Water Techno

An Introduction for Environmental Scientists and Engineers

Water Technology

An Introduction for Environmental Scientists and Engineers

Second Edition

N. F. Gray Ph.D., Sc.D.

Department of Civil, Structural and Environmental Engineering,
Trinity College, University of Dublin

ELSEVIER
BUTTERWORTH
HEINEMANN

OXFORD • AMSTERDAM • BOSTON • LONDON • NEW YORK • PARIS •
SAN DIEGO • SAN FRANCISCO • SINGAPORE • SYDNEY • TOKYO

Elsevier Butterworth-Heinemann
Linacre House, Jordan Hill, Oxford OX2 8DP
30 Corporate Drive, Burlington, MA 01803

First published by Arnold 1999
Reprinted by Butterworth-Heinemann 2002
Second edition 2005

British Library Cataloguing in Publication Data
A catalogue record for this book is available from the British Library

Library of Congress Cataloging-in-Publication Data
A catalogue record for this book is available from the Library of Congress

ISBN 07506 6633 1

For information on all Elsevier Butterworth-Heinemann
publications visit our website at http://books.elsevier.com

Produced and typeset by Charon Tec Pvt. Ltd, Chennai, India
Printed and bound in Great Britain

This book is dedicated to:

Lucy, Catriona and Rebecca
(My best friends)

Pat Smythe
International show jumper
(My hero)

Ginger Rogers
Actress and dancer
Who did everything Fred Astaire did, only backwards and on high heels
(A fellow sucker)

Contents

Part II

Part IV

13. Nature of Wastewater — 349

14. Introduction to Wastewater Treatment — 368

Preface

Traditionally the water industry has relied on specialist engineers, biologists, chemists and microbiologists. However, with new and challenging legislation it has become increasingly necessary for such specialists to develop a broader understanding of the concepts of each other's disciplines. Environmental engineering and environmental science are new hybrids of civil engineering and science, respectively that requires a thorough knowledge of the physico-chemical and biological nature of the aquatic environment, in order to identify and assess impacts, as well as being able to select, design and operate the most appropriate technology to protect both the environment and the health of the general public. Water technology is a rapidly developing area that is truly interdisciplinary in nature.

The European Union has been developing and implementing legislation to protect the aquatic environment for over 30 years. The Water Framework Directive (2000/60/EC) is the last piece of that legislative jigsaw puzzle that puts in place the management structure to ensure that the water quality of lakes, rivers, groundwaters, estuaries, wetlands and coastal waters is protected, and enhanced through proper co-operation and management. The Directive makes this bold statement at the outset:

> *Water is not a commercial product like any other but, rather, a heritage which must be protected, defended and treated as such.*

The Water Framework Directive is a powerful enabling piece of legislation that provides an integrated management base for water resource protection and management including water supply treatment and wastewater treatment. This new edition of *Water Technology* acknowledges the importance of this legislation and uses the concept of River Basin Management throughout.

It must be stressed that *Water Technology* is an introductory textbook that covers the areas of freshwater quality, pollution and management; the treatment, quality and distribution of drinking water; and the treatment and disposal of wastewater. The text is aimed at pure and applied scientists as well as civil and chemical engineers who require an interdisciplinary transitional text to the most important areas of water technology and science. The text, while easily accessible to all disciplines, is particularly designed for students interested in environmental science and engineering who require a sound understanding of the basic concepts that make up this subject. The emphasis of the text is on practical application and the understanding of the processes involved. Special attention has been paid to those areas where an interdisciplinary approach will be advantageous. The text is supported by an extensive web site that allows students to find out more about critical areas, including case studies, more worked examples and self-assessment questions (http://books.elsevier.com/companions/0750666331).

The Brundtland Report and, subsequently, Agenda 21 (Rio Earth Summit, 1992) both identified sustainable development as a critical goal. However, while there appears to be broad agreement for the idea, we still have no consensus as to its precise meaning or its practical application. Environmental scientists, technologists and engineers are going to have to aggressively take hold of the concept of sustainability and develop it into a practical reality. This can only be achieved by different disciplines working together with mutual respect. The ability of the engineer to understand the constraints and limitations imposed by the scientist and *vice versa* is the key to a safer and cleaner environment. Together almost anything is possible.

Nick Gray
Trinity College, Dublin
April 2005

Acknowledgements

I would like to thank all those authors and publishers who have been so helpful and willing to allow me to reproduce copyright material. In particular, I would like to thank Ing. Paolo Romano, General Manager of the Po-Sangone Waste-water Treatment Plant (Torino, Italy), Mr Gerry O'Leary of the Environmental Protection Agency (Ireland) and Mr Alan Bruce, Consultant Environmental Engineer (UK) for permission to reproduce specific material.

I am very grateful to Sarah Hunt of Elsevier Limited for her help and support in producing this second edition.

Part I Water Resources and Ecology

Chapter 1 Basic Considerations in Hydrobiology

1.1 HYDROLOGICAL CYCLE

Water supplies are not pure in the sense that they are devoid of all dissolved chemical compounds like distilled–deionized water but are contaminated by a wide range of trace elements and compounds. In the early days of chemistry, water was known as the universal solvent, due to its ability to slowly dissolve into solution anything it comes into contact with, from gases to rocks. So as rain falls through the atmosphere, flows over and through the Earth's surface, it is constantly dissolving material, forming a chemical record of its passage from the clouds. Therefore, water supplies have a natural variety in quality, which depends largely on its source.

The total volume of water in the world remains constant. What changes is its quality and availability. Water is constantly being recycled, a system known as the water or hydrological cycle. Hydrologists study the chemical and physical nature of water, and its movement both on and below the ground. In terms of total volume, 97.5% of the world's water is saline with 99.99% of this found in the oceans, and the remainder making up the salt lakes. This means that only 2.5% of the volume of water in the world is actually non-saline. However, not all of this freshwater is readily available for use by humans. Some 75% of this freshwater is locked up as ice caps and glaciers, with a further 24% located under ground as groundwater, which means that less than 1% of the total freshwater is found in lakes, rivers and the soil. Only 0.01% of the world's water budget is present in lakes and rivers, with another 0.01% present as soil moisture but unavailable to humans for supply. So while there appears to be lots of water about, there is in reality very little which is readily available for use by humans (Table 1.1). Within the hydrological cycle, the water is constantly moving, driven by solar energy. The sun causes evaporation from the oceans, which forms clouds and precipitation (rainfall). Evaporation also occurs from lakes, rivers and the soil; with plants contributing significant quantities of water by evapotranspiration. While about 80% of precipitation falls back into the oceans, the remainder falls onto land. It is this water that replenishes the soil and groundwater, feeds the streams and lakes, and provides all the water needed by plants, animals and, of course, humans (Fig. 1.1). The cycle is a continuous one and so water is a renewable resource (Franks, 1987). In essence, the

TABLE 1.1 Total volumes of water in the global hydrological cycle

Type of water	Area (km² × 10³)	Volume (km³ × 10³)	Percentage of total water
Atmospheric vapour	510 000	13	0.0001
(water equivalent)	(at sea level)		
World ocean	362 033	1 350 400	97.6
Water in land areas	148 067	–	–
Rivers (average channel storage)	–	1.7	0.0001
Freshwater lakes	825	125	0.0094
Saline lakes; inland seas	700	105	0.0076
Soil moisture; vadose water	131 000	150	0.0108
Biological water	131 000	(Negligible)	–
Groundwater	131 000	7000	0.5060
Ice caps and glaciers	17 000	26 000	1.9250
Total in land areas (rounded)		*33 900*	*2.4590*
Total water, all realms (rounded)		*1 384 000*	*100*
Cyclic water			
Annual evaporation			
From world ocean		445	0.0320
From land areas		71	0.0050
Total		*516*	*0.0370*
Annual precipitation			
On world ocean		412	0.0291
On land areas		104	0.0075
Total		*516*	*0.0370*
Annual outflow from land to sea			
River outflow		29.5	0.0021
Calving, melting and deflation from ice caps		2.5	0.0002
Groundwater outflow		1.5	0.0001
Total		*33.5*	*0.0024*

Reproduced from Charley (1969) with permission of Routledge Publishers Ltd, London.

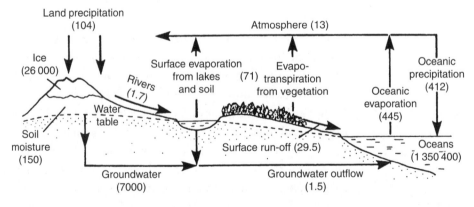

FIGURE 1.1 Hydrological cycle showing the volume of water stored and the amount cycled annually. All volumes are expressed as 10³ km³. (Reproduced from Gray (1994) with permission of John Wiley and Sons Ltd, Chichester.)

more it rains then the greater the flow in the rivers and the higher the water table rises as the underground storage areas (i.e. the aquifers) fill with water as it percolates downwards into the earth. Water supplies depend on rain so when the amount of rain decreases then the volume of water available for supply will decrease, and in cases of severe drought will fall to nothing. Therefore, in order to provide sufficient water for supply all year round, careful management of resources is required.

Nearly all freshwater supplies come from precipitation which falls onto a catchment area. Also known as a watershed or river basin, the catchment is the area of land, often bounded by mountains, from which any water that falls into it will drain into a particular river system. A major river catchment will be made up of many smaller sub-catchments each draining into a tributary of the major river. Each sub-catchment will have different rock and soil types, and each will have different land use activities which also affect water quality. So the water draining from each sub-catchment may be different in terms of chemical quality. As the tributaries enter the main river they mix with water from other sub-catchments upstream, constantly altering the chemical composition of the water. Therefore, water from different areas will have a unique chemical composition.

When precipitation falls into a catchment one of three major fates befalls it:

1. It may remain on the ground as surface moisture and be eventually returned to the atmosphere by evaporation. Alternatively, it may be stored as snow on the surface until the temperature rises sufficiently to melt it. Storage as snow is an important source of drinking water in some regions. For example, throughout Scandinavia lagoons are constructed to collect the run-off from snow as it melts, and this provides the bulk of their drinking water during the summer.
2. Precipitation flows over the surface into small channels becoming surface run-off into streams and lakes (Section 2.1). This is the basis of all surface water supplies with the water eventually evaporating into the atmosphere, percolating into the soil to become groundwater, or continuing as surface flow in rivers back to the sea.
3. The third route is for precipitation to infiltrate the soil and slowly percolate into the ground to become groundwater that is stored in porous sediments and rocks. Groundwater may remain in these porous layers for periods ranging from just a few days to possibly millions of years (Section 2.2). Eventually groundwater is removed by natural upward capillary movement to the soil surface, plant uptake, groundwater seepage into surface rivers, lakes or directly to the sea, or artificially by pumping from wells and boreholes.

Therefore, from a management perspective the hydrological cycle at this point can be represented as:

$$P = E + R + I \tag{1.1}$$

where P is the precipitation, E is the evapotranspiration, R is the run-off and I is the infiltration. In France, approximately 55% of precipitation evaporates, 25% becomes

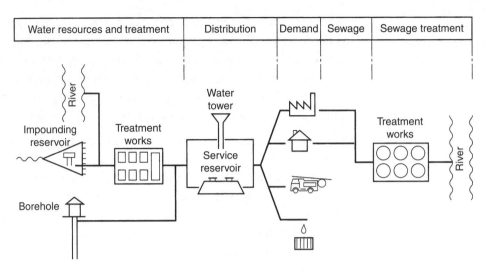

| Water resources and treatment | Distribution | Demand | Sewage | Sewage treatment |

FIGURE 1.2
Water services cycle showing where the water is used by humans during its movements within the hydrological cycle. (Reproduced from Latham (1990) with permission of the Chartered Institution of Water and Environmental Management.)

surface run-off and 20% infiltrates into the ground. However, this will vary enormously between areas.

The water in the oceans, ice caps and aquifers is all ancient, acting as sinks for pollutants. All pollution eventually ends up in the cycle and will ultimately find its way to one of these sinks. Freshwater bodies are often closely inter-connected, forming part of a continuum from rainwater to saline (marine) water, and so may influence each other quite significantly.

People manage the hydrological cycle in order to provide water for industry, agriculture and domestic use. This requires the management of surface and groundwater resources, the treatment and supply of water, its subsequent collection, cleansing and return to the cycle (Fig. 1.2).

1.2 COMPARISON BETWEEN FRESHWATER AND SALINE WATER

Conditions in freshwater differ to saline water in many ways; for example, lower concentrations of salts (especially Na, SO_4, Cl) and smaller volumes of water. Smaller volumes result in a greater range and more rapid fluctuations in temperature, oxygen and other important parameters, making freshwater ecosystems more vulnerable to pollution; the smaller volume and area of freshwater bodies acts as a physical restriction on organisms and restricts the free dispersal of organisms as well as pollutants. In terms of pollution, both lakes and oceans act as sinks with the former having only limited dilution and assimilative capacity.

Freshwaters have a low concentration of salts which presents organisms living in such waters with two problems: first, how to acquire and retain sufficient minimum salts for metabolism; and, second, how to maintain concentrations of salts in their internal fluids that are higher than those in the water (i.e. osmotic and diffusion problems).

TABLE 1.2   Water residence time in inland freshwater bodies

	Hours	Days	Months	Years	10 years	100 years	1000 years
Running waters	Streams	Rivers					
Standing waters			Shallow lakes		Deep lakes		
Groundwaters		Reservoirs Karst aquifers		Alluvial Sedimentary aquifers			Deep aquifers
			Bank filtration				

Reproduced from Meybeck *et al.* (1989) with permission of Blackwell Science Ltd, Oxford.

To some extent this is why the total number of species of organisms that live in freshwater is smaller than the number living wholly on land or in the sea.

1.3 FRESHWATER HABITATS

There are a wide range of freshwater habitats. For example, there are lakes, ponds, turrlochs (temporary ponds), springs, mountain torrents, chalk streams, lowland rivers, large tropical lakes, and a range of man-made habitats such as abandoned peat workings (e.g. the Norfolk Broads) and gravel pits. These are just a few of the wide variety of freshwater bodies that can be found and the hydrological, physico-chemical and biological characteristics vary enormously between them.

In the most basic classification water can be separated into surface and groundwater, with surface waters further split into flowing systems (lotic) and standing systems (lentic). In reality, lotic and lentic systems often grade into each other and may be difficult in practice to differentiate. Lakes generally have an input and an output with what may be a very small throughput. Therefore, these systems can only be separated by considering the relative retention times (Table 1.2). For example, many rivers have been impounded along their length to form reservoirs resulting in lentic–lotic (e.g. River Severn in England) or lotic–lentic–lotic systems (e.g. River Liffey in Ireland).

Rivers can be hydraulically defined as having a unidirectional flow, with generally short retention times measured in days rather than weeks. They have a relatively high average flow velocity of between 0.1 and 1.0 m s^{-1}, although the discharge rate can be highly variable over time, and is a function of climatic and discharge conditions within the watershed. Rivers generally display continuous vertical mixing due to currents and turbulence, although lateral mixing may only occur over considerable distances. Lakes in contrast have a low surface flow velocity of between 0.001 and 0.01 m s^{-1}, with water and element retention times varying from 1 month to >100 years. Currents are multi-directional with diffusion an important factor in dispersion of pollutants. Depending on climatic conditions and depth, lakes can display alternating periods of stratification and vertical mixing. Groundwaters display a steady flow pattern in terms of direction and velocity, resulting in poor mixing. Average flow velocities are very low, between 10^{-10} and $10^{-3}\,\mathrm{m\,s^{-1}}$. This is governed by the porosity and permeability of the aquifer material but results in very long retention periods (Table 1.2). There are a number of transitional forms of water bodies that do fit into the above categories. The most common are flood plains which are seasonal and intermediate between rivers and lakes; reservoirs which are also seasonal depending on water release or use and are intermediate between rivers and lakes; marshes are intermediate between lakes and aquifers; while karst and alluvial aquifers are intermediate between rivers and groundwater because of their relatively rapid flow. The Water Framework Directive not only recognizes surface freshwater and groundwater resources, but also transitional waters such as wetlands, estuaries and coastal waters that are highly influenced by freshwater inputs, in the overall management approach of catchment areas (Section 7.3).

The hydrodynamic characteristics of a water body are dependent primarily on the size of the water body and the climatic and drainage conditions within the catchment area. More specifically, rivers are further characterized by their discharge variability, lakes by their retention time and thermal regime controlling stratification, and groundwaters by their recharge regime (i.e. infiltration through the unsaturated aquifer zone to replenish the water stored in the saturated zone).

Lotic systems have a more open system than lentic ones, with a continuous and rapid throughput of water and nutrients, resulting in a unique flora and fauna (biota). The main advantages of aquatic habitats, and of lotic systems in particular, are that there is no gravity; food comes to the consumer; and that there are no waste disposal problems for the biota.

1.4 THE CATCHMENT AS AREA OF STUDY

Water bodies and lotic systems in particular are not isolated entities, they impinge on the atmosphere and the land. Across these boundaries there is a constant movement of materials and energy. Across the air–water interface gases, water, dissolved nutrients and particles in rain all enter the river. The land–water interface is the main source of material from weathered and eroded rocks, and from dead and decaying plants and animals. Other factors such as transpiration rate of trees, soil and rock type, and land use all affect water quality, therefore it is necessary to study and manage the catchment area or watershed as a whole. This facilitates integrated management of the water cycle within that area. In Ireland, individual catchment areas have detailed water quality management plans (Section 7.2.1), which are operated by the relevant state and semi-state agencies (Fig. 1.3).

Rivers have been managed on a catchment basis for many years although it is relatively recently, with the introduction of geographical information system (GIS) and remote sensing, that it has been possible to integrate land use and groundwater protection in

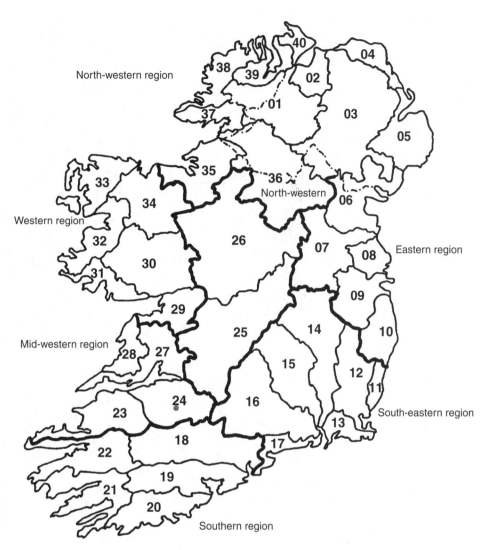

FIGURE 1.3
Water resources
regions and hydro-
metric areas of
Ireland.

water quality management at the catchment level (Chapter 7). Although groundwater resources do not always correspond to river catchment areas and often overlap catchment boundaries. Catchment management in Europe is currently being harmonized through the introduction of the *EU Directive Establishing a Framework for Community Action in the Field of Water Policy* (2000/60/EC). The Water Framework Directive, as it is generally known, proposes a new system of water quality management based on natural river catchments. It provides an integrated approach to the protection, improvement and sustainable use of European water bodies, including lakes, rivers, estuaries, coastal waters and groundwaters. Under this legislation catchments are split into River Basin Districts (RBDs) and then managed through River Basin Management Plans (RBMPs). In Ireland, catchments have been combined to form eight RBDs of which three are International RBDs (i.e. transboundary) (Fig. 1.4). The Water Framework Directive aims to prevent further deterioration of water quality of all water resources

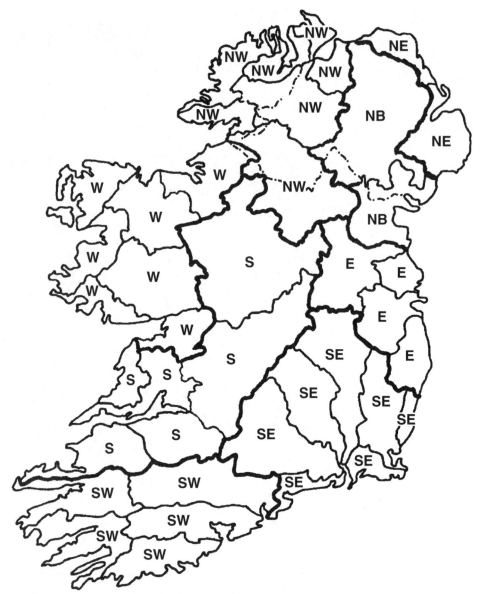

FIGURE 1.4
The newly adopted
International RBDs
(NW: North Western;
NB: Neagh Bann; S:
Shannon) and RBDs
(NE: North Eastern;
E: Eastern; W:
Western; SE: South
Eastern; SW: South
Western) in Ireland.
RBDs are the basic
management unit
used by the Water
Framework Directive
to management
water resources and
are based on river
catchments.

and raise their minimum chemical and ecological status to a newly defined good quality status by 2012. Also by eliminating the discharge of certain hazardous substances to
inland and groundwaters, it is expected that the concentrations of these pollutants in
marine waters will eventually fall to background or zero levels. The Water Framework
Directive is fully explored in Section 7.3.

REFERENCES

Chapman, D. (ed.), 1996 *Water Quality Assessments*, 2nd edn, E. & F. N. Spon, London.
Charley, R. J. (ed.), 1969 *Introduction to Geographical Hydrology*, Routledge, London.

Franks, F., 1987 The hydrologic cycle: turnover, distribution and utilisation of water. In: Lorch W., ed., *Handbook of Water Purification*, Ellis Horwood, Chichester, pp 30–49.

Gray, N. F., 1994 *Drinking Water Quality: Problems and Solutions*, John Wiley and Sons, Chichester.

Latham, B., 1990 *Water Distribution*, Chartered Institution of Water and Environmental Management, London.

Meybeck, M., Chapman, D. and Helmer, R. (eds), 1989 *Global Freshwater Quality: A First Assessment*, Blackwell Science, Oxford.

FURTHER READING

Horne, A. J. and Goldman, C. R., 1994 *Limnology*, McGraw-Hill, New York.

Uhlmann, D., 1979 *Hydrobiology: A Text for Engineers and Scientists*, John Wiley and Sons, Chichester.

Chapter 2 Water Resources

2.1 SURFACE WATERS

Surface water is a general term describing any water body which is found flowing or standing on the surface, such as streams, rivers, ponds, lakes and reservoirs. Surface waters originate from a combination of sources:

(a) *Surface run-off*: rainfall which has fallen onto the surrounding land and that flows directly over the surface into the water body.
(b) *Direct precipitation*: rainfall which falls directly into the water body.
(c) *Interflow*: excess soil moisture which is constantly draining into the water body.
(d) *Water table discharge*: where there is an aquifer below the water body and the water table is high enough, the water will discharge directly from the aquifer into the water body.

The quality and quantity of surface water depends on a combination of climatic and geological factors. The recent pattern of rainfall, for example, is less important in enclosed water bodies, such as lakes and reservoirs, where water is collected over a long period and stored. However, in rivers and streams where the water is in a dynamic state of constant movement, the volume of water is very much dependent on the preceding weather conditions.

2.1.1 RIVERS AND LAKES

Generally in rivers the flow is greater in the winter than the summer due to a greater amount and longer duration of rainfall. Short fluctuations in flow, however, are more dependent on the geology of the catchment. Some catchments yield much higher percentages of the rainfall as stream flow than others. Known as the 'run-off ratio', the rivers of Wales and Scotland can achieve values of up to 80% compared with only 30% in lowland areas in southern England. Hence, while the River Thames, for example, has a vast catchment area of some 9869 km², it has only half of the annual discharge of a river, such as the River Tay, which has a catchment of only $4584\,km^2$. Of course in Scotland there is higher rainfall than in south-east England and also lower evaporation rates due to lower temperatures.

Even a small reduction in the average rainfall in a catchment area, say 20%, may halve the annual discharge from a river. This is why when conditions are only marginally drier than normal a drought situation can so readily develop. It is not always a case of the more it rains the more water there will be in the rivers; groundwater is also an important factor in stream flow. In some areas of England during the dry summer of 1975, despite the rainfall figures, which were way below average, the stream flow in rivers which receive a significant groundwater input were higher than normal due to the excessive storage built-up in the aquifer over the previous wet winter. The drought period in England that started in 1989, where there were three successive dry winters, resulted in a significant reduction in the amount of water stored in aquifers with a subsequent fall in the height of the water table. This resulted in some of the lowest flows on record in a number of south-eastern rivers in England, with sections completely dried up for the first time in living memory. Before these rivers return to normal discharge levels, the aquifers that feed them must be fully replenished and this still had not occurred 8 years later. As groundwater contributes substantially to the base flow of many lowland rivers, any steps taken to protect the quality of groundwaters will also indirectly protect surface waters.

Precipitation carries appreciable amounts of solid material to earth, such as dust, pollen, ash from volcanoes, bacteria, fungal spores and even on occasions, larger organisms. The sea is the major source of many salts found dissolved in rain, such as chloride, sodium, sulphate, magnesium, calcium and potassium ions. Atmospheric discharges from the house and industry also contribute material to clouds which are then brought back to earth in precipitation. These include a wide range of chemicals including organic solvents, and the oxides of nitrogen and sulphur that cause acid rain. The amount and types of impurities in precipitation vary due to location and time of year, and can affect both lakes and rivers. Land use, including urbanization and industrialization, significantly affects water quality, with agriculture having the most profound effect on supplies due to its dispersed and extensive nature (Section 7.4).

The quality and quantity of water in surface waters are also dependent on the geology of the catchment. In general, chalk and limestone catchments result in clear hard waters, while impervious rocks, such as granite, result in turbid soft waters. Turbidity is caused by fine particles, both inorganic and organic in origin, which are too small to readily settle out of suspension and so the water appears cloudy. The reasons for these differences are that rivers in chalk and limestone areas rise as springs or are fed from aquifers through the riverbed. Because appreciable amounts of the water are coming from groundwater resources, the river retains a constant clarity, constant flow and indeed a constant temperature throughout the year, except after the periods of prolonged rainfall. The chemical nature of these rivers is also very stable and rarely

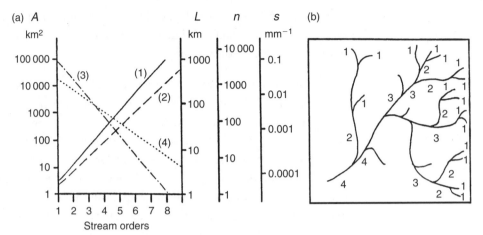

FIGURE 2.1 (a) The relationship between stream orders and hydrological characteristics using a hypothetical example for a stream of order 8: (1) watershed area (A); (2) length of river stretch (L); (3) number of tributaries (n); (4) slope (mm^{-1}). (b) Stream order distribution within a watershed. (Reproduced from Chapman (1996) with permission of the UNESCO, Paris.)

alters from year to year. The water has spent a very long time in the aquifer before entering the river, and during this time dissolves the calcium and magnesium salts comprising the rock, resulting in a hard water with a neutral to alkaline pH. In comparison, soft-water rivers are usually raised as run-off from mountains, so flow is very much linked to rainfall. Such rivers suffer from wide fluctuations in flow rate with sudden floods and droughts. Chemically these rivers are turbid due to all the silt washed into the river with the surface run-off and, because there is little contact with the bed rock, they contain low concentrations of cations, such as calcium and magnesium, which makes the water soft with a neutral to acidic pH. Such rivers often drain upland peaty soils and so the water contains a high concentration of humus material giving the water a clear brown-yellow colour, similar to beer in appearance.

Land areas can be divided into geographical areas drained by a river and its tributaries. Each area is known as a river basin, watershed or catchment. Each river basin is drained by a dendritic network of streams and rivers. They increase in size (order) from very small (1) feeding eventually into the main river channel (4–10) (Fig. 2.1). The stream order is a function of the catchment area feeding that tributary or collection of tributaries, although other parameters, such as average discharge rate, width, etc., are also used to classify rivers (Table 2.1). Stream order is positively correlated to catchment area and the length of the river section, while the number of tributaries within the catchment and slope are negatively correlated with stream order.

Most lakes have an input and an output, and so in some ways they can be considered as very slowly flowing rivers. The very long period of time that water remains in the lake or reservoir ensures that the water becomes cleaner due to bacterial activity removing any organic matter present, and physical flocculation and settlement processes, which remove small particulate material. Storage of water, therefore, improves the

TABLE 2.1 Classification of rivers based on discharge characteristics, drainage area and river width

River size	Average discharge $(m^3 s^{-1})$	Drainage area (km^2)	River width (m)	Stream order[a]
Very large rivers	>10 000	>10^6	>1500	>10
Large rivers	1000–10 000	100 000–10^6	800–1500	7–11
Rivers	100–1000	10 000–100 000	200–800	6–9
Small rivers	10–100	1000–10 000	40–200	4–7
Streams	1–10	100–1000	8–40	3–6
Small streams	0.1–1.0	10–100	1–8	2–5
Brooks	<0.1	<10	<1	1–3

[a]Depending on local conditions.
Reproduced from Chapman (1996) with permission of the UNESCO, Paris.

quality which then minimizes the treatment required before supply ('Introduction to treatment' in Chapter 10). However, this situation is complicated by two factors:

1. In standing waters much larger populations of algae are capable of being supported than in rivers.
2. Deep lakes and reservoirs may become thermally stratified, particularly during the summer months.

Both of these factors can seriously affect water quality.

Thermal stratification is caused by variations in the density of the water in both lakes and reservoirs, although it is mainly a phenomenon of deep lakes. Water is at its densest at 4°C, when it weighs exactly 1000 kg per cubic metre ($kg\,m^{-3}$). However, either side of this temperature water is less dense (999.87 $kg\,m^{-3}$ at 0°C and 999.73 $kg\,m^{-3}$ at 10°C). During the summer, the sun heats the surface of the water reducing its density, so that the colder, denser water remains at the bottom of the lake. As the water continues to heat up then two distinct layers develop. The top layer or epilimnion is much warmer than the lower layer, the hypolimnion. Owing to the differences in density the two layers, separated by a static boundary layer known as the metalimnion or thermocline, do not mix but remain separate (Fig. 2.2).

The epilimnion of lakes and reservoirs is constantly being mixed by the wind and so the whole layer is a uniform temperature. As this water is both warm and exposed to sunlight it provides a very favourable environment for algae. Normally the various nutrients required by the algae for growth, in particular phosphorus and nitrogen, are not present in large quantities (i.e. limiting concentrations). When excess nutrients are present, for example due to agricultural run-off, then massive algal growth may occur. These so-called algal blooms result in vast increases in the quantity of algae in the water, a phenomenon known as eutrophication (Section 6.4). The algae are completely mixed throughout the depth of the epilimnion and in severe cases the water can become highly coloured. Normally this top layer of water is clear and full

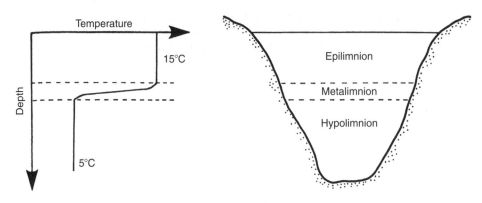

of oxygen, but if eutrophication occurs, then the algae must be removed by treatment. The algae can result in unpleasant tastes in the water even after conventional
water treatment (Section 11.1.1), as well as toxins (Section 6.4.1). Like all plants, algae
release oxygen during the day by photosynthesis, but at night they remove oxygen
from the water during respiration. When eutrophication occurs then the high numbers of algae will severely deplete the oxygen concentration in the water during the
hours of darkness, possibly resulting in fish kills. In contrast, there is little mixing or
movement in the hypolimnion which rapidly becomes deoxygenated and stagnant,
and devoid of the normal aerobic biota. Dead algae and organic matter settling from
the upper layers are degraded in this lower layer of the lake. As the hypolimnion has
no source of oxygen to replace that already used, its water may become completely
devoid of oxygen. Under anaerobic conditions iron, manganese, ammonia, sulphides,
phosphates and silica are all released from sediments in the lake or reservoir into the
water while nitrate is reduced to nitrogen gas. This makes the water unfit for supply
purposes. For example, iron and manganese will result in discoloured water complaints as well as taste complaints. Ammonia interferes with chlorination, depletes the
oxygen faster and acts as a nutrient to encourage eutrophication (as does the phosphorus and silica). Sulphides also deplete the oxygen and interfere with chlorination,
have an awful smell and impart an obnoxious taste to the water (Section 11.1.1).

The metalimnion, the zone separating the two layers, has a tendency to move slowly
to lower depths as the summer progresses due to heat transfer to the lower hypolimnion.
This summer stratification is usually broken up in autumn or early winter as the air
temperature falls and the temperature of the epilimnion declines. This increases the
density of the water making up the epilimnion to a comparable density with the
hypolimnion, making stratification rather unstable. Subsequently, high winds will eventually cause the whole water body to turn over breaking up so that the layers become
mixed. Throughout the rest of the year the whole lake remains completely mixed,
resulting in a significant improvement in water quality. Limited stratification can also
develop during the winter as surface water temperatures approach 0°C while the
lower water temperature remains at 4°C. This winter stratification is broken up in the
spring as the temperature increases and the high winds return. The classification of
lakes is considered in detail in Section 3.8.

2.1.2 RESERVOIRS

As large cities expanded during the last century they relied on local water resources, but as demand grew they were forced to invest in reservoir schemes often quite remote from the point of use. Examples include reservoirs built in Wales, the Pennines and the Lake District to supply major cities, such as Birmingham, Manchester and Liverpool, with water in some cases being pumped over 80 km to consumers. Most are storage reservoirs where all the water collected is used for supply purposes. Such reservoirs are sited in upland areas at the headwaters (source) of rivers. Suitable valleys are flooded by damming the main streams. They can take many years to fill and once brought into use for supply purposes then they must be carefully managed. A balance must be maintained between the water taken out for supply and that being replaced by surface run-off. Normally the surface run-off during the winter far exceeds demand for supply so that the excess water can be stored and used to supplement periods when surface run-off is less than demand from consumers. There is, of course, a finite amount of water in a reservoir, and so often water rationing is required in order to prevent storage reservoirs drying up altogether during dry summers. A major problem is when there is a dry winter, so that the expected excess of water does not occur resulting in the reservoirs not being adequately filled at the beginning of the summer. Under these circumstances water shortages may occur even though the summer is not excessively dry.

The catchment area around reservoirs is normally owned by the water supply company. They impose strict restrictions on farming practice and general land use to ensure that the quality of the water is not threatened by indirect pollution. Restricted access to catchment areas and reservoirs has been relaxed in recent years, although it is still strictly controlled. This controlled access and restrictions on land use has caused much resentment, especially in Wales where as much as 70% of all the water in upland reservoirs is being stored for use in the English midlands. The problem is not only found in Wales, but in England as well. For example, as much as 30% of the Peak District is made up of reservoir catchment areas. Water supply companies want to ensure that the water is kept as clean as possible because water collected in upland reservoirs is of a very high quality. Storage also significantly improves its purity further, and the cleaner the raw water, the cheaper it is to treat to drinking water standard ('Introduction to treatment' in Chapter 10). So restricted access to catchment areas means less likelihood of a reduction in that quality. Conflicts between those who want access to the reservoir or the immediate catchment area for recreation or other purposes, and the water companies who want to supply water to their customers at the lowest price possible, is inevitable.

In order to maximize water availability for supply, hydrologists examine the hydrological cycle within the catchment, measuring rainfall, stream flow and surface run-off, and, where applicable, groundwater supplies. Often they can supplement water abstracted from rivers at periods of very low flow by taking water from other resources, such as groundwater or small storage reservoirs, using these limited resources to top up the primary source of supply at the most critical times. More common is the construction of reservoirs at the headwaters which can then be used to control the flow of the river itself, a process known as compensation. Compensation reservoirs are designed as an integral part of the river system. Water is collected as surface run-off from upland areas and stored during wet periods. The water is then released when needed to ensure that the minimum dry weather flow is maintained downstream to protect the biota, including fisheries, while allowing abstraction to continue. In winter, when the majority of precipitation falls and maximum flows are generated in the river, all the excess water are lost. By storing the excess water by constructing a reservoir and using it to regulate the flow in the river, the output from the catchment area is maximized. A bonus is that such reservoirs can also play an important function in flood prevention. The natural river channel itself is used as the distribution system for the water, unlike supply reservoirs where expensive pipelines or aqueducts are required to transport the water to the point of use. River management is also easier because the majority of water abstracted is returned to the same river. Among the more important rivers in the UK which are compensated are the Dee, Severn and Tees.

Reservoirs are not a new idea and were widely built to control the depth in canals and navigable rivers. Smaller reservoirs, often called header pools, were built to feed mill races to drive waterwheels. Without a reservoir, there are times when after dry spells, where there is negligible discharge from the soil, the only flow in the river will be that coming from groundwater seeping out of the underlying aquifer. Some rivers, rising in areas of permeable rock, may even dry up completely in severe droughts. In many natural rivers the natural minimum flow is about 10% of the average stream flow. Where river regulation is used, this minimum dry weather flow is often doubled, and although this could in theory be increased even further, it would require an enormous reservoir capacity.

Reservoirs are very expensive to construct. However, there are significant advantages to regulating rivers using compensation reservoirs rather than supplying water directly from a reservoir via an aqueduct. With river regulation much more water is available to meet different demands than from the stored volume only. This is because the reservoir is only fed by the upland section of the river contained by the dam; downstream all the water draining into the system is also available. Hence compensation reservoirs are generally much smaller in size and so cheaper to construct.

Reservoirs can only yield a limited supply of water, and so the management of the river system is difficult in order to ensure adequate supplies every year. Owing to the expensive cost of construction, reservoirs are designed to provide adequate supplies for most dry summers. However, it is not cost-effective to build a reservoir large enough to cope with the severest droughts which may only occur once or twice a century. Water is released from the reservoir to ensure the predicted minimum dry weather flow. Where more than one compensation reservoir is available within the catchment area, water will first be released from those that refill quickly. Water regulation is a difficult task requiring operators to make intuitive guesses as to what the weather may do over the next few months. For example, many UK water companies were severely criticized for maintaining water restrictions throughout the winter of 1990–91, in order to replenish reservoirs which failed to completely fill during the previous dry winter. Imposing bans may ensure sufficient supplies for essential uses throughout a dry summer and autumn; however, if it turns out to be a wet summer after all, such restrictions will be deemed to have been very unnecessary indeed by consumers. While water planners have complex computer models to help them predict water-use patterns and so plan the best use of available resources, it is all too often impossible to match supplies with demand. This has mainly been a problem in south-eastern England, where the demand is the greatest due to a high population density and also a high industrial and agricultural demand, but where least rainfall is recorded. Where reservoirs are used to prevent flooding it requires that there is room to store the winter floodwater. This may mean allowing the level of the reservoir to fall deliberately in the autumn and early winter to allow sufficient capacity to contain any floodwater. This is the practice on the Clywedog reservoir on the upper Severn. But if the winter happens to be drier than expected, then the reservoir may be only partially full at the beginning of the summer. Therefore, operating reservoirs and regulating rivers are a delicate art and, as the weather is so unpredictable, decisions made on the best available information many months previously may prove to have been incorrect.

2.2 Groundwater

British groundwater is held in three major aquifer systems, with most of the important aquifers lying south-east of a line joining Newcastle-upon-Tyne and Torquay (Fig. 2.3). An aquifer is an underground water-bearing layer of porous rock through which water can flow after it has passed downwards (infiltration) through the upper layers of soil. On average 7000 million litres (Ml) of water are abstracted from these aquifers each day. Approximately 50% of this vast amount of water comes from Cretaceous chalk aquifers, 35% from Triassic sandstones and the remainder from smaller aquifers, the most important of these being Jurassic limestones. Of course groundwater is not only abstracted directly for supply purposes, it often makes a significant contribution to rivers also used for supply and other uses by discharging into the river as either base flow or springs (Fig. 2.4). The discharge of groundwater into rivers may be permanent or seasonal, depending on the height of the water table within the aquifer. The water table separates the unsaturated zone of the porous rock comprising the aquifer from the saturated zone; in essence it is the height of the water in the aquifer (Fig. 2.5). The water table is measured by determining the level of the water in boreholes and wells. If numerous measurements are taken from wells over a wide area, then the water table can be seen to fluctuate in height depending on the topography and climate conditions. Rainfall replenishes or recharges the water lost or taken from the aquifer and hence raises the level of the water

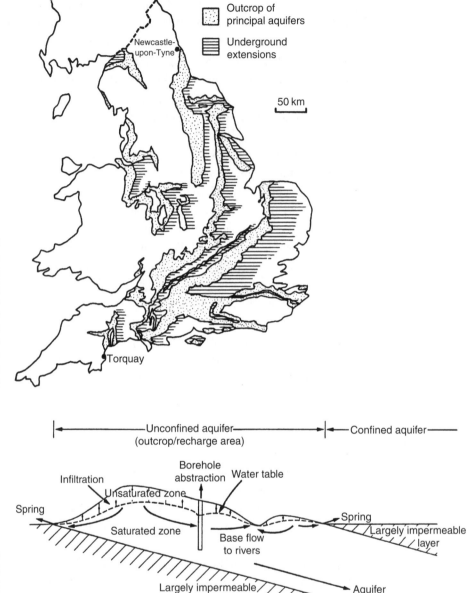

FIGURE 2.3
Location of the principal aquifers in England and Wales. (Reproduced from Open University (1974) with permission of the Open University Press, Milton Keynes.)

FIGURE 2.4
Schematic diagram of groundwater systems. (Reproduced by permission of the British Geological Survey.)

table. So, if the level falls during periods of drought or due to over-abstraction for water supply, then this source of water feeding the river may cease. In periods of severe drought, groundwater may be the only source of water feeding some rivers and so, if the water table falls below the critical level, the river itself could dry up completely.

Aquifers are classified as either confined or unconfined. An unconfined aquifer is one that is recharged where the porous rock is not covered by an impervious layer of

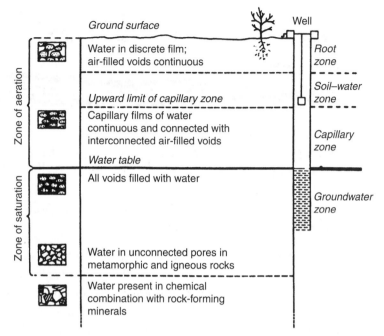

FIGURE 2.5 Cross section through soil and aquifer showing various zones in the soil and rock layers, and their water-bearing capacities. (Reproduced from Open University (1974) with permission of the Open University Press, Milton Keynes.)

soil or other rock. The unsaturated layer of porous rock, which is rich in oxygen, is separated from the saturated water-bearing layer by an interface known as the water table. Where the aquifer is overlain by an impermeable layer, no water can penetrate into the porous rock from the surface; instead water slowly migrates laterally from unconfined areas. This is a confined aquifer. There is no unsaturated zone because all the porous rocks are saturated with water as they are below the water table level, and of course there is no oxygen (Fig. 2.4). Because confined aquifers are sandwiched between two impermeable layers, the water is usually under considerable hydraulic pressure, so that the water will rise to the surface under its own pressure via boreholes and wells, which are known as artesian wells. Artesian wells are well known in parts of Africa and Australia, but are also found on a smaller scale in the British Isles. The most well-known artesian basin is the one on which London stands. This is a chalk aquifer which is fed by unconfined aquifers to the north (Chiltern Hills) and the south (North Downs) (Fig. 2.6). In the late nineteenth century the pressure in the aquifer was such that the fountains in Trafalgar Square were fed by natural artesian flow. But if the artesian pressure is to be maintained, then the water lost by abstraction must be replaced by recharge, in the case of London by infiltration at the exposed edges of the aquifer basin. Continued abstraction in excess of natural recharge has lowered the piezometric surface (i.e. the level that the water in an artesian well will naturally rise to) by about 140 m below its original level in the London basin. A detailed example of an aquifer is given later in this section.

It is from unconfined aquifers that the bulk of the groundwater supplies are abstracted. It is also from these aquifers, in the form of base flow or springs, that a major portion of the flow of some lowland rivers in eastern England arises. These rivers are widely

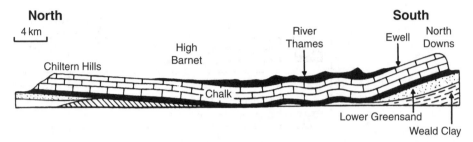

FIGURE 2.6 Cross section through the London artesian basin. (Reproduced from Open University (1974) with permission of the Open University Press, Milton Keynes.)

used for supply purposes and so this source of drinking water is largely dependent on their aquifers; good management is therefore vital.

Groundwater in unconfined aquifers originates mainly as rainfall and so is particularly vulnerable to diffuse sources of pollution, especially agricultural practices and the fallout of atmospheric pollution arising mainly from industry. Elevated chemical and bacterial concentrations in excess of the European Union (EU) limits set in the Drinking Water Directive (Section 8.1) are commonly recorded in isolated wells. This is often a local phenomenon with the source of pollution normally easily identified to a local point source, such as a septic tank, a leaking sewer or farmyard drainage. In the 1970s, there was increasing concern over rising nitrate levels in particular, which often exceeded the EU maximum admissible concentration. This was not an isolated phenomenon, but elevated levels were found throughout all the principal unconfined aquifers in the UK. The areas affected were so large that, clearly, only a diffuse source could be to blame, rather than point sources (Royal Society, 1983). However, it was not until the mid-1970s that the widespread rise in nitrate concentration in groundwater was linked to the major changes in agricultural practices that had occurred in Britain since World War II. The major practice implicated at that time was regular cereal cropping which has to be sustained by the increasing usage of inorganic fertilizer (Section 11.1.3). Trace or micro-organic compounds are another major pollutant in groundwaters. Owing to the small volumes of contaminant involved, once dispersed within the aquifer, it may persist for decades. Many of these compounds originate from spillages or leaking storage tanks. A leak of 1,1,1-trichloroethane (also used as a thinner for typing correction fluids) from an underground storage tank at a factory manufacturing micro-chips for computers in San Jose, California in the USA caused extensive groundwater contamination resulting in serious birth defects, as well as miscarriages and stillbirths, in the community receiving the contaminated drinking water. Agriculture is also a major source of organic chemicals. Unlike spillages or leaks, which are point sources, agricultural-based contamination is a diffuse source (Section 7.4). Many different pesticides are also being reported in drinking water (Section 11.2.1). Other important sources of pollution include landfill and dump sites, impoundment lagoons including slurry pits, disposal of sewage sludge onto land, run-off from roads and mining (OECD, 1989; Mather *et al.*, 1998).

Unlike confined aquifers, unconfined aquifers have both an unsaturated and saturated zone. The unsaturated zone is situated between the land surface and the water table of the aquifer. While it can eliminate some pollutants, the unsaturated zone has the major effect of retarding the movement of most pollutants thereby concealing their occurrence in groundwater supplies for long periods. This is particularly important with major pollution incidents, where it may be many years before the effects of a spillage, or leakage from a storage tank for example, will be detected in the groundwater due to this prolonged migration period.

Most of the aquifers in Britain have relatively thick unsaturated zones. In chalk they vary from 10 to 50 m in thickness, which means that surface-derived pollutants can remain in this zone for decades. An example of this is the recent appearance in some UK groundwater supplies in south-east England of pesticides, such as dichloro-diphenyl-trichloroethane (DDT), which were banned in the early 1970s. Another problem is that the soils generally found above aquifer outcrops are thin and highly permeable, and so allow rapid infiltration of water to the unsaturated zone taking the pollutants with them. Soil bacteria and other soil processes therefore have little opportunity to utilize or remove pollutants. This unsaturated zone is not dry; it does in fact hold large volumes of water under tension in a process matrix, along with varying proportions of air. But below the root zone layer, the movement of this water is predominantly downwards, albeit extremely slowly (Fig. 2.5).

It is in the saturated zone of unconfined aquifers that the water available for abstraction is stored. The volume of water in unconfined aquifers is many times the annual recharge from rainfall. It varies according to rock type and depth; for example in a thin Jurassic limestone the ratio may be up to 3, whereas in a thick porous Triassic sandstone it may exceed 100. The saturated zone also contains a large volume of water which is immobile, locked up in the micro-porous matrices of the rock, especially in chalk Jurassic limestone aquifers.

Where aquifers have become fissured (cracked), the water movement is much more rapid. However, the movement of pollutants through unfissured rock is by diffusion through the largely immobile water which fills the pores which takes considerably longer. This, combined with the time lag in the unsaturated zone, results in only a small percentage of the pollutants in natural circulation within the water-bearing rock being discharged within a few years of their originally infiltrating through agricultural soils. Typical residence times vary but will generally exceed 10–20 years. The deeper the aquifer, the longer this period. Other factors, such as enhanced dilution effects, where large volumes of water are stored, and the nature of the porous rock all affect the retention time of pollutants. An excellent introduction to aquifers and groundwater has been prepared by Price (1991), while Raghunath (1987) has published a more technical and advanced monograph. The management and assessment of the risk of contamination in groundwaters is reviewed by Reichard et al. (1990), while treatment and remediation technologies are dealt with by Nyer (1992).

2.2.1 GROUNDWATER QUALITY

The quality of groundwater depends on a number of factors:

(a) The nature of the rain water, which can vary considerably, especially in terms of acidity due to pollution and the effects of wind-blown spray from the sea which affects coastal areas in particular.
(b) The nature of the existing groundwater which may be tens of thousands of years old.
(c) The nature of the soil through which water must percolate.
(d) The nature of the rock comprising the aquifer.

In general terms groundwater is comprised of a number of major ions which form compounds. These are calcium (Ca^{2+}), magnesium (Mg^{2+}), sodium (Na^+), potassium (K^+), and to a lesser extent iron (Fe^{2+}) and manganese (Mn^{2+}). These are all cations (they have positive charges) which are found in water combined with an anion (which have negative charges) to form compounds sometimes referred to as salts. The major anions are carbonate (CO_3^{2-}), bicarbonate (HCO_3^-), sulphate (SO_4^{2-}) and chloride (Cl^-).

Most aquifers in the British Isles have hard water. Total hardness is made up of carbonate (or temporary) hardness caused by the presence of calcium bicarbonate ($CaHCO_3$) and magnesium bicarbonate ($MgHCO_3$), while non-carbonate (or permanent) hardness is caused by other salts of calcium and magnesium (Section 3.4). It is difficult to generalize, but limestone and chalk contains high concentrations of calcium bicarbonate, while dolomite contains magnesium bicarbonate. Sandstones are often rich in sodium chloride (NaCl), while granite has elevated iron concentrations. The total concentration of ions in groundwater, called the total dissolved solids (TDS), is often an order of magnitude higher than surface waters. The TDS also increases with depth due to less fresh recharge to dilute existing groundwater and the longer period for ions to be dissolved into the groundwater. In very old, deep waters, the concentrations are so high that they are extremely salty. Such high concentrations of salts may result in problems due to over-abstraction or in drought conditions when old saline groundwaters may enter boreholes through upward replacement, or due to saline intrusion into the aquifer from the sea. In Europe, conductivity is used as a replacement for TDS measurement (Section 3.4), and is used routinely to measure the degree of mineralization of groundwaters (Table 2.2).

TABLE 2.2   Mineralization of groundwater can be characterized by conductivity

Conductivity in μS/cm, at 20°C	Mineralization of water
<100	Very weak (granitic terrains)
100–200	Weak
200–400	Slight
400–600	Moderate (limestone terrains)
600–1000	High
>1000	Excessive

In terms of volume of potable water supplied, confined aquifers are a less important
source of groundwater than unconfined aquifers. However, they do contribute substantial volumes of water for supply purposes and, of course, can locally be the major
source of drinking water. Groundwater in confined aquifers is much older and therefore is characterized by a low level of pollutants, especially nitrate and, of course,
micro-organics, including pesticides. Therefore, this source is currently of great interest
for use in diluting water from sources with high pollutant concentrations, a process
known as blending (Section 11.1).

Confined aquifers are generally not used if there is an alternative source of water
because of low yields from boreholes and quality problems, especially high salinity
(sodium and chloride concentrations) in some deep aquifers, excessive iron and/or
manganese, high fluoride and sulphate concentrations, problem gases, such as hydrogen sulphide and carbon dioxide, and the absence of dissolved oxygen. These problems
can be overcome by water treatment, although this, combined with costs for pumping, makes water from confined aquifers comparatively expensive.

An example of a major British aquifer is the Sherwood Sandstone aquifer in the East
Midlands, which is composed of thick red sandstone. It is exposed in the western part
of Nottinghamshire which forms the unconfined aquifer, and dips uniformly to the
east at a slope of approximately 1 in 50, and where it is overlain by Mercia mudstone
forming the confined aquifer (Fig. 2.7). Three groundwater zones have been identified in this aquifer, each of increasing age. In zone 1 the groundwater is predominantly modern and does not exceed a few tens or hundreds of years in age, while in

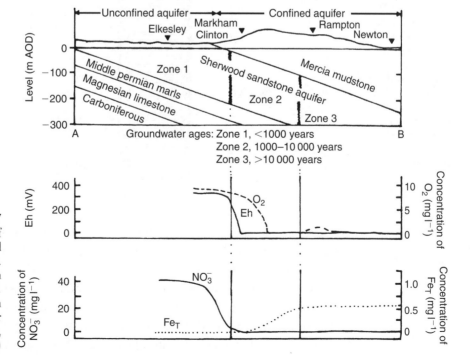

FIGURE 2.7
Hydrochemistry of
the Sherwood
sandstone aquifer.
(Reproduced from
Kendrick *et al.*
(1985) with
permission of Water
Research Centre plc,
Medmenham.)

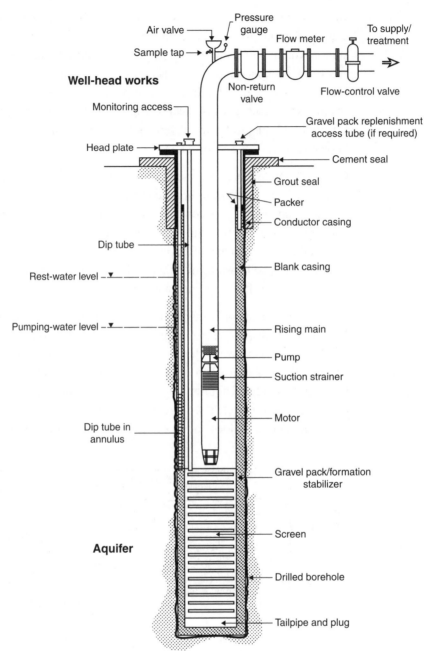

Air valve — Pressure gauge
Sample tap →
Well-head works
Flow meter
To supply/ treatment
Non-return valve
Flow-control valve

Monitoring access —
Gravel pack replenishment access tube (if required)

Head plate —
Cement seal

Grout seal
Packer
Conductor casing

Dip tube —
Blank casing

Rest-water level — ▼ — — — —

Pumping-water level — ▼ — — — —
Rising main

Pump
Suction strainer

Motor

Dip tube in annulus —

Gravel pack/formation stabilizer

Aquifer
Screen

Drilled borehole

Tailpipe and plug

FIGURE 2.8
Typical modern borehole system for groundwater abstraction. (Reproduced from Howsam *et al.* (1995) with permission of the Construction Industry Research and Information Association, London.)

zone 2 the groundwater ranges from 1000 to 10 000 years old. But in zone 3 the groundwater was recharged 10 000–30 000 years ago. In this aquifer the average groundwater velocity is very slow, just 0.7 m per year (Edmonds *et al.*, 1982). The variations in chemical quality across these age zones is typical for unconfined and confined aquifers. This is seen in the other major aquifers, especially the Lincolnshire limestone aquifer in eastern England where the high nitrate concentrations in the youngest groundwater zone is due to the use of artificial fertilizers in this intensively cultivated area.

There are large groundwater resources below some major cities; for example in the UK these include London, Liverpool, Manchester, Birmingham and Coventry. However, these resources are particularly at risk from point sources of pollution from industry, especially solvents and other organic chemicals, as well as more general contaminants from damaged sewers and urban run-off. Such resources are potentially extremely important, although, due to contamination, are often under-exploited (Section 11.1) (Lerner and Tellam, 1993).

The design and construction of boreholes and wells for groundwater abstraction is very technical and requires the expertise of a groundwater hydrologist (Fig. 2.8). Clark (1988) deals comprehensively with the location and installation of water supply boreholes, while the maintenance, monitoring and rehabilitation of such systems is reviewed by Howsam *et al.* (1995) and Detay (1997). A less specialist overview is given by Brassington (1995).

REFERENCES

Brassington, R., 1995 *Finding Water: A Guide to the Construction and Maintenance of Private Water Supplies*, John Wiley and Sons, Chichester.

Chapman, D. (ed.), 1996 *Water Quality Assessments*, 2nd edn, E. & F. N. Spon, London.

Clark, L., 1988 *The Field Guide to Water Wells and Boreholes*, John Wiley and Sons, Chichester.

Detay, M., 1997 *Water Wells: Implementation, Maintenance and Restoration*, John Wiley and Sons, Chichester.

Edmonds, W. M., Bath, A. H. and Miles, D. L., 1982 Hydrochemical evolution of the east Midlands Triassic sandstone aquifer, England, *Geochimica et Cosmochimica Acta*, **46**, 2069.

Gray, N. F., 1994 *Drinking Water Quality: Problems and Solutions*, John Wiley and Sons, Chichester.

Howsam, P., Misstear, B. and Jones, C., 1995 *Monitoring, Maintenance and Rehabilitation of Water Supply Boreholes*, Report 137, Construction Industry Research and Information Association, London.

Kendrick, M. A. P., Clark, L., Baxter, K. M., Fleet, M., James, H. A., Gibson, T. M. and Turrell, M. B., 1985 *Trace Organics in British Aquifers: A Baseline Study*, Technical Report 223, Water Research Centre, Medmenham.

Lerner, D. N. and Tellam, J. H., 1993 The protection of urban ground water from pollution. In: Currie, J. C. and Pepper, A. T., eds., *Water and the Environment*, Ellis Horwood, Chichester, pp 322–37.

Mather, J., Banks, D., Dumpleton, S. and Fermor, M. (eds), 1998 *Groundwater Contaminants and Their Migration*. Geological Society Special Publication No. 128, The Geological Society, London.

Nyer, E. K., 1992 *Groundwater Treatment Technology*, 2nd edn, Van Nostrand Reinhold, New York.

OECD, 1989 *Water Resource Management: Integrated Policies*, Organization for Economic Co-operation and Development, Paris.

Open University, 1974 *The Earth's Physical Resources: 5. Water Resources, S26*, Open University Press, Milton Keynes.

Price, M., 1991 *Introducing Ground Water*, Chapman and Hall, London.

Raghunath, H. M., 1987 *Ground Water*, John Wiley and Sons, New York.

Reichard, E. *et al.*, 1990 *Ground Water Contamination and Risk Assessment*, IAHS Publication 196, International Association of Hydrological Sciences Press, Wallingford, Oxford.

Royal Society, 1983 *The Nitrogen Cycle of the United Kingdom, A Study Group Report*, The Royal Society, London.

FURTHER READING

Brassington, R., 1998 *Field Hydrology*, 2nd edn, John Wiley and Sons, Chichester.

Eriksson, E., 1985 *Principles and Applications of Hydrochemistry*, Chapman and Hall, London.

ReVelle, C., 1997 Water resources: surface water systems. In: Revelle, C. and McGarity, A. E., eds, *Design and Operation of Civil and Environmental Engineering Systems*, John Wiley and Sons, New York, pp 1–40.

Robins, N. S., 1997 The role of ground water as a strategic and economic asset in Northern Ireland. *Journal of the Chartered Institution of Water and Environmental Management*, **11**, 246–50.

Wanielista, M. P., 1990 *Hydrology and Water Quality Control*, John Wiley and Sons, New York.

Yeh, W. W.-G., 1997 Ground water systems. In: Revelle, C. and McGarity, A. E., eds, *Design and Operation of Civil and Environmental Engineering Systems*, John Wiley and Sons, New York, pp 97–164.

Chapter 3 Factors Determining the Distribution of Animals and Plants in Freshwaters

INTRODUCTION

Many factors affect the presence or absence of organisms and plants in freshwaters. These can be categorized as:

(a) natural dispersion;
(b) abiotic factors;
(c) biotic factors.

The biotic factors can be significant locally in determining whether a species occurs or not. Species may be precluded by predation, competition, or lack of suitable food resources or habitat. These factors are considered in Chapter 4. In this chapter the importance of natural dispersion and the abiotic factors are discussed.

3.1 NATURAL DISPERSION

When carrying out biological surveys of surface waters it is always important to ask the question of whether the expected species naturally occur at the site. This is not only a problem if using an American textbook when working in Europe, but significant variations occur within Europe itself. This is perhaps seen most clearly when comparing extreme northern climates with extreme southern climates. However, it is most profound when the British Isles are compared to mainland Europe. The wave of recolonization of animals from Europe after the last ice age had only reached Scotland before the land bridge to Ireland had separated. While this is the primary cause for the lack of snakes in Ireland, it is also true for freshwater fish. Probably only those migratory species, such as salmon (*Salmo salar*), sea trout (*S. trutta*) and eels (*Anguilla anguilla*), that spend part of their life cycle in the sea have colonized the British Isles naturally. Humans have undoubtedly introduced other species. The diversity of freshwater fish in Ireland still remains lower than in Britain, which in turn is less than on the European continent with the number of species increasing towards the equator. At present there are only 35 species of freshwater fish in the UK.

An example of an introduced species is the roach (*Rutilus rutilus*) which was introduced into Ireland by an angler in 1889, who used it as live bait on the River Blackwater, where it rapidly spread throughout the catchment. During the 1930s it was introduced separately into the Foyle catchment from where it was spread by anglers who wanted to improve coarse fishing or accidentally when it was being used as live bait. It is a hardy fish that likes eutrophication. This gave it an advantage over other species during the 1960s and 1970s when Irish rivers became increasingly nutrient enriched, so by 1980 it was widespread throughout the midlands of the country including the Shannon and Boyne systems. Within a decade of its introduction into Lough Neagh, a large shallow lake in Northern Ireland ($32 \times 16\,km^2$), it had become so common that it was seriously affecting the commercial eel fishery.

In Lough Neagh regular fluctuations in the population of the roach have been reported. Numbers fluctuate between 1 and 17 million fish over a 5- to 7-year cycle. This is due to a biotic factor, an intestinal parasite, *Ligula intestinalis*. It infects the young roach (up to six parasites per fish) and can make up to 35% of the body weight of the fish as the parasite grows. This causes pouching so that the lighter sides of the fish are pushed out like a balloon past its camouflaged back making it clearly visible to predators. Roach lay eggs on macrophytes and so need to enter shallow water or streams feeding the lake in order to spawn on the plants. It is during this time that massive predation of the fish occurs by gulls and other sea birds including waterfowl. The parasite makes the fish progressively lethargic, making them easy prey. The parasite matures in the bird, which reinfects the water with eggs via droppings. The life cycle of the parasite involves an intermediate host, copepods, which are very common in the lake throughout the year. As young roach feed off plankton, it is generally the young fish that are infected completing the life cycle of the parasite.

In recent years exotic species have been introduced throughout the British Isles, some of which are having very serious effects on the natural ecology of lakes and rivers through the introduction of new diseases and parasites, elimination of other aquatic species, and the reduction of other fish species through predation and competition. The Chinese mitten crab has managed to invade many British rivers, such as the Thames and Humber. Probably brought over accidentally in the ballast tanks of ships it excavates deep burrows along riverbanks causing them to collapse. The red-eared terrapin was once sold in thousands as pets throughout Europe. It grows into an aggressive reptile over 200 mm in diameter. It is a voracious carnivore that is wiping out other pond and river animals, such as newts and frogs. The pumpkinseed is an ornamental fish from the US, which is now common in rivers throughout southern England. It is believed to eat the eggs of native fish species. Two monsters have also been introduced by anglers and commercial fisheries. The Wels catfish can grow up to 65 kg and has been introduced throughout the Midlands. It is so large and aggressive that it is able to catch waterfowl. Beluga sturgeon have also been introduced into commercial fisheries where they are highly prized by anglers. However, these fish, which can grow up to 1.25 tonnes, have escaped and are now to be found in a number of English rivers. This introduction of species has gone on for centuries. For example, the rainbow trout, originally from the US, is common in rivers throughout Europe, and in the UK has managed to exclude the native brown trout from many southern rivers. Other native species are currently under severe threat with many close to extinction such as the burbot, a miniature catfish once common in eastern rivers, or the vendace which was once common throughout Britain but is now restricted to Bassenthwaite and Derwentwater in the Lake District. This latter species is under threat from the roffe, a small perch-like fish, which feeds on its eggs.

There are many other examples of alien species that have been deliberately or accidentally introduced. These include macrophytes, bankside vegetation, invertebrates as well as fish (Table 3.1). For example, originally from North America the floating pennywort (*Hydrocotyle ranunculoides*) is an ornamental pond escapee that is now common in streams throughout Europe. It can grow up to 300 mm a day forming large floating mats that cause widespread deoxygenation as well as causing localized flooding by blocking drainage ditches. One of the best-known invertebrate invaders is the Zebra mussel (*Dreissena polymorpha*). This small bivalve is originally from the Black Sea but has spread throughout the world. Distributed originally in the ballast water of ships it was first recorded in the Great Lakes in North America in 1985 reaching peak densities of over $300\,000\,m^2$ in some places within just 3 years. Such high densities have had a serious impact on the lake ecology with the mussels representing up to 70% of the invertebrate biomass in some areas (Nalepa *et al.*,

TABLE 3.1   Examples of alien species causing problems in freshwater systems

Generic name	Common name	Areas of serious invasion
Macrophytes		
Alternanthera philoxeroides	Alligator weed	USA
Crassula helmsii	Swamp stonecrop	Europe
Eichhornia crassipes	Water hyacinth	Tropics, Australia
Hydrocotyle ranunculoides	Floating pennywort	Europe, England
Lythrum salicaria	Purple loosestrife	USA
Pistia stratiotes	Water lettuce	Tropics, Africa
Salvinia molesta	Water fern	Tropics, Africa
Bank-side vegetation		
Fallopia japonica	Japaneses knotweed	Almost world-wide
Heracleum mantegazzianum	Giant hogweed	Europe
Impatiens glandulifera	Himalayan balsam	Europe
Invertebrates		
Astacus astacus	Noble crayfish	Europe, England
Astacus leptodactylus	Turkish crayfish	Europe, England
Dreissena polymorpha	Zebra mussel	USA, Europe
Mysis relicta	Opossum shrimp	Europe
Oronectes limosus	Striped crayfish	Europe, England
Pacifastacus leniusculus	Signal crayfish	Europe
Procambarus clarkii	Red swamp crayfish	Europe, England

2000). Apart from altering community structure and causing deoxygenation due to such high biomasses they also block water supply and irrigation intake screens and pipewick, and are widely reported colonizing water distribution mains. The species has also been implicated in summer blooms of the toxic blue green alga *Microcystis* (Section 6.4.1).

3.2 CATCHMENT WATER QUALITY

The catchment has a profound effect on water quality with variation seen between sub-catchments due to different rock geology and soil types. Water arising from hard water catchments has elevated concentrations of calcium, magnesium and carbonate. Some surface waters close to the coast can be affected by wind-blown sea spray resulting in elevated concentrations of sodium, sulphate and chloride. As precipitation falls on the catchment the water dissolves trace amounts of anything it comes into contact with as it flows over plants, soil and through the rock to finally enter the river system. The longer this contact time the more that will be taken into solution. In this way all surface and ground waters have a unique chemistry, reflecting primarily the nature of the geology and soils of the catchment. Table 3.2 compares the quality of a number of different waters. Table 3.3 shows the origin of a number of common elements in a small catchment in the US. It shows that apart from nitrogen, which is utilized by soil micro-organisms, a significant portion of the elements in water are derived from weathered parent rock.

The discharge rate in rivers is governed by many factors. For example, trees can have a significant effect on stream flow due to high water demand from growth and water loss via evapotranspiration. So when trees are either planted or cut down stream flow may be reduced or increased, respectively. Tree planting can also cause acidification by the trees capturing more precipitation especially in coastal areas (Section 6.6). Harvesting of trees also results in increased run-off carrying organic and inorganic material into the water.

In practice, water found in the natural environment is never pure; it always contains some impurities. The chemistry of water is constantly changing. In the atmosphere water present as vapour has an extremely high surface area ensuring maximum uptake of gases both natural and man-made. This process of dissolving chemicals continues as it makes its way through or over the soil into the surface and underground resources. As aquatic organisms use these impurities for metabolism and reproduction, a certain level of contamination of the water is vital to sustain life. The question of how much contamination by these chemicals is acceptable before the water is no longer suitable for these organisms is the basis of pollution chemistry.

There are three important factors affecting community structure in freshwater systems and in lotic environments in particular. These are the substrate, dissolved mineral concentration and dissolved oxygen which is closely linked with the temperature of the water.

TABLE 3.2 Comparison of the water quality from a number of different catchments in the UK with sea water

	Ennerdale, Cumbria	Cambridge tap water (pre-treatment)	Braintree, Essex	Sea water
Sodium	5.8	1.8	34.3	10 720
Potassium		7.5		380
Magnesium	0.72	26.5	15	1320
Calcium	0.8	51[a]	17.5	420
Chloride	5.7	18	407[b]	19 320
Sulphate	4.6	16	96	2700
Carbonate	1.2	123	218	70
	Soft water	Hard water catchment		

[a]Water containing $CaCO_3$ passes through Na^+ soil (Thanet old maritime sands). Ca^+ replaced by Na^+ so the Ca^+ level is reduced.
[b]Usually due to spray from sea, wind-blown sand or pollution.
Reproduced from Open University (1974) with permission of the Open University Press, Milton Keynes.

TABLE 3.3 Origin of certain elements in a small stream

	Precipitation input ($kg\,ha^{-1}\,year^{-1}$)	Stream output ($kg\,ha^{-1}\,year^{-1}$)	Net difference[*] ($kg\,ha^{-1}\,year^{-1}$)
Silicon	Very low	16.4	+16.4
Calcium	2.6	11.7	+9.1
Sulphur	12.7	16.2	+3.5
Sodium	1.5	6.8	+5.3
Magnesium	0.7	2.8	+2.1
Potassium	1.1	1.7	+0.6
Nitrogen	5.8	2.3	−3.5

[*]The net difference is from weathered parent rock.
Reproduced from Open University (1974) with permission of the Open University Press, Milton Keynes.

3.3 RIVER FLOW AND SUBSTRATE

Rivers are one-way flow systems generated by gravity, and downstream there is a trend towards:

(a) decreasing river slope;
(b) increasing depth;
(c) increasing water volume;
(d) increasing discharge rate;
(e) decreasing turbulence.

The most important component governing plant and animal habitats, and so the basic community structure of streams and rivers, is the flow rate. This in turn controls the nature of the riverbed. There is much confusion between discharge rate ($m^3 s^{-1}$) and current velocity ($m s^{-1}$) partially because current velocity has been used to explain the nature of substrate comprising the bed of rivers and streams. This is out-lined in Table 3.4. The theory is based on the observation that in the upper reaches of the river the current velocity is greatest and here the bed is eroded with fine material being washed downstream, with only the coarse material being deposited. The coarse material is occasionally rolled further downstream causing further erosion and dis-turbance of fine material that is also carried in suspension downstream. As the lower reaches of the river are approached the current velocity falls and so the suspended material is finally deposited. This has resulted in the general classification of rivers by current velocity. Rivers with current velocities $>0.3\,m s^{-1}$ are classed as erosive, while rivers with velocities $<0.3\,m s^{-1}$ are classed as depositing.

In reality, the current velocity has little to do with the nature of the substrate, which is controlled by local forces. The confusion is caused by early freshwater ecologists determining current velocity by measuring surface velocity using floats or by a single measurement using a current meter. In practice, current is not uniform in a stream or river, so a single measurement can give an erroneous estimate of velocity of the water as a whole. The float is in essence measuring the velocity of the float, not the water. In addition, settlement rates and the erosive capabilities of a river are related to other factors that are largely independent of current velocity. The rate of flow (or dis-charge rate) equals the cross-sectional area of a stream or pipe multiplied by the mean velocity of the water through it:

$$Q = AV \ m^3 s^{-1} \tag{3.1}$$

where Q is the volume of water to flow past a given point per unit time ($m^3 s^{-1}$), A is the cross-sectional area of the stream, pipe or conduit (m^2) and V is the velocity of fluid travel in terms of distance per unit time ($m s^{-1}$) (Section 9.2.4).

Mean current velocity (V) is influenced by the slope (s), mean flow depth (d), and the resistance to flow offered by the riverbed and its banks (f). Using the simple equation below to describe mean current velocity, where g is the gravity constant, it can be seen

TABLE 3.4   Relationship of current velocity and the nature of the river bed

Current velocity (m s⁻¹)	Nature of bed	Type of habitat
>1.21	Rock	Torrential
>0.91	Heavy shingle	Torrential
>0.60	Light shingle	Non-silted
>0.30	Gravel	Partially silted
>0.20	Sand	Partially silted
>0.12	Silt	Silted
<0.12	Mud	Pond-like

that mean flow velocity increases where the slope (s) is steeper, flow depth (d) is greater and where the resistance to flow (f) is less:

$$V = \sqrt{\frac{8gsd}{f}} \qquad (3.2)$$

Slope is greatest in the headwaters and declines downstream, and if slope was the only determinant of flow then current velocity would decrease with distance from source, which is the original concept. However, as flow depth increases downstream due to tributaries, and the resistance to flow decreases because of depth increasing and the substrate being composed of finer material, both decreasing the frictional resistance, the mean current velocity will have a tendency to increase downstream. So how can the substrate type be explained? There is great variation in the velocity of a river in cross-section, often reaching maximum velocity at the centre and minimum velocity at the banks if macrophytes are present (Fig. 3.1). In fact, the velocity of the water decreases in a logarithmic way from the water surface (Fig. 3.2). Generally the mean velocity of a lotic system is measured at 60% of its depth, which is important if

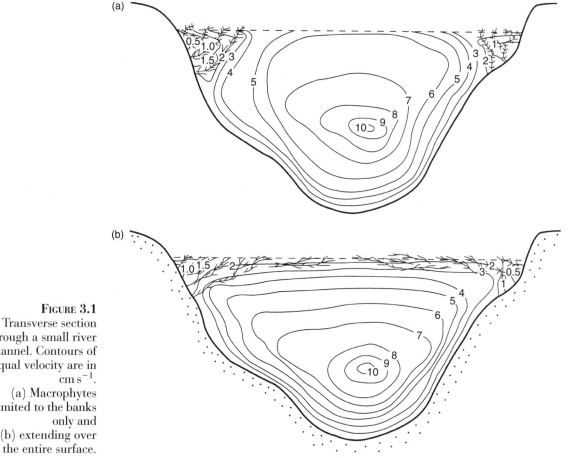

FIGURE 3.1
Transverse section through a small river channel. Contours of equal velocity are in cm s^{-1}.
(a) Macrophytes limited to the banks only and
(b) extending over the entire surface.

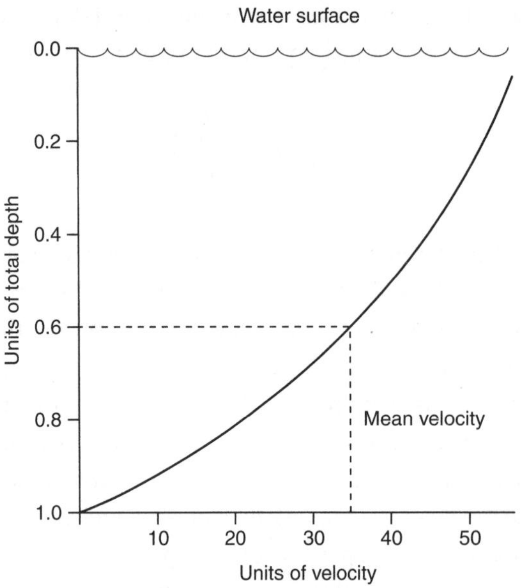

Water surface

FIGURE 3.2
Relationship
between depth and
flow velocity.
(Reproduced from
Townsend (1980)
with permission of
Edward Arnold,
London.)

quantitative samples of drift, planktonic or nektonic (swimming) organisms are being collected. However, the surface or mean velocity of the water is not important to benthic animals or to the determination of the substrate, as these are controlled by local forces known as shear stress or shear velocity.

For an idealized channel, where the roughness of the riverbed can be ignored, the shear stress (u) can be determined as:

$$u = \sqrt{gds} \ \text{m s}^{-1} \tag{3.3}$$

where g is the gravity constant (9.81), d is the flow depth (m) and s is the slope.

In practice, the flow depth and slope vary in opposite manner downstream. As the slope declines far more rapidly than depth increases (100 × decrease in slope and 10 × increase in depth is usual), so the shear stress declines downstream (Townsend, 1980).

EXAMPLE 3.1 Calculate the sheer stress (u) in an ideal channel where: (a) upstream conditions are simulated with a water depth of 0.5 m and a slope of 0.1; (b) downstream conditions have a water depth of 5 m and the slope 0.001.

Using Equation (3.3):

(a) Upstream

$$u = \sqrt{9.81 \times 0.5 \times 0.1} = 0.7 \ \text{m s}^{-1}$$

(b) Downstream

$$u = \sqrt{9.81 \times 5 \times 0.001} = 0.2 \ \text{m s}^{-1}$$

The greater the shear stress exerted on the bed of the river the more likely it is for organisms to be swept away, resulting in an erosional type habitat, and as the shear stress decreases so the riverbed becomes more silted and characteristic of lowland depositional rivers. This gradation of substrate types is not uniform, as slope, depth, mean current velocity and shear stress all alter along the length of rivers. However, unlike slope, shear stress is not constant (Fischer *et al.*, 1979). The effects of turbulence in rivers are considered in more detail in Section 5.3.

3.4 DISSOLVED SOLIDS

Water passes over and through a variety of rocks and soils, and dissolves salts, although rarely in high concentrations in British waters. The nature of the water therefore primarily depends on geology, although other sources such as atmospheric deposition and activities of people (agriculture, drainage, urban run-off, waste disposal, etc.) are also important. These dissolved minerals determine the chemical properties of water such as hardness, acidity, conductivity, which in turn affect the physical properties of water such as colour, taste and odour (Detay, 1997), as well as the capacity of the water to support life.

Conductivity is a measure of water's ability to conduct an electric current. This is linked to the concentration of mineral salts in solution. In fact, conductivity is controlled by the degree to which these salts dissociate into ions, the electrical charge on each ion, their mobility and, of course, the temperature. Conductivity is measured using a specific electrode and is expressed in micro-siemens per centimetre ($\mu S\,cm^{-1}$). The total dissolved solids ($mg\,l^{-1}$) can be calculated by multiplying the conductance by a specific factor (usually between 0.55 and 0.75). The factor increases with increasing sulphate concentration. This factor has to be determined for specific water bodies, but once calculated it remains very stable. Conductivity is widely used for pollution monitoring. Natural rivers and lakes have conductivities between 10 and $1000\,\mu S\,cm^{-1}$. Levels in excess of this certainly indicate pollution, or in groundwater indicate mineralization or salinity problems (Table 2.2). It is possible to approximate water salinity (ppm) from conductivity ($\mu S\,cm^{-1}$) using the relationship:

$$1\text{ ppm} = 1.56\ \mu S\ cm^{-1} \tag{3.4}$$

pH is a measure of the acid balance of a solution and is defined as the negative of the logarithm to the base 10 of the hydrogen ion concentration. So pH 5 is 1000 times more acidic than pH 8. At a given temperature, pH (or hydrogen ion concentration) gives an indication of the acidic or alkaline nature of the water. It is controlled by the dissolved chemical compounds and the biochemical processes in the water. In unpolluted waters pH is primarily controlled by the balance between carbon dioxide, carbonate and bicarbonate ions, as well as natural compounds such as humic and fulvic acids (Fig. 3.3). The pH is normally measured in the field using a meter and an electrode with temperature compensation, as pH is temperature dependent. The pH is also closely linked with corrosion (Section 11.4.1).

Carbon dioxide dissolves readily in water and is closely linked with the chemical processes that determine the acidity and alkalinity of water.

Carbon dioxide is highly soluble, more so at lower temperatures and at higher pressures. It is 40 times more soluble than oxygen (Table 3.5). Although the atmosphere only contains 0.04% CO_2 compared to 21% O_2, CO_2 readily dissolves into cloud vapour due to its high surface area to volume ratio. This forms a weak acid, carbonic acid, making all rain weakly acidic

$$CO_2 + H_2O \rightarrow H_2CO_3 \qquad (3.5)$$

Carbonic acid dissociates to produce free hydrogen ions (H^+) and bicarbonate ions (HCO_3^-)

$$H_2CO_3 \rightarrow H^+ + HCO_3^- \qquad (3.6)$$

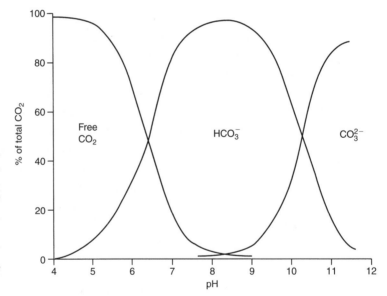

FIGURE 3.3 The relative proportions of different forms of inorganic carbon in relation to the pH of water under normal conditions. (Reproduced from Chapman (1996) with permission of UNESCO, Paris.)

TABLE 3.5 Solubilities of common gases in water

Gas	Solubility $(mg\,l^{-1})$
Carbon dioxide (CO_2)	2318
Hydrogen (H_2)	1.6
Hydrogen sulphide (H_2S)	5112
Methane (CH_4)	32.4
Nitrogen (N_2)	23.3
Oxygen (O_2)	54.3

So the acidity of water is determined by the abundance or, more correctly, the activity of H^+ ions. In pure water, dissociation is limited with only 10^{-7} mol l^{-1} of H^+ ions present (i.e. pH 7). Higher concentrations of H^+ ions make water more acidic while lower concentrations make it more alkaline. For example, at pH 5 water contains 10^{-5} mol l^{-1} of H^+ ions which is 100 times more acidic than at pH 7 and vice versa in alkaline waters. Carbonic acid dissociates readily and has a pH of 5.6. Other chemicals can release or lock up H^+ ions so that the pH is altered as they are dissolved. This is the essence of buffering.

Bicarbonate ions (HCO_3^-) can further dissociate into H^+ and CO_3^{2-}

$$HCO_3^- \rightarrow H^+ + CO_3^{2-} \qquad (3.7)$$

Both HCO_3^- and CO_3^- can further react with water to produce a hydroxide ion (OH^-).

$$HCO_3^- + H_2O \rightarrow H_2CO_3 + OH^- \qquad (3.8)$$

$$CO_3^{2-} + H_2O \rightarrow HCO_3^- + OH^- \qquad (3.9)$$

All three of these ions contribute to the alkalinity of water.

Aquatic plants can use CO_2 or HCO_3^-. During the day uptake of CO_2 may become exhausted so that HCO_3^- is used. Hydroxide ions are secreted replacing the HCO_3^-. Some of the freed CO_2 will be precipitated as $CaCO_3$ (marl) in hard waters and is then permanently lost (Section 20.5.1). This often results in vegetation and debris becoming coated with $CaCO_3$ occasionally resulting in drains and small culverts becoming blocked by the precipitated material. The overall result is an elevated pH (9–10) in water containing actively photosynthesizing macrophytes or algal blooms. At night CO_2 is released and the process is reversed so that the pH returns to normal. In clean water pH is controlled by the balance between CO_2, HCO_3^- and CO_3^{2-} as well as organic acids. So CO_3^{2-}, HCO_3^- and H_2CO_3 are all inorganic forms of CO_2, and their relative contribution to the total CO_2 concentration controls the pH.

Acidity and alkalinity are the base and acid neutralizing capacities of water. If the water has no buffering capacity then these are inter-related with pH. Most natural waters will contain weak acids and bases, so acidity and alkalinity should also be tested with pH. Acidity in water is controlled by the presence of strong mineral acids, weak acids (e.g. carbonic, humic, fulvic) and the hydrolysing salts of metals (e.g. iron and aluminium). It is determined by titration with a strong base up to pH 4 (free acidity) or to pH 8.3 (total acidity). Alkalinity is controlled by the sum of titratable bases (principally carbonate, bicarbonate and hydroxides). Waters of low alkalinity (e.g. <24 mg l^{-1} as $CaCO_3$) have a low buffering capacity making them very susceptible to pH alteration due to atmospheric fallout, acid rain, eutrophication (photosynthesis)

TABLE 3.6   Examples of the arbitrary scales of hardness in use

$mg\,l^{-1}$ as $CaCO_3$	Degree of hardness	$mg\,l^{-1}$ as $CaCO_3$	Degree of hardness
0–75	Soft	<50	Soft
75–150	Moderately hard	50–100	Moderately soft
150–300	Hard	100–150	Slightly hard
300+	Very hard	150–250	Moderately hard
		250–350	Hard
		>350	Excessively hard

and acidic pollutants. Alkalinity is determined by titration using a strong acid to lower the pH to 8.3 for free alkalinity or to pH 4 for total alkalinity.

The hardness of water varies from place to place, reflecting the nature of the geology with which the water has been in contact. In general, surface waters are softer than groundwaters, although there are many extremely soft groundwaters. Hardness is caused by divalent metal cations that can react with certain anions present to form a precipitate. Only divalent cations cause hardness, so Na^+ (monovalent) is not important. Hardness is expressed in mg $CaCO_3\,l^{-1}$. The principal cations (and major anions associated with them) are: Ca^{2+} and (HCO_3^-); Mg^{2+} and (SO_4^{2-}); Sr^{2+} and (Cl^-); Fe^{2+} and (NO_3^{2-}) and finally Mn^{2+} and (SiO_3^{2-}). Strontium, ferrous iron and manganese are usually found in trace amounts in water and so are usually ignored in the calculation of hardness, so that total hardness is taken to be the concentration of calcium and magnesium only. Aluminium and ferric iron could affect hardness but solubility is limited at the pH of natural water so that ionic concentrations are negligible. In hard water Ca^{2+}, Mg^{2+}, SO_4^{2-}, CO_3^- and HCO_3^- ions are more abundant. The separation of water into soft and hard relies on the use of an arbitrary scale (Table 3.6). There are a number of different forms of hardness, apart from total hardness, which is used here, that only measure one component of hardness as defined above. These include calcium hardness, carbonate hardness, magnesium hardness, temporary and permanent hardness (Flanagan, 1988).

Water analysis may not show hardness as a parameter but simply give concentrations for individual ions. It is possible to convert concentrations for the individual ions into a total hardness value expressed as calcium carbonate equivalent by using the equation:

$$\text{Hardness (mg } CaCO_3\,l^{-1}) = \text{Ion (mg } l^{-1}) \times \frac{\text{Equivalent weight of } CaCO_3}{\text{Equivalent weight of ion}} \quad (3.10)$$

EXAMPLE 3.2   For a water sample containing $12\,mg\,l^{-1}$ of Mg^{2+} and $84\,mg\,l^{-1}$ of Ca^{2+}, where the equivalent weights of $CaCO_3$, Mg^{2+} and Ca^{2+} are 50, 12.16 and 20.04, respectively, the hardness is calculated as:

$$\text{Hardness} = \left[12 \times \frac{50}{12.16}\right] + \left[84 \times \frac{50}{20.04}\right] = 259 \text{ mg } CaCO_3\,l^{-1}$$

TABLE 3.7   Examples of the hardness of some Irish rivers

River	$mg\,l^{-1}$ as $CaCO_3$
R. Ward (N. Dublin)	320–415
R. Tolka (W. Dublin)	290–390
R. Suir (Thurles)	155–350[a]
R. Liffey (S. Dublin) Headwater	40–75
R. Liffey (S. Dublin) Lucan	104–332[a]
R. Woodford (Cavan)	74–87
R. Dodder (S. Dublin)	30–210[a]
R. Blackwater (Mallow)	30–140
R. Sullane (Macroom)	20–44
R. Owenmore (Cavan)	20–26
R. Owenduff (Blackpool)	12–48

[a]These rivers indicate systems showing a marked increase
in their hardness with stream length.

The hardness of surface waters varies considerably between catchments, although the hardness generally increases downstream, with high values often due to human activity (Table 3.7). The hardness of water has profound effects on the animals present. For example, the osmotic regulatory problems faced by organisms in very soft water makes them more susceptible to pollution. The presence of Ca^{2+} and Mg^{2+} ions affect the rate of respiration, increasing the rate at lower temperatures and reducing the rate at higher temperatures. Some species require Ca^{2+} for shell construction, and many species of snails, flatworms and leeches are restricted to hard waters where Ca^{2+} ion concentration is above certain concentrations. Thus many insects, which are indifferent to water hardness, are found in greater abundance in soft waters due to reduced competition. Crustaceans, snails, larvae of caddis and chironomids, and most worms are most abundant in very hard waters.

As a general rule rivers with hard waters differ from soft waters due to Ca^{2+} and pH, and as they usually rise as springs and may have springs in the bed, they retain constant clarity, flow and temperature throughout the year. In contrast, soft water rivers are usually raised as surface run-off from mountains and are therefore usually flashy (i.e. rapid variation in discharge rate) resulting in sudden floods and droughts. Such rivers are usually highly turbid with high suspended-solid concentrations during high flows. Soft rivers often have an impoverished fauna with a low productivity, and the water has a high concentration of humic material giving it a characteristic brown colour. The different physical parameters encourage a different fauna, as do the chemical factors.

3.5 DISSOLVED OXYGEN AND TEMPERATURE

Plants and animals are all vital to the cleanliness of rivers, and all aquatic organisms require oxygen. However, water at normal temperatures holds very little oxygen compared to the air. Gas molecules in the atmosphere diffuse or move from an area of high concentration to an area of low concentration. In the same way, oxygen molecules diffuse through the air–water interface into the water where they become dissolved. At the same time oxygen is diffusing in the opposite direction, but when the volume of oxygen diffusing in either direction is equal, then the water is said to be in equilibrium and is therefore saturated with oxygen.

The solubility of oxygen depends on three factors, the temperature, pressure and the concentration of dissolved minerals in the water. Fresh water at 1 atmosphere pressure at 20°C contains $9.08\,g$ of $O_2\ m^{-3}$ and as the temperature increases the saturation concentration (i.e. the amount of oxygen that can dissolve in water) decreases (Table 3.8). An increase in the concentration of dissolved salts lessens the saturation concentration

TABLE 3.8   Variation in the dissolved oxygen saturation concentration of freshwater with temperature at 1 atmosphere pressure

Temperature (°C)	Dissolved oxygen (mg l⁻¹)
0	14.6
1	14.2
2	13.8
3	13.4
4	13.1
5	12.8
6	12.4
7	12.1
8	11.8
9	11.6
10	11.3
11	11.1
12	10.8
13	10.5
14	10.3
15	10.1
16	9.9
17	9.7
18	9.5
19	9.3
20	9.1
21	8.9
22	8.7
23	8.6
24	8.4
25	8.2
26	8.1
27	8.0
28	7.8
29	7.7
30	7.5

of oxygen, which is why sea water has lower saturation concentrations than freshwater at the same temperature and pressure. A decrease in atmospheric pressure causes a decrease in oxygen saturation. Therefore, streams at high altitudes have less oxygen at a certain temperature than lowland streams (Table 3.9). In the British Isles this is of little practical significance, although in respiration experiments, where small changes in the oxygen concentration are being measured, then changes in pressure can be highly significant.

It is very important to know how quickly oxygen dissolves into water, and this depends to a large extent on the level of oxygen in relation to saturation concentration of the water, which is known as the oxygen deficit. For example, water containing $10\,g\,O_2\,m^{-3}$ but with a saturation concentration of $12\,g\,O_2\,m^{-3}$ has a oxygen deficit of $2\,g\,O_2\,m^{-3}$. Since the rate of diffusion is directly proportional to the oxygen deficit, if the same water now contains only $4\,g\,O_2\,m^{-3}$ the oxygen deficit will be $8\,g\,O_2\,m^{-3}$ causing the oxygen to diffuse four times faster.

When water is saturated with oxygen and is equilibrated then it is said to be 100% saturated regardless of the temperature. So for comparative purposes in water monitoring the percentage saturation is normally used. The oxygen saturation of water (%) at temperature A is calculated as:

$$\frac{\text{Concentration of oxygen in water at temperature } A}{\text{Saturation concentration for oxygen at temperature } A} \times 100 \quad (3.11)$$

TABLE 3.9   Correction factor for changes in oxygen concentration due to pressure

Altitude (m)	Pressure (mm)	Factor	Altitude (m)	Pressure (mm)	Factor
0	760	1.00	1300	647	1.17
100	750	1.01	1400	639	1.19
200	741	1.03	1500	631	1.20
300	732	1.04	1600	623	1.22
400	723	1.05	1700	615	1.24
500	714	1.06	1800	608	1.25
600	705	1.08	1900	601	1.26
700	696	1.09	2000	594	1.28
800	687	1.11	2100	587	1.30
900	676	1.12	2200	580	1.31
1000	671	1.13	2300	573	1.33
1100	663	1.15	2400	566	1.34
1200	655	1.16	2500	560	1.36

Divide the saturation concentration (Table 3.8) by the correction factor to give the adjusted saturation concentration at the specific elevation. For example, the saturation concentration of freshwater at 12°C is $10.8\,mg\,l^{-1}$. For an altitude of 600 m this is divided by the correction factor 1.08 to give the saturation concentration of water at 600 m of $10.0\,mg\,l^{-1}$.

In water bodies where there is little mixing, the oxygen must diffuse from the air–water interface into the water setting up an oxygen gradient. This is often the case in lentic systems (Section 3.8). Turbulence causes quicker mixing within the water breaking down the oxygen gradient and ensuring a maximum oxygen deficit, and so increasing the transfer rate. Turbulence is directly related to river gradient and bed roughness. Overall amounts of oxygen transferred depend on the surface area to volume ratio of the water body so that the ratio is larger in a shallow wide river than a narrow deep one. This means that although the oxygen deficit is the same, the shallow wider river will re-oxygenate faster (Section 5.3).

Oxygen concentration in rivers can become supersaturated, up to 200% under conditions of agitation below waterfalls and weirs, and also in bright sunlight where algae is abundant due to photosynthesis. In practice, the elevated oxygen concentration will quickly return to equilibrium by diffusion and excess oxygen will be lost to the atmosphere. The diurnal variation caused by excessive algal growth (Fig. 3.4) in water bodies is considered in detail in Section 6.3. Biologically animals and plants are always using oxygen and so are always increasing the oxygen deficit. Some inorganic substances, such as sulphite (SO_3^{2-}), sulphide (S^{2-}) and iron (Fe^{2+}), take part in chemical reactions which also consume oxygen, thereby increasing the deficit.

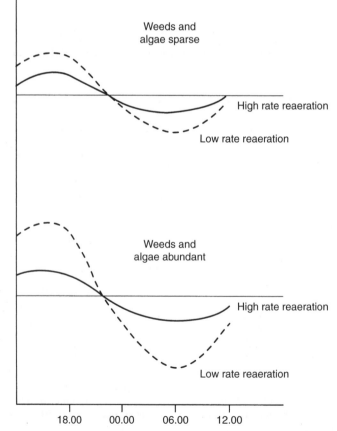

FIGURE 3.4 Diurnal variation of dissolved oxygen in rivers due to eutrophication. Difference in oxygen concentration (y-axis) at different rates of reaeration (Section 5.3) is shown.

3.6 SUSPENDED SOLIDS

The type and concentration of suspended solids controls the turbidity and transparency of water. Suspended solids are insoluble particles, or soluble particles, too large to dissolve quickly, which are too small to settle out of suspension under the prevailing turbulence and temperature. Suspended solids consist of silt, clay, fine particles of inorganic and organic matter, soluble organic compounds, plankton and other micro-organisms. These particles vary in size from 10 nm to 0.1 mm in diameter. In practice, suspended solids are measured as the fraction of these solids retained on a filter paper with a pore size of 0.45 μm. Particles less than 1 μm can remain in suspension indefinitely and are known as colloidal solids. These fine solids impart a cloudy appearance to water known as turbidity. Turbidity is caused by the scattering and absorption of the incident light by the particles present. This affects the transparency or visibility within the water. Turbidity can be caused naturally by surface run-off due to heavy rain, or by seasonal biological activity. Turbidity is also caused by pollution and as such can be used to monitor effluents being discharged to sewers or surface waters, and can be closely correlated with other parameters. Turbidity is measured using nephelometry, which is the measurement of light scattering by the suspended solids. Results are expressed in nephelometric turbidity units (NTU). Owing to problems of flocculation and settlement of particles, as well as precipitation if the pH alters during storage, turbidity should be measured in the field whenever possible.

Erosion, transportation and deposition of particulate matter are a function of a number of factors including current velocity, turbulence, particle size and density (Fig. 3.5). Under extreme flow conditions large material can be mobilized. Very large material either rolls or bounces (saltation) downstream. In an idealized channel

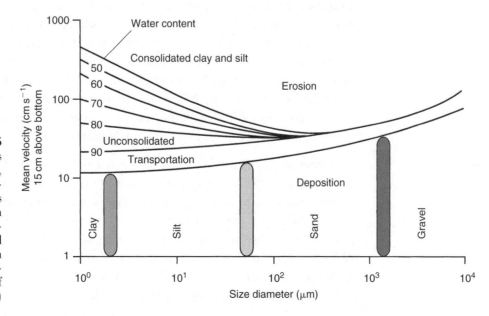

FIGURE 3.5 Velocity curves defining erosion, transport and deposition of sediments of differing grain size and water content. (Reproduced from Chapman (1996) with permission of UNESCO, Paris.)

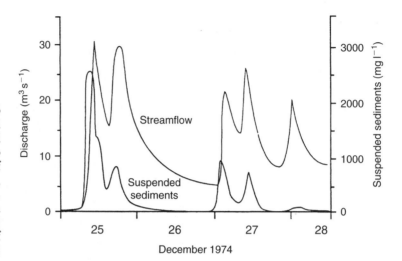

FIGURE 3.6
The temporal
relationship of total
suspended solids to
the hydrography of
the River Exe,
England.
(Reproduced from
Chapman (1996)
with permission of
UNESCO, Paris.)

the vertical profile through the water would show that at high flows the smallest particles are at the top and the largest particles near the bottom of the water column (e.g. clay, silt, sand). The concentration of suspended solids varies with the discharge rate in the river (Fig. 3.6). Sediment is mobilized by an increase in discharge rate, although it quickly becomes exhausted as it is scoured away. Subsequent storms may result in much lower suspended-solid concentrations if insufficient time has passed to allow the sediment to be replenished. The peak suspended-solid concentration generally occurs before the peak in discharge rate occurs, a process known as 'advance'.

3.7 RIVER ZONATION

River zones are classified according to the fish present, and this longitudinal pattern of fish distribution is a reflection of oxygen availability along the length of the river (Table 3.10). There is a strong correlation between dissolved oxygen concentration and zonation in rivers, with the trout zone at the headwaters and the bream zone at the lowland end. In practice, some zones may be repeated along the length of the river, while other zones may be missing. Alternatively a river may comprise of a single zone along its entire length. For example, some short mountainous rivers that quickly discharge into the sea may be entirely trout zone.

The community structure of a water body is dependent on a wide variety of physical and chemical factors. Some of these effects are more influential than others, some are direct while others indirect, affecting organisms by influencing other factors (Fig. 3.7). For example, if the depth of a stream is increased this will reduce light penetration which will suppress both plant growth and photosynthesis, which in turn reduces habitat diversity and primary productivity leading to a reduction in species diversity and species abundance (Section 4.2). Man-made factors (e.g. pollution) can have a profound effect on community structure (Chapter 6).

TABLE 3.10   River zones classified by fish presence

Trout zone
Characteristic species   Trout, bullhead
Oxygen requirements   Fish require much oxygen. Likely to be excluded if concentration
                                          falls below saturation. Minimum DO concentration is 7–8 mg l^{-1}

Minnow zone
Characteristic species   Perch, minnow
Oxygen requirements   Fish do not require 100% saturation. Can live for periods of
                                          weeks at 6–7 mg l^{-1} DO. Need more oxygen for active life

Chub zone
Characteristic species   Roach, pike, chub
Oxygen requirements   Fish require only 60–80% saturation for active life. Can tolerate
                                          very low DO, and live at 3 mg l^{-1} for long periods

Bream zone
Characteristic species   Carp, tench, bream
Oxygen requirements   Fish, survive very low DO concentrations 30–40% for long
                                          periods, down to <1.0 mg l^{-1} for short periods

DO: dissolved oxygen.

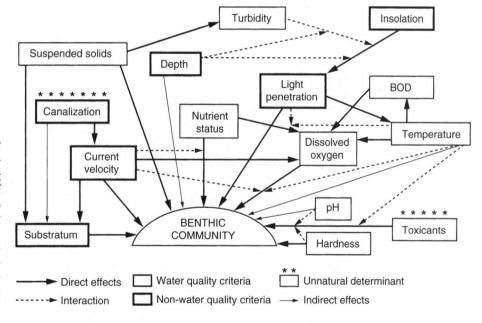

FIGURE 3.7
Sample flow
diagram showing
interactions of
abiotic factors
affecting community
structure.
(Reproduced from
Hawkes (1970) with
permission of John
Wiley and Sons Ltd,
Chichester.)

3.8 LENTIC SYSTEMS

Flowing (lotic) and standing (lentic) waters are different in numerous ways. In deep
ponds and lakes many of the ecologically important physico-chemical parameters,
such as temperature, light intensity and dissolved oxygen, vary in intensity or concentration in vertical directions as distinct depth profiles. While in rivers horizontal
or longitudinal patterns of these parameters develop.

Lentic bodies are not only lakes and ponds, but any hollow or depression that will hold water, even temporarily, can be classified as a lentic system. Among the more unusual lentic systems are puddles, water collected in discarded rubbish bins or hollows in trees. Each of these temporary systems has a unique fauna and flora (Williams, 1987). Lakes can be classified according to their mode of formation:

(a) rock basins are formed by a depression of the landscape due to glacial, tectonic or volcanic activity, dissolution of the bedrock and even the impact of meteorites;

(b) barrier basins are formed when an open channel is blocked by a landslide, larva flow, glacial ice or moraine, for example, ox-bow lakes;

(c) organic basins are those created by humans, animals (e.g. beavers) or vegetation.

The substrate in lakes is generally similar to the sediment found in a very depositional lowland river, being composed of fine material although rarely as rich in organic matter. Sediment arises from a number of different sources in lakes. The main sources are riverine inputs; shoreline erosion; erosion of the lake bed which depends on the strength of local erosional forces; airborne inputs of fine inorganic and organic particulate matter (e.g. pollen); and particles generated within the lake (autochtonous material) which can be either organic (e.g. algae and faecal pellets) or inorganic (e.g. chemical precipitates).

The euphotic zone is that surface layer of the lake that is well illuminated and permits photosynthesis to take place. Transparency in lakes is measured by lowering a Secchi disc, a circular disc 300 mm in diameter and marked in black and white sectors, into the water. The depth at which the disc disappears and just reappears is known as the depth of transparency and is used to measure the euphotic depth in lakes. The lowest limit of light penetration is the compensation point below which is the profundal zone. Heat penetration is closely linked to light intensity which leads to thermal stratification (Fig. 2.2).

Trophic status is widely used to classify lakes using the concentration of phosphorus, the limiting nutrient for algal growth in lentic systems, and determinations of the biomass using chlorophyll concentrations (Table 3.11). Lakes with a low productivity, low biomass and low nutrient concentration (nitrogen and phosphorus) are classed as oligotrophic. These lakes tend to be saturated with oxygen throughout their depth and are dominated by trout and whitefish. Lakes with a high concentration of nutrients and a resultant high biomass production and poor transparency are known as eutrophic. Oxygen concentrations are subject to diurnal fluctuations in the epilimion and may be below $1\,\mathrm{mg\,l^{-1}}$ in the hypolimnion during summer stratification (Figs 2.2 and 3.8). Fish species are restricted to those able to tolerate low dissolved oxygen concentrations, such as perch, roach and bream. Mesotrophic lakes are a transition between the two conditions due to nutrient enrichment. Lake trout are usually eliminated with

TABLE 3.11   Categorization of lakes using nutrient levels, biomass and productivity

Trophic category	Mean total phosphorus (mg m⁻³)	Annual mean chlorophyll a (mg m⁻³)	Chlorophyll a maxima (mg m⁻³)	Annual mean Secchi disc transparency (m)	Secchi disc transparency minima (m)	Minima oxygen (% saturation)ᵃ	Dominant fish
Ultra-oligotrophic	<4.0	<1.0	<2.5	>12.0	>6.0	<90	Trout, whitefish
Oligotrophic	<10.0	<2.5	<8.0	>6.0	>3.0	<80	Trout, whitefish
Mesotrophic	10–35	2.5–8	8–25	6–3	3–1.5	40–89	Whitefish, perch
Eutrophic	35–100	8–25	25–75	3–1.5	1.5–0.7	40–0	Perch, roach
Hypereutrophic	>100.0	>25.0	>75.0	<1.5	<0.7	10–0	Roach, bream

ᵃPercentage saturation in bottom waters depending on mean depth.
Reproduced from Chapman (1996) with permission of UNESCO, Paris.

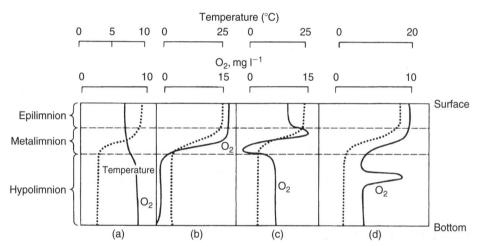

FIGURE 3.8   Types of oxygen distribution with depth. (a) Orthograde curve typical of an unproductive lake. (b) Clinograde curve from a productive lake. (c) Positive and negative heterograde curves. Here photosynthesis from a layer of algae just above the thermocline raises oxygen concentration in the upper part of the metalimnion. Respiration occurring during bacterial decomposition and zooplankton grazing lowers oxygen levels just below the thermocline (metalimnion minimum). (d) Anomalous curves due to the inflow of dense, cool, oxygen-rich stream inflows, which form a discrete layer. In this example, the oxygen-rich stream inflow is in midhypolimnion. (Reproduced from Horne and Goldman (1994) with permission of McGraw-Hill, New York.)

whitefish and perch dominating, and there will be some depression of oxygen levels in the hypolimnion during summer stratification. Two extreme lake classifications are also used, ultra-oligotrophic and hypereutrophic. The former are usually newly formed lakes or isolated lakes with few input routes for nutrients. The later are those with extreme nutrient concentrations which are devoid of oxygen in the hypolimnion during summer stratification. In lentic systems the process of eutrophication is the internal generation of organic matter (autotrophy). However, external inputs of organic matter from the catchment (allotrophy) are also important and can result in lakes rich in humic material (dystrophic lakes). The physical parameters that significantly affect water quality in lakes are retention time, nature of inflow, temperature changes and wind-induced mixing.

REFERENCES

Chapman, D. (ed.), 1996 *Water Quality Assessments*, 2nd edn, E. & F. N. Spon, London.

Detay, M., 1997 *Water Well: Implementation, Maintenance and Restoration*, John Wiley and Sons, Chichester.

Fischer, H. B., List, E. J., Koh, R. C. Y., Imberger, J. and Brooks, N. H., 1979 *Mixing in Inland and Coastal Waters*, Academic Press, New York.

Flanagan, P. J., 1988 *Parameters of Water Quality: Interpretation and Standards*, Environmental Research Unit, Dublin.

Hawkes, H. A., 1970 Invertebrates as indicators as river water quality. In: Evison, L. M. and James, A., eds, *Biological Indicators of Water Quality*, John Wiley and Sons, Chichester, pp 2.1–2.45.

Hawkes, H. A., 1975 River zonation and classification. In: Whitton, B. A., ed., *Studies in Ecology: 2. River Ecology*, Blackwell Scientific Publications, Oxford, pp 312–74.

Horne, A. J. and Goldman, C. R., 1994 *Limnology*, McGraw-Hill, New York.

Nalepa, T. F., Fahnensteil, G. L. and Johengen, T. H., 2000 Impacts of the zebra mussel (*Dreissena polymorpha*) on water quality: a case study in Saginaw Bay, Lake Huron. In: Claudi, R. and Leach, J. H., eds, *Non-indigenous Freshwater Organisms*, Lewis, Boca Raton, FL, pp 255–72.

Open University, 1974 *The Earth's Physical Resources: 5. Water Resources, S26,* Open University Press, Milton Keynes.

Townsend, C. R., 1980 *The Ecology of Streams and Rivers*, Edward Arnold, London.

Williams, D. D., 1987 *The Ecology of Temporary Water*, Croom Helm, London.

FURTHER READING

Horne, A. J. and Goldman, C. R., 1994 *Limnology*, McGraw-Hill, New York.

Peters, N. E., Brickner, O. P. and Kennedy, M. M., 1997 *Water Quality Trends and Geochemical Mass Balance*, John Wiley and Sons, Chichester.

Petts, G. and Calow, P. (eds), 1966 *River Flows and Channel Forms*, Blackwell Science, Oxford.

Whitton, B. A. (ed.), 1975 *Studies in Ecology: 2. River Ecology*, Blackwell Scientific Publications, Oxford.

Chapter 4 Basic Aquatic Ecosystems

4.1 Community Structure

Energy is required to sustain all living systems. On land solar energy is converted to organic molecules by photosynthesis. In aquatic ecosystems the energy comes from two separate sources. As on land some energy is produced by primary producers by photosynthesis (autochthonous input), while the remainder comes from organic and inorganic material swept into the water from land (allochthonous input). In rivers the secondary production (i.e. biomass of invertebrates and fish) is more dependent on the primary production of the surrounding terrestrial zone than the primary production of the river itself.

The basic food cycle in a river is shown in Fig. 4.1. From this, different groups of organisms can be seen, each with a specific feeding regime, forming a food chain comprising a number of different trophic levels. A balance develops between total productivity of living organisms, and the amount of death and decomposition over a period of time and over a large area of clean river. A clean river neither becomes choked nor devoid of living organisms, with the amount of photosynthesis balanced by the amount of respiration. This stability is known as 'ecological equilibrium'. Built into this system is the self-purification capacity of rivers. This is the ability of the system to increase activity at various trophic levels to deal with an increase of organic matter or other energy source into the system. The increase in nutrients or organic matter will therefore produce an increase in the number of organisms in the food chain resulting in the breakdown of equilibrium, the rapid expansion of certain groups and the elimination of others. In practice, unless the pollution input is constant over a long time, only the micro-consumers will be able to respond to increased allochthonous inputs. Nutrients will result in excess algal development (eutrophication) (Section 6.4), while extra soluble organic carbon may cause sewage fungus (Section 6.3). The organisms comprising the various consumer groups have long life cycles, compared to the bacteria, fungi and algae comprising the micro-consumers, and so are unable in the short term to control their development through increased feeding. Owing to the increased demand for oxygen exerted by the micro-organisms and the algae, the higher trophic levels may even become eliminated due to reduced oxygen concentrations in the water. Sewage

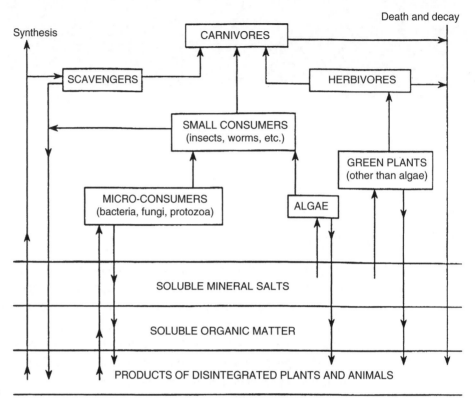

FIGURE 4.1
Simple lotic commu-
nity structure.

treatment utilizes this natural ability of micro-organisms and algae to respond quickly to increases in organic matter and nutrients, and to oxidize them to safe end products (Chapter 15). In practice, the more complex an aquatic ecosystem, the more stable it is. Simplifying the food web of an ecosystem decreases the probability of maintaining ecological equilibrium.

There are five major communities in aquatic systems: phototrophic (autotrophic), heterotrophic, detritivore, herbivore and predator.

4.1.1 PHOTOTROPHIC (AUTOTROPHIC)

Primary producers from either the microalgal or macrophyte communities both produce plant material as the energy base for the rest of the system. Microalgae are present either as a biofilm that grows over the surfaces of solid substrates (periphyton) or as free-living cells suspended in the water (phytoplankton). It is an important food source not only for many macro-invertebrates, but also for fish. Macrophytes are large aquatic plants, although their distribution is dependent on sheer stress, substrate type and depth. In upstream areas, the rocky substrate is unsuitable for root development. The high sheer stress and possible low light intensity due to shading restricts macrophyte development to mosses and liverworts. In contrast, in lowland reaches where the sheer stress is lower, there is high light intensity and a silty substrate for easy rooting, macrophytes are common. However, variability in the water level may reduce extensive macrophyte development to banks, while an increase in turbidity may reduce submerged growth.

4.1.2 HETEROTROPHIC

Bacteria, fungi and protozoa all break down organic matter. They may form a biofilm on submerged surfaces (periphyton) and use dissolved organic matter. They can also grow on organic matter which they decompose, often relying on the water column for nutrients. Periphyton is a major source of food for detritivores, as is the partially degraded organic matter.

4.1.3 DETRITIVORES

These feed on particulate detritus present in the water either as discrete particles or periphyton. Detritivores can be either micro- or macroscopic, but are divided into functional feeding groups based on their mode of food selection and ingestion. Collectors aggregate fine particulate organic matter (FPOM), shredders consume large or coarse particulate organic matter (CPOM), while scrapers (grazers) feed off the thin accumulation of periphyton from solid surfaces.

4.1.4 HERBIVORES

There are very few animals that eat living plant material in aquatic ecosystems. Those that do are specialized as either piercers or shredders. Some macro-invertebrates, such as the hydroptilid caddis fly larvae, pierce algal cells and ingest the contents. While many more species, such as *Lepidoptera* larvae and some adult beetles, eat macrophyte tissue (shredders). Some shredders and scrapers eat both living algal and macrophyte material, and so must be considered both herbivorous and detritivorous.

4.1.5 PREDATORS

Predators can be micro- and macro-invertebrates, fish and even higher vertebrates, such as amphibians, birds and mammals. Generally detritivores are eaten, as are small fish. There may be several predator levels within the food web (e.g. detritivore (chironomid) → predator (*Stenopoda*) → predator (trout) → predator (grey heron)).

An example of a food chain is given in Fig. 4.2. Here the organisms are separated into producers, consumers and decomposers.

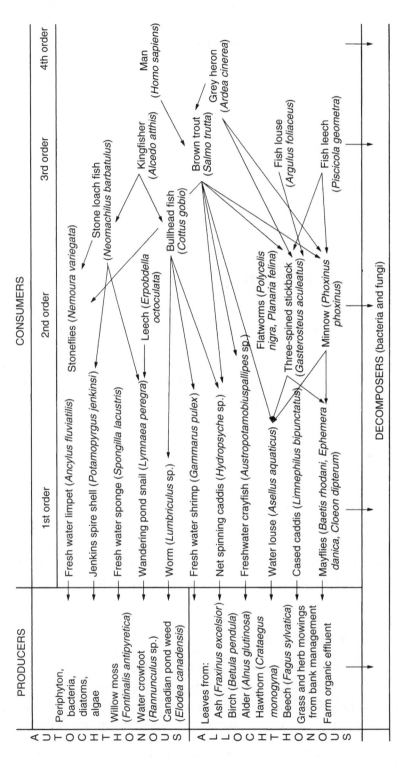

Figure 4.2 Stream food web of the Woodbrook, Loughborough, UK. (Reproduced with permission of the Open University Press, Milton Keynes.)

4.2 ALLOCHTHONOUS AND AUTOCHTHONOUS INPUTS

The contribution of energy from allochthonous and autochthonous inputs (Section 4.1) will vary along the length of a river, and will also vary seasonally. This can be illustrated by examining a theoretical river basin (Fig. 4.3) (Vannote *et al.*, 1980).

4.2.1 UPLAND (SHADED) REGION

The main input of organic matter is plant litter from overhanging trees and bank vegetation (riparian). This input will be seasonal, and must be microbially broken down (i.e. conditioned) prior to being consumed by detritivores, mainly shredders. Some organic matter will be washed downstream, although this will be limited. This is a heterotrophic system with the energy coming largely from outside the stream. It is a net importer of material and energy. This region can be classified as having a low production (P):respiration (R) ratio ($P/R < 1$) or a low phototrophic:heterotrophic ratio.

4.2.2 MIDLAND (UNSHADED) REGION

In this section the river is wider and less shaded resulting in a higher light penetration. This gives a high growth of microalgae on submerged objects, including epiphytes on larger plants and macrophytes. The primary production is controlled by depth and water clarity, being greater if the water is shallow and clear. This is a phototrophic system supported primarily by organic matter fixed *in situ* by photosynthesis. The area may also be

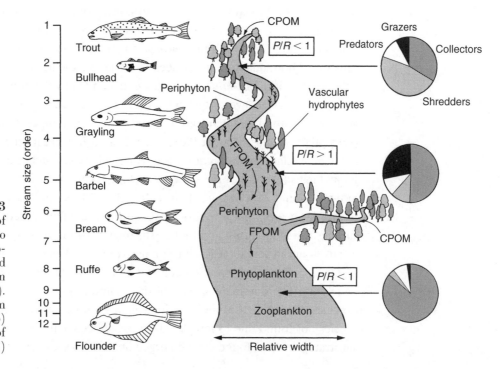

FIGURE 4.3 Classification of rivers according to associated macro-invertebrates and ratios of production (P) to respiration (R). (Reproduced from Chapman (1996) with permission of UNESCO, Paris.)

a net exporter of organic matter. This is an area dominated by grazers, collectors and herbivores. It has a high P/R ratio (>1) and a high phototrophic:heterotrophic ratio.

4.2.3 LOWLAND (UNSHADED) AREA

Further downstream the continuing increase in depth and a higher turbidity reduces photosynthesis. The source of energy is mostly from upstream FPOM. Once again there is a low P/R ratio (<1) and a low phototrophic:heterotrophic ratio.

Rivers are primarily heterotrophic systems with an average of 60% of the biotic energy allochthonous in origin. Lakes in contrast are primarily phototrophic systems with an average of 80% of the biotic energy autochthonous in origin.

4.3 HABITAT ZONATION

Riverine habitats can be loosely classified as either erosional or depositional depending on the dominant substrate particle size (Table 4.1). This also influences the type of organic matter (e.g. coarse or fine), type of primary producers (i.e. vegetation type) which then controls the relative abundance of the specific groups of invertebrates present (Cummins, 1975).

There are two main types of substrate in rivers, stony and silty. Each has its own flora and fauna adapted to living under the prevailing conditions. The substrates are separated by a sandy substrate, which is largely devoid of both plants and animals. Stony substrates are found in erosional areas, where there is a high sheer stress and rapid turbulent flow. Most of the organisms are adapted to this environment. For example, limpets and leeches attach themselves to stones, while *Simulium* and the caseless caddis (e.g. *Hydropsyche*) are anchored by silken webs. Most insect larvae, mayflies, stoneflies and some caddis flies have strong prehensile claws in order to cling to the substrate. Other adaptations are seen, for example the flattened bodies of flatworms and the behaviour of *Gammarus* hiding under and between stones to stop themselves from being washed away. Most members of this habitat require a high dissolved oxygen concentration and many are tolerant of the low temperatures typical of upland (torrential) streams. Stony (riffle) areas are up to 20 times more productive than pool areas in similar stream sections. Stony areas have a much more diverse flora and fauna than silty areas of the same river (Figs 4.4 and 4.5). Silty substrates are found in depositional zones where there is low sheer stress and little turbulence. Unless macrophytes are present to provide an alternative habitat, then only burrowing and surface dwelling forms are found. Burrowing forms include oligochaete worms, certain dipteran larvae (e.g. chironomids, *Psychoda*) and mussels. On the bottom surface the detritivore *Asellus* dominates along with the predatory larvae of the alder fly (*Sialis*). The dominant species of the community is probably determined by the organic content of the mud resulting in a reduced oxygen concentration. Those species not found due to low dissolved oxygen concentration but due to a suitable substrate are often found on submerged objects. To some extent artificial substrates

TABLE 4.1 Characterization of running water habitats on the basis of functional feeding groups based on North American species

Habitat sub-system	Dominant detritus (particle size)	Primary producers	Macro-invertebrates				Fish predators
			Grazers (scrapers)	Shredders (large particle detritivores)	Collectors (fine particle detritivores)	Predators	
Erosional	Coarse (>16 mm): logs, branches, bark, leaves	Diatoms mosses	Mollusca (Ancylidae) Ephemeroptera (Heptageniidae) Trichoptera (Glossosomatidae) Coleoptera (Psephenidae Elmidae)	Plecoptera (Nemouridae, Pteronarcidae, Peltoperlidae, Tipulidae)	Ephemeroptera (Heptageniidae, Baetidae, Siphlonuridae) Trichoptera (Hydropsychidae) Diptera (Simuliidae, Chironomidae-orthocladiinae)	Plecoptera (Perlidae) Megaloptera (Corydalidae)	Cottidae (sculpins) Salmonidae (trouts and chars)
Intermediate	Medium (<16 mm, >1 mm) leaf, bark, twig fragments	Green algae, vascular hydrophytes (e.g. Heteranthera, narrow-leaved Potamogeton)	Mollusca (Sphaeridae, Pleuroceridae Planorbidae)	Trichoptera (Limnephilidae, e.g. Pycnopsyche)	Ephemeroptera (Ephemera)	Odonata (Corduligasteridae, Petalaridae) Plecoptera (Perlidae) Megaloptera (Sialidae)	Etheostominae (darters) Cyprinidae (Rhinichthys)
Depositional	Fine (<1 mm) small terrestrial plant fragments, faeces	Vascular hydrophytes (e.g. Elodea)	Molusca (Physidae, Unionidae) Trichoptera (Phryganeidae)	Trichoptera (Limnephilidae, e.g. Platycentropus)	Oligochaeta Ephemeroptera (Hexagenia, Caenidae) Diptera (Chironomidae: Chironominae)	Odonata (Gomphidae, Agrionidae)	Cyprinidae (Notropis) (Catostomidae) (Catostomus)

Reproduced from Cummins (1975) with permission of Blackwell Science Ltd, Oxford.

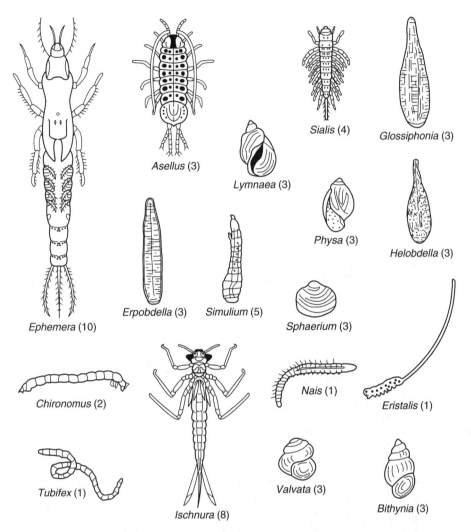

FIGURE 4.4
Macro-invertebrate species associated with depositional substrates. The Biological Monitoring Working Party (BMWP) score (Section 9.3) is given for each species in brackets. (Reproduced from Hellawell (1986) with permission of Kluwer Academic Publishers, Dordrecht, The Netherlands.)

allow comparisons to be made between the water quality of stony and silty reaches (Section 9.2.6). The presence of macrophytes increases animal diversity in silty areas significantly, providing a substrate for algae, *Gammarus*, some caddis, mayflies, *Simulium* and *Nais* worms, all of which are unable to colonize the bottom silt but are able to flourish on plants.

Many attempts have been made to bring together the abiotic and biotic factors that characterize areas of rivers under a single descriptive system. Fish zones are one, while the term rhithron and potamon is the system most preferred by limnologists. The key characteristics of each are given in Table 4.2, while each term can be further broken down into sub-classes (Table 4.3).

Fish are usually used to characterize zones within river systems, with specific species indicative of water quality (Section 3.7). Preferred habitats for fish depend on suitable breeding sites (e.g. a gravel substrate, dense macrophyte growth, rapid flow), minimum oxygen concentration and appropriate food supply. These factors are typical

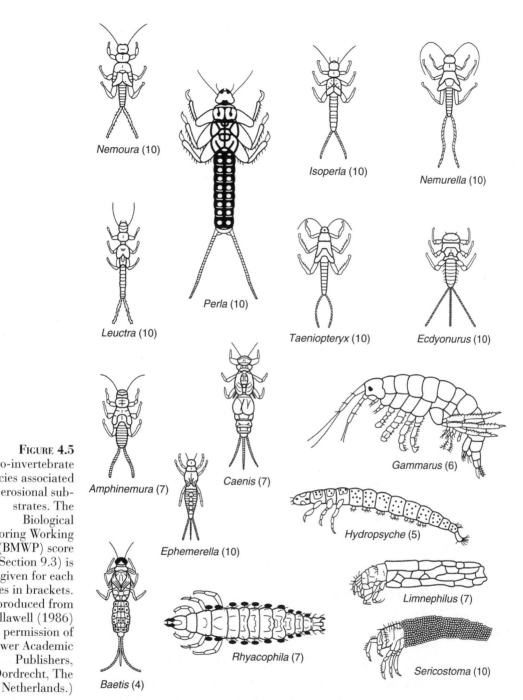

Figure 4.5
Macro-invertebrate species associated with erosional substrates. The Biological Monitoring Working Party (BMWP) score (Section 9.3) is given for each species in brackets. (Reproduced from Hellawell (1986) with permission of Kluwer Academic Publishers, Dordrecht, The Netherlands.)

Nemoura (10)
Perla (10)
Isoperla (10)
Nemurella (10)
Leuctra (10)
Taeniopteryx (10)
Ecdyonurus (10)
Amphinemura (7)
Caenis (7)
Gammarus (6)
Ephemerella (10)
Hydropsyche (5)
Limnephilus (7)
Baetis (4)
Rhyacophila (7)
Sericostoma (10)

of particular river zones; therefore habitat and community zonation are closely linked. For example, trout feed by sight on invertebrates and so require clear water. They have streamline powerful bodies adapted to the rapid turbulent water found in upland zones. In contrast, bream have adapted mouths to suck up the silt and have flattened bodies, and are able to withstand the low dissolved oxygen conditions typical of lowland reaches. A summary is given in Table 4.4.

TABLE 4.2   Comparison of the characteristics of the rhithron and potamon in rivers

Rhithron
Mean monthly temperature rise ≤20°C
Dissolved oxygen concentration always high
High shear stress
High turbulence
Substrate is stony
River contained within channel – rarely flooded
Primary allochthonous system although autochthonous input may be significant
(i.e. mosses, filamentous algae, diatoms)
Major erosional zone

Potamon
Mean monthly temperature rise ≥20°C
Dissolved oxygen concentration deficit may occur
Flow less or non-turbulent
Low shear stress
Substrate sand–silt
Flood plain present
Macrophytes present
Primary autochthonous system although organic debris from upstream
can be significant
Major depositional zone

TABLE 4.3   Classification of rivers

I	Eucrenon	
II	Hypocrenon	
III	Epirhithron	Rhithron
IV	Metarhithron	
V	Hyporhithron	
VI	Epipotamon	
VII	Metapotamon	Potamon
VIII	Hypopotamon[a]	
	↓	
	Downstream	

[a]Extends into estuary.

4.4 LENTIC ECOLOGY

Lakes can be split up into a number of physical zones, the unique character of which is reflected by different flora and fauna. The bottom substrate of a lake, from high water to the bottom of the euphotic zone, is the littoral zone and receives sufficient light for photosynthesis. This is the area of macrophytes, emergent plant species at the edge, floating plants in shallow water and submerged species in deeper water (Fig. 4.6). The substrate in the profundal zone is almost in complete darkness and is known as the sub-littoral zone. As the depth increases, the macrophytes are excluded, leaving an area of open water known as the pelagic zone. These terms are used to describe the type of community structure (e.g. littoral community). More precise

Table 4.4 General characteristics of the faunal zones as defined by river zones

Fish zone	Trout	Minnow	Chub	Bream
Gradient	Very steep	Steep	Gentle	Very gentle
Turbulence	+++	++	+	–
Substrate				
Coarse	+++	++	+	–
Fine	–	+	++	+++
Plants				
Mosses	+	–	–	–
Macrophytes	–	+	++	+++
Invertebrates				
Surface of substrate	+++	++	+	–
Plant living	+	++	+++	+
Burrowing	–	+	++	+++
Dissolved oxygen	+++	+++	++	+

+: low; ++: medium; +++: high; –: not found.

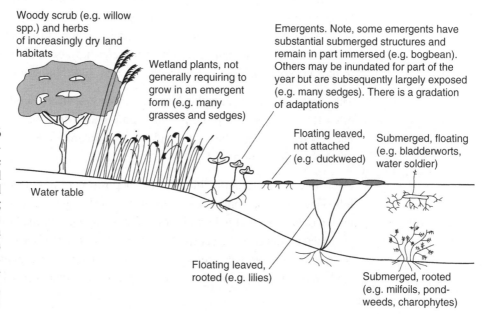

Figure 4.6 Ecological classification of aquatic macrophytes based on position and growth form during the summer. (Reproduced from Jeffries and Mills (1993) with permission of John Wiley and Sons Ltd, Chichester.)

Labels in figure: Woody scrub (e.g. willow spp.) and herbs of increasingly dry land habitats; Wetland plants, not generally requiring to grow in an emergent form (e.g. many grasses and sedges); Emergents. Note, some emergents have substantial submerged structures and remain in part immersed (e.g. bogbean). Others may be inundated for part of the year but are subsequently largely exposed (e.g. many sedges). There is a gradation of adaptations; Floating leaved, not attached (e.g. duckweed); Submerged, floating (e.g. bladderworts, water soldier); Water table; Floating leaved, rooted (e.g. lilies); Submerged, rooted (e.g. milfoils, pondweeds, charophytes)

terms are also used to describe organisms that live in special habitats in lakes. These include:

(a) *Nekton*: free swimming organisms that move easily between zones (e.g. fish).
(b) *Plankton*: free swimming organisms whose movement is largely controlled by the overall water movement in the lake – these are further categorized as zooplankton (animals such as protozoa, rotifers, cladocera) or phytoplankton (algae).
(c) *Seston*: general term to include all small material both living (plankton) and dead.
(d) *Benthos*: animals living on the bottom substrate.

(e) *Interstitial*: microscopic organisms that live within the water-filled pores within the sediment.

(f) *Neuston*: animals living at the surface, epineuston on top and hyponeuston immediately below the surface.

As in lotic systems, the organisms have adapted to the different lentic habitats. For example, low oxygen conditions at depth can be overcome by having haemoglobin to maximize oxygen uptake (e.g. *Chrionomus* midge larvae), large modified external gills (e.g. *Zygoptera* damselfly larvae); many beetles trap an air bubble under wing cases or in body hair, while other invertebrates have long breathing tubes to reach the surface air (e.g. *Nepa cinera* the water scorpion). The problem of staying in the euphotic zones, rather than sinking into the dark profundal zone, is easily achieved by active swimmers, which include the zooplankton which perform a diurnal vertical migration from the surface. Other species have gas vacuoles (e.g. *Chaoboridae* phantom midge larvae), which they can use to adjust their buoyancy so that they can hang motionless in open water. The phantom midge larvae are almost transparent to protect them from predators and hence their common name. While phytoplankton rely mainly on currents and turbulence to keep them in the euphotic zone, many colonial forms have elaborate structures to reduce settlement. Many invertebrates have been able to exploit the surface by having very long legs to displace their weight (e.g. *Hydrmetridae* water measurers).

The community structure within lakes is complicated due to depth, with benthos species often less important than the plankton and nekton species in the pelagic zone. The main food chain would be phytoplankton → zooplankton → fish. This is considered in more detail in Section 18.2.

Over time lakes gradually fill up by the deposition of material washed into them by inflowing streams, from surface run-off or overhanging vegetation. Thus, the distribution of habitats within lakes is gradually changing, with the littoral zone very slowly encroaching towards the centre. With time the entire surface of the lake will become covered by macrophytes eventually being taken over by emergent species and becoming a wetland. Such a transformation can, however, take a very long time.

REFERENCES

Chapman, D. (ed.), 1996 *Water Quality Assessments: A Guide to the Use of Biota, Sediments and Water in Environmental Monitoring*, 2nd edn, E. & F. N. Spon, London.

Cummins, K. W., 1975 Macro-invertebrates. In: Whitton, B. A., ed., *River Ecology. Studies in Ecology 2*, Blackwell Science, Oxford, pp 170–98.

Hellawell, J. M., 1986 *Biological Indicators of Freshwater Pollution and Environmental Management*, Elsevier Applied Science, London.

Jeffries, M. and Mills, D., 1990 *Freshwater Ecology*, Belhaven, London.

Vannote, R. L., Minshall, G. W., Cummins, K. W., Sedell, J. R. and Cushing, C. E., 1980 The river continuum concept. *Canadian Journal of Fish and Aquatic Science*, **37**, 130–7.

FURTHER READING

Abel, P. D., 1989 *Water Pollution Biology*, Ellis Horwood, Chichester.

Horne, A. J. and Goldman, C. R., 1994 *Limnology*, McGraw-Hill, New York.

Moss, B., 1980 *Ecology of Freshwaters*, Blackwell Science, Oxford.

Petts, G. and Calow, P. (eds), 1996 *River Biota*, Blackwell Science, Oxford.

Chapter 5

Micro-organisms and Pollution Control

Introduction

Micro-organisms have a number of vital functions in pollution control. It is the microbial component of aquatic ecosystems that provides the self-purification capacity of natural waters in which micro-organisms respond to organic pollution by increased growth and metabolism. It is essential that the same processes which occur in natural waters are utilized in biological treatment systems to treat wastewater (Section 15.1). Apart from containing food and growth nutrients, wastewater also contains the micro-organisms themselves, and by providing a controlled environment for optimum microbial activity in a treatment unit or reactor, nearly all the organic matter present can be degraded. Micro-organisms utilize the organic matter for the production of energy by cellular respiration and for the synthesis of protein and other cellular components in the manufacture of new cells. This overall reaction of wastewater treatment can be summarized as:

$$\text{Biomass} + \text{Organic matter} + O_2 + NH_4^{2+} + P \rightarrow \text{New cells} + CO_2 + H_2O \quad (5.1)$$

Similar mixed cultures of micro-organisms are used in the assessment of wastewater and effluent strength by the biochemical oxygen demand (BOD) standard 5-day test (BOD_5), in which the oxygen demand exerted by an inoculum of micro-organisms growing in the liquid sample is measured over 5 days in the dark at 20°C to give an estimate of the microbially oxidizible fraction in the wastewater (Section 5.4). Many diseases are caused by waterborne micro-organisms, a number of which are pathogenic to humans. The danger of these diseases being transmitted via wastewater is a constant threat to public health (Chapter 12). Therefore the use of micro-organisms, such as *Escherichia coli*, as indicator organisms to assess the microbial quality of water for drinking, recreation and industrial purposes, as well as in the assessment of wastewater treatment efficiency, is an essential tool in pollution control.

5.1 Nutritional Classification

In freshwaters and in wastewater treatment it is the bacteria that are primarily responsible for the oxidation of organic matter. However, fungi, algae, protozoa (collectively known as the *protista*) and higher organisms all have important secondary roles in the transformation of soluble and colloidal organic matter into biomass (Fig. 4.1). In order to function properly the micro-organisms require a source of energy and carbon for the synthesis of new cells as well as other nutrients and trace elements. The micro-organisms are classified as either heterotrophic or autotrophic according to their source of nutrients. Heterotrophs require organic matter both for energy and as a carbon source for the synthesis of new micro-organisms, while autotrophs do not utilize organic matter but oxidize inorganic compounds for energy and use carbon dioxide as a carbon source.

Heterotrophic bacteria, which are also referred to as saprophytes in older literature, utilize organic matter as a source of energy and carbon for the synthesis of new cells, respiration and mobility. A small amount of energy is also lost as heat during energy transfer reactions. The heterotrophs are subdivided into three groups according to their dependence on free dissolved oxygen, namely aerobic, anaerobic and facultative bacteria.

1. Aerobes require free dissolved oxygen in order to decompose organic material:

$$\text{Organics} + O_2 \xrightarrow{\text{aerobic micro-organisms}} \text{Aerobic micro-organisms} + CO_2 + H_2O + \text{Energy} \quad (5.2)$$

Like all microbial reactions it is autocatalytic, that is, the micro-organisms that are required to carry out the reaction are also produced. Aerobic bacteria predominate in natural watercourses and are largely responsible for the self-purification process (Section 4.1). They are also dominant in the major biological wastewater treatment processes, such as fixed film systems and activated sludge (Chapters 16 and 17). Aerobic processes are biochemically efficient and rapid in comparison with other types of reactions, producing by-products that are usually chemically simple and highly oxidized, such as carbon dioxide and water.

2. Anaerobes oxidize organic matter in the complete absence of dissolved oxygen by utilizing the oxygen bound in other compounds, such as nitrate:

$$\text{Organics} + NO_3 \xrightarrow{\text{anaerobic micro-organisms}} \text{Anaerobic micro-organisms} + CO_2 + N_2 + \text{Energy} \quad (5.3)$$

or sulphate:

$$\text{Organics} + SO_4^{2-} \xrightarrow{\text{anaerobic micro-organisms}} \text{Anaerobic micro-organisms} + CO_2 + H_2S + \text{Energy} \quad (5.4)$$

Anaerobic bacterial activity is found in freshwater and estuarine muds rich in organic matter and at the treatment works in the digestion of sludge. Anaerobic processes are normally biochemically inefficient and generally slow, giving rise to chemically complex by-products that are frequently foul smelling (e.g. organic acids and sulphur-containing compounds) (Table 13.4).

3. Facultative bacteria use free dissolved oxygen when available, but in the absence of oxygen are able to gain energy anaerobically and so are known as facultative aerobes or equally accurately as facultative anaerobes. An example of a facultative bacterium is *Escherchia coli* a common and important coliform; this and other such bacteria are common in both aerobic and anaerobic environments and treatment systems. Often the term obligate is used as a prefix to these categories of heterotrophic bacteria to indicate that they can only grow in the presence (obligate aerobe) or absence (obligate anaerobe) of oxygen.

Autotrophic bacteria cannot utilize organic matter, instead they oxidize inorganic compounds for energy and use carbon dioxide or carbonate as a carbon source. There are a number of autotrophs in the aquatic ecosystem, however, only the nitrifying, sulphur and iron bacteria are particularly important. The nitrifying bacteria oxidize ammonia nitrogen in a two-step reaction, initially to nitrite and finally to nitrate (Equations (15.4) and (15.5)).

Hydrogen sulphide is given off by sulphate-reducing bacteria if the sediment, water or wastewater becomes anaerobic. If this occurs in sewers then the slightly acidic gas is absorbed into condensation water, which collects on the top or crown of the sewer or on the side walls. Here, sulphur bacteria, which are able to tolerate pH levels of 1.0, oxidize the hydrogen sulphide to strong sulphuric acid using atmospheric oxygen:

$$H_2S + Oxygen \xrightarrow{\text{\textit{Thiobacillus thio-oxidans}}} H_2SO_4 + Energy \qquad (5.5)$$

The sulphuric acid reacts with the lime in the concrete to form calcium sulphate, which lacks structural strength. Gradually, the concrete pipe can be weakened so much that it eventually collapses. Crown corrosion is particularly a problem in sewers that are constructed on flat gradients, in warm climates, in sewers receiving heated effluents, with wastewaters with a high sulphur content or in sewers which are inadequately vented. Corrosion-resistant pipe material, such as vitrified clay or polyvinyl chloride (PVC) plastic, prevents corrosion in medium-size sewers, but in larger diameter sewers where concrete is the only possible material, corrosion is reduced by ventilation which expels the hydrogen sulphide and reduces condensation. In exceptional circumstances the wastewater is chlorinated to prevent sulphate-reducing bacteria forming hydrogen sulphide or the sewer is lined with a synthetic corrosion-resistant coating.

Not all species of iron bacteria are strictly autotrophic, however, those that are, which can oxidize inorganic ferrous (Fe^{2+}) iron to the ferric (Fe^{3+}) form as a source of energy.

The bacteria are filamentous and deposit oxidized iron ($Fe(OH)_3$) in their sheath. Included in the group are those that are acidophilic (e.g. *Thiobacillus ferro-oxidans*) and which are also capable of oxidizing sulphur. They mainly occur in iron-rich mine wastewaters. Those that oxidize ferrous iron at a neutral pH include *Sphaerotilus natans* and *Leptothrix ochracea.* These occur in natural waters with a neutral pH and can also occur in biological wastewater treatment units. For example, they are common in percolating filters that treat domestic effluents receiving infiltration water from coal mining areas and so are rich in iron. If the domestic water supply contains dissolved iron, the bacteria can become established in water pipes, forming yellow or reddish-brown slimes and tainting the water as the mature bacteria die (Section 11.1).

5.2 OXYGEN REQUIREMENT OF BACTERIA

Free dissolved oxygen is essential for the aerobic processes of hetero- and autotrophic bacteria. When aerobic organisms utilize organic nutrients they consume dissolved oxygen at the same time. Each molecule of glucose, which is the basic building block of all carbohydrates, requires six molecules of oxygen for complete conversion to carbon dioxide and water by aerobic bacteria:

$$C_6H_{12}O_6 + 6O_2 \xrightarrow{\text{bacteria}} 6CO_2 + 6H_2O \qquad (5.6)$$

There is also a considerable oxygen demand during the nitrification of nitrogenous compounds by autotrophic nitrifying bacteria (Section 15.1).

If the dissolved oxygen is not replaced then aerobic growth will eventually stop when the oxygen is exhausted, allowing only the slow anaerobic processes to continue. Microbial activity is not only oxygen-limited in the case of aerobic micro-organisms, it is also restricted by the availability of adequate supplies of carbon, nutrients such as nitrogen and phosphorus, trace elements and growth factors. It is the actual composition of micro-organisms which controls the nutrient requirements of organisms, but as proteins are composed mainly of carbon, nitrogen and smaller amounts of phosphorus it is these three elements which are essential for microbial growth. The requirements of carbon, nitrogen and phosphorus by microbial cultures in wastewater treatment processes is expressed as a ratio (C:N:P) and if the waste is deficient in any one of these basic components, complete utilization of the wastewater cannot be achieved. The optimum C:N:P ratio for biological oxidation is 100:5:1 (Section 13.1).

Autotrophs derive energy from either sunlight (photosynthetic) or from inorganic oxidation–reduction reactions (chemosynthetic). Chemoautotrophs do not require external sources of energy but utilize the energy from chemical oxidation, while phototrophs require sunlight as an external energy source:

$$6CO_2 + 12H_2O + \text{Light} \xrightarrow{\text{photosynthetic micro-organisms}} C_6H_{12}O_6 + 6O_2 + 6H_2O$$

$$(5.7)$$

Many inorganic ions, mainly metals, are required to ensure that bacterial enzymatic reactions can occur so trace amounts of calcium, magnesium, sodium, potassium, iron, manganese, cobalt, copper, molybdenum and many other elements are required. These are found in adequate amounts in sewage, as are growth factors such as vitamins. However, if any of these materials are deficient or absent, then microbial activity will be restricted or may even stop.

The mixed microbial cultures found in biological wastewater treatment units degrade and subsequently remove colloidal and dissolved organic substances from solution by enzymatic reactions. The enzymes are highly specific, catalysing only a particular reaction and are sensitive to environmental factors, such as temperature, pH and metallic ions. The major types of enzyme-catalysed reactions in wastewater biochemistry are:

(a) oxidation (the addition of oxygen or the removal of hydrogen);
(b) reduction (the addition of hydrogen or the removal of oxygen);
(c) hydrolysis (the addition of water to large molecules which results in their breakdown into smaller molecules);
(d) deamination (the removal of an NH_2 group from an amino acid or amine);
(e) decarboxylation (the removal of carbon dioxide).

5.3 MICROBIAL OXYGEN DEMAND

It is important to know how much oxygen will be required by micro-organisms as they degrade organic matter present in wastewater for two reasons:

1. To ensure that sufficient oxygen is supplied during wastewater treatment so that oxidation is complete.
2. To ensure receiving waters do not become deoxygenated due to the oxygen demand of these micro-organisms, resulting in the death of the natural fauna and flora.

The amount of organic matter that a stream can assimilate is limited by the availability of dissolved oxygen, which is largely determined by the rate oxygen is utilized by microbial oxidation and the rate at which it can be replaced by reaeration and other processes.

5.3.1 SELF-PURIFICATION

The term self-purification is defined as the restoration, by natural processes, of a river's natural clean state following the introduction of a discharge of polluting matter. In natural river systems, organic matter is assimilated by a number of processes which include sedimentation, which is enhanced by mechanical and biological flocculation, chemical oxidation, and the death of enteric and pathogenic micro-organisms by exposure to sunlight. Of course, the assimilative capacity of rivers, that is, the extent to which the river can receive waste without significant deterioration of some quality criteria, usually the dissolved oxygen concentration, varies according to each river due to available dilution, existing quality and self-purification capability. The most important process in self-purification is biochemical oxidation, that is, the aerobic breakdown of organic material by micro-organisms, although oxidation of nitrogen compounds also occurs. Biodegradable organic matter is gradually eliminated in rivers due mainly to bacterial action, by methods very similar to those occurring in wastewater treatment. Complex organic molecules are broken down into simple inorganic molecules in a process requiring oxygen. In this process of self-purification it is the attached micro-organisms, collectively known as periphyton, which are normally responsible for the greatest removal, while the suspended micro-organisms, which are mainly supplied with the discharge, are less important in the removal of organic material. However, while the decomposition of organic waste by micro-organisms is advantageous, the process does remove oxygen from solution and in order to prevent the destruction of the natural fauna and flora, aerobic conditions must be maintained.

Water in lotic systems not only flows over the surface of the stones and finer material that makes up the riverbed, but also through it. This substrate is coated with a mixture of periphyton becoming dominated by bacteria, rather than algae, and other microbiota as the depth increases. As the river flows over the substrate, water is pulled into and down through the material with soluble and colloidal organic matter oxidized by the bacteria, as is ammonia (Fig. 5.1). The action of the substrate is similar to a fixed film reactor used in wastewater treatment (Chapter 16) and like percolating filters is rich in macro-invertebrates feeding on the developed biofilm. The micro- and macro-grazers control the biofilm development ensuring air and water can freely pass through the interstices of the substrate. The water is continuously cycled through the substrate as it travels downstream constantly improving in quality. Depending on the nature and depth of the substrate this can be a major mechanism in the self-purification process.

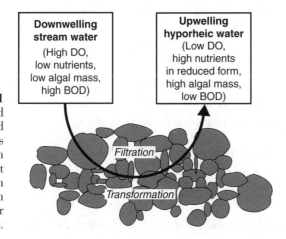

FIGURE 5.1 Water is pulled through the riverbed as it flows downstream with significant improvement in quality. This action increases as the river flow rate increases.

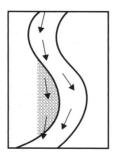

FIGURE 5.2
Gravels and other
deposited material
all act as substrate
for periphyton
development and
in situ the treatment
of river water.

Similar effects are seen as water flows through deposited material or wetlands between meanders in lowland rivers (Fig. 5.2) or through bankside (riparian) vegetation. Periphyton develops on the submerged parts of the vegetation and bacteria develop in close association with the roots of emergent vegetation (Section 18.3.2).

5.3.2 OXYGEN BALANCE

While the dissolved oxygen concentration is affected by factors, such as temperature, BOD_5 and salinity, oxygen depletion is prevented primarily by reaeration, although other sources of oxygen, such as photosynthesis, may also be important under certain conditions. It is important to know how quickly oxygen dissolves into water and, as discussed in Section 3.5, this depends to a large extent on the concentration of oxygen already in solution in relation to the saturation concentration, that is, the oxygen deficit. In general terms, the greater the organic load to the river the greater the response in terms of microbial activity, resulting in a larger demand for the available dissolved oxygen.

5.3.3 REAERATION

Oxygen diffuses continuously over the air–water interface in both directions. In the water, the concentration of oxygen will eventually become uniform due to mixing or, in the absence of mixing, by molecular diffusion. The rate of diffusion is proportional to the concentration gradient, which has been described by Flick's law as:

$$\frac{dM}{dt} = K_d A \cdot \frac{dC}{dx} \tag{5.8}$$

where M is the mass transfer in time t (mass-transfer rate), K_d the diffusion coefficient, A the cross-sectional area across which transfer occurs and C the concentration, and x the distance of transfer (concentration gradient).

If a uniform concentration gradient is assumed, then:

$$\frac{dM}{dt} = K_d A \cdot \frac{C_s - C_t}{x} \tag{5.9}$$

where C is the concentration at saturation (C_s) and after time t (C_t).

The equation can be solved as:

$$C_t = C_s - 0.811(C_s - C_0)\left[e^{-K_d} + \left(\frac{1}{9}\right)e^{-9K_d} + \left(\frac{1}{25}\right)e^{-25K_d} + \cdots\right] \quad (5.10)$$

where C_0 is the concentration after time 0 and,

$$K_d = \frac{K_d \pi^2 t}{4x^2} \quad (5.11)$$

The diffusion coefficient K_d can be expressed in mm^2s^{-1} or cm^2s^{-1}. The K_d for oxygen in water at 20°C is $1.86 \times 10^{-3}mm^2s^{-1}$.

Aeration in time or distance can be expressed as:

$$\frac{dC_t}{dt} = K_2(C_s - C_t) \quad (5.12)$$

integrating with limit $C_t = C_0$ at $t = 0$:

$$\int_{C_0}^{C_t} \frac{dC_t}{C_s - C_t} = K_2 \int_{t_0}^{t} dt \quad (5.13)$$

i.e.

$$\log_e\left(\frac{C_s - C_t}{C_s - C_0}\right) = -K_2 t \quad (5.14)$$

If D_t and D_0 are the dissolved oxygen deficit at times t and 0, respectively, and K_2 is the reaeration constant, then:

$$\log_e\left(\frac{D_t}{D_0}\right) = -K_2 t \quad (5.15)$$

thus,

$$D_t = D_0 e^{-k_2 t} \quad (5.16)$$

A more useful parameter than the reaeration constant (K_2) is the exchange coefficient f. The exchange coefficient, also known as the entry or exit coefficient, is the mass of oxygen transferred across unit area of interface in unit time per unit concentration deficit:

$$f = \frac{K_d}{x} \tag{5.17}$$

$$\frac{dM}{dt} = fA(C_s - C_t) \tag{5.18}$$

and if a finite volume of water (V) is assumed, then:

$$\frac{dC_t}{dt} = \frac{dM}{dt}\frac{1}{V} = \frac{fA}{V}(C_s - C_t) \tag{5.19}$$

that is,

$$\frac{dC_t}{dt} = K_2(C_s - C_t) \tag{5.20}$$

where

$$K_2 = f\frac{A}{V} \tag{5.21}$$

$$f = K_2\frac{V}{A} = K_2 h \tag{5.22}$$

where V is the volume of water below interface, A is the area of the air–water interface and h is the mean water depth.

The exchange coefficient f is expressed in units of velocity (mm h^{-1}) and at 20°C in British rivers it can be estimated by the formula:

$$f = 7.82 \times 10^4 U^{0.67} H^{-0.85} \tag{5.23}$$

where U is the water velocity (m s^{-1}) and H the mean depth (mm). Typical values for f range from 20 for a sluggish polluted lowland river to over 1000 for a turbulent unpolluted upland stream. Values for the exchange coefficient for various aeration systems expressed in cm h^{-1} are summarized in Table 5.1.

In the US, $K_L a$ (the oxygen transfer rate) is used instead of K_2. In Europe, K_2 is used primarily for river reaeration while $K_L a$ is used in modelling oxygen transfer in wastewater operation (Section 17.3).

TABLE 5.1   Typical values for the exchange coefficient f

Aeration system	$f\ (cm\ h^{-1})$
Stagnant water	0.4–0.6
Water flowing at $0.4\,m\,min^{-1}$ in a small channel	
Water polluted by sewage	0.4
Clean water	0.5
Water flowing at $0.6\,m\,min^{-1}$ in a channel	1
Polluted water in dock and tidal basin	1–3
Sluggish polluted river	2
Sluggish clean water about 51 mm deep	4
Thames estuary under average conditions	5.5
Water flowing at $10.06\,m\,min^{-1}$ in a small channel	7.5
The open sea	13
Water flowing at $14.94\,m\,min^{-1}$ in a channel	30
Turbulent lakeland beck	30–200
Water flowing down a $30°$ slope	70–300

A rise in temperature can increase the rate of reaeration and vice versa. The reaeration rate constant (K_2) can be related to temperature (T) by:

$$K_{2(T)} = K_{2(20)}1.047^{(T-20)} \tag{5.24}$$

In general terms, an increase in temperature of 1°C will result in an increase in the exchange coefficient f of about 2%.

A number of physical factors affect reaeration. The transfer of oxygen at the air–water interface results in the surface layer of water becoming saturated with oxygen. If the water is turbulent, as is the case in upland streams, the saturated surface layer will be broken up and mixing will ensure that reaeration is rapid. When no mixing occurs, as in a small pond to take an extreme example, then oxygen has to diffuse throughout the body of the water. In some cases the diffusion rate may be too slow to satisfy the microbial oxygen demand so that anaerobic conditions may occur at depth. In rivers, velocity, depth, slope, channel irregularity and temperature will all affect the rate of reaeration. To increase the rate of aeration and speed up the self-purification process, weirs can be built below discharges. More recently, floating aerators have been employed on rivers during periods of high temperature when the dissolved oxygen concentration has fallen dangerously low. More sensitive rivers, containing salmonid fish, have been protected by pumping pure oxygen from barges into the river at times of particular stress. This technique has been developed specifically to control deoxygenation caused by accidental discharges of pollutants. The use of a compressor with a perforated rubber hose has also been successfully employed. In emergencies, for example where a deoxygenated plug of water is moving downstream, the local fire brigade has been able to prevent total deoxygenation by using the powerful pumps on their tenders to recirculate as much water as possible, with the water returned to the river via high-pressure hoses. The main advantage of this method is that the fire crews can make their way slowly downstream, keeping abreast of the toxic plug. Hydrogen peroxide is also used in emergencies for rapid reaeration in larger rivers.

5.3.4 THE OXYGEN-SAG CURVE

When an organic effluent is discharged into a stream it exerts a biochemical oxygen demand with the processes of oxygen consumption and atmospheric reaeration proceeding simultaneously. While other processes, such as photosynthesis, sedimentation and oxidation, of the bottom deposits can also affect oxygen concentration; oxygen consumption and reaeration are the primary processes affecting oxygen status.

In many cases the oxygen demand will initially exceed the reaeration rate, so the dissolved oxygen concentration will fall downstream of the outfall (discharge point). As the rate of diffusion across the air–water interface is directly proportional to the oxygen deficit, if the rate of consumption lowers the oxygen concentration, the oxygen mass-transfer rate will increase. At some point downstream the rate of reaeration and consumption become equal and the oxygen concentration stops declining. This is the critical point of the curve where the oxygen deficit is greatest (D_c) and the dissolved oxygen concentration is lowest (Fig. 5.3). Thereafter reaeration predominates and the dissolved oxygen concentration rises to approach saturation. The characteristic curve, which results from plotting dissolved oxygen against time or distance downstream, is known as the oxygen-sag curve. The long tail associated with the recovery phase of the curve is due to the rate of mass transfer of oxygen. As the river's dissolved oxygen concentration recovers, the oxygen deficit is reduced and, as the rate of mass transfer is proportional to the oxygen deficit, the rate of reaeration slows, thus extending the curve.

The shape of the curve remains more or less the same except that the critical point will vary according to the strength of the organic input. It is possible for the dissolved oxygen concentration to be reduced to zero and an anaerobic or septic zone to be formed. Deoxygenation is generally a slow process, so the critical point may occur some

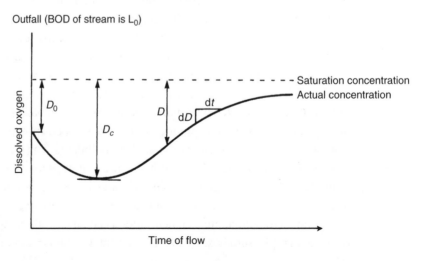

FIGURE 5.3
Dissolved oxygen-sag curve.

considerable distance downstream of the outfall. The degree of deoxygenation not only depends on the strength of the discharge, but also on dilution, BOD of the receiving water, nature of the organic material in terms of availability and biodegradability, temperature, reaeration rate, dissolved oxygen concentration of the receiving water and the nature of the microbial community of the river.

The oxygen-sag curve can be expressed mathematically for idealized conditions in terms of the initial oxygen demand, the initial dissolved oxygen concentration in the river and the rate constants for oxygen consumption (K_1) and reaeration (K_2). These mathematical formulations were derived by Streeter and Phelps when working on the Ohio River (Streeter and Phelps, 1925). This large river had long uniform stretches between pollution discharges, also relatively little photosynthesis, so the only major factors affecting the oxygen status were oxygen consumption and reaeration. They considered that the rate of biochemical oxidation of the organic matter was proportional to the remaining concentration of unoxidized organic material, typified by the first-order reaction curve (Fig. 5.4). Assuming first-order kinetics, the oxygen demand with no aeration can be represented as:

$$\frac{dL_t}{dt} = -K_1 L_t \tag{5.25}$$

$$L_t = L_0 - D_t \cdot d(L_0 - D_t) = -dD_t \tag{5.26}$$

Thus,

$$\frac{dD_t}{dt} = -K_1 L_t \tag{5.27}$$

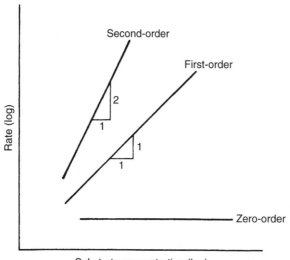

FIGURE 5.4 Log plots of reaction rates.

Reaeration with no oxygen demand:

$$\frac{dC_t}{dt} = K_2(C_s - C_t) \qquad (5.28)$$

Therefore,

$$\frac{d(C_s - C_t)}{dt} = K_2 D_t \qquad (5.29)$$

So,

$$\frac{dD_t}{dt} = -K_2 D_t \qquad (5.30)$$

It is possible to express both demand and aeration in terms of change in the oxygen deficit (dD_t)/dt. So for simultaneous oxygen demand and reaeration:

$$\frac{dD_t}{dt} = K_1 L_t - K_2 D_t \qquad (5.31)$$

where D is the dissolved oxygen deficit at time t (D_t), L is the ultimate BOD at time t (L_t) or initially (L_0), K_1 the BOD reaction rate constant and K_2 the reaeration rate constant.

Providing oxygen is not a limiting factor, the oxygen demand is not dependent on the oxygen deficit. So, by substituting L_t according to the equation:

$$L_t = L_0 e^{-K_1 t} \qquad (5.32)$$

$$\frac{dD_t}{dt} = K_1 L e^{K_1 t} - K_2 D_t \qquad (5.33)$$

When this equation is integrated with limit $D_t = D_0$ when $t = 0$:

$$D_t = \frac{K_1 L_0}{K_2 - K_1} (e^{K_1 t} - e^{-K_2 t}) + D_0 e^{-K_2 t} \qquad (5.34)$$

which is the well-known Streeter and Phelps equation.

By changing to base 10 ($K = 0.4343k$):

$$D_t = \frac{K_1 L_0}{K_2 - K_1} (10^{K_1 t} - 10^{-K_2 t}) + D_0 10^{-K_2 t} \qquad (5.35)$$

The minimum dissolved oxygen concentration (the critical point), which occurs at maximum oxygen deficit D_t when $dD_t/d_t = 0$, is given by:

$$t_c = \frac{1}{K_2 - K_1} \cdot \log_e \left\{ \frac{K_2}{K_1} \left[1 - \frac{(K_2 - K_1)D_0}{K_1 L_0} \right] \right\} \tag{5.36}$$

where the critical (maximum) deficit (D_c) occurs at time t_c.

Both K_1 and K_2 in the model are assumed to be constant; however, while K_1 is measured by running a BOD determination in the laboratory (Section 5.4), it may vary with time. The K_2 value will vary from reach to reach within the river and must be measured in the field. Both these constants are temperature functions and so temperature effects must be taken into consideration. For domestic sewage, K_1 approximates to 0.1 at 20°C while K_2, which is mainly a function of turbulence, can be assessed using the equation:

$$K_2 = \frac{(K_d U)^{1/2}}{H^{3/2}} \tag{5.37}$$

where K_2 is the reaeration coefficient (base e) per hour, K_d the diffusion coefficient of oxygen into water, U is the velocity of flow and H the depth. Only approximate values can be obtained. Low values represent deep, slow-moving rivers and high values shallow fast-flowing upland streams. In reality, K_2 is at best a crude estimate, and often an assumed value can have severe effects on the predictive estimate. The measurement of K_1 is even more critical.

The oxygen-sag curve can be more accurately assessed by providing a third point. This is provided by the point of inflexion, where the net rate of aeration is at a maximum when $(d^2 D_t/dt^2) = 0$:

$$t_i = \frac{1}{K_2 - K_1} \cdot \log_e \left\{ \left(\frac{K_2}{K_1} \right)^2 \left[1 - \frac{(K_2 - K_1)D_0}{K_1 L} \right] \right\} \tag{5.38}$$

where the inflexion deficit (D_i) occurs at time t_i.

So, it is now possible to construct the oxygen-sag curve and to predict the minimum oxygen concentration downstream of a point discharge of organic waste, such as sewage (Fig. 5.3).

While the Streeter and Phelps model provides an extremely useful basis for the study of the sequence of events that occur in an organically polluted river, it must be applied with care. This is especially so in rivers where conditions change frequently, there is appreciable photosynthesis, deposition of debris and sediment or discharges of inhibitory or toxic substances. The model is only valid for a single pollution discharge and where there is no dilution from tributaries. Where these occur the river must be split up into discrete sections according to changes in flow or discharge, so that each section can be treated as an individual case and the model applied. The output data from one section provides the input data for the next, and in this way the entire river system can be covered to provide an overall calculation. This type of model is the basis of predictive water quality models with many other variables, such as benthic and nitrogenous oxygen demand, salinity and temperature also included.

A final word of caution on the use of the Streeter and Phelps model. The model assumes that the flow does not vary over time, that the organic matter is distributed uniformly across the stream's cross-section and that there is no longitudinal mixing, with the effects of algae and bottom sediments unimportant and so neglected in the equation. In reality, however, the dissolved oxygen-sag curve can be affected by other factors apart from microbial oxygen demand and reaeration rate. Among those worthy of further consideration are photosynthesis, with the addition of oxygen during the day and the uptake of oxygen by plant respiration at night; benthic oxygen demand, the removal of oxygen by gases released from the sediments and the release of soluble organic material from the sediments which have an oxygen demand; and finally, the input of oxidizable material from surface water. These inputs and the dissolved oxygen are constantly being redistributed within the water column by longitudinal mixing. Some of these factors can be easily predicted and so built into the existing model, whereas other factors are less quantifiable. The use of water quality models is discussed further in Section 9.4.2.

EXAMPLE 5.1 A treated sewage effluent discharged to a river with a mean flow velocity (U) of $0.1\,\mathrm{m\,s}^{-1}$ raises the BOD immediately below the mixing zone from 2.6 to $4.5\,\mathrm{mg\,l}^{-1}$. The upstream dissolved oxygen concentration is $9.5\,\mathrm{mg\,l}^{-1}$ at a water temperature of 12°C. The constants are derived from Table 5.2 where the deoxygenation constant (K_1) for treated wastewater is taken as $0.15\,\mathrm{day}^{-1}$ and the reaeration constant (K_2) for an average velocity river is taken as $0.5\,\mathrm{day}^{-1}$. Calculate the time (t_c) when the minimum dissolved oxygen concentration will occur, the distance downstream (x_c) where this will occur, and finally the minimum dissolved oxygen concentration.

TABLE 5.2 Typical deoxygenation (K_1) and reaeration (K_2) constants

Type of water	K_1 (to base e) day^{-1}	Type of water	K_2 (at 20°C) day^{-1}
Untreated wastewater	0.35–0.70	Small ponds, backwaters	0.10–0.23
Treated wastewater	0.10–0.25	Sluggish streams, large lakes	0.23–0.35
Polluted river water	0.10–0.25	Low-velocity rivers	0.35–0.46
		Average-velocity rivers	0.46–0.69
		Fast rivers	0.69–1.15
		Rapid rivers	>1.15

(a) Calculate initial oxygen deficit (D_0):

$$D_0 = \text{Saturation oxygen concentration at } 12°C \\ - \text{Actual oxygen concentration} \tag{5.39}$$

$$10.8 - 9.5 = 1.3 \text{ mg l}^{-1}$$

(b) Calculate ultimate oxygen demand (L_0):

$$L_0 = \frac{\text{BOD}_5}{1 - e^{-K_1 t}} \tag{5.40}$$

$$L_0 = \frac{4.5}{1 - e^{-0.15 \times 5}} = \frac{4.5}{0.528} = 8.52 \text{ mg l}^{-1}$$

(c) Calculate time before minimum dissolved oxygen concentration occurs (t_c) using Equation (5.36):

$$t_c = \frac{1}{0.5 - 0.15} \cdot \ln \left\{ \frac{0.5}{0.15} \left[1 - \frac{1.3(0.5 - 0.15)}{0.15 \times 8.52} \right] \right\}$$

$$t_c = 2.18 \text{ days.}$$

(d) Distance downstream for minimum dissolved oxygen concentration to occur (x_c). Remember to multiply by 86 400 to convert flow to m day^{-1} and then by 10^{-3} to convert m to km (i.e. 86.4).

$$x_c = U t_c \tag{5.41}$$
$$x_c = 0.1 \times 2.18 \times 86.4 = 18.8 \text{ km}$$

(e) Calculate the maximum oxygen deficit (D_c):

$$D_c = \frac{K_1}{K_2} L_0 e^{-K_1 t_c} \tag{5.42}$$

$$D_c = \frac{0.15}{0.5} \times 8.52 \times e^{-0.15 \times 2.18} = 1.84 \text{ mg l}^{-1}$$

(f) Calculate the minimum dissolved oxygen concentration (D_{min}) where the saturation concentration at $12°C$ (D_{sat}) is 10.8 mg l^{-1}:

$$D_{min} = D_{sat} - D_c = 10.8 - 1.84 = 8.96 \text{ mg l}^{-1}$$

5.4 THE BOD TEST

There are two widely used measures of oxygen demand: chemical oxygen demand (COD) which measures the organic content in terms of biodegradable and non-biodegradable compounds, and the BOD test, which measures only the biodegradable fraction of the wastewaters by monitoring the assimilation of organic material by aerobic micro-organisms over a set period under strictly controlled conditions, usually 5 days at 20°C in the dark. The COD will always be higher than the BOD as the former includes substances that are chemically oxidized as well as biologically oxidized. The ratio of COD:BOD provides a useful guide to the proportion of organic material present in wastewaters that is biodegradable, although some polysaccharides, such as cellulose, can only be degraded anaerobically and so will not be included in the BOD estimation. The COD:BOD relationship varies from 1.25 to 2.50 for organic wastewaters depending on the waste being analysed. The ratio increases with each stage of biological treatment as biodegradable matter is consumed but non-biodegradable organics remain and are oxidized in the COD test. After biological treatment the COD:BOD of sewage can be >12:1. The relationship remains fairly constant for specific wastes, although the correlation is much poorer when the COD values are $<100\,\mathrm{mg\,O_2\,l^{-1}}$.

The COD of domestic wastewater can be approximated by:

$$COD = 1.64 \times BOD + 11.36 \tag{5.43}$$

Not all the substrate within the BOD bottle will be oxidized to CO_2, some will be converted to new cells. So, if a simple organic source like glucose is oxidized both chemically and biologically, there will be a discrepancy. For example, the COD test will predict an oxygen consumption of $192\,\mathrm{g\,O_2\,mol^{-1}}$ of glucose compared with only $150\,\mathrm{g\,O_2\,mol^{-1}}$ using the BOD test. So the BOD test does not give a measure of the total oxidizable matter present in wastewaters, because of the presence of considerable quantities of carbonaceous matter resistant to biological oxidation. However, it does indicate the potential possessed by a wastewater for deoxygenating a river or stream.

Complete breakdown of even the most biodegradable wastes can take several weeks, so during the 5-day test only a proportion of the organic material will be broken down. Some organic materials, such as cellulose, can remain virtually unaffected by aerobic micro-organisms, only being broken down anaerobically. When the organic fraction has been aerobically broken down as completely as possible, the oxygen consumed is termed the ultimate BOD or ultimate oxygen demand.

The test can incorporate two distinct stages forming the characteristic BOD curve (Fig. 5.5). Stage I is the basic curve and represents the carbonaceous material which can take up to 3 weeks to be fully degraded at 20°C, while in stage II the nitrogenous material is oxidized. The oxygen demand from nitrification is only important in wastewaters. In raw wastewaters, nitrification only becomes a significant source of

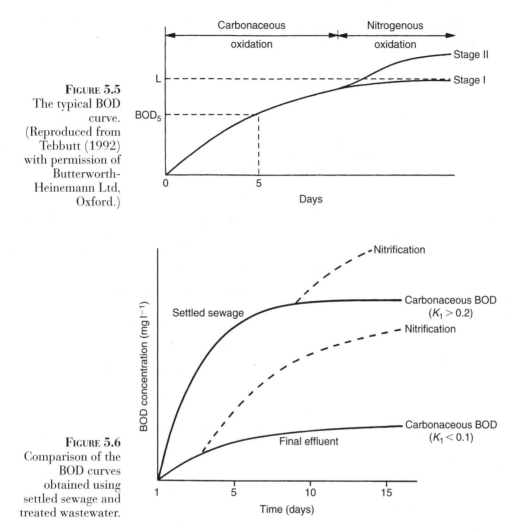

FIGURE 5.5
The typical BOD curve. (Reproduced from Tebbutt (1992) with permission of Butterworth-Heinemann Ltd, Oxford.)

FIGURE 5.6
Comparison of the BOD curves obtained using settled sewage and treated wastewater.

oxygen demand after 8–10 days, while in partially treated effluents nitrification can dominate the oxygen demand after just a few days (Fig. 5.6).

The BOD_5 measures only the readily assimilable organic material (carbonaceous) with the low molecular weight carbohydrates in particular being primarily utilized. The BOD_5 gives a far more reliable estimation of the possible oxygen demand that a waste will have on a river than the COD test, which also measures the more refractory (non-biodegradable) compounds. Owing to the similarity between the self-purification and wastewater treatment process, the BOD test has been widely used as a measure of organic strength of river water and effluents.

With domestic wastewaters, only 60–70% of the total carbonaceous BOD is measured within 5 days at 20°C (BOD_5), with only the most biodegradable fraction utilized. For most materials, an incubation period of about 20 days (BOD_{20}) is required for complete breakdown, even though some more recalcitrant organic compounds, such as certain polysaccharides, will not have been degraded even then. The test is essentially the oxidation of carbonaceous matter:

$$C_xH_yO_z \rightarrow CO_2 + H_2O \tag{5.44}$$

However, while this first stage may be the only component of the BOD curve, often a second stage is present, nitrification. Where necessary this second stage can be suppressed by the addition of allythiourea (ATU) to the sample. The addition of $0.5\,mg\,l^{-1}$ ATU prevents the onset of nitrification for a period of up to 9 days with no effect on carbonaceous oxidation. Theoretically, $3.4\,g$ of oxygen is required by *Nitrosomonas* to oxidize $1\,g$ of ammonia to nitrite and a further $1.14\,g$ of oxygen by *Nitrobacter* to oxidize $1\,g$ of nitrite to nitrate. The extent of nitrification in the BOD test can be easily measured by incubating a parallel set of samples, one with and one without nitrification suppressed, the difference being the nitrogenous oxygen demand. The ultimate oxygen demand (L) can be estimated from the BOD_5 with ATU ($BOD_{5(ATU)}$) value as:

$$L = (1.5 \times BOD_{5(ATU)}) + (4.6 \times NH_3\text{–}N) \tag{5.45}$$

Glucose ($C_6H_{12}O_6$) is used as a reference for the BOD test and is also useful for examining the stoichiometry of the test. Glucose is completely oxidized as:

$$C_6H_{12}O_6 + 6O_2 \rightarrow 6CO_2 + 6H_2O \tag{5.46}$$

For complete oxidation a glucose solution of $300\,mg\,l^{-1}$ concentration will require $320\,mg\,l^{-1}$ of oxygen at 20°C. However, using the standard 5-day BOD_5 test only $224\,mg$ of oxygen is utilized, with complete oxidation taking longer than 5 days. So the BOD_5 only measures part of the total oxygen demand of any waste, and in this case:

$$\frac{BOD_5}{BOD_{20}} = \frac{224}{300} = 70\%$$

The BOD has been traditionally modelled as a continuous first-order reaction, so that the rate of breakdown of carbonaceous material is proportional to the amount of material remaining. In this type of reaction the rate of breakdown is at first rapid when the organic content is high, but gets progressively slower as the organic material is utilized. This can be expressed as:

$$\frac{dL}{dt} = -K_1L \tag{5.47}$$

where K_1 is the BOD reaction rate constant and L the ultimate BOD (carbonaceous only).

This integrates to:

$$L_t = L_0 e^{-K_1 t} \qquad (5.48)$$

where the initial BOD (L_0) is L_t after time t. The amount of oxygen consumed during the BOD test period (Y) is:

$$Y = L_0 - L_t \qquad (5.49)$$

So,

$$Y = L_0(1 - e^{-K_1 t}) \qquad (5.50)$$

or using base 10:

$$Y = L_0(1 - 10^{-K_1 t}) \qquad (5.51)$$

Thus, for a test where 65% of the carbonaceous material is broken down within the 5 days, K_1 will equal $0.223\,\text{day}^{-1}$ and the removal rate is approximately 20% per day. Therefore, 95% removal will take 13 days and 99% removal 21 days, although adherence to the relationship between K_1 at base e and base 10 is:

$$K_e = 2.303\,K_{10} \qquad (5.52)$$

It is convention to quote K_1 to the base 10. The rate constant K_1 varies according to the quantity and nature of the organic matter present, the temperature and the type of micro-organisms in the wastewater. This can be best illustrated by considering the way in which micro-organisms utilize the available organic material present. Essentially two reactions take place within a BOD bottle, a rapid synthesis reaction in which there is a rapid consumption of oxygen due to the high concentration of available organics, which is characteristic of raw wastewaters or effluents high in low molecular weight carbohydrates, followed by a slower endogenous metabolism (Fig. 5.7). In treated effluents most of the organics originally present in the wastewater have been removed, so oxygen is consumed at the lower endogenous rate. Therefore, the greater rate of reaction due to the concentration of assimilable organic material, the larger the K_1 value. The average BOD rate constant at 20°C ranges from 0.04–0.08 for rivers with low pollution, 0.06–0.10 for biologically treated effluents, 0.12–0.22 for partially treated effluents and those using high-rate systems, to 0.15–0.28 for untreated wastewaters. It is possible for samples with different reaction rates to have the same BOD_5 (Fig. 5.8).

The rate constant K_1 and the ultimate BOD (L) are traditionally calculated using the Thomas graphical method. The procedure is based on the function:

$$\left(\frac{t}{y}\right)^{1/3} = (2.3\,K_1 L)^{-1/3} + \frac{K_1^{2/3}}{3.43\,L^{1/3}} \cdot t \qquad (5.53)$$

where y is the BOD exerted in time t, K_1 the reaction rate constant (base 10) and L the ultimate BOD. This equation forms a straight line with $(t/y)^{1/3}$ plotted as a function of

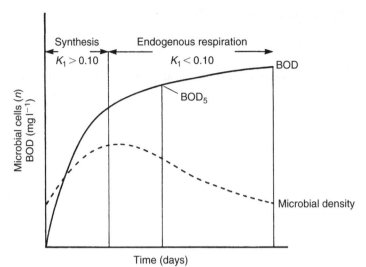

FIGURE 5.7
Reactions which
occur in the BOD
bottle.

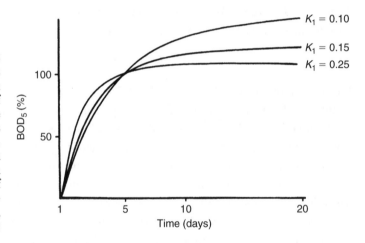

FIGURE 5.8
The effect of various
rate constants in
the calculation of
the same BOD. The
result is expressed as
a percentage.
(Reproduced from
Tebbutt (1992) with
permission of
Butterworth-
Heinemann Ltd,
Oxford.)

time t. The slope $K_1^{2/3}/3.43\,L^{1/3}$ and the intercept $(2.3\,K_1L)^{-1/3}$ of the line of best fit of the data is used to calculate K_1 and L.

Using the form $Z = a + bt$ for the straight line where $Z = (t/y)^{1/3}$ and $b = K_1^{2/3}/3.43\,L^{1/3}$:

$$K_1 = 2.61\left(\frac{b}{a}\right) \tag{5.54}$$

$$L = \frac{1}{2.3\,K_1a^3} \tag{5.55}$$

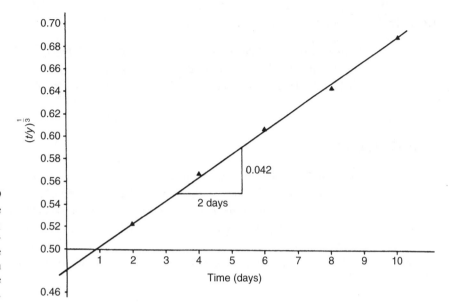

FIGURE 5.9
Determination of the
BOD constant K_1
(reaction rate con-
stant) and L (the
ultimate BOD) from
BOD data using the
Thomas method.

EXAMPLE 5.2 Over a 10-day period the BOD was measured every second day. From this data $(t/y)^{1/3}$ can be calculated:

t (day^{-1})	2	4	6	8	10
y (mg l^{-1})	14	22	27	30	32
$(t/y)^{1/3}$	0.523	0.567	0.606	0.644	0.679

The graph of $(t/y)^{1/3}$ is plotted against t (Fig. 5.9) and from this the intercept can be measured ($a = 0.481$) and slope b calculated:

$$\text{Slope } b = \frac{0.042}{2} = 0.021$$

From these values the rate reaction rate K_1 and the ultimate BOD (L) can be estimated:

$$K_1 = 2.61\left(\frac{b}{a}\right)$$

$$= 2.61\left(\frac{0.021}{0.481}\right) = 0.114$$

$$L = \frac{1}{2.3K_1a^3}$$

$$= \frac{1}{2.3(0.114)(0.481)^3} = 34.3 \text{ mg l}^{-1}$$

REFERENCES

Streeter, H. W. and Phelps, E. B., 1925 *A Study of the Pollution and Natural Purification of the Ohio River*, US Public Health Bulletin 146, Washington, DC.

Tebbutt, T. H. Y., 1992 *Principles of Water Quality Control*, Pergamon Press, London.

FURTHER READING

Britton, G., 1994 *Waste Water Microbiology*, Wiley-Liss, New York.

Hromadka, T. V., McCuen, R. H., De Vries, J. J. and Durbin, T. J., 1993 *Computer Methods in Environmental and Water Resource Engineering*, Lighthouse Publications, California.

Nemerow, N. L., 1991 *Stream, Lake, Estuary and Ocean Pollution*, Van Nostrand Reinhold, New York.

Welch, E. B., 1980 *Ecological Effects of Waste Water*, Cambridge University Press, Cambridge.

INTRODUCTION

Even the most well-treated effluent from a sewage works is not as clean as river water and will contain a wide variety of substances including partially oxidized organic and inorganic material. Whenever an effluent enters a receiving water there will always be some change in water quality and a resultant change in the community structure.

The effect of pollutants on river communities depends on:

(a) the type of pollutant;
(b) its concentration in the water;
(c) the length of exposure to the community.

These effects can be summarized according to their physico-chemical or biological nature as:

(a) the addition of toxic substances;
(b) the addition of suspended solids;
(c) deoxygenation;
(d) the addition of non-toxic salts;
(e) heating the water;
(f) the effect on the buffering system;
(g) the addition of human, animal and plant pathogens.

These effects rarely occur singly and often all together (e.g. sewage). Therefore, most discharges to water are classed as multi-factor pollutants. Each of these major effects is considered below except for faecal contamination or contamination by other pathogens (g) which is examined in Chapter 12. The introduction of alien plant and animal species can also be considered as pollution (Section 3.1). The effects of climate change are explored in Section 23.1.

6.1 TOXIC SUBSTANCES

Toxic substances, which include heavy metals and trace organics, decrease in concentration downstream after discharge due to an increase in flow volume from tributaries, by precipitation and adsorption. Metals are lost from solution by precipitation as the pH changes (Table 6.1). Many organic poisons are degraded, while some harmless organics are broken down to produce toxic substances, such as ammonia and sulphides, resulting in an increase in toxicity and a reduction in dissolved oxygen further downstream. Other toxic material is adsorbed onto suspended and other particulate matter, and eventually settles out of suspension. The toxicity of heavy metals can be listed in order of decreasing toxicity as $Hg > Cd > Cu > Zn > Ni > Pb > Cr > Al > Co$, although this is only approximate as the vulnerability of species to individual metals varies. Toxicity also varies according to environmental conditions that control the chemical speciation of the metals (Campbell and Stokes, 1985; Tessier and Turner, 1995).

Adsorption is primarily controlled by the surface area of particles; therefore, adsorption capacity is inversely proportional to particle size. In practice, the finest sediments are the richest in trace elements. Pollutants and nutrients associated with particulate matter can be partitioned into different phases or forms (speciation). The major forms in which pollutants and nutrients occur in particulate matter are, in terms of the most to least reactive:

(a) adsorbed (electrostatically or specifically) onto mineral particles;
(b) bound to the organic matter (e.g. organic debris and humic substances);
(c) bound to carbonates;
(d) bound to sulphides;
(e) occluded in Fe and Mn oxides which occur commonly as coatings on particles;
(f) within a mineral lattice (e.g. calcium phosphate for P, copper oxide or sulphide for Cu);
(g) in silicates and other non-alterable minerals.

TABLE 6.1 Maximum pH values for the precipitation of metal ions as hydroxides or other salts

Metal	Minimum pH as hydroxide	Other salts
Sn	4.2	
Fe (III)	4.3	
Al	5.2	
Pb (II)	6.3	6.0
Cu (II)	7.2	5.3
Zn	8.4	7.0
Ni	9.3	
Fe (II)	9.5	
Cd	9.7	
Mn (II)	10.6	

Under non-polluted conditions the majority of inorganic compounds, for example trace elements and phosphorus, are found in categories (e), (f) and (g). While under polluted conditions discharged compounds are mainly adsorbed onto mineral particles (a) and/or bound to organic matter (b). The majority of synthetic organic compounds are found in the adsorbed fractions. Particulate organic matter has a very high surface area and so a very high adsorption capacity. Therefore, the concentration of pollutants in sediments is often proportional to the amount of organic matter present or to the amount of carbon adsorbed onto mineral surfaces.

Toxicity of metals is reduced in waters rich in humic acids (humic and fulvic acids) as they become bound to the organic compounds, a process known as chelation. While these organic compounds remain in solution the metals are essentially unavailable biologically. Metals can also react with organic compounds to form toxic organo-metal complexes. Hardness also plays an important part in metal toxicity, which varies depending on the concentration of calcium in the water. For example, the higher the concentration of Ca^{2+}, the lower the toxicity of Hg, Pb, Cu and Zn. This is reflected in the EU Freshwater Fish Directive (Table 6.2).

Toxic compounds are rarely present on their own so that the response of organisms to individual pollutants is often different when other pollutants or compounds are present. Normally the combined effects of two or more toxic compounds is additive, for example Zn + Cd, but they can also increase (synergism) or decrease (antagonism) the toxic effects of the compounds when acting independently. For example, Cu is synergistic when present with Zn, Cd or Hg, while Ca is antagonistic with Pb, Zn and Al (Table 6.2). Detergents act synergistically with all pollutants as they reduce the surface tension on the gill membranes in freshwater organisms, especially fish, which increases the permeability of metals and other toxic compounds. This is shown in Fig. 6.1 where the response of 1 unit of pollutant A in the absence of pollutant B is 1.0 (i.e. its lethal concentration 50 (LC_{50}) concentration) and the response of 1 unit

TABLE 6.2   The permissible concentration (I) of Zn and the recommended concentration (G) of Cu in rivers of different hardness in order to sustain fish life

	Hardness of water mg $CaCO_3$ l^{-1}			
	10	50	100	500
Total zinc (mg l^{-1})				
Salmonid waters	0.03	0.2	0.3	0.5
Cyprinid waters	0.3	0.7	1.0	2.0
Total copper (mg l^{-1})				
Salmonid waters	0.005[a]	0.022	0.04	0.112
Cyprinid waters				

[a] The presence of fish in waters containing higher concentrations of copper at this hardness may indicate a predominance of dissolved organo-cupric complexes.
Adapted from the EU Freshwater Fish Directive (78/659/EEC), reproduced with permission of the European Commission, Luxembourg.

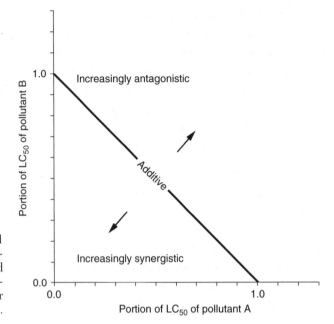

FIGURE 6.1
Schematic descrip-
tion of the combined
effects of two pollu-
tants. See text for
explanation.

of B is also 1.0 in the absence of A. If the combined response of A + B falls on the diagonal line, for example 0.2A + 0.8B, the effect is additive. If, however, the response falls below the diagonal line, for example 0.3A + 0.5B, the effect is synergistic showing that A + B increases the combined toxicity effect. If the response is above the line, for example 0.6A + 0.8B, then it is antagonistic with A + B decreasing the combined toxicity effect.

Toxicity is generally classified as either acute (i.e. a large dose over a short duration resulting in a rapid critical response that is often lethal) or chronic (i.e. a low dose over a long duration which is normally non-lethal but may eventually cause permanent disabilitating effects). These effects may be behavioural as well as physiological. Toxicity can also be cumulative with effects brought about by successive exposures.

There are a wide variety of methods for assessing toxicity. These usually involve the exposure of test organisms, such as zooplankton, macro-invertebrates or fish, to a range of polluting concentrations or mixtures of pollutants over prolonged time periods. The most widely adopted toxicity assessment procedure is the measurement of lethal concentration (LC) where toxicity of a pollutant is expressed as the percentage of organisms killed at a particular concentration over a set time period. So the 96-h LC_{50} is the concentration of pollutant that results in 50% mortality of the test organism over a test period of 96 h. The effective concentration (EC) is used as an alternative to LC when death is not used as the end point of the test, but another effect such as loss of movement or developmental abnormalities. The LC_{50} varies between organisms due to sensitivity, which in turn is affected by factors such as sex, age or acclimation (i.e. acquired tolerance). For example, the common macro-invertebrate indicators *Gammarus pulex*, *Asellus aquaticus* and *Chironomus riparius* show difference sensitivities to common pollutants. The reported 96-h LC_{50} test values for Cd are 0.03, 0.60, 200 mg l^{-1}; phenol 69, 180 and 240 mg l^{-1}; ammonia 2.05, 2.30 and 1.65 mg l^{-1}, respectively.

The US EPA has listed 129 priority compounds that pose a serious risk to aquatic habitats. Similarly, the EU has placed toxic compounds into two priority lists under the Dangerous Substances Directive (70/464/EEC). List I, the black list, includes those compounds that must not be discharged to aquatic habitats. The criteria for inclusion into List I is based on compound toxicity, persistence and bioaccumulation potential. List II, the grey list, compounds are less dangerous but can result in chronic toxicity (Table 8.16). This control of such compounds will eventually come under the control of the EU Water Framework Directive (2000/60/EC) (Sections 7.3 and 8.3). In the UK it is estimated that sewage contains traces of up to 60 000 different chemicals. The effects of toxic compounds on the treatment of wastewaters are examined in Section 15.3.

Pollutants, especially metals and pesticides, are readily accumulated in organisms. Biomagnification is where pollutants increase in concentration through the food chain with maximum concentrations found in the top carnivores (e.g. mercury and organochlorine pesticides). For example, the concentration of polychlorinated biphenyls (PCB) in the Firth of Clyde in Scotland during October 1989 was 0.01 $\mu g\,l^{-1}$, yet the concentration in shellfish was 200–800 $\mu g\,l^{-1}$, a concentration factor of up to 80 000. During the same period concentration factors of up to 300 000 were recorded for dieldrin and dichloro-diphenyl-trichloroethane (DDT). Bioaccumulation is independent of trophic level with uptake normally directly from the water (e.g. Cd and Zn). While metal concentrations in many algae and bacteria are directly related to metal concentration in the water, this is not usually the case with macro-invertebrates or fish. Pollutants can be taken up via food or water, and this varies according to the pollutant and the organism. Burrowing macro-invertebrates tend to obtain Cd and Zn from ingested food, while surface dwellers such as *Asellus* and *Gammarus* generally take up these metals directly from the water. Bioaccumulation and biomagnification of metals can affect the whole food chain resulting in high concentrations in shellfish or fish that could be eaten by humans. The most well-known example biomagnification is the poisoning of the inhabitants of Minamata, a small town on the West coast of Kyushu Island (Japan), by methyl mercury that had accumulated in the fish used as the staple diet of the islanders (Laws, 1993). The mercury originated from wastewater containing mercuric sulphate, which was used as a catalyst in the production of polyvinyl chloride (PVC) at a local factory. While 43 deaths were recorded between 1953 and 56, the discharge continued until 1988 with serious mercury pollution still present. To date, 18 000 people are thought to have been affected by mercuric poisoning from the factory. Mercury is usually discharged in its inorganic form (Hg^{2+}) and is converted to the highly toxic methyl mercury ($Hg(CH_3)_2$) by microbial action. It is the methyl form that is more readily absorbed by tissues and dissolves in fat, with biomagnification reported as high as 3.6×10^6 in fish from Minamata Bay.

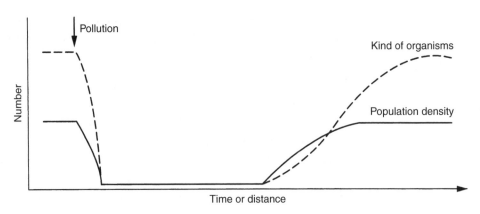

FIGURE 6.2
Generalized effect of
a toxic discharge on
lotic freshwater
community struc-
ture. (Reproduced
from UNESCO
(1978) with permis-
sion of UNESCO
and WHO.)

While it is difficult to generalize, the overall effect of toxic compounds on the biota is a rapid reduction of both species diversity and abundance leading to complete elim-ination of all species if toxic concentrations are sufficient. Recovery is generally slow with tolerant species returning first, often at elevated population densities due to a lack of competition (Fig. 6.2). Each species shows a different tolerance to different toxic compounds, while indirect effects can also be significant. For example, if a sen-sitive plant species is eliminated then tolerant macro-invertebrate species dependent on it will also be excluded. Other pollutants included in this category are used and waste pharmaceutical compounds and endocrine disrupters, both of which are com-mon in wastewater effluents (Chapter 13).

6.2 SUSPENDED SOLIDS

Solids discharged to water settle out downstream, the rate depending on the size of the particles and the turbulence. Settlement especially occurs where the turbulence is lowest (e.g. behind weirs and in pools). Suspended and settleable solids are produced by a wide range of industries (Table 6.3), and can be classed as either inert or oxidizable (degradable).

6.2.1 INERT SOLIDS

Inert solids come from mining activity, coal washing, construction sites, particularly road and bridge construction, river channeling and dredging. The solids accumulate on the riverbed and gradually extend further downstream. Floods may extend such solids even further. The primary (direct) effects are as follows:

1. The inert particles clog the feeding mechanisms of filter feeders and the gills of aquatic animals, which are both, eliminated if the effect is prolonged.
2. Stony erosional substrates become coated with the solids, smothering the fauna and replacing them with silty dwelling species (Section 4.3).

TABLE 6.3   Examples of typical suspended solids concentrations in rivers and effluents

Source	Suspended solids concentrations (mg l⁻¹)
Industrial	
China clay	500–100 000
Paper manufacture	200–3000
Tannery	2000–5000
Meat packing	1000–3000
Cannery	200–2500
Sugar beet	800–5000
Sewer	
Storm water overflow	500–3000
Raw sewage	300–800
Treated sewage	
Primary sewage treatment only	200–350
Secondary sewage	30

FIGURE 6.3
Generalized effect of inert solids on lotic freshwater community structure. (Reproduced from UNESCO (1978) with permission of UNESCO and WHO.)

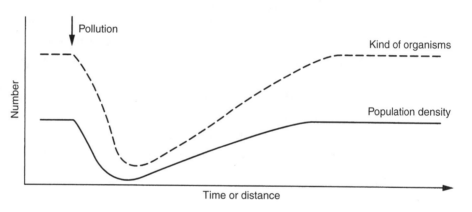

3. With the increased turbidity light penetration is reduced depressing photosynthesis and reducing primary productivity. This affects the whole food web.

There are also numerous secondary (indirect) effects. These include:

(a) the elimination of predators that feed on eliminated organisms (i.e. a food shortage);

(b) the possible loss of macrophytes will result in a loss of shelter, reduction in habitat diversity and an increased loss of animals due to greater predation and scouring.

The general effect is a suppression of both species diversity and abundance, as the diversity of habitats is reduced. There is a selective increase in a few species that are adapted to survive the stress imposed by the solids (Fig. 6.3). However, due to the lack of organic matter in the sediment, abundance of burrowing organisms will be low. Fines from washing gravel and road chippings can extend silt deltas in rivers. While this can have a positive effect in some rivers by encouraging reed development including alder/willow, the effects are generally undesirable.

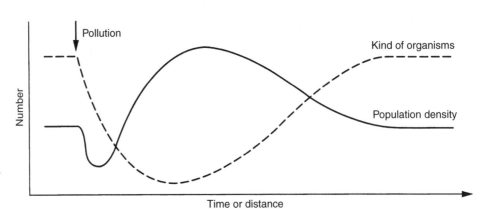

FIGURE 6.4
Generalized effect of
an organic non-toxic
pollution on lotic
freshwater commu-
nity structure.
(Reproduced from
UNESCO (1978)
with permission of
UNESCO and
WHO.)

6.2.2 OXIDIZABLE SOLIDS

These settle out in the same way as inert solids but once deposited they decompose. Gradually a balance results between the rate of settlement and decomposition, and so the effect decreases downstream. However, as the rate of decomposition is very slow, the area affected can be quite large. The solids blanket the substrate and undergo anaerobic decomposition releasing methane, sulphide, ammonia and other toxic compounds (Fig. 19.1). Irregular areas of deposition can occur further downstream at dams, weirs and pools, resulting in localized areas of high oxygen demand, even though at this point the rest of the river is unaffected. Sensitive benthic organisms are rapidly eliminated and, due to the increased organic matter, are replaced by high population densities of a few species tolerant of low oxygen silty conditions (Fig. 6.4).

6.3 DEOXYGENATION

Oxygen is not very soluble in water and the concentration available to the biota is a function between oxygen demand in the water, the water temperature and the rate of diffusion. This is explained in detail in Section 3.5. The survival of the biota is dependent on the extent to which their oxygen demand can be matched by the availability of oxygen from the environment (Abel, 1989). Oxygen is an important limiting factor for all freshwater organisms, and for fish in particular (Table 3.8). Inputs of readily oxidizable organic matter cause the dissolved oxygen concentration to fall downstream due to a gradual increase in bacterial activity (Fig. 6.5). As the oxygen deficit increases then the rate of transfer increases; however, when the rate of oxygen consumption by the bacteria is greater than that diffusing from the atmosphere, the dissolved oxygen concentration falls and may reach anaerobic conditions. Recovery occurs once this is reversed, but as the oxygen deficit decreases so will the rate of oxygen transfer. This slows the rate of recovery resulting in the typical shape of oxygen-sag curve (Fig. 6.5). While deoxygenation is generally due to bacterial decomposition it can also be caused by autotrophic bacterial action. This is examined in detail in Section 5.3. Foam from non-biodegradable detergents or the presence of oil on the surface of water can significantly reduce the oxygen transfer rate. A thick film of oil

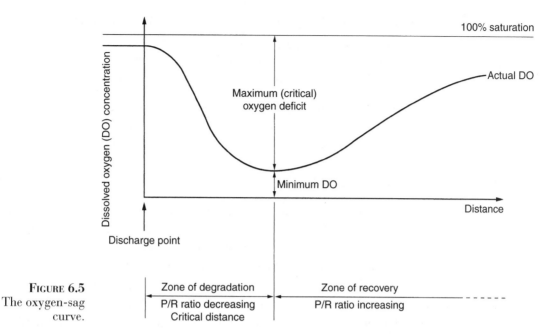

FIGURE 6.5
The oxygen-sag
curve.

TABLE 6.4 Typical BOD concentration ranges expected in rivers and effluents

Source	BOD (mg l⁻¹)
Natural background levels expected	
Upland streams	0.5–2.0
Lowland streams	2.0–5.0
Large lowland rivers	3.0–7.0
Sewage	
Raw/crude sewage	200–800
Treated effluent	3–50
Agricultural	
Pig slurry	25 000–35 000
Poultry	24 000–70 000
Silage	60 000
Processing	
Abattoir	650–2500
Meat processing	2000–3000
Sugar beet	3500–5000
Dairy	
Milk manufacture	300–2000
Cheese manufacture	1500–2000
Breweries	500–1300

can eliminate oxygen transfer altogether (Section 13.1). Once anaerobic conditions prevail the oxygen demand falls off, but reduced substances that are produced and that build up further downstream cause a further oxygen demand.

The effect on the biota depends on the oxygen demand (Table 6.4) and the subsequent reduction in dissolved oxygen concentration in the water. The first response to the increase in soluble organic matter is rapid bacterial growth. This, and the development of sewage fungus, which is a massive development of the periphyton (mainly filamentous bacteria, fungus and protozoa) that smothers the riverbed as a thick wool-like growth, is responsible for the high oxygen demand (Gray, 1985).

The presence of sewage fungus is the most obvious visual sign of organic pollution in lotic systems, and in contrast to its name it can be caused by any source of organic enrichment and is usually comprised of filamentous bacteria. The main sewage fungus forming organisms are given in Table 6.5, while its impact on lotic ecosystems is summarized in Fig. 6.6.

If a severe deficiency occurs most of normal clean water will be eliminated and replaced by large population densities of tolerant species. The response of the biota in lotic systems to organic pollution is summarized in Fig. 6.7. The affected river can be broken down into various zones describing the impact and recovery, although the length of each zone depends on the amount of organic matter in the effluent and the rate of reaeration. Therefore, a fast-flowing turbulent river recovers over a shorter distance than a slower non-turbulent river (Table 6.6). Downstream of the outfall the biochemical oxygen demand (BOD) concentration decreases, the oxygen concentration increases so that the least sensitive organisms reappear first (i.e. *Tubificidae* → *Chironomus* → *Asellus*)

TABLE 6.5   The occurrence of the commonest slime forming organisms expressed a percentage of the total sites examined in the UK (Curtis and Harrington, 1971) and Ireland

	UK			Ireland		
	Dominant	*Secondary*	*Total*	*Dominant*	*Secondary*	*Total*
Bacteria						
Beggiatoa alba	6.3	21.4	27.7	5.5	23.3	28.8
Flavobacterium spp.	3.1	37.1	40.2	0	0	0
Sphaerotilus natans	52.1	37.1	89.2	52.8	23.3	76.1
Zoogloea spp.	58.5	34.0	92.5	11.1	43.3	54.4
Fungi						
Fusarium aquaeductuum	1.9	0	1.9	5.5	0	5.5
Geotrichium candidum	4.4	3.1	7.5	0	0	0
Leptomitus lacteus	3.1	0.6	3.7	22.2	3.3	25.5
Protozoa						
Carchesium polypinum	6.3	10.1	16.4	2.8	0	2.8
Algae						
Stigeoclonium tenue	3.1	7.6	10.7	0	3.3	3.3

Reproduced from Gray (1985) with permission of Cambridge Philosophical Society, Cambridge.

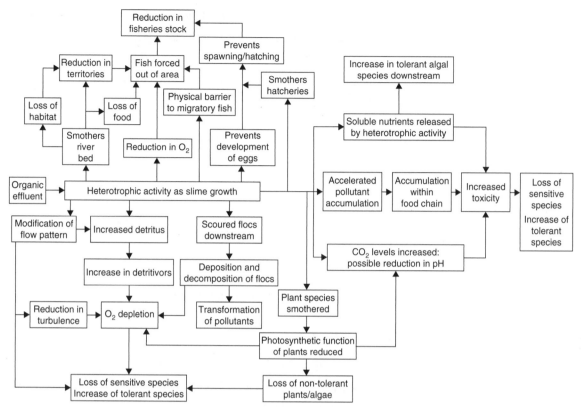

FIGURE 6.6 Summary of the major pathways and subsequent effects of sewage fungus in lotic ecosystems.

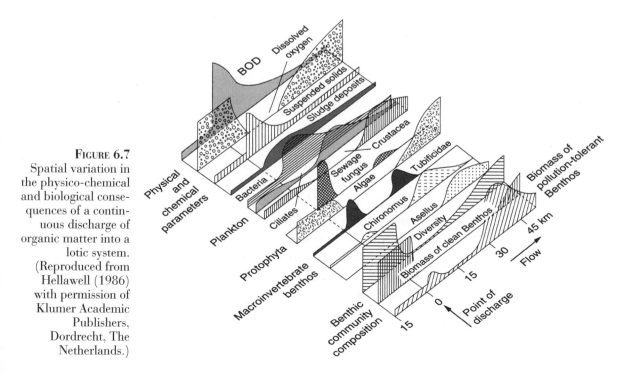

FIGURE 6.7
Spatial variation in the physico-chemical and biological consequences of a continuous discharge of organic matter into a lotic system. (Reproduced from Hellawell (1986) with permission of Klumer Academic Publishers, Dordrecht, The Netherlands.)

TABLE 6.6 Summary of effects of an organic discharge on a lotic (river) system showing various pollutional zones as seen in Fig. 6.4

| Zone | Clean water | Organic pollution | | Recovery | Clean water |
		Gross pollution	Lesser pollution		
Dissolved oxygen	High	Very low	Low	Moderate	High
BOD	Low	Very high	Moderate to high	Moderate	Low
Species diversity	High	Very low	Low	Moderate	High
Key organism	Clean water fauna/flora	Free bacteria Tubificidae	Sewage fungus Ciliates *Chironomus*	Filamentous algae *Asellus*	Clean water fauna/flora

(Section 8.2). Often if less tolerant predators or competitors are absent then exceptionally abundant population levels occur where it first appears.

6.4 ADDITION OF NON-TOXIC SALTS

The effect of non-toxic salts can be split into two categories. Salinization results in an increase in conductivity and salinity due to anions such as Cl^- and SO_4^{2-} and cations such as K^+ and Na^+. Eutrophication is due to increases in P and N that increase plant biomass.

The main sources of mineral salts associated with salanization are:

(a) mining wastewaters;
(b) industrial wastewaters;
(c) increased evaporation or evapotranspiration in the catchment, although this is mainly a phenomenon of sub-arid regions only.

Among the expected salts that occur as major point sources are Cl^-, K^+ and Na^+ from potash and salt mines; SO_4^{2-} from iron and coal mines; specific ions from industrial and mining sources with Na^+ and CO_3^- common industrial wastes. Other minor point and dispersed sources (principally Na^+, Cl^-, Ca^{2+}, SO_4^{2-}) that may be locally important include domestic wastewater, atmospheric pollution, de-icing salts and fertilizer run-off. Conductivity is used to monitor such wastes. Non-essential ions are not removed by sewage treatment or water treatment. Therefore, any salts discharged to sewer are discharged with the final effluent after treatment. Urine contains 1% NaCl and this builds up in surface water when it is recycled as drinking water. This is why some recycled waters have a dull flat taste that is characteristic of waters with a high Cl^- concentration before they begin to taste salty. The concentration of these salts can only be reduced by dilution and assimilation downstream.

Nutrients that cause eutrophication are principally N and P. These come from fertilizer use, waste tip leachate and as the breakdown products of sewage (NO^{3-}, SO_4^{2-}, PO_4^-). The major source of phosphate in wastewaters is from the widespread use of linear alkylate sulphonate (LAS) detergents that employ sodium tripolyphosphate builders (Section 13.1). Other components of detergent formulations are toxic, such as boron from perborate additives (Table 13.3), and can adversely affect crops when contaminated surface water is used for irrigation. The ratio of N:P in water can identify which nutrient is most likely to be limiting to eutrophication. If the N:P ratio is >16:1 then P is most likely to be the nutrient limiting algal development, while at ratios <16:1 then N is probably limiting. These nutrients cause increased algal growth leading to eutrophication and a possible fall in dissolved oxygen. Eutrophication in lakes is readily modelled (Golterman, 1991). Vollenweider (1975) proposed a simple input–output model (Equation 6.1) assuming steady-state conditions with nutrient loading and flushing rates remaining constant with time and with uniform mixing:

$$P = \frac{L}{z(r_s - r_f)} \text{ g m}^{-3} \tag{6.1}$$

where P is the total phosphorus concentration (g m^{-3}) measured at time of complete mixing just prior to stratification in spring, L is the annual loading of phosphorus per unit area ($\text{g m}^{-2} \text{ year}^{-1}$), z is the mean depth (m), r_s is the sedimentation rate coefficient (i.e. the fraction of phosphorus lost per annum to the sediment) and r_f is the hydraulic flushing rate (i.e. the number of times the entire lake water volume is replaced annually). Using this model permissible loadings for nutrients can be calculated.

Unless nutrients are discharged directly to lotic systems (e.g. agricultural run-off) the algal (eutrophication) zone normally occurs some distance downstream of an organic waste discharge for two reasons. First, organic matter has to be broken down to release nutrients and secondly, reduced light penetration due to a high turbidity caused by bacteria. In England, over 70% of nitrates and 40% of phosphates in rivers originate from agricultural land, with 29% of rivers in 2002 having a nitrate concentration >30 mg l^{-1}. These nutrient rich catchments are associated with areas of intensive arable farming (Fig. 8.2). River nitrate concentrations show definite seasonal trends being low in summer and high in winter, especially during times of high surface run-off. The worse scenario for nitrate pollution is heavy rainfall after a dry summer when nitrogen has been able to accumulate in the soil. While nitrate concentrations in rivers appear to be leveling off, it will be many decades before any reversal in nitrate concentration is seen in aquifers with long retention times. Remediation is discussed in Sections 8.4 and 11.1.3.

The effect of eutrophication in lotic systems depends on flow velocity and stream order. For example, in small streams and rivers eutrophication promotes macrophyte growth or periphyton including the development of filamentous algae. In medium-sized

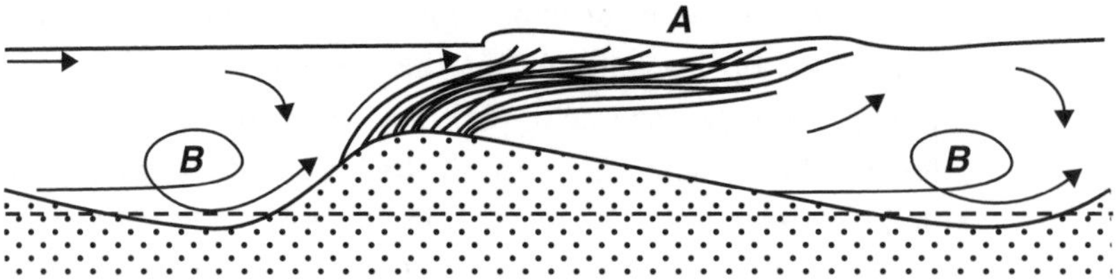

FIGURE 6.8 Macrophyte and filamentous algal growth in rivers results in a number of physical effects including A raising the water level by enhancing settlement of solids and B causing hydraulic scouring upstream and downstream of the growth. (Reproduced from Uhlmann (1979) with permission of John Wiley and Sons Ltd, Chichester.)

rivers periphyton is promoted by increased nutrients and, to a lesser extent, macrophyte growth. In large rivers phytoplankton dominates rather than macrophytes. Moderate increases in primary productivity as plant growth increases the abundance of herbivorous and detritivorous animals leading to an increase in overall productivity, including fish, in the system as well as increasing overall species diversity (Section 4.2). The main effects of eutrophication are:

(a) blanketing of substrata resulting in a reduction in habitat diversity;
(b) reduction of oxygen concentration and light penetration in water;
(c) species diversity is often reduced with different dominant species;
(d) salmonids are eliminated;
(e) agal blooms can result in the production of algal toxins which are harmful to the biota and animals drinking the water (Section 6.4.1);
(f) algal blooms cause discoloration of the water and production of offensive tastes and odours (Section 11.1);
(g) as the plant material decays it causes organic pollution with increased oxygen demands;
(h) increased macrophyte and filamentous algae will impede water flow, act as a physical barrier for fish migration and reduce the recreational value of rivers. Reducing the flow rate raises the water level leading to a higher water table in adjacent fields. Macrophytes also cause increased sedimentation within the vegetation as well as scouring the riverbed both upstream and downstream of the macrophyte growth (Fig. 6.8).

The importance of these effects varies between lotic and lentic systems, with stratification a major factor in eutrophication in lentic systems, which is explored in Sections 2.1 and 3.8. Lakes gradually increase their nutrient levels over time with the nutrients entering from diffuse surface run-off or from point sources of pollution. These nutrients are constantly recycled within the lake (Fig. 6.9). Phytoplankton growth is restricted to the epilimnion (Fig. 2.2). During winter low temperatures and low light intensity restricts phytoplankton development, but with the increasing temperature and light intensity in the spring the phytoplankton growth increases rapidly. The development of algae is limited by nutrient concentrations in the water, which is

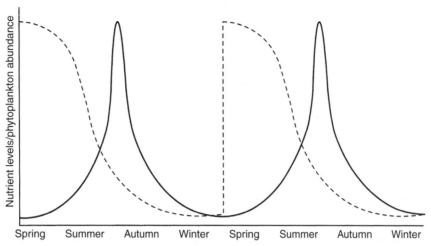

FIGURE 6.9   Seasonal cycles of nutrients and phytoplankton on the surface layer of lakes. Nutrient concentrations (dotted line) decline as the phytoplankton develops (solid line). Stratification leads to nutrient depletion in the surface layer resulting in a decline in phytoplankton density with senescent cells falling into the hypolimnion, where they decay releasing nutrients. With the breakdown of stratification in winter these nutrients are redistributed allowing a new cycle to begin.
(Reproduced from Abel (1989) with permission of Addison Wesley Longman Ltd, Harlow.)

gradually depleted by algal growth. This leads to a levelling off of the density of plant cells in the water. There is a continuous cycle of death and decay of the phytoplankton, which sink slowly through the hypolimnion to the bottom of the lake where they decompose. This process accelerates as the temperature and light intensity fall with the onset of winter. Inorganic nutrients released during their decomposition remain locked in the hypolimnion due to stratification. In winter, however, stratification fails due to the epilimnion cooling and becoming denser than the lower hypolimnion resulting in the lake becoming completely mixed, redistributing nutrients ready for the following spring.

Construction of reservoirs and locks may reduce flow velocities in a river, resulting in severe eutrophication where nutrient assimilation is enhanced by longer retention times. Eutrophication leads to daily (diel) variations in both oxygen concentration and pH. During the night photosynthesis is less than respiration (P < R), so both the oxygen concentration and pH fall. During the day photosynthesis exceeds respiration (P > R), so that the oxygen concentration and pH both rise (Fig. 6.10). If the river is also organically polluted total deoxygenation may occur during the night period. In practice, the lowest oxygen concentrations occur several hours after sunrise (6 a.m–9 a.m), while the highest oxygen concentrations occur mid to late evening (8 p.m–12 a.m). The pH shift is due to all the CO_2 being stripped from the water and other effects on the bicarbonate–carbonate buffering system of water (Section 3.4). Another effect of the rise in pH is the increased toxicity of ammonia as it changes from the ionized to the unionized form (Table 9.25 and Fig. 6.11). The biological assessment of eutrophication has been reviewed by Kelly and Whitton (1998).

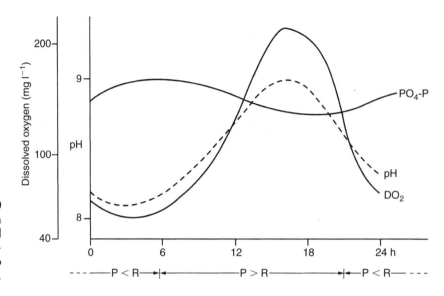

FIGURE 6.10
Daily variation of pH, dissolved oxygen and phosphate due to eutrophication.

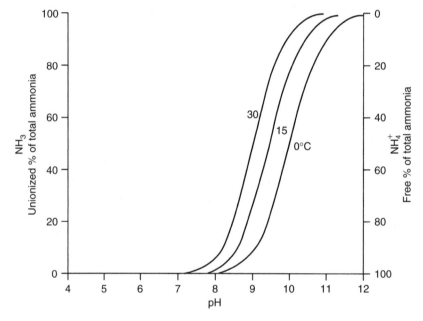

FIGURE 6.11
The general variation between the proportion of unionized to free ammonia at varying pH and temperatures. (Reproduced from Chapman (1996) with permission of UNESCO, Paris.)

6.4.1 ALGAL TOXINS

The algal species characteristic of oligotrophic and eutrophic lakes are given in Table 6.7. Blue-green algae are unique micro-organisms having some of the properties of both algae and bacteria. They are capable of photosynthesis and some species can fix nitrogen, but despite their name they belong to a group of true bacteria known as cyanobacteria with over 1000 known species. In lentic systems planktonic forms of blue-green algae can increase to very high densities often forming a surface scum downwind of the bloom due to the presence of air-filled sacks making up as much as 20% of the cell, making them very buoyant. These algae can produce chemicals that

TABLE 6.7 Characteristic algal associations of oligotrophic and eutrophic lakes

	Algal group	*Example*
Oligotrophic lakes	Picoplankton (often small cyanobacteria)	*Synechococcus*
	Desmid plankton	*Staurdesmus, Staurastrum*
	Chrysophycean plankton	*Dinobryon*
Mesotrophic lakes	Diatom plankton	*Cyclotella, Tabellaria*
	Dinoflagellate plankton	*Peridinium, Ceratium*
	Chlorococcal plankton	*Oocystis, Eudorina*
Eutrophic lakes	Diatom plankton	*Asterionella, Fragilaria Stephanodiscus, Melosira*
	Dinollagellate plankton	*Peridinium, Glenodinium*
	Chlorococcal plankton	*Scenedesmus*
	Cyanobacterial plankton	*Aphanizomenon, Anabaena, Microcystis*

Reproduced from Mason (1996) with permission of Addison Wesley Longman Ltd, Harlow.

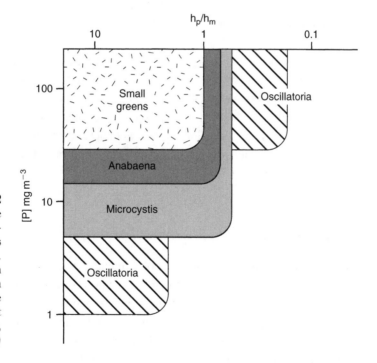

FIGURE 6.12 Factors affecting the distribution of blue-green algae in lakes and reservoirs. (Reproduced from NRA (1990) with permission of the Environment Agency, Peterborough.)

are toxic to all mammals including humans (NRA, 1990). They are found mainly in eutrophic or hypertrophic lakes where they can form blooms at water temperatures ($>20°C$) where there is a high P concentration; although growth is limited at low temperatures ($<10°C$) and low phosphate concentrations ($<0.4\,\mu g\,l^{-1}$). Typical distribution of blue-green algae is related to light, mixing and the availability of phosphorus as shown in Fig. 6.12. Here, the ratio of the depth of the fully mixed zone of the lake (h_m) to the photosynthetic depth (h_p) is plotted against the phosphorus concentration so that genera characteristic of oligotrophic lakes appear at the base of the figure. Different species of *Oscillatoria* predominate at extremes, while other types of planktonic algae such as *Microcystis* compete with other groups in the more favourable environment (NRA, 1990).

Cyanobacterial poisoning is caused by contact with water containing blue-green algae, consumption of fish taken from contaminated waters or drinking water with cyanobacterial toxins (Hunter, 1991). The toxins can kill fish and other biota, farm animals and waterfowl. The effects on humans is very serious causing a wide range of symptoms, normally liver damage, although mortalities are rare (Codd *et al.*, 1989; Falconer, 1999). The toxins produced by blue-green algae are varied but include:

(a) neurotoxins (mainly alkaloids) produced by species of the genera of *Anabaena*, *Aphanizomenon* and *Oscillatoria*, that cause paralysis of the skeletal and respiratory muscles resulting in death in as little as 5 min;
(b) hepatotoxins (microcystin – peptides) produced by species of the genera of *Microcystis*, *Oscillatoria*, *Anabaena* and *Nodularia*, that mainly cause severe and often fatal liver damage.

All algal toxins are highly toxic and difficult to control. In practice, water with blue-green algal blooms are not used for supply purposes, although algal toxins can be removed from drinking water by granular activated carbon (Section 20.6). The WHO has proposed a guideline value of $1 \, \mu g l^{-1}$ for microcystin-LR in drinking water (Falconer, 1999). Not all blooms of cyanobacteria result in the release of toxins. However, algal toxins are difficult to monitor and detect, making prevention very difficult (Lawton and Codd, 1991).

6.5 HEATING OF THE WATER

Many effluents may be warmer or cooler than the receiving water; however, the main problem is waste heat from cooling towers or power generation. Heat added to rivers is quickly dissipated downstream. Heat loss depends on turbulence, dilution, water velocity and temperature difference (Fig. 6.13). In the US, river temperatures $\geqslant 30°C$ are not uncommon, whereas in the British Isles river temperatures above 24°C are rare. So the problem of dissipating heat from discharges varies significantly from place to place.

Elevation of water temperature causes a number of direct responses:

(a) heat stress or death of sensitive species;
(b) enhanced micro-organism respiration;
(c) increased toxicity of many poisons;
(d) the attraction–repulsion of mobile species.

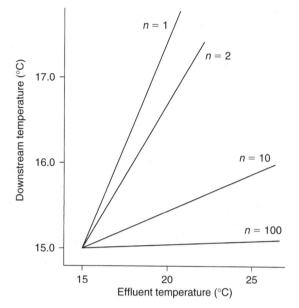

FIGURE 6.13
Effect of dilution
and effluent temper-
ature on the down-
stream temperature
of a river with a
temperature of
15°C, where n is the
dilution factor.

The increased temperature also lowers dissolved gas solubility. In terms of oxygen availability sensitive species will be excluded, especially if the BOD is elevated due to the higher micro-organism respiration rate. Nitrogen is also a problem, coming out of solution as the temperature rises causing gas bubble disease in fish and macro-invertebrates, which is normally fatal. The abstraction and disposal of cooling water can cause other problems. For example, the abstraction of cooling water causes a local increase in water velocity that can attract and damage species. As another example, anti-fouling biocides are used to control the development of biofilms within the cooling systems. Chlorine is generally used, although the pesticides are also employed. Not only do these kill any organisms passing through the system, they can also be discharged, resulting in a toxic effect on the receiving water biota. Thermal pollution is not only caused by heated effluents and cooling water from power stations. Water from other water resources such as ground water, reservoirs and lakes used for flow compensation in rivers may also be either warmer or colder than the receiving water causing a significant temperature change. Warmer water may not readily mix with colder denser water and so form a surface plume. This will reduce oxygen transfer to the colder water beneath, although the benthic organisms are protected from any sudden changes in temperature (Abel, 1989).

Overall the effects can be summarized as the loss of temperature-sensitive species, especially salmonids, an increase in the rate of biological activity and, therefore, an increase in BOD, and finally a lower oxygen saturation concentration in the water. Climate change due to global warming is causing a slow but steady rise in the mean temperature of European rivers and lakes. This is causing concern as it is resulting in significant changes in the composition of the biota of surface waters (Section 23.2).

6.6 The Effect on the Buffering System

The buffering system of freshwaters has already been described in Section 3.4. There are two major causes of acidification of freshwater systems, acid rain and acid mine drainage.

6.6.1 ACID RAIN

When sulphur dioxide (SO_2) and the oxides of nitrogen (NO_x), mainly from motor exhausts and the combustion of fossil fuel, reach the atmosphere they react with water vapour being oxidized to form sulphuric and nitric acids. This is then carried to earth as rain or snow (wet deposition) or in particulate form (dry deposition). Uncontaminated rainwater in the atmosphere is in equilibrium with CO_2 and has a pH of 5.6 (Section 3.4). However, most rainfall in Europe has a pH less than 4.5 due to acid formation, with an average 70% of the acidity due to sulphuric acid. Acid rain lowers the pH of surface waters by falling into surface waters, as surface run-off within the catchment or by the generation of acidity within the soils of the catchment. The impact on surface waters is linked closely to catchment geology and the acid neutralizing capability (ANC) of the water. Soft waters are more at risk from acidification due to their poor acid neutralizing capacity (i.e. alkalinity $<24\,\mathrm{mg\,l^{-1}}$ as $CaCO_3$). The buffering capacity of lakes can be calculated as:

$$\text{ANC} = \Sigma \text{ base cations} - \Sigma \text{ strong acid anions } (\mu\text{eq l}^{-1}) \qquad (6.2)$$

$$= (Ca^{2+} + Mg^{2+} + Na^+ + K^+) - (SO_4^{2-} + NO_3^- + Cl) \qquad (6.3)$$

This scale can be used to classify lakes as to their sensitivity to acidification:

ANC (μeq l^{-1})	Sensitivity
0	Acidified
<1–40	Very sensitive
>40–200	Sensitive
>200	Insensitive

Soil acidification is a major factor in water acidification. In the soil the acids in precipitation dissociate into their respective ions (H^+, SO_4^{2-}, NO_3). As the dissociated H^+ ions increase, cation exchange in the soil increases with Na^+, K^+ and Mg^{2+} displaced from exchange sites on soil particles and transported into surface waters with SO_4^{2-} effectively transferring acidity from the soil to surface water. The acid precipitation also mobilizes metals, including aluminium ions (Al^{3+}), where the pH is <4.5. Trees, in particular conifers, are very effective at scavenging dry deposition and pollution held in mist and fog (occult deposition) causing increased acidity in the precipitation falling under the vegetation. In coastal areas oceanic aerosols are also effectively captured by the vegetation resulting in increased sulphate deposition and release of H^+ ions. The effect of acidification by acid rain is best seen in lakes. The effect on pH is dependent on the concentration of bicarbonate ions in the water. Initially the bicarbonate buffers the input of acids according to:

$$H^+ + HCO_3^+ \rightarrow H_2O + CO_2 \tag{6.4}$$

The pH remains above 6, and the flora and fauna is unaffected. Lakes where there is enough bicarbonate buffer to neutralize the acidity are classed as bicarbonate lakes (Fig. 6.14). Where the bicarbonate is exhausted the pH may collapse resulting in fish kills. If this is a seasonal or occasional event then the lake will recover (transitional lake). If the loss of alkalinity is complete then the water will retain a low but stable pH of <5, although metal levels including aluminium (Fig. 20.2) will be elevated (acid lake). The general biological effect is the rapid elimination of sensitive species as the pH falls and metal contamination rises, in particular aluminium, with fish particularly sensitive. Most waters sensitive to acidity are already classified as soft which naturally have a low species diversity of macro-invertebrates. Few species can tolerate such conditions, especially in a low calcium environment. Ephemeroptera are particularly sensitive to acidification.

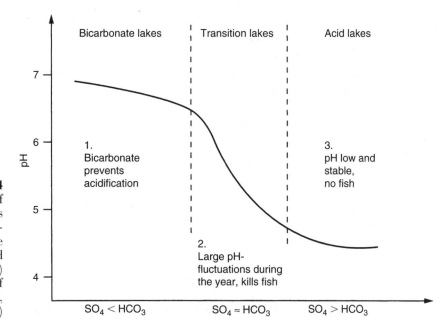

FIGURE 6.14 Classification of acidification of lakes using the ratio of sulphate to bicarbonate ions. (Reproduced from Mason (2002) with permission of Prentice Hall Ltd, London.)

6.6.2 ACID MINE DRAINAGE

Acid mine drainage occurs whenever mineral ores are mined. It is caused by the oxidation of a sulphide ore, usually pyrite (FeS_2). The most commonly associated minerals are pyritic sulphur, copper, zinc, silver, gold, lead and uranium. The most frequently mined sulphides are pyrite (FeS_2), chalcopyrite ($CuFeS_2$), sphalerite (ZnS) and galena (PbS). However, in Europe the commonest source of acid mine drainage is coal mining and storage.

The oxidation of pyrite is a chemical process although chemoautotrophic bacteria are also closely involved. Pyrite is oxidized in the presence of air and water to ferrous sulphate and sulphuric acid:

$$2FeS_2 + 2H_2O + 7O_2 \rightarrow FeSO_4 + H_2SO_4 \tag{6.5}$$

The ferrous sulphate is further oxidized to ferric sulphate. This step is catalysed by iron-oxidizing bacteria, such as *Thiobacillus ferro-oxidans*:

$$4FeSO_4 + O_2 + 2H_2SO_4 \rightarrow 2Fe_2(SO_4)_3 + 2H_2O \tag{6.6}$$

The ferric sulphate then reacts with water to produce ferric hydroxide and more acid:

$$Fe_2(SO_4)_3 + 6H_2O \rightarrow 2Fe(OH)_3 + 3H_2SO_4 \tag{6.7}$$

In practice, the chemical reactions are very slow. The presence of bacteria, such as *T. ferro-oxidans* (iron oxidizing) and *T. thio-oxidans* (sulphur oxidizing), greatly accelerate the process due to enzymatic activity. These bacteria grow at a very low pH, do not require organic matter and so rapidly dominate the flora. They utilize pyrites as an energy source, so their enzymes can accelerate the rate of oxidation by up to 1 million times. The bacteria are collectively known as acidophilic (acid loving) chemoautotrophs.

The oxidation of pyrite is in fact a four-step reaction:

$$(1) \qquad FeS_2 + \tfrac{1}{2}O_2 + H_2O \rightarrow Fe^{2+} + 2SO_4^{2-} + 2H^+ \tag{6.8}$$

$$(2) \qquad Fe^{2+} + \tfrac{1}{4}O_2 + H^+ \rightarrow Fe^{3+} + \tfrac{1}{2}H_2O \tag{6.9}$$

$$(3) \qquad Fe^{3+} + 3H_2O \rightarrow Fe(OH)_3 + 3H^+ \tag{6.10}$$

$$(4) \qquad FeS_2 + 14Fe^{3+} + 8H_2O \rightarrow 15Fe^{2+} + SO_4^{2-} + 16H^+ \tag{6.11}$$

Step (2) is the rate-limiting reaction, the oxidation of ferrous iron (iron II) to ferric iron (iron III). There is a propagation cycle between steps (2) and (4) where Fe^{3+} one of the products of step (2), acts as an oxidant of pyrite in step (4), and the Fe^{2+} produced by this reaction can be used as a reductant in step (2). Step (2), the oxidation of ferrous sulphate to ferric sulphate, is normally catalysed by iron-oxidizing bacteria. These bacteria keep the ratio of Fe^{3+} to Fe^{2+} in the Fe^{3+} oxidizing step (4), high. In this way the bacteria are ensuring that more pyrite is oxidized to produce more Fe^{2+} to drive step (2).

Pyrite is only broken down in the presence of water and oxygen. Pyrites can remain in their reduced form in an undisturbed state so long as they are anaerobic. Only a few cases of naturally occurring acid streams have been recorded, most occur as a result of mining activity. The major factor is the surface area available for oxidation of pyrite, as this determines the overall rate of reaction.

Acid mine drainage is a multi-factor pollutant (Fig. 6.15). The low pH adversely affects the biota and takes into solution heavy metals which are toxic. Iron is toxic in its own right at elevated concentrations. As the drainage water is diluted the pH rises and ferric hydroxide flocs are formed. These not only cause turbidity but settle out of suspension forming a layer of precipitate (ochre). There are four main pollutional effects: acidity, heavy metals, suspended solids (i.e. ferric hydroxide precipitate) and salinization (i.e. sulphate) (Kelly, 1989). The ecological effects are summarized by Gray (1997).

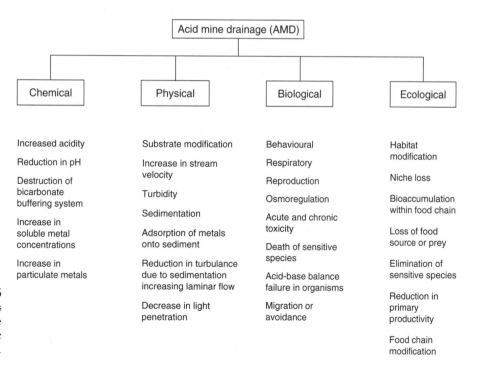

FIGURE 6.15 The major effects of acid mine drainage on a lotic (river) system.

TABLE 6.8   Fish kills in Ireland 1988–1990

Source	1988	1989	1990	Cause
Agriculture	26	26	23	Organic
Deoxgenation[a]	0	25	2	Deoxygenation
Industrial	7	14	14	Organic/toxic
Eutrophication	4	12	4	nutrients
Civil works	3	12	1	Inert solids
Sewage	0	8	1	Organic
Water works	3	4	2	Inert solids/toxic
Unknown	7	10	5	
Total	50	111	52	

[a] Due to low flows.
Reproduced from Moriarty (1991) with permission of Roinn Na Mara, Dublin.

6.7 FISH KILLS

The causes of fish kills are similar throughout Europe, although the more industrialized a country then the more important industrial pollution is as a causative factor in fish mortality. The causes of fish kills in Ireland between 1988 and 1990 are summarized in Table 6.8 (Moriarty, 1991), and the category of pollutant this involves is shown (i.e. organic, inert solids, etc.). Over half of the reported kills took place in eastern and southern fishery areas (i.e. the areas of high population density and most industry), with agriculture and industry the most serious causes. Sewage treatment plants are rarely a problem except during exceptionally low flows when the available dilution may become inadequate. Nutrients come from slurry spreading, agricultural fertilizers and domestic sewage. The majority of fish kills occur in the summer months of May–August, with a peak in June and July. There are very few kills in the winter from October to March. The significant increase in fish kills in 1989 was due mainly to low water levels resulting in lower dilution. Increased sunshine and temperature that year resulted in an increase in eutrophication and made fish much more susceptible to low dissolved oxygen conditions.

6.8 CONCLUSION

Discharges can make rivers more silty, more acid or alkaline, less oxygenated, warmer, harder, saltier and richer in nutrients, but still a natural river so long as these parameters remain within the bounds of normal variability. All rivers and streams have the capacity of self-purification and polluted water will eventually regain its natural quality with time. Many of the more persistent pollutants are removed finally by precipitation and/or adsorption. Where there is a steady input of pollutant the river will show various permanent changes in chemical, physical and biological quality downstream.

REFERENCES

Abel, P. D., 1989 *Water Pollution Biology*, Ellis Horwood, Chichester.

Campbell, P. G. C. and Stokes, P. M., 1985 Acidification and toxicity of metals to aquatic biota. *Canadian Journal of Fish and Aquatic Science*, **42**, 2034–49.

Chapman, D. (ed.), 1996 *Water Quality Assessments: A Guide to the Use of Biota, Sediments and Water in Environmental Monitoring*, 2nd edn, E. & F. N. Spon, London.

Codd, G. A., Bell, S. G. and Brooks, W. P., 1989 Cyanobacterial toxins in water, *Water Science and Technology*, **21** (3), 1–13.

Curtis, E. J. C. and Harrington, D. W., 1971 The occurrence of sewage fungus in rivers in the United Kingdom. *Water Research*, **5**, 281–90.

Ellis, K. V., 1989 *Surface Water Pollution and its Control*, Macmillan, London.

Falconer, I. R., 1999 An overview of problems caused by toxic blue-green algae (cyanobacteria) in drinking and recreational water. *Environmental Toxicology*, **14**, 5–12.

Golterman, H. L., 1991 Reflections on post-OECD eutrophication models. *Hydrobiologia*, **218**, 167–76.

Gray, N. F., 1985 Heterotrophic slimes in flowing waters, *Biological Reviews of the Cambridge Philosophical Society*, **60**, 499–548.

Gray, N. F., 1997 Environmental impact and remediation of acid mine drainage: a management problem. *Environmental Geology*, **30**, 62–71.

Hellawell, J. M., 1986 *Biological Indicators of Freshwater Pollution and Environmental Management*, Elsevier Applied Science, London.

Hunter, P. R., 1991 An introduction to the biology, ecology and potential public health significance of the blue-green algae. *PHLS Microbiology Digest*, **8**, 13–15.

Kelly, M., 1989 *Mining and the Freshwater Environment*, Elsevier Applied Science, London.

Kelly, M. G. and Whitton, B. A., 1998 Biological monitoring of eutrophication in rivers. *Hydrobiologia*, **384**, 55–67.

Laws, E. A., 1993 *Aquatic Pollution*, John Wiley and Sons Inc., New York.

Lawton, L. A. and Codd, G. A., 1991 Cyanobacterial (blue-green algal) toxins and their significance in the UK and European waters. *Journal of the Chartered Institution of Water and Environmental Management*, **5**, 460–5.

Mason, C. F., 2002 *Biology of Freshwater Pollution*, 4th edn, Prentice Hall, London.

Moriarty, C., 1991 *Fish Kills in Ireland 1990*, Fishery leaflet 149, Roinn Na Mara, Dublin.

National Rivers Authority (NRA), 1990 *Toxic Blue-green Algae*, Water Quality Series No. 12, National Rivers Authority (Anglian Region), Peterborough.

Tessier, A. and Turner, D. R., 1995 *Metal Speciation and Bioavailability in Aquatic Systems*, John Wiley and Sons, Chichester.

UNESCO, 1978 *Water Quality Surveys*. Studies in Hydrobiology: 23, UNESCO and WHO, Geneva.

Vollenweider, R. A., 1975 Input–output models with special reference to the phosphorus loading concept in limnology, *Schweizerische Zeitschrift für Hydrologie*, **37**, 53–84.

FURTHER READING

Cushing, C. E. (ed.), 1997 *Freshwater Ecosystems and Climate Change in North America: A Regional Assessment*, John Wiley and Sons, New York.

Nemerow, N. L., 1991 *Stream, Lake, Estuary and Ocean Pollution*, 2nd edn, Van Nostrand Reinhold, New York.

Redman, M., Winder, L. and Merrington, G., 1999 *Agricultural Pollution: Problems and Solutions*, E. & F. N. Spon, London.

Steinberg, C. E. W. and Wright, R. F., 1994 *Acidification of Freshwater Ecosystems: Implications for the Future*, John Wiley and Sons, Chichester.

Uhlmann, D., 1979 *Hydrobiology: A Text for Engineers and Scientists*, John Wiley and Sons, Chichester.

Welch, E. B., 1986 *Ecological Effects of Wastewater*, Cambridge University Press, Cambridge.

Part II Water Quality Management and Assessment

INTRODUCTION

Prior to the introduction of the European Union (EU) Water Framework Directive (WFD) (Section 7.2), water resources were managed on an individual basis, with little coherent policy in terms of how water bodies interacted with each other, or how land-based activities affected them. The concept of catchment management emerged in the 1970s and led to the formation of catchment-based water authorities in the UK. This allowed improved management of water resources, especially the regulation of flow in major rivers. River Catchment Management Plans were developed in Ireland in the early 1980s and were used to set water quality standards based on use criteria. However, other water bodies such as lakes, groundwaters and coastal waters were largely ignored within such plans leading to these resources being largely unmanaged and unmonitored. The WFD provides a unique management structure that deals with the whole water cycle and isolates the water cycle from contamination by hazardous substances by eliminating them at source. Using an updated and streamlined group of Directives, water is effectively protected from all land-based sources of pollution for the first time, preventing hazardous substances from being lost into the marine environment. Setting emission limits through the Urban Wastewater Treatment and Integrated Pollution Prevention Control Directives controls point sources of pollution; while diffuse pollution is controlled by the Nitrates Directive and the application, through the WFD, of best environmental practices (Section 7.4). The development of catchment management is examined in Sections 7.1 and 7.2.

7.1 BASIC MANAGEMENT PROGRAMME FOR RIVERS

The objective of river management is to balance the interests of users with the development of the resource, while at the same time improving and preserving environmental quality. The ideal situation is the optimal utilization of resources without deterioration of their natural quality. Uses can be classified as either consumptive (e.g. irrigation, water supply, waste disposal, fishery production, etc.) or non-consumptive (e.g. aesthetic, ecological, scientific, etc.). It is the former uses that reduce the quality or availability of water for the non-consumptive functions, requiring

TABLE 7.1   Major anthropogenic activities affecting river systems

Supra-catchment effects
Acid deposition
Inter-basin transfers

Catchment land use change
Afforestation and deforestation
Urbanization
Agricultural development
Land drainage/flood protection

Corridor engineering
Removal of riparian vegetation
Flow regulation (dams, channelization, weirs, etc.)
Dredging and mining

Instream impacts
Organic and inorganic pollution
Thermal pollution
Abstraction
Navigation
Exploitation of native species
Introduction of alien species

Reproduced form Boon *et al.* (1992) with permission of John Wiley and Sons Ltd, Chichester.

conservation action to be taken. The fundamental problem of water management is to find an acceptable balance between use and conservation of a system. Owing to the growth in population and the increasing demand for water, many of the anthropogenic activities affecting freshwater systems are increasing (Table 7.1).

Water quality is very hard to define and to a great extent extremely subjective. It is not simply a case of the cleaner or purer the better. For example, distilled water is extremely pure chemically and so its quality can be considered as being high as it contains no toxicants or pollutants, yet it is unsuitable for potable use and it lacks the trace elements necessary for freshwater biota. Water quality can only be defined in relation to some potential use for which the limiting concentrations of various parameters can be identified. This approach makes particular sense as concern for quality is normally related to some practical need (e.g. drinking, fishing, agriculture, etc.).

There are a variety of uses for water each requiring their own set of specific quality requirements (criteria). These can be categorized into simple groups, such as:

(I) those requiring water of highest quality and free of pathogens, uses include: drinking water supply, salmonid fishery, swimming and certain industrial processes such as food processing;

(II) those requiring water of lesser quality but still free from toxins and pathogens, uses include: coarse fishery, amenity and recreation such as boating, also agricultural irrigation and certain industries;

(III) where quality is unimportant, just quantity; uses include: cooling water and navigation.

This classification system was originally proposed by the World Health Organization (IHD-WHO, 1978) but is now seen as being rather crude for modern management purposes. Most countries have developed their own classifications linking water quality to potential uses (Tables 9.20 and 7.2). The EU has also produced a wide range of usage-based Directives on water quality (e.g. Surface Water Directive, Freshwater Fish Directive, etc.) and these are considered in detail in Chapter 8. The main criteria for use are freedom from pathogens, non-toxicity, and consistency of quality and consistency in the quantity of water (Table 7.3).

No European rivers are designated as mere effluent carriers; all regulatory authorities accept that a minimum standard for all waters is that it should be suitable for recreation (excluding swimming) and other group uses. Where rivers are required for supply or group (I) uses, then a greater effort is expended in achieving the high quality required and in maintaining it. With the introduction of the WFD, this approach must change with all water bodies, including rivers, maintained to at least Good Quality Status. Where rivers are so severely damaged that improvement would require a very large investment, then the current quality status must not be allowed to deteriorate further, with the long-term objective to bring them up to the minimum Good Quality Status.

Group (I) uses are those most often protected through EU legislation. River classification in Europe is largely based on the former UK system where rivers are classified according to their water quality ranging from good (Classes 1A and 1B) to bad (Class 4) quality (Table 7.2). This is achieved largely on the basis of chemical criteria and in particular biochemical oxygen demand (BOD) and dissolved oxygen (DO) concentration. Most European countries have a high proportion of river channels with good water quality that are classed as Grade 1A and 1B. The less industrialized the country the greater the proportion of rivers in these top groups. France and Germany both have less than 45% of Grade 1 rivers with the overall situation in many countries slowly deteriorating, especially in terms of rising nutrient levels (Newman et al., 1992).

In water management, decisions are based on the comparison of water quality data with criteria and standards. Criteria are scientific requirements on which a decision or judgement may be based concerning the suitability of water quality to support a designated use; that is, the determination of the basic water quality requirements for a particular use. Sets of criteria exist for five categories of specific water use:

1. Raw water sources for drinking, supply.
2. Public recreational waters, aesthetics.
3. Agricultural supply.
4. Industrial supply.
5. Preservation of freshwater, estuarine and marine ecosystems.

With time this information is refined and extended. Objectives are the levels of parameters to be attained within a water quality management programme, often

TABLE 7.2   The former UK river quality classification system[a] is linked to water use

River Class	Quality criteria	Remarks	Current potential uses
	Class limiting criteria (95th percentile)		
1A Good quality	(i) DO saturation greater than 80% (ii) BOD not greater than 3 mg l^{-1} (iii) Ammonia not greater than 0.4 mg l^{-1} (iv) Where the water is abstracted for drinking water, it complies with requirements for A2[b] water (v) Non-toxic to fish	(i) Average BOD probably not greater than 1.5 mg l^{-1} (ii) Visible evidence of pollution should be absent	(i) Water of high quality suitable for potable supply abstractions and for all other abstractions (ii) Game or other high-class fisheries (iii) High amenity value
1B Good quality	(i) DO greater than 60% saturation (ii) BOD not greater than 5 mg l^{-1} (iii) Ammonia not greater than 0.9 mg l^{-1} (iv) Where water is abstracted for drinking water, it complies with the requirements for A2[b] water (v) Non-toxic to fish	(i) Average BOD probably not greater than 2 mg l^{-1} (ii) Average ammonia probably not greater than 0.5 mg l^{-1} (iii) Visible evidence of pollution should be absent (iv) Waters of high quality which cannot be placed in Class 1A because of the high proportion of high-quality effluent present or because of the effect of physical factors such as canalization, low gradient or eutrophication (v) Classes 1A and 1B together are essentially the Class 1 of the River Pollution Survey	Water of less high quality than Class 1A but usable for substantially the same purposes
2 Fair quality	(i) DO greater than 40% saturation (ii) BOD not greater than 9 mg l^{-1}	(i) Average BOD probably not greater than 5 mg l^{-1}	(i) Waters suitable for potable supply after advanced treatment

Class			
	(iii) Where water is abstracter for drinking water it complies with the requirements for A3[b] water (iv) Non-toxic to fish in EIFAC terms (or best estimates if EIFAC figures not available)	(ii) Similar to Class 2 of River Pollution Survey (iii) Water not showing physical signs of pollution other than humic colouration and a little foaming below weirs	(ii) Supporting reasonably good coarse fisheries (iii) Moderate amenity value
3 Poor quality	(i) DO greater than 10% saturation (ii) Not likely to be anaerobic (iii) BOD not greater than 17 mg l^{-1}. This may not apply if there is a high degree of reaeration	Similar to Class 3 of River Pollution Survey	Waters which are polluted to an extent that fish are absent or only sporadically present. May be used for low grade industrial abstraction purposes. Considerable potential for further use if cleaned up
4 Bad quality	Waters which are inferior to Class 3 in terms of DO and likely to be anaerobic at times	Similar to Class 4 of River Pollution Survey	Waters which are grossly polluted and are likely to cause nuisance
X	DO greater than 10% saturation		Insignificant watercourses and ditches not usable where the objective is simply to prevent nuisance developing

Notes:
(i) Under extreme weather conditions (e.g. flood, drought, freeze-up), or when dominated by plant growth, or by aquatic plant decay, rivers usually in Classes 1, 2 and 3 may have BODs and DO levels, or ammonia content outside the stated levels for those classes. When this occurs the cause should be stated along with analytical results.

(ii) The BOD determinations refer to 5-day carbonaceous BOD with allylthiourea. Ammonia figures are expressed as NH$_4$.

(iii) In most instances the chemical classification given above will be suitable. However, the basis of the classification is restricted to a finite number of chemical determinants and there may be a few cases where the presence of a chemical substance other than those used in the classification markedly reduces the quality of the water. In such cases, the quality classification of the water should be down graded on the basis of biota actually present, and the reasons stated.

[a]The river classification system is now to be superseded by the new Environment Agency GQA scheme (Section 9.5).

[b]European Economic Community (EEC) Category A2 and A3 requirements are those specified in the EEC Council Directive of 16 June 1975 concerning the quality of surface water intended for abstraction of drinking water in the Member State (Table 8.5).

Reproduced with permission of Her Majesty's Stationery Office, London.

TABLE 7.3   Four main requirements of water in relation to use

Use	Freedom from pathogens	Non-toxic	Consistency	
			Quality	Quantity
Domestic	×	×	×	×
Industrial				
process	–	–	×	×
cooling	–	–	×	
Agricultural	–	×	×	×
Navigation	–	–	–	×
Fisheries	–	×	×	–
Waste disposal	–	–	–	×
Sport				
swimming	×	×	–	×
boating	×	–	–	×
skiing	×	×	–	×

referred to as water quality goals or guidelines. Standards are legally prescribed limits for discharges adopted by governments or other legal authorities. These have been based on technical feasibility, cost–benefit and risk–benefit analysis. Standards are used to achieve objectives which in turn are based on the critical assessment of national priorities, such as cost, population trends, present and projected water usage, industrialization and economic resources (IHD-WHO, 1978). Water quality standards have been set by various organizations including the WHO, the EU and the US Environmental Protection Agency (USEPA). The Directives currently affecting water quality in Europe are summarized in Table 8.1.

Water pollution legislation allows consents (or discharge licences) to be set for dischargers to both receiving waters and sewers. Consents can include details on the nature and composition of an effluent, its temperature, volume or rate of discharge, or any factor deemed important (e.g. times of discharge in tidal areas). These consents are legally enforceable and can lead to prosecution under the relevant legislation. In deciding a consent condition a discharge must not make any appreciable change in quality. This is calculated using mass balance, waste assimilative capacity and, increasingly, water quality models (Section 7.1). Consents are based on low flow conditions when any resultant pollution will be at a maximum. Ammonia is an exception, as it is not oxidized so quickly in winter so conditions may be based on winter conditions. Of course there will also be a safety margin built in. Like any other planning decision, dischargers can normally appeal against a consent or licence. They are renewed usually every 5 years, sometimes annually or when a factory expands or its discharge significantly changes due to work practices. Once consents are agreed then a routine sampling programme must be initiated to monitor both the discharge and the receiving water quality. The monitoring may be chemical, biological or both (Section 9.2).

Complex river systems have led to an ever-increasing use of systems modelling. Using base-line data, extensive models have been constructed to estimate assimilative capacities and use demands (Section 7.2). The increasing complexity of co-ordinating discharge licences and protecting water quality to attain the environmental quality objectives (EQO) has led to the development of water quality (catchment or basin) management plans.

7.2 WATER QUALITY AND CATCHMENT MANAGEMENT PLANS

7.2.1 WATER QUALITY MANAGEMENT PLANS

In Ireland, prior to the introduction of the WFD, the Water Pollution Act 1977 enabled a number of county councils to work together to produce water quality management plans (WQMP), based on catchments and not administrative boundaries. These provided a legal document, similar to a county development plan, which defined management objectives and set water quality objectives and standards. They had a draft stage to allow for public comment or if necessary an enquiry, but once adopted a WQMP became law. The primary aim of a WQMP was to ensure that water quality was maintained in a satisfactory condition and where necessary improved thereby: safeguarding public health; catering for the abstraction of increasing quantities of water for domestic, industrial and agricultural purposes; catering for the needs of commercial, game and coarse fish (as appropriate); and catering for the relevant water-based amenities and recreational requirements. Such plans covered a period of 20 years and were to be reviewed and, where necessary, revised at least every 5 years (McCumiskey, 1982).

There are six major stages in the development of a WQMP for a river catchment:

1. To decide on the uses of a particular river (i.e. group designation).
2. To decide on the water quality conditions necessary in the river to support the uses decided at stage 1 (i.e. water quality criteria).
3. To assess the effect of existing discharges on a river and to attempt a forecast of future effluents.
4. To decide upon the standards that are required for each effluent discharge in order to leave a river with the necessary quality (i.e. water quality objectives).
5. To produce consent (discharge) conditions which will include the standards decided under stage 4.
6. To initiate a sampling programme which will both ensure that discharges comply with the above standards and also indicate from the river water quality whether revised effluent standards are necessary.

These steps are expanded into a full methodology in Table 7.4 (Fig. 7.1). The major factor affecting long-term water quality is the initial setting of standards that are dependent on the designated uses identified and their water quality criteria. In Ireland all rivers for which a WQMP were formulated have been classified as salmonid. Therefore, the standards adopted have all been the same (Table 7.5). Those catchments that have also been designated as salmonid under the EU Freshwater Fish Directive use the EQO in Table 8.13.

TABLE 7.4 The major steps in the production of a WQMP for a river (visualized in a flow diagram in Fig. 7.1)

1. Identify beneficial water uses (existing and future) to be protected
2. Compile all available data on water quality
3. Examine the data in (2) and determine the characteristic elements which mainly determine water quality at selected control sections along main channel and major tributaries
4. Select water quality standards according to (1). Due reference to EU Directives and other standards
5. Compute assimilative capacities at key locations
6. Compute population projections for catchment (20 years)
7. Compute generated waste loads and loads discharged in catchment
8. Compute projections of generated loads over projected period
9. Identify existing and future abstractions
10. Estimate water quality at key locations as a result of (a) existing and (b) future waste loads. Identify the level of waste-load reduction required
11. Review all possible options in relation to the treatment of existing and projected wastewater loads prior to discharge and determine the reserve capacity of the river at key sections in relation to these options
12. Select the appropriate treatment options for existing wastewater discharges (public and private) so that receiving water standards can be met and a suitable reserve capacity for future developments maintained
13. Determine the main priorities for capital investment in both public and private wastewater treatment facilities
14. Outline the general procedures that should be adopted for the laying down of effluent emission standards for future wastewater discharges

7.2.2 CATCHMENT MANAGEMENT

The management of water resources cannot be isolated from the use of the land within the catchment. Therefore, the WQMP approach outlined above is being replaced with a more integrated approach to river management where land use is more closely controlled in order to minimize the impact of diffuse pollution, especially from agriculture and forestry. This is reflected by the introduction of new legislation, such as the Nitrates Directive (91/676/EEC), resulting in new codes of practices for agriculture and the designation of nitrate-sensitive areas (Sections 8.4 and 11.1.3) (Table 7.6). Other important considerations include the effect of small treatment plants (e.g. septic tanks) (Chapter 22) causing localized nutrient, trace organic and pathogen pollution of surface waters either directly or via groundwater, and the problems caused by urban and rural run-off on water quality (Section 7.4). More attention is also paid to other water uses, such as conservation and recreation. Geographical information systems (GIS) are likely to become increasingly important to the management of large systems. Its primary function is to analyse and present information quickly and effectively allowing management decisions to be made much faster than with traditional mapping technology. The use of GIS in conjunction with CORINE land cover maps allows potential risk areas within the catchment to be identified. One effect is to isolate diffuse sources from

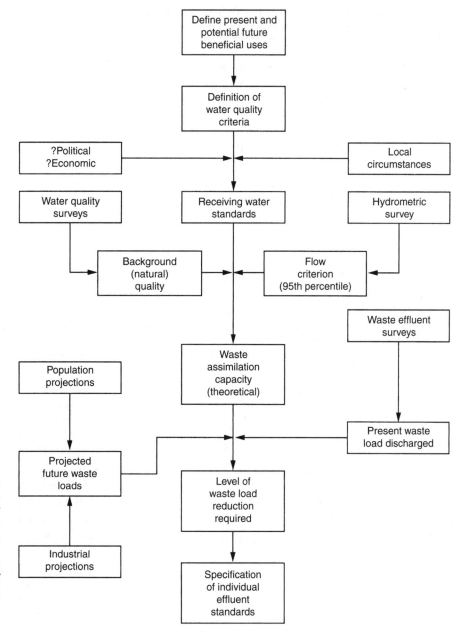

FIGURE 7.1
Flow diagram
showing the major
steps in the forma-
tion of a WQMP.
(Reproduced from
McCumiskey (1982)
with permission of
the Environmental
Protection Agency,
Dublin.)

individual farms or other land-based activities and turn them into point sources so that they can be treated.

River channelization for flood prevention is widespread and involves the straightening of river bends and the removal of natural obstructions. This results in a straight river channel devoid of bankside trees and other large vegetation, a uniform substrate with a very low habitat diversity. The presence of riparian vegetation, especially mature trees, shades the river and suppresses macrophyte growth. However, engineered rivers require

TABLE 7.5   Water quality standards used in Irish WQMP for salmonid rivers (percentile values based on 95th percentile flow)

Dissolved oxygen
$\geq 4 \, \text{mg} \, \text{l}^{-1}$ for 99.9% of the time
$\geq 6 \, \text{mg} \, \text{l}^{-1}$ for 95% of the time
$\geq 9 \, \text{mg} \, \text{l}^{-1}$ for 50% of the time

Biochemical oxygen demand
$\leq 5 \, \text{mg} \, \text{l}^{-1}$ for 95% of the time
$\leq 3 \, \text{mg} \, \text{l}^{-1}$ for 50% of the time

Ammonia
$\leq 0.02 \, \text{mg} \, \text{l}^{-1}$ for 95% of the time (unionized)
$\leq 0.05 \, \text{mg} \, \text{l}^{-1}$ for 95% of the time (total ammonia)

Oxidized nitrogen
$\leq 11 \, \text{mg} \, \text{l}^{-1}$ for 95% of the time

Orthophosphate
$\leq 0.2 \, \text{mg} \, \text{l}^{-1}$ for 95% of the time
$\leq 0.1 \, \text{mg} \, \text{l}^{-1}$ for 50% of the time

TABLE 7.6   Key actions of a Code of Practice to control nitrate leaching from agricultural soils

Reduction in leaching during the wet seasons	Crop rotations, soil winter cover, catch crops
Nutrient management plans	Balancing application with soil supply and crop needs, frequent manure and soil analysis, general limitations per crop for both mineral and organic nitrogen application rates
Appropriate application	Sufficient manure storage, application only when the crop needs nutrients, good spreading practices.
Buffer zones	Use of non-fertilized grass strips and hedges along watercourses and ditches to remove nitrates
Appropriate land use	Avoidance of steeply sloping land, highly permeable soils and land bordering vulnerable aquifers and surface waters

regular maintenance to control macrophyte development either using machinery, which is very damaging to the riparian zone, or by the application of herbicides. Channelization results in a large reduction in species diversity and abundance, especially of vertebrate species, such as fish, otters and water voles. In the UK, the Environment Agency has introduced an integrated management approach to prevent such damage to wildlife. This involves carrying out river corridor surveys using a standard methodology to evaluate the conservation value of each section of a river (500 m) and to identify the best engineering solution to any problems with minimal disturbance (NRA, 1993). An example of how riverbeds can be rehabilitated or improved by

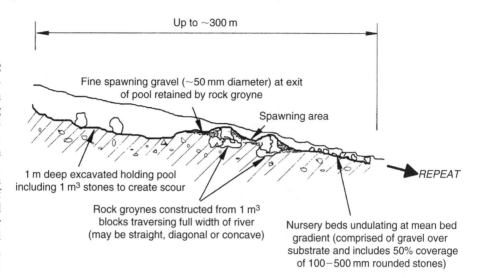

FIGURE 7.2
Schematic longitudi-
nal section through
a riverbed showing
typical reha-
bilitation action.
(Reproduced from
Ferguson *et al.*
(1998) with
permission of
the Chartered
Institution of Water
and Environmental
Management,
London.)

simple engineering techniques is illustrated in Fig. 7.2 (Ferguson *et al.*, 1998). The loss of flood plains and marginal wetlands due to land drainage and channelization of rivers is a major factor in the reduction of fish stocks in lowland rivers. Major restoration projects have been carried out on a number of such rivers (Brookes and Shields, 1996). In Denmark, a 25 km stretch of the River Brede, and a further 3 km of tributaries, which had been channelled 40–50 years previously, were re-meandered to improve flood control and the ecological quality of the river (Fig. 7.3). Small lakes, ponds and wetlands were created by raising the level of the substrate in the re-meandered sections of the river increasing the hydraulic contact between the river and the surrounding meadows. Gravel was introduced to create spawning grounds and shallow backwaters, often referred to as fry ponds, to allow more young fish to mature (Neilsen, 2002). These are also being constructed on lowland rivers throughout the east and south-east of England and have been particularly successful in providing cover to allow fish to feed and hide during development. The restoration project on the River Brede has significantly improved water storage in the lower catchment by recreating flood plains thereby reducing downstream flooding. With the rise in sea level due to climate change, such actions are increasingly important in helping to minimize the effect of flood incidents.

The Environment Agency introduced catchment management planning in the early 1990s and defines the process as:

A catchment management plan treats a river, together with the land, tributaries and the underground water connected with it, as a complete unit, or catchment. The plan sets out a common vision for the river catchment, reached after consultation with all interested parties. It identifies objectives and lists actions for conservation, recreation and amenity, as well as including all of the other functions of the Agency. Catchment management plans are the primary method by which the Agency are able to fully understand and plan any changes

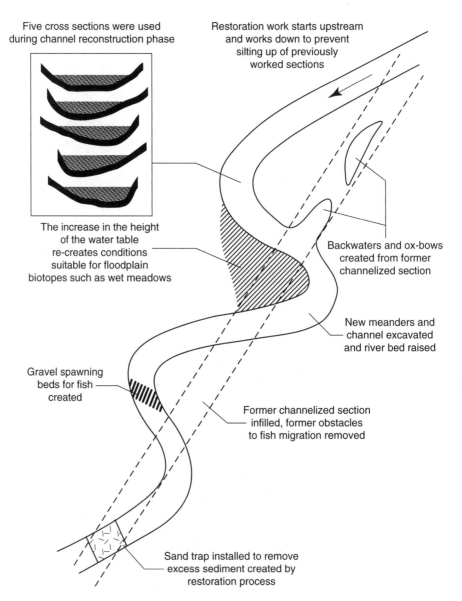

Five cross sections were used during channel reconstruction phase

Restoration work starts upstream and works down to prevent silting up of previously worked sections

The increase in the height of the water table re-creates conditions suitable for floodplain biotopes such as wet meadows

Backwaters and ox-bows created from former channelized section

New meanders and channel excavated and river bed raised

Gravel spawning beds for fish created

Former channelized section infilled, former obstacles to fish migration removed

Sand trap installed to remove excess sediment created by restoration process

FIGURE 7.3
River restoration on the River Brede, Denmark. (Reproduced from Neilsen (2002) with permission of the Chartered Institution of Water and Environmental Management, London.)

that are envisaged to a particular river. By using these plans, all sites along a river can be seen within a much larger context, ensuring that any changes elsewhere are neither harmful nor impact on existing recreational usage or on the general environment (NRA, 1994).

The fundamental aim in catchment planning is to conserve, enhance and, where appropriate, restore the total river environment through effective land and resource planning across the total catchment area. This is the basis of sustainability, and those rivers that are currently under greatest stress are those where development and land use within the catchment have ignored their potential impact on the river system. The introduction of the WFD has taken the concepts of catchment management and created an integrated structure for the management, conservation and improvement of all water resources within river basins. This pioneering legislation will ensure sustainable use of water, preserve ecological quality and eliminate the transfer of hazardous pollutants from land-based activities into the marine environment through the hydrological cycle.

7.3 THE WATER FRAMEWORK DIRECTIVE

7.3.1 INTRODUCTION

The European Community (EC) WFD (2000/60/EC) provides an integrated approach to the protection, improvement and sustainable use of Europe's rivers, lakes, estuaries, coastal waters and groundwater. The main aims of the Directive are to prevent further deterioration, protect and enhance the status of aquatic ecosystems and associated wetlands; to promote the sustainable consumption of water; to reduce pollution of waters from priority substances; to prevent the deterioration in the status, and to progressively reduce pollution, of groundwater; and to reduce the effects of floods and droughts. The WFD is unique as it deals with all water resources, including both surface and ground waters, fresh and saline. Estuaries and coastal waters are included as the volume of freshwater flowing into them generally affects their equilibrium. Wetlands are highlighted in the Directive as, apart from being a threatened habitat, they play a vital role in the protection of water resources. The overall objectives of the WFD are to provide a sufficient supply of good quality surface and ground waters to provide for sustainable, balanced and equitable water use throughout the Member States. Also to significantly reduce pollution in both ground and surface waters that have continued to show an overall reduction in quality over the past decades even though legislation has been in place to protect them. Finally, to protect the marine environment by reducing the concentration of pollutants to near background values for naturally occurring substances and to zero for man-made synthetic substances.

The Directive makes this bold statement at the outset:

> Water is not a commercial product like any other but, rather, a heritage which must be protected, defended and treated as such.

The Directive was published in its final form on 22 December 2000 and harmonizes and updates existing EU water legislation, replacing many existing Directives (Fig. 7.4; Table 8.2). While it provides quality standards for water quality, it is unusual in that it also deals with water policy and management. The Directive does this through provisions dealing with organizational and procedural aspects of water management and by formalizing the concept of integrated river basin management. Although catchment management has been used for several decades, the WFD uses a holistic approach to managing the water cycle for the first time, integrating all

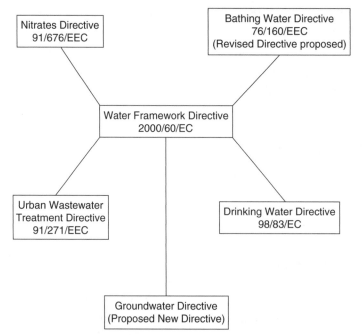

FIGURE 7.4 The WFD is supported by other key Directives that provide an integrated approach to the control of point and diffuse pollution, and the protection of drinking water.

aspects of management and control. The concept of river basin planning, by eliminating both regional and national boundaries, is expected to have a significant effect on water quality, especially within the expanded EU by the control of transboundary problems.

EU countries introduced the Directive into domestic legislation during 2003, with draft River Basin Management Plans (RBMPs) required by 2008 and, after public consultation, finalized and published by 2009. These plans must be fully operational by 2012 with the overall requirement to achieve *good ecological and good chemical status* in all waters by 2015, unless there are grounds for derogation. There is also a general *no deterioration* provision in the Directive to prevent the deterioration in status of any water body within the EU. The Directive also requires the reduction and ultimate elimination of priority hazardous substances (PHSs) (Table 8.17) and the reduction of priority substances to below set quality standards by this date (Section 8.3). A timetable for its full implementation is given in Table 7.7 with subsequent reviews of RBMPs carried out every 6 years. Implementation of the WFD is complex and will require much new development in monitoring and management techniques. While overall standards and timescale for implementation are strictly fixed, there is a large degree of flexibility on how Member States deliver individual RBMPs. The main steps are summarized in Fig. 7.5. In the UK, the Environment Agency has been appointed as the competent authority to carry out the implementation of the Directive. Like other Member States it has established a Water Framework Directive Programme to deliver all the scientific, technical and managerial requirements so that RBMPs can be produced by 2008.

TABLE 7.7 Timetable for the implementation of the WFD

Year	Requirement
2003	Transpose Directive into domestic law (Article 24) Identify RBDs and the competent authorities (Article 3) that will be empowered to implement the Directive
2004	Complete first characterization (Article 5) and assessment of impacts on RBDs. Complete first economic analysis of water use Establish a register of protected areas in each RBD (Articles 6 and 7)
2005	Identify significant upward trends in groundwater and establish trend reversal (Article 17)
2006	Establish environmental monitoring programmes (Article 8) Publish a work programme for producing the first RBMPs (Article 14). Establish environmental quality standards for priority substances and controls on principle sources (Article 16)
2007	Publish an interim overview of the significant water management issues in each RBD for general consultation (Article 14)
2008	Publish draft RBMPs for consultation (Article 14)
2009	Finalize and publish first RBMPs (Article 13) Finalize programme of measures to meet objectives (Article 11)
2010	Ensure water pricing policies are in place (Article 9)
2012	Ensure all measures are fully operational (Article 11) Publish timetable and work programme for second RBMPs Report progress in implementing measures (Article 15)
2013	Review characterization and impact assessment for RBDs Review economic analysis of water use Publish an interim overview of the significant water management issues
2014	Publish second draft RBMPs for consultation
2015	Achieve environmental objectives specified in first RBMPs (Article 4) Finalize and publish second RBMP with revised Programme of Measures
2021	Achieve environmental objectives specified in second RBMPs Publish third RBMPs
2027	Achieve environmental objectives specified in third RBMPs Publish fourth RBMPs.

The WFD is designed to allow further integration of the protection and sustainable use of water into other areas of community policy dealing with energy, transport, agriculture, fisheries, regional policy and tourism.

7.3.2 RIVER BASIN DISTRICTS

The WFD requires the identification of River Basin Districts (RBDs). These districts are areas of land and sea, made up of one or more river basins together with their associated groundwaters and coastal waters. RBDs are determined using existing hydrological river basin (catchment) boundaries. Small river basins may be joined with larger river basins, or joined with neighbouring small basins to create a RBD.

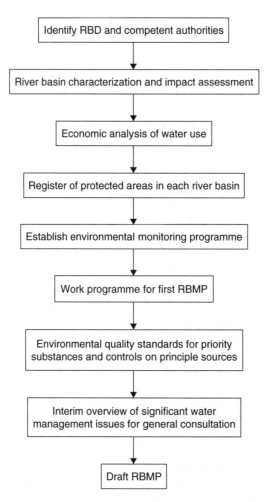

Identify RBD and competent authorities

River basin characterization and impact assessment

Economic analysis of water use

Register of protected areas in each river basin

Establish environmental monitoring programme

Work programme for first RBMP

Environmental quality standards for priority substances and controls on principle sources

Interim overview of significant water management issues for general consultation

Draft RBMP

FIGURE 7.5
Key implementation steps of the WFD.

Groundwaters often do not follow a particular river basin and so are assigned to the nearest or most appropriate RBD.

Environmental objectives, established by the Directive, are co-ordinated for the whole of the RBD, and reported to the European Commission as such. Coastal waters, out to 1 nautical mile from the shoreline, are assigned to the nearest or most appropriate RBD.

There are nine RBDs identified in England and Wales, and an additional two cross-border RBDs with Scotland (Northumbria RBD and the Solway Tweed RBD). In Ireland, there are a total of eight RBDs of which seven are in the Republic of Ireland and four are in Northern Ireland, three of which are international or transboundary RBDs (Fig. 1.4).

7.3.3 RIVER BASIN CHARACTERIZATION

Under the Directive RBDs, which can comprise a single or number of individual catchment areas, must be identified by June 2004. Although characterization of RBDs will

TABLE 7.8   Main descriptors used to characterize rivers in the preparation of RBMPs

Rivers	Lakes
Obligatory factors	
Altitude	Altitude
Latitude	Latitude
Longitude	Longitude
Geology	Geology
Size	Size
Optional factors	
Distance from source	Mean water depth
Energy of flow (function of flow and gradient)	Lake shape
Mean water width	Residence time
Mean water depth	Mean air temperature
Mean water slope	Air temperature range
Form and shape of main river bed	Mixing characteristics (e.g. monomictic, dimictic, polymictic)
River discharge category	Acid neutralizing capacity
Valley shape	Background nutrient status
Transport of solids	Mean substratum composition
Acid neutralizing capacity	Water level fluctuation
Mean substratum composition	
Chloride	
Air temperature range	
Mean air temperature	
Precipitation	

Note: Annex II of the WFD lists the prescribed physical and chemical factors needed to characterize water bodies in RBDs. This table is for rivers and lakes, but specific characterization factors are listed in the annex for all water bodies.

continue up to 2009, when the first RBMPs are published, a preliminary characterization of both surface and groundwaters is required by December 2004. This will include an analysis of the characteristics of all aquatic systems, a review of the impact of human activity on the status of the water bodies, an economic analysis of water use and a register of protected areas in each RBD (Table 8.4). From this characterization, appropriate environmental standards will be set with monitoring effort targeted towards those water bodies at the greatest risk of failing to meet their stated objectives, and a programme of measures will be established to ensure that water bodies reach the environmental objectives. Characterization is outlined in Annex II of the Directive and requires detailed morphological analysis. This must be done using GIS with the individual river basin making up each RBD clearly defined. Basic descriptions for surface waters are listed in Table 7.8. A RBMP should include the following:

1. General description of the characterization of the RBD (Article 5; Annex II).
2. Summary of the significant pressures and impact of human activity on the status of surface and ground waters.

3. Identification and mapping of protected areas (Article 6; Annex IV).
4. A map of the monitoring stations and the results of the monitoring programmes, in map form, for all surface waters (ecological and chemical), groundwaters (chemical and quantitative) and protected areas (Article 8; Annex V).
5. List of environmental objectives for all water bodies (Article 4).
6. A summary of the economic analysis of water use (Article 5; Annex III).
7. Summary of programme of measures/actions to be taken in order to achieve the objectives listed in 5, above (Article 11).
8. A register of more detailed programmes and management plans for specified areas or issues within RBD.
9. Summary of the public information and consultation measures taken, and the effect this had on the RBMP as a result.
10. A list of competent authorities.
11. Contact points and procedures for obtaining background documentation and information (Article 14).

7.3.4 IMPLEMENTATION OF THE WATER FRAMEWORK DIRECTIVE IN THE UK

The Environment Agency is sole competent authority charged with the Directive's implementation in England and Wales. In Scotland and Northern Ireland the Scottish Environment Protection Agency and the Environment and Heritage Service are the competent authorities, respectively.

The UK Technical Advisory Group (UK TAG) has been established to provide technical advice and solutions on issues that are relevant to all parts of the UK, including the Republic of Ireland. UK TAG consists of members from the Environment Agency (England and Wales's representatives), Environment and Heritage Services (Northern Ireland), Scottish Environment Protection Agency (SEPA), English Nature (EN), Scottish Natural Heritage (SNH), Environmental Protection Agency (Republic of Ireland), Countryside Council for Wales (CCW), and Scottish and Northern Ireland Forum for Environmental Research (SNIFFER). The Technical Advisory Group is made up of a number of sub-groups covering groundwater, rivers, lakes or transitional waters (estuaries and coastal waters). Each one of these categories of water resources is very different requiring specialist expertise in order to develop appropriate and consistent assessment and classification methods. Apart from also providing technical advice to the UK authorities, UK TAG is also co-ordinating the UK input into the development of European Guidance and methods, the common implementation strategy (CIS).

While the implementation of the Directive is the responsibility of individual European Member States, it was recognized that there was a need to establish a CIS to ensure that the implementation was both coherent and harmonious across the EU. The aim of the CIS is to achieve a common understanding of the technical and scientific implications of the Directive. It is divided into four key activity modules:

1. Information sharing.
2. Developing guidance on technical issues.
3. Information and data management.
4. Application, testing and validation.

Almost 20 Expert Advisory Panels have so far been established, which produce guidance documents, recommendations for operational methods and other supporting information. The documents produced are informal in character and not legally binding, but minimize any risks associated with the application of the Directive and ensure rapid implementation.

7.3.5 RIVER BASIN PLANNING

The WFD encourages the prudent and rationale use of water through a policy of protection and sustainable use. The precautionary principle is stressed in the management of RBDs, with preventive action paramount and, where necessary, remedial action taken at source ensuring that the polluter pays. A key element in river basin planning is that decisions should be taken as close as possible to the locations where water is affected or used.

The main output of the WFD is the RBMP that must be produced for each RBD by 2009 (Table 7.7). These are similar to County Development or Management Plans produced by planning authorities and will be the basis for the integrated management of water. Many of the basic principles used by planning authorities will also apply to the development of RBMPs and it will require the co-operation of a much greater number of organizations and bodies outside the immediate area of water management (e.g. land use planning) to deliver the objectives set out in the Directive. The involvement of stakeholders in both the development and public consultation of RBMPs is seen as vital to the successful implementation of the objectives of the WFD.

The process of River Basin Planning is not only to facilitate the production of RBMPs, but also to integrate and manage all those stakeholders involved in some way in the water cycle as it affects the operation of the RBMP. This involves the collection and analysis of river basin data and producing management solutions in order to achieve the objectives of the WFD within the appropriate timescale. The planning process is followed by implementation of the management solutions that together are referred to as river basin management, which is the practical operation of the WFD at river basin level. So the RBMP is the final output of the river basin management process, which is itself a continuous process, with new plans produced every 6 years (Table 7.7). Thus the RBMP is the main reporting mechanism to both the Commission and the public, being the official record of the current status of water bodies within the RBD, with a summary of what measures are planned to meet objectives.

There is no procedural information in the Directive on how Member States should carry out either river basin planning or management, although the CIS has published a guidance document *Best Practices in River Basin Planning*. This identifies nine discrete components in the river basin planning process:

1. Assessment of current status and preliminary gap analysis (i.e. identification of areas where improvement is required and the assessment of the necessary management solutions).
2. Setting environmental objectives.
3. Establishing monitoring programmes.
4. Gap analysis.
5. Setting up programmes of measures.
6. Development of RBMPs.
7. Implementation of programmes of measures.
8. Evaluation.
9. Public information and consultation.

Component 9 reflects the importance the EU places on transparency and the active involvement of interested parties. A key aspect to the development of RBMPs is how they will interact with other regulatory drivers. Some existing and proposed legislation, such as the Directive on public access to information, the Strategic Environmental Assessment Directive, the Groundwater Daughter Directive and the Proposed Environmental Liability Directive, may have important ramifications for river basin management.

7.3.6 ENVIRONMENTAL MONITORING

The WFD requires the development of new monitoring and classification systems for surface and ground waters by 25 December 2006. Although the Environment Agency has been preparing new quality assessment procedures for England and Wales for many years (Section 9.5), the final methodologies have yet to be agreed. To assist in this task the Environment Agency has set up the Environmental Monitoring Classification and Assessment Reporting Project (EMCAR) to develop monitoring, classification, assessment and reporting protocols. These will include techniques to monitor all the biological, hazardous chemical and other pollutants, hydrological, morphological and physico-chemical elements identified in the Directive; specific monitoring systems for wetlands and protected areas; appropriate classification and assessment methods including surveillance, operational and investigative monitoring; and mechanisms to collate all monitoring data generated by the monitoring networks established. All the adopted procedures are required to be scientifically robust and risk-based.

The Directive identifies three types of monitoring. *Surveillance* is carried out in order to validate the risk assessments, detect long-term trends, assess impacts, and in order to design the monitoring strategy. *Operational monitoring* is used to classify those

TABLE 7.9 Quality elements used to characterize surface and ground waters in the preparation of RBMPs

	Rivers	Lakes
Biological elements	Composition and abundance of aquatic flora Composition and abundance of benthic invertebrate fauna Composition, abundance and age structure of fish fauna	Composition, abundance and biomass of phytoplankton Composition and abundance of other aquatic flora Composition and abundance of benthic invertebrate fauna Composition, abundance and age structure of fish fauna
Hydromorphological elements supporting the biological elements	Hydrological regime *Quantity and dynamic of water flow* *Connection to groundwater bodies* River continuity Morphological conditions *River depth and width variation* *Structure and substrate of the river bed* *Structure of the riparian zone*	Hydrological regime *Quantity and dynamic of water flow* *Residence time* *Connection to groundwater bodies* Morphological conditions *Lake depth variation* *Quantity, structure and substrate of lake bed* *Structure of lake shore*
Chemical and physico-chemical elements supporting the biological elements	General *Thermal conditions* *Oxygenation conditions* *Salinity* *Acidification status* *Nutrient conditions*	General *Transparency* *Thermal conditions* *Oxygenation conditions* *Salinity* *Acidification status* *Nutrient conditions*
Specific pollutants	Pollution by all priority substances identified as being discharged into the body of water Pollution by other substances identified as being discharged in significant quantities into the body of water	Pollution by all priority substances identified as being discharged into the body of water Pollution by other substances identified as being discharged in significant quantities into the body of water

water bodies that are at risk of failing good status. Finally, *Investigative monitoring* enables the cause and effects of a failure to be ascertained when it is not clear. All waters will be assigned to a new ecological classification system.

The quality elements required for the classification of ecological and chemical quality status are listed in Annex V of the WFD for surface and ground waters (Table 7.9).

7.3.7 ECOLOGICAL CLASSIFICATION SYSTEM FOR SURFACE AND GROUND WATERS

A new approach to water quality classification has been adopted for aquatic systems. Surface waters (i.e. rivers, lakes, estuaries and coastal waters) are assessed using biological (plankton and phytobenthos, macrophytes, invertebrates and fish) elements, hydromorphology, physio-chemical (including pollutants) elements and PHSs and other pollutants. There are five quality classes: High, Good, Moderate, Poor and Bad Status (see Table 7.10). The overall status is determined by the worst element for a

TABLE 7.10 General definitions of ecological status

High status	Good status	Moderate status	Poor status	Bad status
There are no, or only very minor, anthropogenic alterations to the values of the physico-chemical and hydromorphological quality elements for the surface water body type from those normally associated with that type under undisturbed conditions.	The values of the biological quality elements for the surface water body type show low levels of distortion resulting from human activity, but deviate only slightly from those normally associated with the surface water body type under undisturbed conditions.	The values of the biological quality elements for the surface water body type deviate moderately from those normally associated with the surface water body type under undisturbed conditions. The values show moderate signs of distortion resulting from human activity and are significantly more disturbed than under conditions of good status.	Waters showing evidence of major alterations to the values of the biological quality elements for the surface water body type and in which the relevant biological communities deviate substantially from those normally associated with the surface water body type under undisturbed conditions, shall be classified as bad.	Waters showing evidence of severe alterations to the values of the biological quality elements for the surface water body type and in which large portions of the relevant biological communities normally associated with the surface water-body type under undisturbed conditions are absent, shall be classified as bad.
The values of the biological quality elements for the surface water body reflect those normally associated with that type under undisturbed conditions, and show no, or only very minor, evidence of distortion.		Waters achieving a status below moderate shall classified as poor or bad.		
These are the type-specific conditions and communities.				

particular site. Only two classifications Good or Bad Status are used in groundwater assessment, which is based on both chemistry and quantity of water. Protection of water status will provide economic benefits at the RBD level by safeguarding and developing the potential uses of water. For example the protection of both inland and coastal fish populations.

While chemical and ecological quality status are used as defining terms of water quality, often only ecological status is used as it describes the quality of the structure and functioning of aquatic ecosystems, which is, of course, dependent on chemical water status, and is the ultimate objective. However, drinking water quality may in some cases require higher than the minimum Good Ecological Status required for all water bodies under the Directive. The general definition of status is given in Table 7.11 for all surface waters, while specific criteria are outlined for each specific water body in Annex V (Table 7.12).

7.3.8 PRIORITY AND OTHER SPECIFIC POLLUTING SUBSTANCES

For nearly 30 years the Dangerous Substances Directive (76/464/EEC) has been the primary regulatory legislation in Europe controlling the discharge of a wide range of chemical pollutants to surface and ground waters. The WFD will eventually take over the provisions of this Directive, which will be repealed (Section 8.3). A list of chemical polluting substances is given in Annex X of the WFD, some of which are identified as being PHSs (Table 8.17). The WFD aims to eliminate PHSs at source by preventing inputs via discharges, emissions and other losses, including accidents. The Directive requires that management plans to deal with accidental pollution and incidents are in place.

The environmental objectives in RBMPs include chemical status for both surface and ground waters, which are based on achievement of environmental quality standards. Chemical status includes priority substances and other pollutants (i.e. any substance liable to cause pollution including those listed in Annex VIII and X) (Table 7.13). The Directive requires environmental quality standards for priority substances and controls on their principle sources to be established by the end of 2006 for each RBD. Member States have set up National Expert Working groups to assist with developing lists of dangerous substances. For example, in Ireland their Expert Group has put forward 161 candidate substances and named 33 priority substances that have definite relevance in the context of the implementation of the Dangerous Substances and WFDs. These candidate substances will be subjected to a screening process, developed by the Expert Group, to assess their relevance in the Irish aquatic environment, in order to achieve a progressive reduction in hazardous substances in water.

7.3.9 CONCLUSION

Throughout Europe water resources are coming under increasing pressure from the continuous growth in demand for sufficient quantities of good quality water for all

TABLE 7.11 General definitions of ecological status classifications given in the WFD, Annex V, based on biological, chemical and hydromorphological quality elements for all surface waters (refer Table 7.10 for general definition of ecological quality)

Element	High status	Good status	Moderate status
Biological quality elements			
Phytoplankton	The taxonomic composition of phytoplankton corresponds totally or nearly totally to undisturbed conditions.	There are slight changes in the composition and abundance of planktonic taxa compared to the type-specific communities. Such changes do not indicate any accelerated growth of algae resulting in undesirable disturbances to the balance of organisms present in the water body or to the physico-chemical quality of the water or sediment.	The composition of planktonic taxa differs moderately from the type-specific communities.
	The average phytoplankton abundance is wholly consistent with the type-specific physico-chemical conditions and is not such as to significantly alter the type-specific transparency conditions.		Abundance is moderately disturbed and may be such as to produce a significant undesirable disturbance in the values of other biological and physico-chemical quality elements.
	Planktonic blooms occur at a frequency and intensity which is consistent with the type-specific physico-chemical conditions.	A slight increase in the frequency and intensity of the type-specific planktonic blooms may occur.	A moderate increase in the frequency and intensity of planktonic blooms may occur. Persistent blooms may occur during summer months.
Macrophytes and phytobenthos	The taxonomic composition corresponds totally or nearly totally to undisturbed conditions.	There are slight changes in the composition and abundance of macrophytic and phytobenthic taxa compared to the type-specific communities. Such changes do not indicate any accelerated growth of phytobenthos or higher forms of plant life resulting in undesirable disturbances to the balance of organisms present in the water body or to the physico-chemical quality of the water or sediment.	The composition of macrophytic and pyhtobenthic taxa differs moderately from the type-specific community and is significantly more distorted than that at good status.
	There are no detectable changes in the average macrophytic and the average phytobenthic abundance.		Moderate changes in the average macrophytic and the average phytobenthic abundance are evident.
		The phytobenthic community is not adversely affected by bacterial tufts and coats present due to anthropogenic activity.	The phytobenthic community may be interfered with and in some areas, displaced by bacterial tufts and coats present as a result of anthropogenic activities.
Benthic invertebrate fauna	The taxonomic composition and abundance correspond totally or nearly totally to undisturbed conditions.	There are slight changes in the composition and abundance of invertebrate taxa from the type-specific communities.	The composition and abundance of invertebrate taxa differ moderately from the type-specific communities.

	High status	Good status	Moderate status
	The ratio of disturbance-sensitive taxa to insensitive taxa shows no signs of alteration from undisturbed levels. The level of diversity of invertebrate taxa shows no sign of alteration from undisturbed levels.	The ratio of disturbance-sensitive taxa to insensitive taxa shows slight signs of alteration from type-specific levels. The level of diversity of invertebrate taxa shows slight signs of alteration from type-specific levels.	Major taxonomic groups of the type-specific community are absent. The ratio of disturbance-sensitive taxa to insensitive taxa, and the level of diversity, are substantially lower than the type-specific level and significantly lower than for good status.
Fish fauna	Species composition and abundance correspond totally or nearly totally to undisturbed conditions. All the type-specific disturbance-sensitive species are present. The age structures of fish communities show little sign of anthropogenic disturbance and are not indicative of a failure in the reproduction or development of any particular species.	There are slight changes in species composition and abundance from the type-specific communities attributable to anthropogenic impacts on physico-chemical and hydromorphological quality elements. The age structures of the fish communities show signs of disturbance attributable to anthropogenic impacts on physico-chemical or hydromorphological quality elements, and, in a few instances, are indicative of a failure in the reproduction or development of a particular species, to the extent that some age classes may be missing.	The composition and abundance of fish species differ moderately from the type-specific communities attributable to anthropogenic impacts on physico-chemical or hydromorphological quality elements. The age structure of the fish communities shows major signs of anthropogenic disturbance, to the extent that a moderate proportion of the type specific species are absent or of very low abundance.

Hydromorphological quality elements

	High status	Good status	Moderate status
Hydrological regime	The quantity and dynamics of flow, and the resultant connection to groundwaters, reflect totally, or nearly totally, undisturbed conditions.	Conditions consistent with the achievement of the values specified above for the biological quality elements.	Conditions consistent with the achievement of the values specified above for the biological quality elements.
River continuity	The continuity of the river is not disturbed by anthropogenic activities and allows undisturbed migration of aquatic organisms and sediment transport.	Conditions consistent with the achievement of the values specified above for the biological quality elements.	Conditions consistent with the achievement of the values specified above for the biological quality elements.
Morphological conditions	Channel patterns, width and depth variations, flow velocities, substrate conditions and both the structure	Conditions consistent with the achievement of the values specified above for the biological quality elements.	Conditions consistent with the achievement of the values specified above for the biological quality elements.

(Continued)

TABLE 7.11 (Continued)

Element	High status	Good status	Moderate status
	and condition of the riparian zones correspond totally or nearly totally to undisturbed conditions.		
Physico-chemical quality elements[a]			
General conditions	The values of the physico-chemical elements correspond totally or nearly totally to undisturbed conditions.	Temperature, oxygen balance, pH, acid neutralizing capacity and salinity do not reach levels outside the range established so as to ensure the functioning of the type specific ecosystem and the achievement of the values specified above for the biological quality elements.	Conditions consistent with the achievement of the values specified above for the biological quality elements.
	Nutrient concentrations remain within the range normally associated with undisturbed conditions.		
	Levels of salinity, pH, oxygen balance, acid neutralizing capacity and temperature do not show signs of anthropogenic disturbance and remain within the range normally associated with undisturbed conditions.	Nutrient concentrations do not exceed the levels established so as to ensure the functioning of the ecosystem and the achievement of the values specified above for the biological quality elements.	
Specific synthetic pollutants	Concentrations close to zero and at least below the limits of detection of the most advanced analytical techniques in general use.	Concentrations not in excess of the standards set in accordance with the procedure detailed in Section 1.2.6 of the WFD without prejudice to Directive 91/414/EC and Directive 98/8/EC. (<EQS)	Conditions consistent with the achievement of the values specified above for the biological quality elements.
Specific non-synthetic pollutants	Concentrations remain within the range normally associated with undisturbed conditions (bgl).	Concentrations not in excess of the standards set in accordance with the procedure detailed in Section 1.2.6[a] of the WFD without prejudice to Directive 91/414/EC and Directive 98/8/EC. (<EQS)	Conditions consistent with the achievement of the values specified above for the biological quality elements.

bgl: background level; EQS: environmental quality standard.
[a]Application of the standards derived under this protocol shall not require reduction of pollutant concentrations below background levels: (EQS > bgl).

Table 7.12 Example of the detailed definitions of ecological status classifications given in the WFD, Annex V, based on biological, chemical and hydromorphological quality elements for coastal waters (refer Table 7.10 for general definition of ecological quality)

Element	High status	Good status	Moderate status
Biological quality elements			
Phytoplankton	The composition and abundance of phytoplanktonic taxa are consistent with undisturbed conditions.	The composition and abundance of phytoplanktonic taxa show slight signs of disturbance.	The composition and abundance of planktonic taxa show signs of moderate disturbance.
	The average phytoplankton biomass is consistent with the type-specific physico-chemical conditions and is not such as to significantly alter the type-specific transparency conditions.	There are slight changes in biomass compared to type-specific conditions. Such changes do not indicate any accelerated growth of algae resulting in undesirable disturbance to the balance of organisms present in the water body or to the quality of the water.	Algal biomass is substantially outside the range associated with type-specific conditions, and is such as to impact upon other biological quality elements.
	Planktonic blooms occur at a frequency and intensity which is consistent with the type specific physico-chemical conditions.	A slight increase in the frequency and intensity of the type-specific planktonic blooms may occur.	A moderate increase in the frequency and intensity of planktonic blooms may occur. Persistent blooms may occur during summer months.
Macroalgae and angiosperms	All disturbance-sensitive macroalgal and angiosperm taxa associated with undisturbed conditions are present.	Most disturbance-sensitive macroalgal and angiosperm taxa associated with undisturbed conditions are present.	A moderate number of the disturbance-sensitive macroalgal and angiosperm taxa associated with undisturbed conditions are absent.
	The levels of macroalgal cover and angiosperm abundance are consistent with undisturbed conditions.	The level of macroalgal cover and angiosperm abundance show slight signs of disturbance.	Macroalgal cover and angiosperm abundance is moderately disturbed and may besuch as to result in an undesirable disturbance to the balance of organisms present in the water body.
Benthic invertebrate fauna	The level of diversity and abundance of invertebrate taxa is within the range normally associated with undisturbed conditions.	The level of diversity and abundance of invertebrate taxa is slightly outside the range associated with the type-specific conditions.	The level of diversity and abundance of invertebrate taxa is moderately outside the range associated with the type-specific conditions.
	All the disturbance-sensitive taxa associated with undisturbed conditions are present.	Most of the sensitive taxa of the type-specific communities are present.	Taxa indicative of pollution are present. Many of the sensitive taxa of the type-specific communities are absent.

(Continued)

TABLE 7.12 (Continued)

Element	High status	Good status	Moderate status
Hydromorphological quality elements			
Tidal regime	The freshwater flow regime and the direction and speed of dominant currents correspond totally or nearly totally to undisturbed conditions.	Conditions consistent with the achievement of the values specified above for the biological quality elements.	Conditions consistent with the achievement of the values specified above for the biological quality elements.
Morphological conditions	The depth variation, structure and substrate of the coastal bed, and both the structure and condition of the inter-tidal zones correspond totally or nearly totally to the undisturbed conditions.	Conditions consistent with the achievement of the values specified above for the biological quality elements.	Conditions consistent with the achievement of the values specified above for the biological quality elements.
Physico-chemical quality elements[a]			
General conditions	The physico-chemical elements correspond totally or nearly totally to undisturbed conditions.	Temperature, oxygenation conditions and transparency do not reach levels outside the ranges established so as to ensure the functioning of the ecosystem and the achievement of the values specified above for the biological quality elements.	Conditions consistent with the achievement of the values specified above for the biological quality elements.
	Nutrient concentrations remain within the range normally associated with undisturbed conditions.	Nutrient concentrations do not exceed the levels established so as to ensure the functioning of the ecosystem and the achievement of the values specified above for the biological quality elements.	
	Temperature, oxygen balance and transparency do not show signs of anthropogenic disturbance and remain within the range normally associated with undisturbed conditions.		
Specific synthetic pollutants	Concentrations close to zero and at least below the limits of detection of the most advanced analytical techniques in general use.	Concentrations not in excess of the standards set in accordance with the procedure detailed in Section 1.2.6 (WFD) without prejudice to Directive 91/414/EC and Directive 98/8/EC. (<EQS)	Conditions consistent with the achievement of the values specified above for the biological quality elements.
Specific non-synthetic pollutants	Concentrations remain within the range normally associated with undisturbed conditions (bgl).	Concentrations not in excess of the standards set in accordance with the procedure detailed in Section 1.2.6[a] (WFD) without prejudice to Directive 91/414/EC and Directive 98/8/EC. (<EQS)	Conditions consistent with the achievement of the values specified above for the biological quality elements.

bgl: background level; EQS: environmental quality standard.

[a] Application of the standards derived under this protocol shall not require reduction of pollutant concentrations below background levels: (EQS > bgl).

TABLE 7.13   List of Annex VIII substances that are considered to be the main pollutants in water

1. Organohalogen compounds and substances which may form such compounds in the aquatic environment
2. Organophosphorous compounds
3. Organotin compounds
4. Substances and preparations, or the breakdown products of such, which have been proved to possess carcinogenic or mutagenic properties or properties which may affect steroidogenic, thyroid, reproduction or other endocrine-related functions in or via the aquatic environment
5. Persistent hydrocarbons and persistent and bioaccumulable organic toxic substances
6. Cyanides
7. Metals and their compounds
8. Arsenic and its compounds
9. Biocides and plant protection products
10. Materials in suspension
11. Substances which contribute to eutrophication (in particular, nitrates and phosphates)
12. Substances which have an unfavourable influence on the oxygen balance, (and can be measured using parameters such as BOD, COD, etc.)

purposes. The WFD is the EU response to protect Community waters in both qualitative and quantitative terms. It is the most far-reaching piece of environmental legislation ever introduced by the EU and will change the way in which water is perceived
and managed in Europe forever. Water is not an isolated entity and is continuously
being modified. So, in order to protect it all discharges that may eventually find their
way into the water cycle need to be controlled and regulated. The WFD along with
its associated Directives puts into place the management and legislative structure to
achieve this.

7.4 DIFFUSE POLLUTION

The most difficult pollution inputs to control are not those discharged at a single
location (point sources) but those that enter surface waters over a wide area (diffuse
or non-point sources) (Fig. 7.6). Diffuse pollution can be caused by atmospheric deposition or contaminated sediments, but more commonly are caused through land use
activities resulting in widespread run-off or infiltration (e.g. agriculture, forestry and
large-scale construction) (Table 7.14). Agriculture is the major cause of diffuse pollution in Europe resulting in the run-off of nutrients (mainly as nitrogen), pesticides
and suspended solids causing turbidity. This is considered in detail in Section 11.1,
while the effect of afforestation is discussed below.

7.4.1 REMEDIAL ACTION

The nature of diffuse pollution determines the potential of options for its control
(D'Arcy et al., 1998). There are a wide variety of remedial actions that can be taken

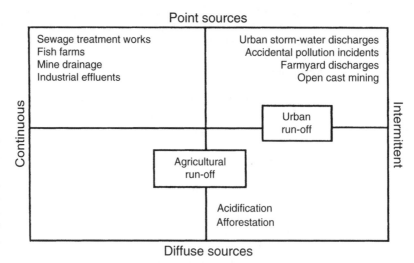

FIGURE 7.6
Water pollution
sources.
(Reproduced from
British Ecological
Society (1990) with
permission of the
British Ecological
Society, London.)

TABLE 7.14 Major diffuse sources of pollution in the UK and the problems they cause

Pollutant	Example sources	Environmental problem
Oil and other hydrocarbons	Car maintenance. Disposal of waste oils. Spills from storage and handling. Road run-off. Industrial run-off	Toxicity. Contamination of urban stream sediments. Groundwater contamination. Nuisance (surface waters). Taste (potable supplies)
Pesticides	Municipal application to control roadside weeds. Agriculture. Private properties?	Toxicity. Contamination of potable supplies
Sediment	Run-off from aerable land. Upland erosion. Forestry. Urban run-off. Construction industry	Destruction of gravel riffles. Sedimentation of atural pools and ponds. Costs to abstractors (e.g. fish farms, potable supplies)
Organic wastes	Agricultural wastes (slurry, silage, liquor, surplus crops). Sewage sludge. Industrial wastes for land application	Oxygen demand. Nutrient enrichment
Faecal pathogens	Failures of conventional sewage systems (wrong connections in separate sewer systems). Dog faeces in towns and cities. Application of organic wastes to farmland	Health risks. Non-compliance with recreational water standards
Nitrogen	Agricultural fertilizers. Atmospheric deposition	Eutrophication (especially coastal waters). Contamination of potable supplies (rivers and groundwaters). Acidification
Phosphorus	Soil erosion. Agricultural fertilizers. Contamination of urban run-off (detergents, organic material)	Eutrophication of freshwaters: ecological degradation; blue-green algae; increased filtration costs for potable reservoirs/rivers
Trace metals	Urban run-off. Industrial and sewage sludges applied to land. Contaminated groundwater	Toxicity
Iron	Water table rebound following mining (especially coal)	Toxicity. Aesthetic nuisance
Solvents	Cleaning plant on industrial yards, wrong connections of trade effluent to surface water	Toxicity. Contamination of potable supplies (rivers and groundwaters)

Reproduced from D'Arcy et al. (1998) with permission of Elsevier Science Ltd, Oxford.

to reduce the effects of diffuse pollution. Structural remedies range from fencing off riverbanks to complex engineering works (Table 7.15). The management of the riparian (bankside) zones is the most important factor in surface water conservation, especially of rivers, and reflects the shift in attitude from just the river channel to overall consideration of the catchment as a whole. Specific attention is now given to the aquatic–terrestrial interface, which in most rivers is not only the primary source of its productivity, but also the source of detrimental inputs such as nutrients and toxicants (Boon *et al.*, 1992). Bankside vegetation provides habitat diversity, allochthonous organic matter (Section 4.2), and helps to regulate water temperature, light, surface run-off and interflow from the adjoining land, erosion of banks, nutrient transfer and, most importantly, helps to control flooding downstream. Planting

TABLE 7.15 Structural solutions to diffused pollution prevention

Practice	Definition	Management objective	% Reduction of nutrient loads
Terraces, diversions	Earthen ridge or channel across slope	Intercept run-off; decrease run-off velocity; minimize soil erosion	Terraces, 90% adsorbed and 30% dissolved nutrients. Diversions, 20–45% adsorbed nutrients
Sediment basins, detention ponds	Artificial depression which may drain or release water to land and may be vegetated	Collect and store run-off; detain sediments and nutrients; increase uptake of nutrients; recycle water	Sediment basins, 40% total P; 30% adsorbed N; Detention ponds, +40% total P; higher if vegetated
Animal waste management systems	Any system which collects and stores animal wastes: lagoons, storage tanks, composters	Biological treatment (lagoons); separates clean and soiled waters; reduce volume and odour of waste (composter); delay land spreading (tanks and ponds)	Combined, 21% total P, 62% total N; individual practices, range of 10–69% dissolved P, 32–91% total N
Fencing	Barricade to exclude livestock from sensitive areas	Minimize erosion; protect habitats; protect water quality and public health	Depends on type of fence and if vegetated; range of 50–90% total P
Streambank stabilization	Structural repair to streambanks, usual to combine wetland vegetation measures	Minimize bank erosion; control stream temperature, sediment and flows	Combined wetland riparian buffers, 50–70% total P; Combined wetland grass filter strips, 30–90% total P
Rock-reed water treatment filters	Long shallow microbial filter system of reed beds in rock	Remove heavy metals/toxins from run-off; reduce BOD of effluents	Depends on vegetation
Wetland rehabilitation and development	Enhance natural capacity of wetlands to trap, store and process pollutants	Increase nutrient uptake; immobilize P; detain and treat run-off; minimize sediment transport	10–70% P; 40–80% N, depending on location in hydrological unit; 80–90% sediment removal

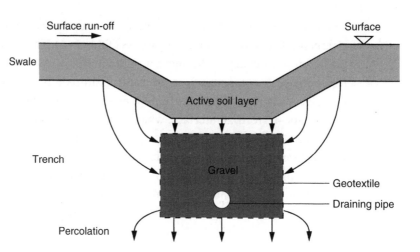

FIGURE 7.7
Cross-section
through a
swale infiltration
trench system
(Mulden–Rigolen
System®).
(Reproduced from
Sieker (1998) with
permission of
Elsevier Science Ltd,
Oxford.)

conifers and poor forestry practice exerts significant chemical and physical influences on surface waters causing acidification in base-poor catchments, increases nutrients (especially phosphorus) and sediment (especially during deforestation). The use of riparian strips of broadleaves and grassland to protect streams has been highly effective in counteracting the effects of afforestation on freshwaters. However, the degree to which macro-invertebrates benefit from riparian management in streams is restricted by low pH. The management of riparian zones is generally significant at a local level but is unable to buffer the effects of acidification, which can only be tackled at a supra-catchment level. The management of riparian zones is, however, the first step within a wider management plan. Riparian buffer zones are highly effective in nutrient and sediment removal and are increasingly used to protect all types of water resources. Swales, wide grassed drainage ditches, are employed throughout the world to intercept surface run-off from urban areas, especially highways and agricultural land. They have a maximum depth of 0.3 m with gently sloping sides (1:3) to allow for easy maintenance (Fig. 7.7).

The run-off is retained in the swale encouraging percolation as well as treatment through settlement and adsorption. Water is also lost through evapotranspiration. To be effective the active soil layer must have a minimum permeability of $10^{-5}\,\mathrm{m\,s^{-1}}$ to ensure infiltration is complete within 3 h after the end of a storm. So the swale offers only short-term storage of storm water while the infiltration trench beneath it offers longer-term storage. The depth of water in the gravel infiltration trench can be controlled if necessary by opening the control throttle of the drainage pipe. A simple overflow device in the manhole ensures that the water level never exceeds the top level of the infiltration trench (Fig. 7.8) (Sieker, 1998). Constructed wetlands are also used in combination with swales to intercept and treat urban and highway run-off before discharge to surface waters resulting in significant reductions of suspended solids, nutrients, trace organic compounds and heavy metals (IAWQ, 1997; Shutes

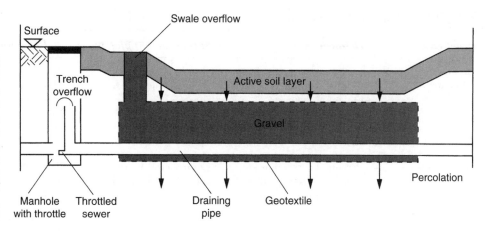

FIGURE 7.8
Longitudinal
section through a
swale infiltration
trench system
(Mulden–Rigolen
System®).
(Reproduced from
Sieker (1998) with
permission of
Elsevier Science Ltd,
Oxford.)

et al., 1997). Improved management practices on farms and of the riparian zones can also bring about significant water quality improvement (Table 7.16).

7.4.2 AFFORESTATION AND WATER QUALITY

Approximately 40% of water supplies in the UK come from upland catchment areas where the water quality is becoming increasingly vulnerable due to changing land use pressures, such as forestry, grazing and recreation. Of these, forestry has the greatest potential impact. Up to the early 1980s forestry development included little consideration for its impact on water quality. Potential problems include changes in water yield and flow rates, increased turbidity, colour and sedimentation, contamination from chemicals and oil, nutrient enrichment and acidification. Forestry development can affect streams, lakes and reservoirs.

Soil and stream erosion, sedimentation and pollution can result in an increase in the cost of drinking water treatment, as well as causing damage to fisheries, wildlife and even roads. Poor management practices identified by Stretton (1998) include:

(a) ploughing at right angles to contours;
(b) ploughing through natural watercourses;
(c) site roads fording natural watercourses;
(d) inadequate use of culvert and bridge structures;
(e) very steep sides to road drainage;
(f) no re-seeding of disturbed ground;
(g) use of unstable material as road dressing;
(h) poor design of interception ditches.

Environmental impact studies are now a normal requirement for new forestry developments and a Code of Practice (HMSO, 1993) has been adopted to help managers

TABLE 7.16 Managerial solutions to diffused pollution prevention

Practice	Definition	Management objective	% Reduction of nutrient loads
Managerial			
Filter strips	Closely growing vegetation between agricultural land and watercourse	Remove sediment from run-off; increase nutrient uptake by plants	Most effective at removal of sediment and sediment bound N; 5–50% total P and ortho-P; may be ineffective at P removal during growing season
Field borders	Perennial vegetation at edge of fields, regardless of proximity to water	Control erosion; reduce velocity of overland flow	Perennial grasses most effective for N removal; mainly for sediment removal
Riparian buffer zones	Maintain vegetation zone along watercourse as ecosystem	Filter sediments with low vegetation; increase nutrient uptake with high vegetation	50–75% total P; 80–90% total N; 80–90% sediment removal
Range and pasture management	Grazing rotation, seeding, brush management, reduce stocking rates	Minimize soil losses; protect vegetation cover	Up to 80% N, P and sediment
Nutrient management planning	Based on limiting nutrient concept, plan for precise nutrient application rates	Limit nutrient in run-off; reduce fertilizer consumption; control animal wastes usage	20–90% total N and P mainly for soluble types of nutrients
Cropping practices			
Conservation tillage	Soil only tilled to prepare seedbed, must leave 30% minimum of soil surface with crop residue after planting	Minimize soil erosion and loss; minimize pesticides and nutrient levels in run-off	35–85% total P; 50–80% total N
Crop rotation	Change crops grown on a field	Minimize soil erosion and loss; improve soil structure; minimize insecticide use by disruption of insect life cycles	Approximately 30% total P; approximately 50% total N
Cover crops	Close-growing crops that cover soil during critical erosion period	Replacement of commercial fertilizers with legumes for N; minimize soil erosion and loss	30–50% total P; certain crops annual reduction: 30% total P, 50% total N
Contour ploughing, strip-cropping	Ploughing and crop rows follow field contours across slope; alternate row and close-grown crops	Minimize soil erosion and loss	Strip-cropping: up to 50% total P and 75% sediment

Reproduced with permission of Carolina State University.

and operators to protect water quality. The key elements of good management are as follows.

Catchment assessment

Trees are very effective scavengers of aerial pollutants causing acidification. Critical load maps are available to identify areas where the natural buffering capacity of soils is inadequate and where afforestation may lead to acidification. Sensitive areas in the UK include parts of central and southwest Scotland, Cumbria, the Pennines, and central and north Wales. Any new forestry development will need to have catchment-based assessments made to measure the vulnerability of surface waters.

Ground preparation

Cultivation and drainage should be designed to ensure minimum run-off from land. This includes ploughing along contours, avoiding natural watercourses and installing cross-drains to prevent cultivation channels carrying large quantities of water from higher ground. Silt traps are also required at the ends of drains where there is a risk of erosion. Natural watercourses must never be diverted into channels or drains or be diverted into another catchment. Drain maintenance or ground preparation should be done at a time to avoid spawning or when salmonid eggs and alevins are living in the gravel of adjacent watercourses.

Protection of watercourses by buffer areas

Areas of undisturbed land are left between the planted area and the watercourse. This filters out sediment and also helps to reduce nutrients from entering the water. Buffer strips are more effective if they have vigorous vegetative growth, and are gently sloping and wide. Exact details of design are given in the code of practice, but the width of the buffer area depends largely on stream size. For example, if channel width is 1, 2 or >2 m then the buffer area should be 5, 10 or 20 m wide on either side of the stream. Where the site is in a sensitive area with easily eroded soils then these buffer areas should be doubled. This may make certain upland areas with significant lengths of watercourse commercially unviable for forestry.

Riparian vegetation management

The riparian zone, the land immediately adjoining the aquatic zone, is particularly important in downstream sections of watercourses. Open or partially shaded conditions are necessary to maintain the bankside vegetation that is so important for wildlife, as well preventing erosion of the banks. Half the length of the watercourse should be exposed to direct sunlight while the remainder can be under partially shaded conditions from trees and shrubs. Trees such as conifers, beech and oak cause too much shading and should only be used occasionally. Trees such as willow, birch, rowan, ash and hazel all have lighter foliage and are ideal for creating partial shade. While alder is particularly successful on riverbanks it can cause considerable shade and may contribute to acidification, so it should be used only sparingly. Grazing animals must be excluded to prevent bank erosion or damage to bankside vegetation.

Road construction and maintenance

The construction of forest roads can be a major cause of damage to watercourses and to fish in particular. For that reason all major road works should be restricted to the period May to September. All roads should be well clear of riparian or buffer zones. Any watercourse intercepted by a road must be culverted or bridged. Roadside drains should not be used to intercept watercourses or to transport large volumes of water from higher ground. Sediment traps should be used to remove solids and these must be maintained. Embankments must be reseeded to prevent erosion. Roads must be inspected regularly, especially after heavy rain. Metalliferous or sulphide-rich material should not be used for road construction due to leaching and acidification.

Felling

Harvesting trees, especially clear felling, causes a considerable threat to watercourses. Large increases in sedimentation occurs which can smother aquatic habitats, damage spawning ground, prevent fish movement as well as physically blocking streams. Bark and other woody debris can also enter the river causing both sedimentation and deoxygenation problems. Soil disturbance must be kept to a minimum, especially from dragging timber, and if necessary brash mats must be used to protect sensitive soils from vehicle damage. New channels and sediment traps to intercept surface run-off may also be required.

Pesticides and fertilizers

There are separate Codes of Practice covering the use of pesticides (HMSO, 1989) and in the UK consultation with the Water Regulatory Authority is legally required before any application of pesticide is made near water or from the air. In sensitive areas, where there are standing waters, fertilizer should be applied by hand and not by aerial application to prevent the risk of eutrophication.

Chemicals and oils

Careful storage and use of these materials is vital, and contingency plans should be drawn up to deal with spillages.

7.4.3 CONCLUSION

The adoption of the best management practices (BMP) outlined in Tables 7.15 and 7.16 for the control of diffuse pollution is the most cost-effective way to tackle diffuse pollution. The use of grass swales, buffer zones and strips, infiltration systems, wet ponds and storm water wetlands can significantly improve the appearance of the landscape by providing attractive natural solutions for the collection and treatment of run-off, as well as achieving substantial water quality protection and improvement. The adoption of BMP for diffuse pollution control is now widespread in the US, Australia, Japan and many European countries.

REFERENCES

Boon, P. J., Calow, P. and Petts, G. E. (eds), 1992 *River Conservation and Management*, John Wiley and Sons, Chichester.

British Ecological Society, 1990 *River Water Quality*, Field Studies Council, Shrewsbury.

Brookes, A. and Shields, F. D. (eds), 1996 *River channel restoration-guiding principles for sustainable projects.* John Wiley and Sons, Chichester.

D'Arcy, B. J., Usman, F., Griffiths, D. and Chatfield, P., 1998 Initiatives to treat diffuse pollution in the UK. *Water Science and Technology*, **38** (10), 131–8.

Environment Agency, 1998 *The State of the Environment in England and Wales: Freshwaters*, The Stationery Office, London.

Ferguson, R. M., Lyness, J. F., Myers, W. R. C. and O'Sullivan, J. J., 1998 Hydraulic performance of environmental features in rivers. *Journal of the Chartered Institution of Water and Environmental Management*, **12**, 286–281.

HMSO, 1989 *Provisional Code of Practice for the Use of Pesticides in Forestry*, Forestry Commission Occasional Paper: 21, Her Majesty's Stationery Office, London.

HMSO, 1993 *Forests and Water Guidelines*, Her Majesty's Stationery Office, London.

IAWQ, 1997 *Nature's Way: Best Management Practice in the Control of Diffuse Pollution*, IAWQ Video, International Association of Water Quality, London.

IHD-WHO, 1978 *Water Quality Surveys*, International Hydrological Decade and World Health Organization Working Group on the Quality of Water, UNESCO, Geneva.

McCumiskey, L. M., 1982 Water Quality Management Plans – what is required. *Irish Journal of Environmental Science*, **2** (1), 44–50.

McGarity, A. E., 1997 Water quality management. In: ReVelle, C. and McGarity, A. E., eds, *Design and Operation of Civil and Environmental Engineering Systems*, John Wiley and Sons, New York, pp 41–96.

Neilsen, M., 2002 Lowland stream restoration in Denmark: background and examples. *Journal of the Chartered Institution of Water and Environmental Management*, **16**, 189–93.

Newman, P. J., Piavaux, M. A. and Sweeting, R. A., 1992 *River Water Quality: Ecological Assessment and Control*, Office for Official Publications of the European Communities, Luxembourg.

NRA, 1993 *River Corridor Manual for Surveyors*, National Rivers Authority, Bristol.

NRA, 1994 *Guidance Notes for Local Planning Authorities on the Methods of Protecting the Water Environment Through Development Plans*, National Rivers Authority, Bristol.

Shutes, R. B. E., Revitt, D. M., Mungar, A. S. and Scholes, L. N. L., 1997 Design of wetland systems for the treatment of urban run-off. *Water Quality International* (March/April), 35–8.

Sieker, F., 1998 On-site storm-water management as an alternative to conventional sewer systems: a new concept spreading in Germany. *Water Science and Technology*, **38** (10), 65–71.

Stretton, C., 1998 Managing the threats of afforestation in reservoir catchments. *Water Science and Technology*, **37** (2), 361–8.

FURTHER READING

Biswas, A. K. (ed.), 1997 *Water Resources: Environmental Planning, Management and Development*, McGraw-Hill, New York.

Brizga, S. O. and Findlayson, B. L., 2000 *River Management: The Australasian Experience*, John Wiley and Sons, London.

Cosgrove, D. and Petts, G. (eds), 1990 *Water, Engineering and Landscape*, Belhaven Press, London.

DeWaal, L., Wade, M. and Large, A., 1998 *Rehabilitation of Rivers: Principles and Implementation*, John Wiley and Sons, Chichester.

Kirby, C. and White, W. R. (eds), 1994 *Integrated River Basin Development*, John Wiley and Sons, Chichester.

Harper, D. M. and Ferguson, R. M. (eds), 1995 *The Ecological Basis for River Management*, John Wiley and Sons, Chichester.

Heathcote, I. W., 1998 *Integrated Watershed Management: Principles and Practice*, John Wiley and Sons, New York.

Mitchell, B. (ed.), 1990 *Integrated Water Management*, Belhaven Press, London.

Novotny, V. and Olem, H., 1994 *Water Quality: Prevention, Identification and Management of Diffuse Pollution*, Van Nostrand Reinhold, New York.

Novotny, V., 1999 Diffuse pollution from agriculture – a worldwide outlook. *Water Science and Technology*, **39** (3), 1–13.

NRA, 1993 *Catchment Management Planning Guidelines*, National Rivers Authority, Bristol.

Petts, G. and Calow, P., 1996 *River Restoration*, Blackwell Science, Oxford.

Sheldon, C. and Yoxon, M., 1999 *Installing Environmental Management Systems: A Step by Step Guide*, Earthscan, London.

Schiechtl, H. M. and Stern, R., 1997 *Water Bioengineering Techniques for Watercourse Bank and Shoreline Protection*, Blackwell Science, Oxford.

Chapter 8 Water Quality and Regulation

INTRODUCTION

Environmental legislation relating to water quality is based largely on quality standards relating to suitability for a specific use, the protection of receiving waters or emission limits on discharges. Standards are usually mandatory with maximum permissible concentrations based on health criteria or environmental quality standards (EQSs) (see Section 7.1). Table 8.1 lists the Directives concerning the aquatic environment that govern legislation in countries (Member States) comprising the European Union. However, European legislation relating to water policy has significantly changed since the introduction of the Water Framework Directive (WFD) (2000/60/EC). Many of the earlier Directives, on which surface water and groundwater management has been based, will be integrated into the new Directive giving one coherent policy document controlling and protecting water resources on a catchment basis (Table 8.2). Acting as the pivotal legislation, the WFD will be supported by a New Groundwater Directive (COM(2003)550 Final), a revised Bathing Water Directive (76/160/EEC; COM(2002)581 Final), and the existing Nitrates (91/676/EEC), Urban Wastewater Treatment (91/271/EEC) and Drinking Water (98/83/EC) Directives. The Sewage Sludge Disposal (86/278/EEC) and the Integrated Pollution Prevention and Control (IPPC) Directive (96/61/EEC), along with a number of associated Directives remain relevant to water quality but are outside those concerned directly with water policy. The Directive controlling sewage sludge disposal to agricultural land (86/278/EEC) is discussed in Chapter 21.

In most Directives, guide (G) and imperative or mandatory (I) values are given. The G-values are those which Member States should be working towards in the long term. In most cases nationally adopted limit values are the I-values, although occasionally more stringent values are set. The key US Federal environmental legislation relating to water quality is summarized in Table 8.3.

TABLE 8.1   EU environmental policy with regard to inland waters by year of introduction

1973
Council Directive on the approximation of the laws of the Member States relating to detergents (73/404/EEC)
Council Directive on the control of biodegradability of anionic surfactants (73/405/EEC)

1975
Council Directive concerning the quality required of surface water intended for the abstraction of drinking water in the Member States (75/440/EEC)

1976
Council Directive concerning the quality of bathing waters (76/160/EEC)
Council Directive on pollution caused by certain dangerous substances discharged into the aquatic environment (76/464/EEC)

1977
Council decision establishing a common procedure for the exchange of information on the quality of surface in the Community (77/795/EEC)

1978
Council Directive on titanium oxide waste (78/178/EEC)
Council Directive on quality of freshwaters needing protection or improvement in order to support fish life (78/659/EEC)

1979
Council Directive concerning the methods of measurement and frequencies of sampling and analysis of surface water intended for the abstraction of drinking water in the Member States (79/869/EEC)
Council Directive on the quality required for shellfish waters (79/923/EEC)

1980
Council Directive on the protection of groundwater against pollution caused by certain dangerous substances (80/68/EEC)
Council Directive on the approximation of the laws of the Member States relating to the exploitation and marketing of natural mineral waters (80/777/EEC)
Council Directive relating to the quality of water intended for human consumption (80/778/EEC)

1982
Council Directive on limit values and quality objectives for mercury discharges by the chlor-alkali electrolysis industry (82/176/EEC)
Council Directive on the testing of the biodegradability of non-ionic surfactants (82/242/EEC)
Council Directive on the monitoring of waste from the titanium oxide industry (82/883/EEC)

1983
Council Directive on limit values and quality objectives for cadmium discharges (83/513/EEC)

1984
Council Directive on limit values and quality objectives for discharges by sectors other than the chlor-alkali electrolysis industry (84/156/EEC)
Council Directive on limit values and quality objectives for discharges of hexachlorocyclohexane (84/491/EEC)

1985
Council Directive on the assessment of the effects of certain public and private projects on the environment (85/337/EEC)

TABLE 8.1 (*Continued*)

1986
Council Directive on the limit values and quality objectives for discharge of certain dangerous substances included in List I of the Annex to Directive 76/464/EEC (86/280/EEC)
1987
Council Directive on the prevention and reduction of environmental pollution by asbestos (87/217/EEC)
1988
Council Directive amending Annex II to the Directive 86/280/EEC on limit values and quality objectives for discharges of certain dangerous substances included in List I of the Annex to Directive 76/464/EEC (88/347/EEC)
1990
Council Directive amending Annex II to the Directive 86/280/EEC on limit values and quality objectives for discharges of certain dangerous substances included in List I of the Annex to Directive 76/464/EEC (90/415/EEC)
1991
Council Directive concerning urban wastewater treatment (91/271/EEC)
Council Directive concerning the protection of waters against pollution caused by nitrates from agricultural sources (91/676/EEC)
1992
Council Directive on pollution by waste from the titanium oxide industry (92/112/EEC)
1996
Council Directive on IPPC (96/61/EEC)
Council Directive amending Directive 80/777/EEC on the approximation of the laws of the Member States relating to the exploitation and marketing of natural mineral waters
1998
Council Directive on the quality of water intended for human consumption (98/83/EEC)
2000
Council Directive establishing a framework for community action in the field of Water Policy (00/60/EEC)

TABLE 8.2 Existing EU legislation that will be repealed by the WFD. The transitional period for this legislation varies between 7 and 13 years after the publishing of the WFD on 22 December 2000.

To be repealed with effect from the year 2007
75/440/EEC Directive concerning the quality required of surface water intended for the abstraction of drinking water in Member States
77/795/EEC Council Decision establishing a common procedure for the exchange of information on the quality of surface freshwater in the Community
79/869/EEC Directive concerning the methods of measurement and frequencies of sampling and analysis of surface water intended for the abstraction of drinking waters in the Member States

To be repealed with effect from the year 2013
76/464/EEC Directive on pollution caused by certain dangerous substances discharged into the aquatic environment
78/659/EEC Directive on the quality of freshwaters needing protection in order to support fish life
79/923/EEC Directive on the quality required by shellfish waters
80/68/EEC Directive on the protection of groundwater against pollution caused by certain dangerous substances

TABLE 8.3   Major US Federal water quality legislation

Name of Act	Abbreviation	Date	Comments
River and Harbors Act		1889	Protects waters from pollution
Federal Water Pollution Control Act	FWPCA	1948	Grants for feasibility studies
Amendments to FWPCA	FWPCA	1956	Grants for feasibility studies
Water Quality Act	WQA	1965	Enacted water quality standards
Amendments to FWPCA – now called the Clean Water Act	CWA	1972	Massive grants for wastewater treatment and collection
Safe Drinking Water Act	SDWA	1974	EPA sets standards for drinking water quality
Fishery Conservation and Management Act	FCMA	1976	
Soil and Water Resources Conservation Act	SWCRA	1977	
EPA Wetlands Regulations		1989	
Amendments to Safe Drinking Water Act	SDWA	1986	EPA publishes lists of contaminants found in public water supplies (85 by 1992)
Amendments to CWA	CWA	1987	Regulation of toxics and sewage sludge in wastewaters and non-point source (NPS) pollution control grants
Surface Water Treatment Rules	SWTR	1989	
National Pollution Discharge Elimination System	NPDES	1990	Permits for industrial and municipal stormwater discharges
Amendments to CWA	CWA	1992	65 000 chemicals regulated
Standards for the use and disposal of sewage sludge	49 CFR 503	1993	Sewage sludge standards
Amendments to CWA	CWA HR 961	1995	Incentives for watershed management, a new wetlands permitting programme based on ecological values, etc.

8.1 POTABLE WATERS

Water used for the abstraction of drinking water is now covered by the WFD (Article 7), replacing the Surface Water Directive, which will be repealed in 2007 (Table 8.2). Under the WFD water bodies used for the abstraction of water intended for human consumption, and either providing $>10\,m^3\,d^{-1}$ or serving more than 50 people, are designated under Article 6 as protected areas. Likewise, those water bodies intended for abstraction in the future are also classified as protected areas (Table 8.4). The Directive requires that these waters receive sufficient treatment to ensure that the water supplied conforms to the Drinking Water Directive (98/83/EC). Member States are required to protect these waters to avoid any deterioration in their quality in

TABLE 8.4   Areas within River Basin Districts (RBDs) requiring special protection under the WFD

Protected areas include the following:
- Areas designated for the abstraction of water intended for human consumption
- Areas that may be designated in the future for the abstraction of water intended for human consumption
- Areas designated for the protection of economically significant aquatic species
- Bodies of water designated as recreational waters, including area designated as bathing waters under Directive (76/160/EEC)
- Nutrient-sensitive areas including areas designated as vulnerable zones under Directive 91/676/EEC and areas designated as sensitive areas under Directive 91/271/EEC
- Areas designated for the protection of habitats of species where the maintenance or improvement of the status of the water is an important factor in their protection, including relevant Natura 2000 sites designated under Directives 92/43/EEC (last amended 97/62/EC) and 79/409/EEC (last amended 97/49/EC)

order to minimize the degree of treatment required. This includes establishing, where necessary, protection zones around water bodies to safeguard quality.

Until it is repealed, surface waters intended for supply purposes must be classified under EU Directive (75/440/EEC). There are three categories (A_1, A_2 or A_3) based on raw water quality using 46 parameters (Table 8.5). For each category there is a mandatory minimum degree of treatment required. These are:

A_1 simple physical treatment and disinfection (e.g. rapid filtration and chlorination);

A_2 normal physical treatment, chemical treatment and disinfection (e.g. pre-chlorination, coagulation, flocculation, decantation, filtration and final chlorination);

A_3 intensive physical and chemical treatment, extended treatment and disinfection (e.g. chlorination to break point, coagulation, flocculation, decantation, filtration, adsorption (activated carbon) and disinfection (ozone or final chlorination)) (Section 10.1).

The methods and sampling frequency for this Directive were not published until 1979 (79/869/EEC), with sampling taking place at designated sites where water is abstracted. Surface water that falls outside the mandatory limits for A_3 waters is normally excluded for use, although it can be blended with better quality water prior to treatment. The Surface Water Directive will continue in force until 2012 when all measures in the first River Basin Management Plans (RBMPs) become fully operational.

Only one Directive deals with drinking water quality, the EU Directive relating to the quality of water intended for human consumption (98/83/EC), approved 3 November 1998. It also covers water used in food production or in processing or marketing products intended for human consumption. This replaces the earlier Drinking Water Directive (80/778/EEC) which had been originally proposed in 1975 and was both scientifically and technically out of date. The publication of the revised World Health Organization (WHO) drinking water guidelines in 1993 led to the revision of the

TABLE 8.5 Standards for potable water abstractions set by the EC Surface Water Directive (75/440/EEC)

Treatment type	A_1		A_2		A_3	
Parameter (mg/l except where noted)	Guide limit	Mandatory limit	Guide limit	Mandatory limit	Guide limit	Mandatory limit
pH units	6.5–8.5		5.5–9.0		5.5–9.0	
Colour units	10	20	50	100	50	200
Suspended solids	25					
Temperature (°C)	22	25	22	25	22	25
Conductivity (µS/cm)	1000		1000		1000	
Odour (DN[a])	3		10		20	
Nitrate (as NO_3)	25	50		50		50
Fluoride	0.7–1.0	1.5	0.7–1.7		0.7–1.7	
Iron (soluble)	0.1	0.3	1.0	2.0	1.0	
Manganese	0.05		0.1		1.0	
Copper	0.02	0.05	0.05		1.0	
Zinc	0.5	3.0	1.0	5.0	1.0	5.0
Boron	1.0		1.0		1.0	
Arsenic	0.01	0.05		0.05	0.05	0.1
Cadmium	0.001	0.005	0.001	0.005	0.001	0.005
Chromium (total)		0.05		0.05		0.05
Lead		0.05		0.05		0.05
Selenium		0.01		0.01		0.01
Mercury	0.0005	0.001	0.0005	0.001	0.0005	0.001
Barium		0.1		1.0		1.0
Cyanide		0.05		0.05		0.05
Sulphate	150	250	150	250	150	250
Chloride	200		200		200	
MBAS	0.2		0.2		0.5	
Phosphate (as P_2O_5)	0.4		0.7		0.7	
Phenol		0.001	0.001	0.005	0.01	0.1
Hydrocarbons (ether soluble)		0.05		0.2	0.5	1.0
PAH[b]		0.0002		0.0002		0.001
Pesticides		0.001		0.0025		0.005
COD					30	
BOD (with ATU[c])	<3		<5		<7	
DO[d] per cent saturation	>70		>50		>30	
Nitrogen (Kjeldahl)	1		2		3	
Ammonia (as NH_4)	0.05		1	1.5	2	4
Total coliforms/100 ml	50		5000		50 000	
Faecal coliforms/ 100 ml	20		2000		20 000	
Faecal streptococci/ 100 ml	20		1000		10 000	
Salmonella	absent in 5 l		absent in 1 l			

Mandatory levels 95% compliance, 5% not complying should not exceed 150% of mandatory level. [a]DN: dilution number; [b]PAH: polycyclic aromatic hydrocarbons; [c]ATU: allythiourea; [d]DO: dissolved oxygen.
Reproduced with permission of the European Commission, Luxembourg.

TABLE 8.6 The quality parameters listed in the new EC Drinking Water Directive (98/83/EEC)

Parameter	Parametric value
Part A: Microbiological parameters	
Escherichia coli	0/100 ml
Enterococci	0/100 ml
In water offered for sale in bottles or containers	
E. coli	0/250 ml
Enterococci	0/250 ml
Pseudomonas aeruginosa	0/250 ml
Colony count at 22°C	100/ml
Colony count at 37°C	20/ml
Part B: Chemical parameters	
Acrylamide	$0.1\,\mu g\,l^{-1}$
Antimony	$5.0\,\mu g\,l^{-1}$
Arsenic	$10\,\mu g\,l^{-1}$
Benzene	$1.0\,\mu g\,l^{-1}$
Benzo(*a*)pyrene	$0.01\,\mu g\,l^{-1}$
Boron	$1.0\,mg\,l^{-1}$
Bromate	$10\,\mu g\,l^{-1}$
Cadmium	$5.0\,\mu g\,l^{-1}$
Chromium	$50\,\mu g\,l^{-1}$
Copper	$2.0\,mg\,l^{-1}$
Cyanide	$50\,\mu g\,l^{-1}$
1,2-Dichloroethane	$3.0\,\mu g\,l^{-1}$
Epichlorohydrin	$0.1\,\mu g\,l^{-1}$
Fluoride	$1.5\,mg\,l^{-1}$
Lead	$10\,\mu g\,l^{-1}$
Mercury	$1.0\,\mu g\,l^{-1}$
Nickel	$20\,\mu g\,l^{-1}$
Nitrate	$50\,mg\,l^{-1}$
Nitrite	$0.5\,mg\,l^{-1}$
Pesticides[a,b]	$0.1\,\mu g\,l^{-1}$
Pesticides (total)[a]	$0.5\,\mu g\,l^{-1}$
Polycyclic aromatic hydrocarbons[a]	$0.1\,\mu g\,l^{-1}$
Selenium	$10\,\mu g\,l^{-1}$
Tetrachloroethene plus trichloroethane	$10\,\mu g\,l^{-1}$
Trihalomethanes (total)[a]	$100\,\mu g\,l^{-1}$
Vinyl chloride	$0.5\,\mu g\,l^{-1}$

[a]Relates to specified compounds in Directive 98/83/EEC.
[b]For aldrin, dieldrin, heptachlor and heptachlor epoxide the parametric value is $0.03\,\mu g\,l^{-1}$.
Reproduced with permission of the European Commission, Luxembourg.

European standards. The new Directive has been significantly altered focusing on parameters that reflect current health and environmental concerns. This involves reducing the number of listed parameters from 66 to 48, 50 for bottled water, of which 15 are new parameters (Tables 8.6 and 8.7). It also introduces a more modern management approach to potable water quality control and assessment. The Directive has been incorporated into law in Britain and Ireland through the Water

TABLE 8.7 The indicator parameters (Part C) in the new EC Drinking Water Directive (98/83/EC) used for check monitoring

Parameter	Parametric value	Notes
Physico-chemical		
Aluminium	$200\,\mu g\,l^{-1}$	
Ammonium	$0.5\,mg\,l^{-1}$	
Chloride	$250\,mg\,l^{-1}$	Water should not be aggressive
Clostridium perfringens	0/100 ml	From, or affected by, surface water only
Colour		Acceptable to consumers and no abnormal change
Conductivity	$2500\,\mu S\,cm^{-1}$	Water should not be aggressive
Hydrogen ion concentration	pH $\geqslant 6.5$, $\leqslant 9.5$	Water should not be aggressive. Minimum values for bottled waters not $\leqslant$pH 4.5
Iron	$200\,\mu g\,l^{-1}$	
Manganese	$50\,\mu g\,l^{-1}$	
Odour		Acceptable to consumers and no abnormal change
Oxidizability	$5.0\,mg\,O_2\,l^{-1}$	Not required if TOC used
Sulphate	$250\,mg\,l^{-1}$	Water should not be aggressive
Sodium	$200\,mg\,l^{-1}$	
Taste		Acceptable to consumers and no abnormal change
Colony count at 22°C	0/100 ml	
Coliform bacteria	0/100 ml	For bottled waters 0/250ml
TOC		No abnormal change. Only for flows $>10\,000\,m^3\,day^{-1}$
Turbidity		Acceptable to consumers and no abnormal change. Normally <1.0 NTU
Radioactivity		
Tritium	$100\,Bq\,l^{-1}$	
Total indicative dose	$0.1\,mSv\,year^{-1}$	

TOC: total organic carbon; NTU: nephelometric turbidity unit.
Reproduced with permission of the European Commission, Luxembourg.

Supply (Water Quality) Regulations 2000 and the European Communities (Drinking Water) Regulations 2000, respectively. Natural mineral waters and medicinal waters are excluded (i.e. Natural Mineral Waters Directive 80/777/EEC), while bottled water not legally classified as natural mineral water (e.g. spring waters) are required to conform to the Drinking Water Directive.

The WHO guidelines for drinking water are used universally and are the basis for both EU and USA legislation (WHO, 1993a, 2004). The guidelines are revised when required to reflect the most recent toxicological and scientific evidence. The latest guidelines published on 21 September 2004 include microbiological (Table 8.8), physico-chemical (Table 8.9) and radiological parameters (Table 8.10). While health-based guidance levels for almost 200 individual radionuclides in drinking water have been published in the 2004 WHO guidelines, the process of identifying and quantifying

TABLE 8.8   WHO drinking water guide values for microbial quality[a]

Organisms	Guideline value
All water directly intended for drinking	
E. coli or thermotolerant coliform bacteria[b,c]	Must not be detectable in any 100-ml sample
Treated water entering the distribution system	
E. coli or thermotolerant coliform bacteria[b]	Must not be detectable in any 100-ml sample
Treated water in the distribution system	
E. coli or thermotolerant coliform bacteria[b]	Must not be detectable in any 100-ml sample

[a]Immediate investigative action must be taken if *E. coli* are detected.

[b]Although *E. coli* is the more precise indicator of faecal pollution, the count of thermotolerant coliform bacteria is an acceptable alternative. If necessary, proper confirmatory tests must be carried out. Total coliform bacteria are not acceptable indicators of the sanitary quality of water supplies, particularly in tropical areas, where many bacteria of no sanitary significance occur in almost all untreated supplies.

[c]It is recognized that in the great majority of rural water supplies, especially in developing countries, faecal contamination is widespread. Especially under these conditions, medium-term targets for the progressive improvement of water supplies should be set.

Reproduced from WHO (2004) with permission of the World Health Organization, Geneva.

TABLE 8.9   WHO drinking water guide values for chemicals of significance to health

Chemical	Guideline value[a] (mg/l)	Remarks
Acrylamide	0.0005[b]	
Alachlor	0.02[b]	
Aldicarb	0.01	Applies to aldicarb sulphoxide and aldicarb sulphone
Aldrin and dieldrin	0.00003	For combined aldrin plus dieldrin
Antimony	0.02	
Arsenic	0.01 (P)	
Atrazine	0.002	
Barium	0.7	
Benzene	0.01[b]	
Benzo[a]pyrene	0.0007[b]	
Boron	0.5 (T)	
Bromate	0.01[b] (A, T)	
Bromodichloromethane	0.06[b]	
Bromoform	0.1	
Cadmium	0.003	
Carbofuran	0.007	
Carbon tetrachloride	0.004	
Chloral hydrate (trichloroacetaldehyde)	0.01 (P)	
Chlorate	0.7 (D)	
Chlordane	0.0002	
Chlorine	5 (C)	For effective disinfection, there should be a residual concentration of free chlorine of ⩾0.5 mg/l after at least 30 min contact time at pH <8.0

(Continued)

TABLE 8.9 (*Continued*)

Chemical	Guideline value[a] (mg/l)	Remarks
Chlorite	0.7 (D)	
Chloroform	0.2	
Chlorotoluron	0.03	
Chlorpyrifos	0.03	
Chromium	0.05 (P)	For total chromium
Copper	2	Staining of laundry and sanitary ware may occur below guideline value
Cyanazine	0.0006	
Cyanide	0.07	
Cyanogen chloride	0.07	For cyanide as total cyanogenic compounds
2,4-D (2,4-dichlorophenoxyacetic acid)	0.03	Applies to free acid
2,4-DB	0.09	
DDT and metabolites	0.001	
Di(2-ethylhexyl)phthalate	0.008	
Dibromoacetonitrile	0.07	
Dibromochloromethane	0.1	
1,2-Dibromo-3-chloropropane	0.001[b]	
1,2-Dibromoethane	0.0004[b] (P)	
Dichloroacetate	0.05 (T, D)	
Dichloroacetonitrile	0.02 (P)	
Dichlorobenzene, 1,2-	1 (C)	
Dichlorobenzene, 1,4-	0.3 (C)	
Dichloroethane, 1,2-	0.03[b]	
Dichloroethene, 1,1-	0.03	
Dichloroethene, 1,2-	0.05	
Dichloromethane	0.02	
1,2-Dichloropropane (1,2-DCP)	0.04 (P)	
1,3-Dichloropropene	0.02[b]	
Dichlorprop	0.1	
Dimethoate	0.006	
Edetic acid (EDTA)	0.6	Applies to the free acid
Endrin	0.0006	
Epichlorohydrin	0.0004 (P)	
Ethylbenzene	0.3 (C)	
Fenoprop	0.009	
Fluoride	1.5	Volume of water consumed and intake from other sources should be considered when setting national standards
Formaldehyde	0.9	
Hexachlorobutadiene	0.0006	
Isoproturon	0.009	
Lead	0.01	
Lindane	0.002	
Manganese	0.4 (C)	
MCPA	0.002	
Mecoprop	0.01	
Mercury	0.001	For total mercury (inorganic plus organic)

TABLE 8.9   (Continued)

Chemical	Guideline value[a] (mg/l)	Remarks
Methoxychlor	0.02	
Metolachlor	0.01	
Microcystin-LR	0.001 (P)	For total Microcystin-LR (free plus cell-bound)
Molinate	0.006	
Molybdenum	0.07	
Monochloramine	3	
Monochloroacetate	0.02	
Nickel	0.02 (P)	
Nitrate (as NO_3^-)	50	Short-term exposure
Nitrilotriacetic acid (NTA)	0.2	
Nitrite (as NO_2^-)	3	Short-term exposure
	0.2 (P)	Long-term exposure
Pendimethalin	0.02	
Pentachlorophenol	0.009[b] (P)	
Pyriproxyfen	0.3	
Selenium	0.01	
Simazine	0.002	
Styrene	0.02 (C)	
2,4,5-T	0.009	
Terbuthylazine	0.007	
Tetrachloroethene	0.04	
Toluene	0.7 (C)	
Trichloroacetate	0.2	
Trichloroethene	0.07 (P)	
Trichlorophenol, 2,4,6-	0.2[b] (C)	
Trifluralin	0.02	
Trihalomethanes		The sum of the ratio of the concentration of each to its respective guideline value should not exceed 1
Uranium	0.015 (P, T)	Only chemical aspects of uranium addressed
Vinyl chloride	0.0003[b]	
Xylenes	0.5 (C)	

[a]P: provisional guideline value, as there is evidence of a hazard, but the available information on health effects is limited; T: provisional guideline value because calculated guideline value is below the level that can be achieved through practical treatment methods, source protection, etc.; A: provisional guideline value because calculated guideline value is below the achievable quantification level; D: provisional guideline value because disinfection is likely to result in the guideline value being exceeded; C: concentrations of the substance at or below the health-based guideline value may affect the appearance, taste or odour of the water, leading to consumer complaints.

[b]For substances that are considered to be carcinogenic, the guideline value is the concentration in drinking water associated with an upper-bound excess lifetime cancer risk of 10^{-5} (one additional cancer per 100 000 of the population ingesting drinking water containing the substance at the guideline value for 70 years). Concentrations associated with upper-bound estimated excess lifetime cancer risks of 10^{-4} and 10^{-6} can be calculated by multiplying and dividing, respectively, the guideline value by 10.

Reproduced from WHO (2004) with permission of the World Health Organization, Geneva.

individual radioactive species requires sophisticated and expensive analysis. For this reason, drinking water is screened using gross alpha and beta activity with further action only required if these activity levels are exceeded. Table 8.11 lists those chemicals that are not harmful to health at the concentrations normally encountered in drinking water. Some may, however, result in consumer complaints.

TABLE 8.10   WHO drinking water guide values for radioactive components

Screening value (Bq/l)	Remarks
Gross α-activity 0.5 Gross β-activity 1	If a screening value is exceeded, more detailed radionuclide analysis is necessary. Higher values do not necessarily imply that the water is unsuitable for human consumption

Reproduced from WHO (2004) with permission of the World Health Organization, Geneva.

TABLE 8.11   Health-related guide values have not been set by the WHO for a number of chemicals that are not considered hazardous in concentrations normally found in drinking water. Some compounds may give rise to consumer complaints

Chemicals excluded from guideline value derivation

Chemical	Reason for exclusion
Amitraz	Degrades rapidly in the environment and is not expected to occur at measurable concentrations in drinking water supplies
Beryllium	Unlikely to occur in drinking water
Chlorobenzilate	Unlikely to occur in drinking water
Chlorothalonil	Unlikely to occur in drinking water
Cypermethrin	Unlikely to occur in drinking water
Diazinon	Unlikely to occur in drinking water
Dinoseb	Unlikely to occur in drinking water
Ethylene thiourea	Unlikely to occur in drinking water
Fenamiphos	Unlikely to occur in drinking water
Formothion	Unlikely to occur in drinking water
Hexachlorocyclohexanes (mixed isomers)	Unlikely to occur in drinking water
MCPB	Unlikely to occur in drinking water
Methamidophos	Unlikely to occur in drinking water
Methomyl	Unlikely to occur in drinking water
Mirex	Unlikely to occur in drinking water
Monocrotophos	Has been withdrawn from use in many countries and is unlikely to occur in drinking water
Oxamyl	Unlikely to occur in drinking water
Phorate	Unlikely to occur in drinking water
Propoxur	Unlikely to occur in drinking water
Pyridate	Not persistent and only rarely found in drinking water
Quintozene	Unlikely to occur in drinking water
Toxaphene	Unlikely to occur in drinking water
Triazophos	Unlikely to occur in drinking water
Tributyltin oxide	Unlikely to occur in drinking water
Trichlorfon	Unlikely to occur in drinking water

TABLE 8.11 (*Continued*)

Chemicals for which guideline values have not been established

Chemical	Reason for not establishing a guideline value
Aluminium	Owing to limitations in the animal data as a model for humans and the uncertainty surrounding the human data, a health-based guideline value cannot be derived; however, practicable levels based on optimization of the coagulation process in drinking water plants using aluminium-based coagulants are derived: 0.1 mg/l or less in large water treatment facilities, and 0.2 mg/l or less in small facilities
Ammonia	Occurs in drinking water at concentrations well below those at which toxic effects may occur
Asbestos	No consistent evidence that ingested asbestos is hazardous to health
Bentazone	Occurs in drinking water at concentrations well below those at which toxic effects may occur
Bromochloroacetate	Available data inadequate to permit derivation of health-based guideline value
Bromochloroacetonitrile	Available data inadequate to permit derivation of health-based guideline value
Chloride	Not of health concern at levels found in drinking water[a]
Chlorine dioxide	Guideline value not established because of the rapid breakdown of chlorine dioxide and because the chlorite provisional guideline value is adequately protective for potential toxicity from chlorine dioxide
Chloroacetones	Available data inadequate to permit derivation of health-based guideline values for any of the chloroacetones
Chlorophenol, 2-	Available data inadequate to permit derivation of health-based guideline value
Chloropicrin	Available data inadequate to permit derivation of health-based guideline value
Dialkyltins	Available data inadequate to permit derivation of health-based guideline values for any of the dialkyltins
Dibromoacetate	Available data inadequate to permit derivation of health-based guideline value
Dichloramine	Available data inadequate to permit derivation of health-based guideline value
Dichlorobenzene, 1,3-	Toxicological data are insufficient to permit derivation of health-based guideline value
Dichloroethane, 1,1-	Very limited database on toxicity and carcinogenicity
Dichlorophenol, 2,4-	Available data inadequate to permit derivation of health-based guideline value
Dichloropropane, 1,3-	Data insufficient to permit derivation of health-based guideline value
Di(2-ethylhexyl) adipate	Occurs in drinking water at concentrations well below those at which toxic effects may occur
Diquat	Rarely found in drinking water, but may be used as an aquatic herbicide for the control of free-floating and submerged aquatic weeds in ponds, lakes and irrigation ditches
Endosulfan	Occurs in drinking water at concentrations well below those at which toxic effects may occur
Fenitrothion	Occurs in drinking water at concentrations well below those at which toxic effects may occur
Flouoranthene	Occurs in drinking water at concentrations well below those at which toxic effects may occur
Glyphosate and AMPA	Occurs in drinking water at concentrations well below those at which toxic effects may occur
Hardness	Not of health concern at levels found in drinking water[a]
Heptachlor and heptachlor epoxide	Occurs in drinking water at concentrations well below those at which toxic effects may occur
Hexachlorobenzene	Occurs in drinking water at concentrations well below those at which toxic effects may occur
Hydrogen sulphide	Not of health concern at levels found in drinking water[a]
Inorganic tin	Occurs in drinking water at concentrations well below those at which toxic effects may occur
Iodine	Available data inadequate to permit derivation of health-based guideline value, and lifetime exposure to iodine through water disinfection is unlikely
Iron	Not of health concern at concentrations normally observed in drinking water, and taste and appearance of water are affected below the health-based value
Malathion	Occurs in drinking water at concentrations well below those at which toxic effects may occur
Methyl parathion	Occurs in drinking water at concentrations well below those at which toxic effects may occur

(*Continued*)

TABLE 8.11 (*Continued*)

Chemical	Reason for not establishing a guideline value
Monobromoacetate	Available data inadequate to permit derivation of health-based guideline value
Monochlorobenzene	Occurs in drinking water at concentrations well below those at which toxic effects may occur, and health-based value would far exceed lowest reported taste and odour threshold
MX	Occurs in drinking water at concentrations well below those at which toxic effects may occur
Parathion	Occurs in drinking water at concentrations well below those at which toxic effects may occur
Permethrin	Occurs in drinking water at concentrations well below those at which toxic effects may occur
pH	Not of health concern at levels found in drinking water[b]
Phenylphenol, 2- and its sodium salt	Occurs in drinking water at concentrations well below those at which toxic effects may occur
Propanil	Readily transformed into metabolites that are more toxic; a guideline value for the parent compound is considered inappropriate, and there are inadequate data to enable the derivation of guideline values for the metabolites
Silver	Available data inadequate to permit derivation of health-based guideline value
Sodium	Not of health concern at levels found in drinking water[a]
Sulphate	Not of health concern at levels found in drinking water[a]
TDS	Not of health concern at levels found in drinking water[a]
Trichloramine	Available data inadequate to permit derivation of health-based guideline value
Trichloroacetonitrile	Available data inadequate to permit derivation of health-based guideline value
Trichlorobenzenes (total)	Occurs in drinking water at concentrations well below those at which toxic effects may occur, and health-based value would exceed lowest reported odour threshold
Trichloroethane, 1,1,1-	Occurs in drinking water at concentrations well below those at which toxic effects may occur
Zinc	Not of health concern at concentrations normally observed in drinking water[a]

[a]May affect acceptability of drinking water (see Chapter 11).
[b]An important operational water quality parameter.

The key changes in water quality standards introduced by the revised Drinking Water Directive are:

(a) Faecal coliforms are replaced by *Escherichia coli*, and *Pseudomonas aeruginosa* is to be measured in bottled water.

(b) Antimony is reduced from 10 to $5\,\mu g\,l^{-1}$.

(c) Lead is reduced from 50 to $10\,\mu g\,l^{-1}$. A 15-year transition period is allowed for the replacement of lead distribution pipes.

(d) Nickel (a precursor for eczema) is reduced from 50 to $20\,\mu g\,l^{-1}$.

(e) Disinfection by products and certain flocculants are also included (e.g. trihalomethanes, trichloroethene and tetracholoroethene, bromate, acrylamide, etc.).

(f) Copper is reduced from 3 to $2\,mg\,l^{-1}$.

(g) Maximum permissible concentrations for individual and total pesticides are retained at $0.1\,\mu g\,l^{-1}$ and $0.5\,\mu g\,l^{-1}$, respectively, with more stringent standards introduced for certain pesticides (i.e. $0.03\,\mu g\,l^{-1}$).

The Directive entered into force on 25 December 1998 with Member States required to transpose the Directive into national legislation within 2 years and full compliance by 25 December 2003. Exceptions have been made for bromate, trihalomethanes and lead for which full compliance is required by 25 December 2008 for the first two parameters and 25 December 2013 for lead. Interim values of 25, 150 and $25\,\mu g\,l^{-1}$, respectively, are to be achieved by 25 December 2003 for these parameters. The new parameters are listed under three parts: Part A. Microbiological; Part B. Chemical (Table 8.6); Part C. Indicator parameters (Table 8.7).

Monitoring of compliance is done at two levels. Regular check monitoring using the parameters listed in Part C (Table 8.7) ensures that the basic organoleptic and microbial quality is maintained, while audit monitoring is done less frequently to ensure compliance with the listed quality parameters in the Directive. Sampling is now done exclusively at consumers' taps and the water must be free from any micro-organisms, parasite or substance which, in numbers or concentrations, constitute a potential danger to human health, as well as meeting the minimum quality standards listed. The Directive introduces some new management functions to ensure better quality drinking water is supplied. All parametric values are to be regularly reviewed and where necessary strengthening them in accordance with the latest available scientific knowledge. Two principle sources of information will be the WHO Water Quality Guidelines and the Scientific Committee on Toxicology and Ecotoxicology. The Directive also seeks to increase transparency and safety for consumers. This is achieved by (i) ensuring that compliance is at the point of use; (ii) obligation on suppliers to report on quality; (iii) an obligation on suppliers to inform the consumer on drinking water quality and measures that they can take to comply with the requirements of the Directive when the non-compliance is because of the domestic distribution system (e.g. internal pipes, plumbing, etc.).

The Drinking Water Directive only covers public water supplies while private supplies are excluded (e.g. springs, wells, etc.). The UK Water Industry Act 1991 defines a private water supply as any supply of water not provided by a statutorily appointed water undertaker. In England and Wales there are over 50 000 private supplies supplying about 350 000 people for domestic purposes. Of these, approximately 30 000 supplies serve a single dwelling. The actual number of people consuming private water supplies at some time is very much greater due to their use at hotels, schools and other isolated locations. The Private Water Supplies Regulations 1991 require local authorities to monitor such supplies to protect public health and, where necessary, demand improvements to be made. However, in practice, private supplies within the EU fall outside the regulatory system.

Drinking water standards in the US arise form the Safe Drinking Water Act 1974 and its subsequent amendments (Gray, 1994). Standards are split into two categories: primary and secondary. Primary drinking water standards are mandatory and cover those parameters considered to be potentially harmful to health. They are broken down into clarity, microbiological, organic contaminants, inorganic contaminants and radionuclides (Table 8.12). Secondary drinking water standards are non-mandatory and cover parameters that are not harmful to health but that affect the aesthetic quality of drinking water. These include colour, odour, chloride, copper, foaming

TABLE 8.12 US Primary Drinking Water Standards

Contaminants	Health effects	MCL (mg/l)	Sources
Organic chemicals			
Acrylamide	Probable cancer nervous system	TT[a]	Flocculants in sewage/waste-water treatment
Alachlor	Probable cancer	0.002	Herbicide on corn and soybeans; under review for cancellation
Aldicarb	Nervous system	0.003	Insecticide on cotton, potatoes, restricted in many areas due to groundwater contamination
Aldicarb sulphone	Nervous system	0.002	Degraded from aldicarb by plants
Altrazine	Reproductive and cardiac	0.003	Widely used herbicides on corn and on non-crop land
Benzene	Cancer	0.005	Fuel (leaking tanks); solvent commonly used in manufacture of industrial chemicals, pharmaceuticals, pesticides, paints and plastics
Carbofuran	Nervous system and reproductive system	0.04	Soil fumigant/insecticide on corn/cotton; restricted in some areas
Carbon tetrachloride	Possible cancer	0.005	Commonly used in cleaning agents, industrial wastes from manufacture of coolants
Chlordane	Probable cancer	0.002	Soil insecticide for termite control on corn, potatoes; most uses cancelled in 1980
2.4-D	Liver, kidney, nervous system	0.07	Herbicide for wheat, corn, rangelands
Dibromochloropropane (DBCP)	Probable cancer	0.0002	Soil fumigant on soybeans cotton; cancelled 1977
Dichlorobenzene *p*-	Possible cancer	0.075	Used in insecticides, mothballs, air deodorizers
Dichlorobenzene *o*-	Nervous system lung, liver, kidney	0.6	Industrial solvent; chemical manufacturing
Dichloroethane (1,2-)	Possible cancer	0.005	Used in manufacture of insecticides, gasoline
Dichloroethylene (1,1-)	Liver/kidney effects	0.7	Used in manufacture of plastics, dyes, perfumes, paints, synthetic organic chemicals (SOCs)
Dichloroethylene (*cis*-1,2-)	Nervous system, liver, circulatory	0.07	Industrial extraction solvent
Dichloroethylene (*trans*-1,2-)	Nervous system liver, circulatory	0.01	Industrial extraction solvent
Dichloropropane (1,2-)	Probable cancer, liver, lungs, kidney	0.005	Soil fumigant, industrial solvent
Endrin	Nervous system/kidney effects	0.0002	Insecticides used on cotton, small grains, orchards (cancelled)
Epichlorohydrin	Probable cancer, liver, kidneys, lungs	TT[a]	Epoxy resins and coatings, flocculants used in treatment
Ethylbenzene	Kidney, liver, nervous system	0.7	Present in gasoline and insecticides; chemical manufacturing
Ethylene dibromide (EDB)	Probable cancer	0.00005	Gasoline additive; soil fumigant, solvent cancelled in 1984; limited uses continue
Heptachlor	Probable cancer	0.0004	Insecticide on corn: cancelled in 1983 for all but termite control
Heptachlor epoxide	Probable cancer	0.0002	Soil and water organisms convert heptachlor to the epoxide
Lindane	Nervous system, liver, kidney	0.0002	Insecticide for seed/lumber/livestock pest control: most uses restricted in 1983
Methoxychlor	Nervous system, liver, kidney	0.04	Insecticide on alfalfa, livestock

Contaminant	MCL	Potential health effects	Sources of contaminant
Monochlorobenzene	0.1	Kidney, liver, nervous system	Pesticide manufacturing; metal cleaner; industrial solvent
Pentachlorophenol	0.001	Probable cancer, liver, kidney	Wood preservative and herbicide; non-wood uses banned in 1987
Polychlorinated byphenyls (PCBs)	0.005	Probable cancer	Electrical transformers, plasticizers; banned in 1979
Styrene	0.1	Liver, nervous system	Plastic manufacturing; resins used in water treatment equipment
Tetrachloroethylene	0.005	Probable cancer	Dry-cleaning/industrial solvent
Toluene	1	Kidney, nervous system, lung	Chemical manufacturing; gasoline additive; industrial solvent
Toxaphene	0.003	Probable cancer	Insecticide/herbicide for cotton, soybeans; cancelled in 1982
2-4-5-TP (Silvex)	0.05	Nervous system, liver, kidney	Herbicide on rangelands, sugar cane, golf courses; cancelled in 1983
Total trihalomethanes (TTMH) (chloroform, bromoform, bromodichloromethane, dibromochloromethane)	0.1	Cancer risk	Primarily formed when surface water containing organic matter is treated with chlorine
Trichloroethane (1,1,1)	0.2	Nervous system problems	Used in manufacture of food wrappings, synthetic fibres
Trichloroethylene	0.005	Possible cancer	Waste from disposal of dry-cleaning materials and manufacturing of pesticides, paints, waxes and varnishes, paint stripper, metal degreaser
Vinyl chloride	0.002	Cancer risk	Polyvinyl chloride pipes and solvents used to joint them; industrial waste from manufacture of plastics and synthetic rubber
Xylenes	10	Liver, kidney, nervous system	Paint/ink solvent; gasoline refining by-product, components of detergents
Inorganic chemicals			
Arsenic	0.01	Dermal and nervous system toxicity effects	Geological, pesticide residues, industrial waste and smelter operations
Asbestos	7 MFL	Benign tumours	Natural mineral deposits; also in asbestos/cement pipe
Barium	2	Circulatory system	Natural mineral deposits; oil/gas drilling operations; paint and other industrial uses
Cadmium	0.005	Kidney	Natural mineral deposits; metal finishing; corrosion product plumbing
Chromium	0.1	Liver/kidney, skin and digestive system	Natural mineral deposits; metal finishing, textile, tanning and leather industries
Copper	TTa	Stomach and intestinal distress; Wilson's disease	Corrosion of interior household and building pipes
Fluoride	4	Skeletal damage	Geological; additive to drinking water; toothpaste; foods processed with fluorinated water
Lead	TT[a]	Central and peripheral nervous system damage; kidney; highly toxic to infants and pregnant women	Corrosion of lead solder and brass faucets and fixtures; corrosion of lead service lines

(Continued)

Table 8.12 (*Continued*)

Contaminants	Health effects	MCL (mg/l)	Sources
Mercury	Kidney, nervous system	0.002	Industrial/chemical manufacturing; fungicide; natural mineral deposits
Nitrate	Methaemoglobinaemia, 'blue-baby syndrome'	10	Fertilizers, feedlots, sewage, naturally in soil, mineral deposits
Nitrite	Methaemoglobinaemia, 'blue-baby syndrome'	1	Unstable, rapidly converted to nitrate prohibited in working metal fluids
Total (nitrate and nitrite)	Not applicable	10	Not applicable
Selenium	Nervous system	0.05	Natural mineral deposits; by-product of copper mining/smelting
Uranium	Cancer	0.03	Radioactive waste, geological/natural
Radionuclides			
Beta particle and photon activity	Cancer	4 mrem/year	Radioactive waste, uranium deposits, nuclear facilities
Gross alpha particle activity	Cancer	15 pCl/l	Radioactive waste, uranium deposits, geological/natural
Radium 226/228	Bone cancer	5 pCl/l	Radioactive waste, geological/natural
Microbiological			
Giardia Lamblia	Stomach cramps, intestinal distress (Giardiasis)	TT[a]	Human and animal faecal matter
Legionella	Legionnaires' disease (pneumonia), Pontiac fever	TT[a]	Water aerosols such as vegetable misters
Total coliforms	Not necessarily disease-causing themselves, coliforms can be indicators of organisms that can cause gastroenteric infections, dysentery, hepatitis, typhoid fever, cholera, and other. Also coliforms interfere with disinfection		Human and animal faecal matter
Turbidity	Interferes with disinfection	0.5–1.0 NTU[b]	Erosion, run-off, discharges
Viruses	Gastroenteritis (intestinal distress)	TT[a]	Human and animal faecal matter
Other substances			
Sodium	Possible increase in blood pressure in susceptible individuals	None (20 mg/l reporting level)	Geological, road salting

[a]Treatment technique requirement in effect.
[b]NTU, on nephelometric turbidity unit.
Reproduced from USEPA(1995) with permission of the US Environmental Protection Agency, Washington, DC.

agents, iron, manganese, sulphate, zinc, conductivity, total dissolved solids (TDS), pH, hardness, sodium, calcium, potassium, magnesium, boron and nitrite.

The Office of Drinking Water at the US Environmental Protection Agency (USEPA) uses two standards for primary parameters. The maximum contaminant level (MCL) is the enforceable standard, while the MCL guide (MCLG) is a non-enforceable standard set at a level at which no known or anticipated adverse health effect occurs. For example, for all known carcinogens the USEPA sets the MCLG at zero as a matter of policy, based on the theory that there exists some risk of cancer, albeit very small, at any exposure level. The MCL is set as close to the MCLG as possible taking technological and cost factors into considerations. The primary drinking water standards must be complied with by all States, although in practice many States set more stringent values for certain parameters. The standards cover public supplies only, which are defined as those serving 25 or more people or from which at least 15 service connections are taken (American Water Works Association, 1990).

8.2 SURFACE WATERS NOT USED PRIMARILY FOR SUPPLY

The ecological and chemical status of surface waters will in the future be protected and managed by setting environmental quality and emission standards in RBMPs under the WFD. Five ecological status classifications are used for water quality of surface waters, high, good, moderate, poor and bad. Ecological status is based on a wide range of biological, physico-chemical and hydromorphological elements (i.e. parameters) with detailed descriptions of status classification for each surface water type, that is lakes, rivers, transitional waters and coastal waters, given in Annex V of the Directive. Guidance is given for each major quality element as to what comprises a particular status for each water type, although specific environmental quality and emission standards are set by Member States for each River Basin District (RBD) for water, sediment and biota. In setting EQSs for polluting substances (Section 8.3) acute and chronic toxicity test data should be obtained for each type of water body. From this safety factors will be set to protect the biota from dangerous substances. Guidance for toxicity assessment is given in the WFD (Annex V). Areas designated as protected areas within RBDs have been identified as requiring special protection, which in practice will mean more stringent or special standards may be required. A register of protected areas is required for all water bodies within an RBD (Table 8.4).

There is a transitional period until 2007 and then 2013 while RBMPs are adopted and become fully operational, respectively (Table 7.7). Existing legislation that is to be integrated into the WFD stays in force until it is repealed (Table 8.2). So during this transitional period surface water quality is maintained by the Directives described below.

The Freshwater Fish Directive (78/649/EEC) specifies 14 physico-chemical parameters in order to maintain water quality suitable for supporting and improving fish life (Table 8.13). The Directive does not apply to all surface waters, with Member States

TABLE 8.13 The standards specified for salmonid and cyprinid waters under the EC Freshwater Fish Directive (78/656/EEC)

Parameter	Salmonid		Cyprinid	
	G	I	G	I
Temperature (°C) (where there is an thermal discharge)	Temperature at edge of mixing zone must not exceed the unaffected temperature by more than:			
		1.5		3
	The temperature must not exceed:			
		21.5		28
		10		10
	The 10°C limit applies to breeding periods when needed			
Dissolved oxygen (mg/l O_2)	50% >9 100% >7	50% >9 When <6, must prove not harmful to fish population	50% >8 100% >5	50% >7 When <4, must prove not harmful to fish population
pH		6–9		6–9
Suspended solids (mg/l)	<25		<25	
BOD (mg/l)	<3		<6	
Total phosphorus	No G or I standards applicable			
Nitrites (mg/l NO_2)	<0.01		<0.03	
Phenolic compounds (mg/l C_6H_5OH)	Must not adversely affect fish flavour			
Petroleum hydrocarbons	Must not be present visibly, detectable by taste of fish, harmful to fish			
Non-ionized ammonia (mg/l NH_3)	<0.005	<0.025	<0.005	<0.025
Total ammonium (mg/l NH_4)	<0.04	<1	<0.2	<1
Total residual chlorine (mg/l HOCl)		<0.005		<0.005
Total zinc (mg/l Zn)	At water hardness >100 mg $CaCO_3$/l			
		<0.3		<1.0
	There are also limit values for hardness between 10 and 500 mg $CaCO_3$/l*			
Dissolved copper (mg/l Cu)	At water hardness >100 mg $CaCO_3$/l			
	<0.04		<0.04	
	There are also limit values at hardness between 10 and 500 mg $CaCO_3$/l*			

*Table 6.2
Reproduced with permission of the European Commission, Luxembourg.

designating specific sections of rivers or standing waters as either salmonid waters (suitable for salmon or trout) or cyprinid waters (suitable for roach, bream, chub, etc.). The mandatory (I) values are more stringent for salmonids reflecting their need for cleaner water (Chapter 7). The toxicity of heavy metals, in this case copper and zinc, varies with water hardness resulting in a variable standard. For example, in salmonid waters the I-value for zinc varies from 0.03 to 0.5 mg l^{-1} at water hardness of 10 and

$500\,\mathrm{mg}\,l^{-1}$ as $CaCO_3$, respectively (Table 6.2). In 1996 there were 2288 stretches of designated rivers in the UK, covering some 1221 km (Environment Agency, 1998). The problem of eutrophication in surface waters is now an issue dealt with under the WFD. However, both the Nitrate Directive (Section 8.4) and the Urban Wastewater Treatment Directive (Section 8.3) contain provisions for the control of eutrophication.

The Bathing Water Directive (76/160/EEC) sets microbiological and physico-chemical standards (both G and I) for inland and coastal waters designated by Member States as bathing areas (Table 8.14). However, since the introduction of this Directive 25 years ago epidemiological knowledge has significantly progressed making the currents standards inappropriate. The Directive has also been widely criticized for its inflexibility, inappropriate fixed standards, and poor management procedures, especially in the way bathing waters are designated (Gray, 2004). A revised Directive was proposed in October 2002 (COM(2002)581 Final) which provides improved health standards, more efficient management, including the active involvement of the public, and provides Member States greater flexibility in the way they implement the Directive. The proposed Directive is also fully integrated with the WFD.

In the 1976 Directive three microbial parameters were used, total coliforms, faecal coliforms and faecal streptococci. Two new microbial indicators, intestinal enterococci (IE) and *Escherichia coli* (EC), which provide better correlations between faecal pollution and health impacts in recreational waters, will replace these. Based on epidemiological studies a dose response relationship has been established between contamination risk and the 95th percentile value of IE for contracting either gastro-enteritis or upper respiratory illness by bathing in microbially contaminated water (Fig. 8.1). The 95th percentile value is calculated by taking the $\log_{10}$ value of all bacterial enumerations in the data sequence for the bathing water, calculating the arithmetic mean of the $\log_{10}$ values (μ) and then calculating the standard deviation (σ). The 95th percentile value is calculated using the equation:

$$\text{95th percentile} = \text{antilog}(\mu + 1.65\,\sigma) \tag{8.1}$$

A ratio of EC:IE of 2–3:1 can also be used to assess risk using this relationship. From these studies legally binding standards have been proposed in the new Directive. Good quality values for EC and EI, which are mandatory, and excellent quality values, which are guide values only, have been set (Table 8.15). It can be seen in Fig. 8.1 that the mandatory standard for good standard bathing waters gives rise to a risk of contracting either gastroenteritis or upper respiratory illness of 5% and 2.5%, respectively. These percentages drop to 3% and 1%, respectively, when excellent quality standards apply. Although these risk assessments are based on repetitive exposure to these levels of contamination, children and those with suppressed immune systems are at the greatest risk. One of the innovations of the proposed Directive is that microbial data should be made more widely available (i.e. via the Internet as well as at the site) and more quickly allowing bathers to make more objective decisions. The proposed limit values are a significant improvement over the

TABLE 8.14 EU Bathing Water (76/110/EEC) quality standards

Parameters	G (90th percentile)	I (95th percentile)	Minimum sampling frequency	Method of analysis and inspection
Microbiological				
Total coliforms/100 ml	500	10 000	Fortnightly[b]	Fermentation on multiple tubes. Subculturing of the positive tubes on a confirmation medium. Count according to most probable number (MPN) or membrane filtration and culture on an appropriate medium such as Tergitol lactose agar. 0.4% Teepol broth, subculturing and identification of the colonies
Faecal coliforms/100 ml	100	2000	Fortnightly[b]	In the case of total and faecal coliforms, the incubation temperature is variable according to whether total or faecal coliforms are being investigated
Faecal streptococci/100 ml	100	–	–[c]	Litsky method. Count according to MPN or filtration on membrane. Culture on an appropriate medium
Salmonella/l	–	0	–[c]	Concentration by membrane filtration. Inoculation on a standard medium. Enrichment – subculturing on isolating agar – identification
Entero viruses PFU/10l	–	0	–[c]	Concentration by filtration, flocculation or centrifuging and confirmation
Physico-chemical				
pH	–	6–9[a]	–[c]	Electrometry with calibration at pH 7 and 9
Colour	–	No abnormal change in colour[a]	Fortnightly[b]	Visual inspection or photometry with standards on the Pt. Co scale
Mineral oils mg/l	–	No film visible on the surface of the water and no odour	–[c]	Visual and olfactory inspection or extraction using an adequate volume and weighing the dry residue
Surface-active substances reacting with methylene blue mg/l (lauryl-sulphate)	–	No lasting foam	Fortnightly[b]	Visual inspection or absorption spectrometry with methylene blue
	≤0.3	–	–[c]	

Parameters	G	I	Minimum sampling frequency	Methods of analysis
Phenols (phenol indices) mg/l C_6H_5OH	– / ≤0.005	No specific odour / ≤0.005	Fortnightly[b]	Verification of the absence of specific odour due to phenol or absorption spectrophotometry 4-aminoantipyrine (4 AAP) method
Transparency m	–[c]	1[a]	Fortnightly[b]	Secchi's disc
Dissolved oxygen % saturation O_2	80–120	–	–[c]	Winkler's method or electrometric method (oxygen meter)
Tarry residues and floating materials such as wood, plastic articles, bottles, containers of glass, plastic, rubber or any other substance. Waste or splinters	Absence	–	Fortnightly[b]	Visual inspection
Ammonia mg/l NH_4	–	–	–[d]	Absorption spectrometry, Nessler's method or indophenol blue method
Nitrogen Kjeldahl mg/l N	–	–	–[d]	Kjeldahl method
Other substances regarded as indications of pollution				
Pesticides (parathion, HCH deldrin) mg/l	–	–	–[c]	Extraction with appropriate solvents and chromatographic determination
Heavy metals such as:				
Arsenic mg/l As	–	–	–[c]	} Atomic absorption possibly preceded by extraction (for As, Cd, Cr VI, Pb, Hg)
Cadmium mg/l Cd	–	–	–[c]	
Chrome mg/l CrVI	–	–	–[c]	
Lead mg/l Pb	–	–	–[c]	
Mercury mg/l Hg	–	–	–[c]	
Cyanides mg/l Cn	–	–	–[c]	Absorption spectrophotometry using a specific reagent
Nitrates mg/l NO_3	–	–	–[c]	Absorption spectrophotometry using a specific reagent
Phosphates PO_4	–	–	–[c]	Absorption spectrophotometry using a specific reagent

G: guide; 1: mandatory.

[a] Provision exists for exceeding the limit in the event of geographical or meteorological conditions.

[b] When a sample taken in the previous years produced results which are appreciably better than those in this table and when no new factor likely to lower the quality of the water has appeared, the competent authorities may reduce the sampling by a factor of 2.

[c] Concentration to be checked by the competent authorities when an inspection in the bathing area shows that the substance may be present or that the quality of the water has deteriorated.

[d] These parameters must be checked by the competent authorities when there is a tendency towards the eutrophication of the water.

Reproduced with permission of the European Commission, Luxembourg.

FIGURE 8.1
Risk of contracting gastroenteritis (GI) and upper respiratory illnesses (RI) based on bacteriological quality as measured by IE. (Reproduced from the proposed EU Bathing Water quality standards (COM(2002)581 Final) with permission of the European Commission, Luxembourg.)

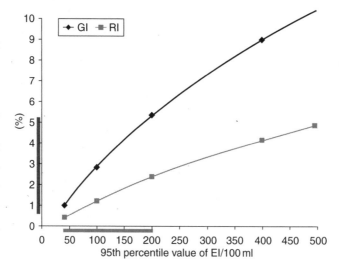

TABLE 8.15   Proposed EU Bathing Water quality standards (COM(2002)581 Final)

	Microbiological parameters	Excellent quality	Good quality	Methods of analysis
1	Intestinal enterococci (IE) in cfu/100 ml	100[a]	200[a]	ISO 7899
2	E. coli (EC) in cfu/100 ml	250[a]	500[a]	ISO 9308-1
3	Phytoplankton blooms or macroalgae proliferation[b]		Negative result on tests	Microscopic monitoring[c] toxicity tests[d], visual inspection

	Physico-chemical parameters	Excellent Quality	Good Quality	Methods of Inspection
4	Mineral oils	–	No film visible on the surface of the water and no odour	Visual and olfactory inspection
5	Tarry residues and floating materials such as wood, plastic, glass, rubber or any other waste substance	–	Absence	Visual inspection
6	pH[e]	–	6–9. No unexplainable variations	Electrometry with calibration on pH 7 and pH 9

[a]Based upon 95th percentile evaluation.
[b]Only for sites which have been revealed physically sensitive to specific toxic blooms (e.g. dinophysis, alexandrium, blue algae).
[c]Determination and counting of cells.
[d]Mouse test, skin test or by direct toxin dosage in plankton cells or water.
[e]Only for freshwaters.
Reproduced with permission of the European Commission, Luxembourg.

original Directive where the I (mandatory) standard implies a risk of 12–15% and the G (guide) standard implies a 5% risk of contracting gastroenteritis.

Another innovation of the proposed Directive is water quality at bathing sites will be classified on the basis of a 3-year trend and not on the basis of 1 year's results, as is currently the case. While this means that the classification will be less susceptible to bad weather conditions or one-off incidents, it also ensures a more accurate assessment of microbial quality overall. Where the quality of water is consistently good over a 3-year period the monitoring frequency may be reduced. Bathing water profiles will also be required in order to identify potential sources of contamination whether these originate from surface water or land. This is a basic management plan for the bathing area and must be updated every 3 years for excellent quality waters, every 2 years where quality is good and annually where the water is poor (i.e. fails). Modelling techniques are to be employed where bathing sites have a history of poor water quality, such as at times of unusually severe weather conditions, and preventive measures should be taken to close the bathing site when such weather conditions are predicted. The 17 physico-chemical parameters have been removed from the new Directive, as the chemical status of waters will in the future be carried out under the provisions of the WFD, which requires that good standards should be achieved for all waters. Bathing waters will, however, be required to be free from phytoplankton blooms or excessive macroalgal development, oil, tarry residues and floating litter (Table 8.15). Freshwater bathing areas are also required to have a neutral to alkaline pH (pH 6–9). The use of RBMPs will provide an integrated approach upstream that will improve the quality of downstream bathing areas.

Coastal and estuarine water quality is also controlled by the Shellfish Directive (79/923/EEC), which specifies 12 physico-chemical and microbiological water quality parameters. Also, where areas are designated for the protection of habitats or species the maintenance or improvement of the status of water may be an important factor in their protection. This includes the relevant Natura 2000 sites designated under the Habitats Directive (92/43/EEC) and the Birds Directive (79/409/EEC). Such areas are classified as protected areas in RBMPs under the WFD, requiring specific monitoring and control measures.

8.3 WASTEWATERS

The Dangerous Substances Directive (76/464/EEC) requires licensing, monitoring and control of a wide range of listed substances discharged to the aquatic environment. List I (black list) substances have been selected mainly on the basis of their toxicity, persistence and potential for bioaccumulation. Those that are rapidly converted into substances that are biologically harmless are excluded. List II (grey list) substances are considered to be less toxic, or the effects of which are confined to a limited area which is dependent on the characteristics and location of the water into which they are discharged (Table 8.16).

TABLE 8.16   List I and List II substances covered by the EC Dangerous Substances Directive
(76/464/EEC)

List I (black list)

Organohalogen compounds and substances which may form such compounds in the aquatic environment

Organophosphorus compounds

Organotin compounds

Substances, the carcinogenic activity of which is exhibited in or by the aquatic environment (substances in List II which are carcinogenic are included here)

Mercury and its compounds

Cadmium and its compounds

Persistent mineral oils and hydrocarbons of petroleum

Persistent synthetic substances

List II (grey list)

The following metalloids/metals and their compounds:

Zinc, copper, nickel, chromium, lead, selenium, arsenic, antimony, molybdenum, titanium, tin, barium, beryllium, boron, uranium, vanadium, cobalt, thalium, tellurium, silver

Biocides and their derivatives not appearing in List I

Substances which have a deleterious effect on the taste and/or smell of products for human consumption derived from the aquatic environment and compounds liable to give rise to such substances in water

Toxic or persistent organic compounds of silicon and substances which may give rise to such compounds in water, excluding those which are biologically harmless or are rapidly converted in water to harmless substances

Inorganic compounds of phosphorus and elemental phosphorus

Non-persistent mineral oils and hydrocarbons of petroleum origin

Cyanides, fluorides

Certain substances which may have an adverse effect on the oxygen balance, particularly ammonia and nitrites

Reproduced with permission of the European Commission, Luxembourg.

Under Article 16 of the WFD a new list of priority substances has been drawn up which has become Annex X of the Directive. This list will eventually replace the existing Dangerous Substances Directive (76/464/EEC) and aims to protect surface water from the chemical pollutants that represent the most dangerous and damaging substances threatening European waters. The substances in the list include selected chemicals, plant protection products, biocides, metals and other groups such as PAH that are mainly by products of incineration, and polybrominated biphenylethers (PBDE), which are used as flame retardants (Table 8.17). The list, which was adopted in November 2003 (Decision No. 2455/2001/EC), contains 33 individual or groups of substances of which 11 have been identified as priority hazardous substances. All priority hazardous substances must be removed from all discharges or emissions within a reasonable time that must not exceed the year 2023. Fourteen further listed substances are subject to review for inclusion into the priority hazardous substances category (Table 8.17).

TABLE 8.17 Priority substances listed in Annex X of the WFD

	CAS number	EU number	Name of priority substance	Identified as priority hazardous substance
(1)	15972-60-8	240-110-8	Alachlor	
(2)	120-12-7	204-371-1	Anthracene	(Possible)[c]
(3)	1912-24-9	217-617-8	Atrazine	(Possible)[c]
(4)	71-43-2	200-753-7	Benzene	
(5)	n.a.	n.a.	Brominated diphenylethers[b]	Yes[d]
(6)	7440-43-9	231-152-8	Cadmium and its compounds	Yes
(7)	85535-84-8	287-476-5	C_{10-13}-chloroalkanes[b]	Yes
(8)	470-90-6	207-432-0	Chlorfenvinphos	
(9)	2921-88-2	220-864-4	Chlorpyrifos	(Possible)[c]
(10)	107-06-2	203-458-1	1,2-Dichloroethane	
(11)	75-09-2	200-838-9	Dichloromethane	
(12)	117-81-7	204-211-0	Di(2-ethylhexyl)phthalate	(Possible)[c]
(13)	330-54-1	206-354-4	Diuron	(Possible)[c]
(14)	115-29-7	204-079-4	Endosulfan (alpha-endosulfan)	(Possible)[c]
(15)	959-98-8	n.a.		
(16)	206-44-0	205-912-4	Fluoranthene[e]	
(17)	118-74-1	204-273-9	Hexachlorobenzene	Yes
(18)	87-68-3	201-765-5	Hexachlorobutadiene	Yes
	608-73-1	210-158-9	Hexachlorocyclohexane (gamma-isomer, Lindane)	Yes
(19)	58-89-9	200-401-2		
(20)	34123-59-6	251-835-4	Isoproturon	(Possible)[c]
(21)	7439-92-1	231-100-4	Lead and its compounds	(Possible)[c]
(22)	7439-97-6	231-106-7	Mercury and its compounds	Yes
(23)	91-20-3	202-049-5	Naphthalene	(Possible)[c]
(24)	7440-02-0	231-111-4	Nickel and its compounds	
(25)	25154-52-3	246-672-0	Nonylphenols (4-(para)-nonylphenol)	Yes
	104-40-5	203-199-4		
(26)	1806-26-4	217-302-5	Octylphenols (para-tert-octylphenol)	(Possible)[c]
(27)	140-66-9	n.a.		
(28)	608-93-5	210-172-5	Pentachlorobenzene	Yes
	87-86-5	201-778-6	Pentachlorophenol	(Possible)[c]
	n.a.	n.a.	PAHs	Yes
	50-32-8	200-028-5	(Benzo(a)pyrene),	
	205-99-2	205-911-9	(Benzo(b)fluoroanthene),	

(Continued)

TABLE 8.17 (Continued)

	CAS number	EU number	Name of priority substance	Identified as priority hazardous substance
	191-24-2	205-883-8	(Benzo(g,h,i)perylene),	
	207-08-9	205-916-6	(Benzo(k)fluoroanthene),	
	193-39-5	205-893-2	(Indeno(1,2,3-cd)pyrene)	(Possible)[c]
(29)	122-34-9	204-535-2	Simazine	Yes
(30)	688-73-3	211-704-4	Tributyltin compounds	
	36643-28-4	n.a.	(Tributyltin-cation)	
(31)	12002-48-1	234-413-4	Trichlorobenzenes	(Possible)[c]
	120-82-1	204-428-0	(1,2,4-Trichlorobenzene)	
(32)	67-66-3	200-663-8	Trichloromethane (Chloroform)	
(33)	1582-09-8	216-428-8	Trifluralin	(Possible)[c]

[a]Where groups of substances have been selected, typical individual representatives are listed as indicative parameters (in brackets and without number). The establishment of controls will be targeted to these individual substances, without prejudicing the inclusion of other individual representatives, where appropriate.

[b]These groups of substances normally include a considerable number of individual compounds. At present, appropriate indicative parameters cannot be given.

[c]This priority substance is subject to a review for identification as possible 'priority hazardous substance'. The Commission will make a proposal to the European Parliament and Council for its final classification not later than 12 months after adoption of this list. The timetable laid down in Article 16 of Directive 2000/60/EC for the Commission's proposals of controls is not affected by this review.

[d]Only Pentabromobiphenylether (CAS number 32534-81-9).

[e]Fluoranthene is on the list as an indicator of other, more dangerous PAHs.

There is a transitional period during which Annex X of the WFD will replace the Dangerous Substances Directive with the old Directive repealed in 2013. Article 16 of the WFD lays down new procedures for the identification of substances and new control measures at a river basin level. As part of these procedures community wide emission controls and quality standards for all priority substances are being prepared to ensure good chemical surface water status by 2015.

Member States are in the process of establishing EQSs for surface and groundwaters. These will be used during the production of RBMPs as maximum permissible concentrations in waters receiving discharges containing such compounds (Table 8.18). Emission limit values (ELVs) and quality objectives for a number of substances were set in amendments to the Directive (76/464/EEC) and unless changed in the RBMPs they remain in force. These include the Mercury Discharges Directive (82/176/EEC), Cadmium Discharges Directive (83/513/EEC), the Mercury Directive (84/156/EEC), Hexachlorocyclohexane Directive (84/491/EEC) and the Dangerous Substances Discharges Directive (86/280/EEC).

The Urban Wastewater Treatment Directive (91/271/EEC) makes secondary treatment mandatory for sewered domestic wastewaters and also biodegradable industrial (e.g. food processing) wastewaters. Minimum effluent standards have been set at biochemical oxygen demand (BOD) $25 \, \mathrm{mg} \, l^{-1}$, chemical oxygen demand (COD) $125 \, \mathrm{mg} \, l^{-1}$ and suspended solids $35 \, \mathrm{mg}^{-1}$. Those receiving waters that are considered to be at risk from eutrophication are classified as sensitive so that discharges require more stringent treatment to bring nutrient concentrations of effluents down to a maximum total phosphorus concentration of $2 \, \mathrm{mg} \, l^{-1}$ as phosphorus and a total nitrogen concentration of $10–15 \, \mathrm{mg} \, l^{-1}$ as nitrogen (Table 14.3). Owing to the cost of nutrient removal, the designation of receiving waters as sensitive has significant cost implications for Member States. Strict completion dates have been set by the Commission for the provision of minimum treatment for wastewaters entering freshwater, estuaries and coastal waters. For example, full secondary treatment including nutrient removal for all discharges to sensitive waters with a population equivalent (pe) $>10\,000$ must have been completed by the end of 1998. By the 31 December 2005 all wastewaters from population centres $<2000 \, \mathrm{pe}$ discharged to freshwaters, and $<10\,000 \, \mathrm{pe}$ to coastal waters must have sufficient treatment to allow receiving waters to meet EQSs, while populations centres larger than these require secondary treatment (Fig. 14.1). The Directive also requires significant changes in the disposal of sewage sludge including:

(a) that sludge arising from wastewater treatment shall be reused whenever possible and that disposal routes shall minimize adverse effects on the environment;
(b) competent authorities shall ensure that before 31 December 1998 that the disposal of sludge from wastewater treatment plants is subject to general rules (i.e. codes of practice) or legislation;

Table 8.18 British EQSs for List I and List II substances

List I substances	Statutory EQS[a] ($\mu g/l$)	Number of discharges	List II substances	Operational EQS[b] ($\mu g/l$)	Measured as
Mercury and compounds	1	752	Lead	10	AD
			Chromium	20	AD
Cadmium and compounds	5	2196	Zinc	75	AT
			Copper	10	AD
Hexachlorocyclohexane (all isomers)	0.1	123	Nickel	150	AD
			Arsenic[c]	50	AD
DDT (all isomers)	0.025	15	Boron	2000	AD
DDT (pp isomers)	0.01	1	Iron	1000	AT
Pentachlorophenol	2	88	pH	6.0–9.0	AD
Carbon tetrachloride	12	51	Vanadium	0	AT
Aldrin	0.01	35	Tributylin[c]	0.02	MT
Dieldrin	0.01	58	Triphenyltin[c]	0.02	MT
Endrin	0.005	37	PCSD	0.05	PT
Isodrin	0.005	7	Cyfluthrin	0.001	PT
Hexachlorobenzene	0.03	20	Sulcofuron	25	PT
Hexachlorobutadiene	0.1	14	Flucofuron	1	PT
Chloroform	12	73	Permethrin	0.01	PT
Trichloroethylene	10	48	Atrazine and simazine[c]	2	A
Tetrachloroethylene	10	51	Azinphos-methyl[c]	0.01	A
Trichlorobenzene	0.4	31	Dichlorvos	0.001	A
1,2-dichloroethane	10	87	Endosulphan[c]	0.003	A
			Fenitrothion	0.01	A
			Malathion[c]	0.01	A
			Trifluralin[c]	0.1	A
			Diazinon	0.01	A
			Propetamphos	0.01	A
			Cypermthrin	0.0001	A
			Isoproturon	2.0	A

A: annual average; P: 95% of samples; D: dissolved; T: total; M: maximum.

[a]Standards are all annual mean concentrations.

[b]Standards quoted for metals are for the protection of sensitive aquatic life at hardness 100–150 mg/l $CaCO_3$, alternative standards may be found in DoE circular 7/89.

[c]Standards for these substances are from the Surface Waters (Dangerous Substances) (Classification) Regulations 1997, SI 2560 in which case these are now statutory.

Reproduced from Environment Agency (1998) with permission of the Environment Agency, London.

(c) the disposal of sludge to surface waters by dumping from ships or discharge from pipelines or other means shall be phased out by 31 December 1998;

(d) that the total amount of toxic, persistent or bioaccumable material in sewage sludge is progressively reduced.

This wide scoping legislation is considered in more detail in Chapters 14 and 21. The disposal options for sewage sludge is further limited if it contains metals or listed substances, which may categorize it as a hazardous waste under the EU Directive on Hazardous Waste (91/689/EEC).

8.4 GROUNDWATERS

In principle both surface and ground waters are both renewable natural resources, but due to the natural time lag in its formation and renewal, groundwater is more complex to manage. As the impact of human activity, for example nitrates, pesticides, and industrial organic compounds, can last for relatively long periods in groundwater, it is imperative that pollutants should be prevented from entering aquifers in the first place. In order to achieve protective and remedial goals for groundwaters, long-term planning and extended timetables for specific achievements in quality are needed.

The existing Groundwater Directive (80/68/EEC) protects groundwater against direct or indirect pollution caused by certain substances listed in the Dangerous Substances Directive (76/464/EEC) through management controls. List I substances must be prevented from entering groundwaters, while pollution by List II substances must be carefully controlled. The Directive requires strict hydrological and environmental impact assessments to be carried out before listed substances can be licensed for disposal where they may enter an aquifer. However, the Directive does not set any clear quality objectives, nor does it require comprehensive monitoring. The WFD requires all waters in the EU to be of good status, and requires specific measures to be taken to prevent and control groundwater pollution and achieve good groundwater chemical status. These measures have to include criteria and methods for assessing the chemical status of groundwater and for identifying trends in pollution of groundwater bodies. To achieve these objectives a new Groundwater Directive was proposed on 19 September 2003 (COM(2003)550 Final) to protect groundwater from pollution. This introduces, for the first time, quality objectives, obliging Member States to monitor and assess groundwater quality on the basis of common criteria and to identify and reverse trends in groundwater pollution. The proposed Directive will also ensure that groundwater quality is monitored and evaluated across Europe in a harmonized way. There is also a serious lack of understanding of how groundwater becomes polluted and about overall chemical quality of European aquifers. This new Directive will address these deficiencies so that European groundwaters can be safeguarded from pollution in the future, improved where necessary and allow aquifers to be used sustainably. There will be a transitional period for its implementation with the original Groundwater Directive (80/68/EEC) repealed by 22 December 2013.

Compliance with good chemical status will be based on a comparison of the monitoring data collected for each aquifer within RBDs with quality standards. Quality standards covering nitrates, plant protection and biocidal products are already prescribed in existing in EU legislation. These set threshold values (i.e. maximum permissible concentrations) in groundwater for a number of pollutants. For polluting substances not covered by existing EU legislation, the proposed Directive requires Member States to establish threshold values by 22 June 2006. This poses a major technical problem, as groundwater characteristics are so variable across the EU, requiring threshold values to be defined at national, river basin or groundwater body level. This makes the setting of quality criteria very flexible, taking account of local characteristics and allowing for further improvements. Monitoring and implementation of the Directive will be through RBMPs set up under the WFD.

Two threshold values are given in the Directive. Nitrate at $50\,mg\,l^{-1}$, except those groundwaters identified under the Nitrate Directive (91/676/EC) in nitrate vulnerable zones (NVZs) where more stringent threshold values are set under the WFD. Also, active ingredients in pesticides, including their relevant metabolites, degradation and reaction products are set at $1\,\mu g\,l^{-1}$, based on the Plant Protection Products Directive (91/414/EC) and the Biocidal Products Directive (98/8/EC). Groundwater protection also features in other European legislation and policy documents: Drinking Water Directive (80/778/EC as amended 98/83/EC); Landfill Directive (99/3/EC), Construction Products Directive (89/106/EEC), Integrated Pollution Prevention and Control Directive (96/61/EC) and the Communication Towards a Thematic Strategy for Soil Protection (COM(2002)179 Final). Annex III of the proposed Groundwater Directive contains a list of eight substances and ions for which Member States must set threshold values. This is seen as the minimum number of substances for which quality standards must be set and it is expected that individual lists of threshold values will contain many more polluting substances. Those substances listed in Annex III that can occur naturally as well as being anthropogenic in origin, include ammonium, arsenic, cadmium, chloride, lead and sulphate. Two synthetic substances that do not naturally occur in groundwaters, trichloroethylene and tetrachloroethylene, are also listed.

An interesting aspect of the new Directive is the identification and reversal of pollution trends. Where a significant and sustained upward trend in a pollutant concentration is identified action must be taken to reverse this trend. The Directive defines trends in both time and environmental significance, from which a starting point for remedial action can be identified to reverse the trend. Environmental significance is defined as the point at which the concentration of a pollutant starts to threaten or worsen the quality of groundwater. It is set at 75% of the quality standard or the threshold value defined by Member States. Upward trends are established over very long monitoring periods (5–15 years), with reversal of trends having to be demonstrated over a minimum of 10 years and a maximum of 30 years. The protocol for the establishment of upward sustained trends and trend reversal is outlined in Annex IV.

In order to protect groundwater it is important to maintain control over indirect discharges of hazardous substances and the new Directive contains provisions that prohibit or limit such discharges. By including quality objectives, the effects of the discharges can be monitored and future risks can be assessed. Monitoring results obtained through the application of the proposed Directive will be used to design the measures to prevent or limit pollution of groundwater. The proposed action required to achieve good quality groundwater status must be published by 2008 in the Draft RBMP (Table 7.7). Measures to prevent or limit pollution of water, including groundwater, must be operational under the WFD by 2012.

The Nitrate Directive was set up to protect surface freshwaters, estuaries, coastal waters and groundwaters from pollution by nitrates arising from agricultural sources (91/676/EEC). The objective is to reduce the problem of eutrophication in surface waters and to limit the concentration of nitrate in drinking water from both surface and ground water sources. The legislation requires Member States to control pollution by:

(a) Encouraging good agricultural practice through the provision of codes of practice (Section 11.1). This also includes the application of fertilizers and manures, including sewage sludge.
(b) Identifying vulnerable aquifers and setting up action programmes. Where there is a risk of ground or surface waters exceeding a nitrate concentration of $50\,\mathrm{mg\,l^{-1}}$ those aquifers or catchments must be designated as NVZs. An action plan is required for each NVZ to reduce the input of nitrate and phosphate from both diffuse and point sources. For diffuse sources this is achieved mainly through the introduction of codes of practice. Some 55% of England is now designated as NVZs (Fig. 8.2) compared to 100% in Austria, Denmark, Germany, Luxembourg and the Netherlands (EC, 1998). Under the Urban Wastewater Directive, an NVZ can also be designated as a sensitive area, requiring wastewater treatment plants within these catchment areas to incorporate nutrient removal technologies.
(c) Monitoring nitrate levels and trends in surface and ground waters.

The Directive does not itself set mandatory standards but relies on other legislation. The Surface Water Directive (75/440/EEC) specifies a mandatory limit of $50\,\mathrm{mg\,l^{-1}}$ as NO_3 in 95% of samples, while the Drinking Water Directive (80/778/EEC) sets the same mandatory value for human consumption via water. So both the Urban Waste Water Treatment Directive (91/271/EEC) and the Nitrate Directive identify this level as being the critical concentration for nitrate requiring pollution control action. The Nitrate Directive forms an important part of European water policy and remains in force alongside the WFD (Fig. 7.4).

8.5 REGULATION

The regulation of water resources varies from country to country and even within the European Union, where the national legislation is controlled by EU Directives, significant differences exist. The regulatory mechanisms to implement the ever-increasing volume of environmental legislation is constantly evolving. In the past decade the structure of the regulatory bodies in most countries has significantly changed, and this is especially the case in the British Isles. The management of water resources, the disposal of wastewaters, and the provision of drinking water in England and Wales is split between the private water undertakers and a number of regulatory bodies.

Ten water service companies supply drinking water to consumers and treat their sewage. A further 29 small water supply companies also provide localized water supplies but

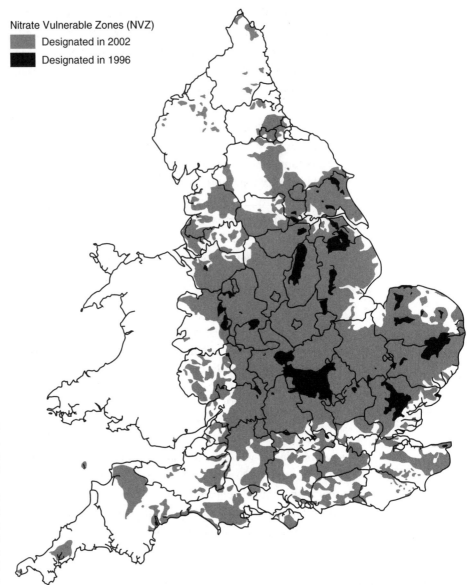

Nitrate Vulnerable Zones (NVZ)
Designated in 2002
Designated in 1996

FIGURE 8.2
NVZs as designated in England and Wales in 1996 and extended in 2002. (Reproduced from Environment Agency (1998) with permission of the Ministry of Agriculture, Fisheries and Food, London.)

rely on the major service companies to provide sewage treatment (Table 8.19). The Drinking Water Inspectorate regulates the quality of drinking water supplied, although the wholesomeness of the water is the sole responsibility of the privatized water companies. The standards of service delivered by the water companies, including the price of water, are regulated by the Office of Water Services (OFWAT) (Gray, 1994).

The Environment Act 1995 created the Environment Agency for England and Wales which carries out the functions of the former National Rivers Authority (NRA) and Her Majesty's Inspectorate of Pollution (HMIP); and the Scottish Environment Protection Agency (SEPA) which now carries out the functions of the River Purification Authorities and the HMIP. The principal aim of the Environment

TABLE 8.19 Water undertakers in England and Wales by region

Anglia region	*South west region*
Anglian Water Services Ltd	South West Water Services Ltd
Cambridge Water Company	
East Anglian Water Company	*Thames region*
Essex Water Company	Thames Water Utilities Ltd
Tendring Hundred Water Company	Colne Valley Water plc
	East Surrey Water plc
Northumbria region	Lee Valley Water plc
Northumbrian Water Ltd	Mid-Southern Water Company
Hartlepool's Water Company	North Surrey Water Company
Newcastle and Gateshead Water Company	Rickmansworth Water plc
Sunderland and South Shields Water Company	Sutton District Water plc
North west region	*Wales region*
North West Water Ltd	Dŵr Cymru (Welsh Water)
	Chester Waterworks Company
Severn Trent region	Wrexham and East Denbighshire
Severn Trent Water Ltd	Water Company
East Worcestershire Waterworks Company	
South Staffordshire Waterworks Company	*Wessex region*
	Wessex Water Services Ltd
Southern region	Bournemouth and District Water Company
Southern Water Services Ltd	Bristol Waterworks Company
Eastbourne Water Company	West Hampshire Water Company
Folkestone Water Company	
Mid-Kent Water Company	*Yorkshire region*
Mid-Sussex Water Company	Yorkshire Water Services Ltd
Portsmouth Water plc	York Waterworks plc
West Kent Water Company	

Each region has sewage and wastewater treatment facilities provided by a single water service company, although there may be several water supply companies.

Agency as stated in Section 4 of the Act is 'to protect or enhance the environment taken as a whole, as to make the contribution towards attaining the objective of achieving sustainable development'. The Agency is responsible for controlling water pollution, regulating waste and overseeing integrated pollution control (IPC) licensing. It also designates water protection zones, protects groundwater and sets water quality objectives. So it protects and improves the aquatic environment by catchment management and setting consents for all discharges including those from industry, agriculture, water company sewage treatment plants and local authorities. The consents can specify discharge limits for any number of parameters including the rate of effluent discharge. The Agency also levies an annual charge on dischargers based on the volume and content of the discharge and the nature of the receiving water. The setting of new environmental quality objectives through RBMPs for surface waters will tighten existing consent conditions even further (Table 8.18). As well as its pollution control functions, the Agency also has the responsibility to: issue licences for water abstraction; promote the conservation and enhancement of freshwaters; to promote the recreational use of freshwater; improve and develop fisheries and regulate them; to issue flood warnings and the provision of defences to reduce the risk of sea and river flooding; the issuing of land drainage consents and many other tasks.

8.6 INTEGRATED POLLUTION CONTROL (IPC)

The Integrated Pollution Prevention and Control (IPPC) Directive (96/61/EEC) was adopted in September 1996 and will impose a system of IPC throughout the European Union by 1999. IPC is a major advance in pollution control in that all discharges and environmental effects to water, air and land are considered together with the best practicable environmental option (BPEO) selected for disposal. In this way pollution problems are solved rather than transferred from one part of the environment to another. In the past licensing of one environmental media (i.e. air, water or land) created an incentive to release emissions to another. IPC also minimizes the risk of emissions crossing over into other environmental media after discharge (e.g. acid rain, landfill leachate). There is only one licence issued under IPC covering all aspects of gaseous, liquid, solid waste and noise emissions, so that the operator has to only make one application as well as ensuring consistency between conditions attached to the licence in relation to the different environmental media. In Britain IPC was introduced by the Environmental Protection Act 1990. It is operated in England and Wales by the Environment Agency and applies to the most complex and polluting industries and substances (e.g. large chemical works, power stations, etc.). The Agency issues guidance for such processes to ensure that the BPEO is carried out. The aim of IPC is to minimize the release of listed substances and to render substances that are released harmless using best available techniques not entailing excessive cost (BATNEEC). The objective of the guidance notes is to identify the types of techniques that will be used by the Agency to define BATNEEC for a particular process. The BATNEEC identified is then used as a base for setting ELVs. Unlike previous practice, in the identification of BATNEEC emphasis is placed on pollution prevention techniques such as cleaner technologies and waste minimization rather than end-of-pipe treatment. Other factors for improving emission quality include in-plant changes, raw material substitution, process recycling, improved material handling and storage practices. Apart from the installation of equipment and new operational procedures to reduce emissions, BATNEEC also necessitates the adoption of an on-going programme of environmental management and control which should focus on continuing improvements aimed at prevention, elimination and progressive reduction of emissions.

The selection of BATNEEC for a particular process takes into account:

(a) the current state of technical knowledge;
(b) the requirements of environmental protection;

(c) the application of measures for these purposes which do not entail excessive costs, having regard to the risk of significant environmental pollution.

For existing facilities the Agency considers:

(a) the nature, extent and effect of the emissions concerned;
(b) the nature and age of the existing facilities connected with the activity and the period during which the facilities are likely to be used or to continue in operation;
(c) the costs which would be incurred in improving or replacing these existing facilities in relation to the economic situation of the industrial sector of the process considered.

So while BATNEEC guidelines are the basis for setting licence emission standards, other factors such as site-specific environmental and technical data as well as plant financial data are also taken into account. In Ireland similar IPC licensing procedures are operated by the Environmental Protection Agency (EPA), and like the Environment Agency in England and Wales public registers of all licences are maintained. The processes requiring IPC licences in Ireland are listed in Table 8.20.

TABLE 8.20   Industries for which an IPC licence is required in Ireland

Minerals and other materials
The extraction, production and processing of raw asbestos
The extraction of aluminium oxide from an ore
The extraction and processing (including size reduction, grading and heating) of minerals within the meaning of the Minerals Development Acts 1940 to 1979, and storage of related mineral waste
The extraction of peat in the course of business which involves an area exceeding 50 ha

Energy
The production of energy in combustion plant, the rated thermal input of which is equal to or greater than 50 MW other than any such plant which makes direct use of the products of combustion in a manufacturing process
The burning of any fuel in a boiler or furnace with a nominal heat output exceeding 50 MW

Metals
The initial melting or production of iron and steel
The processing of iron and steel in forges, drawing plants and rolling mills where the production area exceeds $500 \, m^2$.
The production, recovery, processing or use of ferrous metals in foundries having melting installations with a total capacity exceeding 5 tonnes
The production, recovery or processing of non-ferrous metals, their compounds or other alloys including antimony, arsenic, beryllium, chromium, lead, magnesium, manganese, phosphorus, selenium, cadmium or mercury by thermal, chemical or electrolytic means in installations with a batch capacity exceeding 0.5 tonnes
The reaction of aluminium or its alloys with chlorine or its compounds
The roasting, sintering or calcining of metallic ores in plants with a capacity exceeding 1000 tonnes $year^{-1}$
Swaging by explosives where the production area exceeds $100 \, m^2$
The pressing, drawing and stamping of large castings where the production area exceeds $500 m^2$

(Continued)

TABLE 8.20 *(Continued)*

Boilermaking and the manufacture of reservoirs, tanks and other sheet metal containers where the production area exceeds $500\,m^2$

Mineral fibres and glass
The processing of asbestos and the manufacture and processing of asbestos-based products
The manufacture of glass fibre or mineral fibre
The production of glass (ordinary and special) in plants with a capacity exceeding 5000 tonnes $year^{-1}$
The production of industrial diamonds

Chemicals
The manufacture of chemicals in an integrated chemical installation
The manufacture of olefins and their derivatives or of monomers and polymers, including styrene and vinyl chloride
The manufacture, by way of chemical reaction processes, of organic or organo-metallic chemical products other than those specified in the row above
The manufacture of inorganic chemicals
The manufacture of artificial fertilizers
The manufacture of pesticides, pharmaceuticals or veterinary products and their intermediates
The manufacture of paints, varnishes, resins, inks, dyes, pigments or elastomers where the production capacity exceeds 1000 l per week
The formulation of pesticides
The chemical manufacture of glues, bonding agents and adhesives
The manufacture of vitamins involving the use of heavy metals
The storage in quantities exceeding the values shown, of any one or more of the following chemicals (others than as part of any other activity) – methyl acrylate (20 tonnes); acrylonitrile (20 tonnes); toluene di-isocyanate (20 tonnes); anhydrous ammonia (100 tonnes); anhydrous hydrogen fluoride (1 tonne)

Intensive agriculture
The rearing of poultry in installations, whether within the same complex or within 100 m of that complex, where the capacity exceeds 100 000 units have the following equivalents:
   1 broiler = 1 unit
   1 layer, turkey or other fowl = 2 units
The rearing of pigs in installations, whether within the same complex or within 100 m of that complex where the capacity exceeds 1000 units on gley soils or 3000 units on other soils and where units have the following equivalents:
   1 pig = 1 unit
   1 sow = 10 units

Food and drink
The manufacture of vegetable and animal oils and fats where the capacity for processing raw materials exceeds 40 tonnes day^{-1}
The manufacture of dairy products where the processing capacity exceeds 50 million gallons of milk equivalent per year
Commercial brewing and distilling, and malting in installations where the production capacity exceeds 100 000 tonnes $year^{-1}$
The slaughter of animals in installations where the daily capacity exceeds 1500 units and where units have the following equivalents:
1 sheep = 1 unit
1 pig = 2 units
1 head of cattle = 5 units
The manufacture of fish-meal and fish-oil
The manufacture of sugar
The rendering of animal by-products

TABLE 8.20 *(Continued)*

Wood, paper, textiles and leather
The manufacture of paper pulp, paper or board (including fibreboard particle board and plywood) in installations with a production capacity equal to or exceeding 25 000 tonnes of product per year
The manufacture of bleached pulp
The treatment or protection of wood, involving the use of preservatives, with a capacity exceeding 10 tonnes day^{-1}
The manufacture of synthetic fibres
The dyeing, treatment or finishing (including mothproofing and fireproofing) of fibres or textiles (including carpet) where the capacity exceeds 1 tonne day^{-1} of fibre, yarn to textile material
The fell-mongering of hides and tanning of leather in installations where the capacity exceeds 100 skins per day

Fossil fuels
The extraction, other than offshore extraction, of petroleum, natural gas, coal or bituminous shale
The handling or storage of crude petroleum
The refining of petroleum or gas
The pyrolysis, carbonization, gasification, liquefaction, dry distillation, partial oxidation or heat treatment of coal lignite, oil or bituminous shale, other carbonaceous materials or mixtures of any of these in installations with a processing capacity exceeding 500 tonnes day^{-1}

Cement
The production of cement

Waste
The incineration of hazardous waste
The incineration of hospital waste
The incineration of waste other than that mentioned in the two rows above in plants with a capacity exceeding 1 tonne h^{-1}
The use of heat for the manufacture of fuel from waste

Surface coatings
Operations involving coating with organo-tin compounds
The manufacture or use of coating materials in processes with a capacity to make or use at least 10 tonnes $year^{-1}$ of organic solvents, and powder coating manufacture with a capacity to produce at least 50 tonnes $year^{-1}$
Electroplating operations

Other activities
The testing of engines, turbines or reactors where the floor area exceeds 500 m^2
The manufacture of integrated circuits and printed circuit boards
The production of lime in a kiln
The manufacture of coarse ceramics including refractory bricks, stoneware pipes, facing and floor bricks and roof tiles

Since 1994 new companies require an IPC licence before commencing operation, while licensing for existing companies have been slowly phased in on a sectoral basis to coincide with the 1999 deadline in the IPPC Directive.

REFERENCES

American Water Works Association, 1990 *Water Quality and Treatment: A Handbook of Community Water Supplies*, 4th edn, McGraw-Hill, New York.
EC, 1998 *Measures Taken Persuant to Council Directive 91/676/EEC Concerning the Protection of Waters Against Pollution Caused by Nitrates from Agricultural Sources*, Report of the

Commission to the Council and European Parliament, Office for Official Publications of the European Communities, Luxembourg.

Environment Agency, 1998 *The State of the Environment of England and Wales: Fresh Waters*, The Stationery Office, London.

Environmental Protection Agency, 1994 *Integrated Pollution Control Licensing: Guide to Implementation and Enforcement in Ireland*, Environmental Protection Agency, Dublin.

Gray, N. F. 1994 *Drinking Water Quality: Problems and Solutions*, John Wiley and Sons, Chichester.

Gray, N. F. 2004 *Biology of Wastewater Treatment*, 2nd edn, Imperial College Press, London.

US Environmental Protection Agency, 1995 *National Primary Drinking Water Standards*, US Environmental Protection Agency, Washington, DC.

WHO, 1993a *Guidelines for Drinking Water Quality: 1. Recommendations*, 2nd edn, World Health Organization, Geneva.

WHO, 1993b *Guidelines for Drinking Water Quality: 2. Health Criteria and Other Supporting Information*, World Health Organization, Geneva.

WHO, 2004 *Guidelines for Drinking Water Quality: 1. Recommendations*, 3rd edn, World Health Organization, Geneva.

FURTHER READING

Cook, H. F. 1998 *The Protection and Conservation of Water Resources*, John Wiley and Sons, Chichester.

Johnstone, D. W. M. and Horan, N. J. 1994 Standards, costs and benefits: an international perspective, *Journal of the Institution of Water and Environmental Management*, **8**, 450–8.

Petts, J. and Eduljee, G. 1994 *Environmental Impact Assessment for Waste Treatment and Disposal Facilities*, John Wiley and Sons, Chichester.

9.1 PHYSICO-CHEMICAL AND BIOLOGICAL SURVEILLANCE

Water quality assessment is critical for pollution control and the protection of surface and ground waters. Water quality rarely remains static, so quality data are needed specifically because:

(a) quality varies in space;
(b) quality varies in time;
(c) waste loads vary at different points in the system;
(d) for effluent description;
(e) for setting consents, mass balance calculations and river modelling.

The main factors affecting variation in quality are dilution, water temperature causing variation in biological activity and oxygen solubility, and seasonal changes in waste inputs.

The selection of parameters for water quality assessment is dependent on the type of receiving water, the nature of the discharges into the receiving water, water use and any legal designation relating to the system. The key parameters for physico-chemical assessment of a selection of different water systems and effluents are given below (this does not include all the parameters that are required to be monitored under various legislation (Chapter 8)):

(a) river: biochemical oxygen demand (BOD), oxygen, temperature, NH_3, Cl, PO_4, NO_3, etc.;
(b) lake: oxygen (at various depths), temperature (at various depths), PO_4, NO_3, SiO_2, Fe, Mn, Na, K, etc.;
(c) estuary: oxygen, temperature, BOD, suspended solids;
(d) drinking water: coliforms, Fe, Mn, toxic metals, pesticides, etc.;

TABLE 9.1 Comparison of attributes of physico-chemical and biological assessment of water quality

Attribute	Physico-chemical	Biological
Precision (how much pollution)	Good	Poor (non-quantitative in terms of pollution)
Discrimination (what kind of pollution)	Good	Poor (a general response to all pollutants)
Reliability (how representative is a limited number)	Poor	Good (takes account of temporal or long-term single or effects of pollution samples)
Measure of effects	No (not quantitative in that sense)	Yes
Cost	Can be high	Moderate

(e) effluents:
  (i) biodegradable – sewage, agricultural, food processing: BOD, chemical oxygen demand (COD), suspended solids, NH_3, PO_4, etc.,
  (ii) industrial – toxic: COD, BOD, suspended solids, NH_3, metals and/or other toxic compounds;
(f) general characteristics – typing water source: hardness, alkalinity, pH, colour, conductivity, Fe, Cl, Na, K, silica, SO_4, temperature, etc.

A choice must be made between selecting physico-chemical or biological parameters for assessment based on factors, such as the information gained and cost-effectiveness (Table 9.1). For routine monitoring a combination of both approaches is preferable to either alone. However, some quantitative chemical data will generally be required in order to check compliance with consents, compliance to standards laid down in Directives, especially the Drinking Water and Dangerous Substances Directives, and in order to make prosecutions.

The problem of interpreting biological and chemical data has led to the wide-scale adoption of biological (Section 9.3) and chemical (Section 9.4) indices in water quality assessment.

9.2 SAMPLING SURFACE WATERS

9.2.1 SAMPLING PROGRAMMES

The different types of sampling programmes employed in water quality assessment are defined in Table 9.2 and may vary in scope from the examination of a single parameter on a single occasion to continuous multi-parameter surveillance. Before any fieldwork is undertaken the objectives of the sampling programme must be defined and, combined with a preliminary survey of existing data and other material, a suitable sampling strategy designed. Figure 9.1 outlines a simple protocol for the assessment of water quality in surface waters.

TABLE 9.2   Definitions of sampling programmes

Survey
A survey is a series of intensive, standardized observations designed to measure specific parameters for a specific purpose and is limited to a short sampling period (e.g. an EIA)

Surveillance
Surveillance is used to describe a continued programme of surveys carried out systematically over a longer time to provide a series of observations relative to control or management (e.g. national assessment of river or lake water quality)

Monitoring
Monitoring is continuous standardized surveillance undertaken to ensure that previously formulated standards are being met (e.g. assessment by industry or a regulator to ensure compliance to consent conditions)

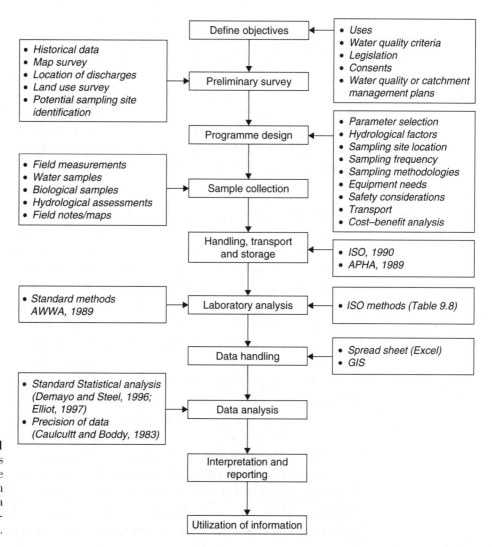

FIGURE 9.1
Main components and, in italics, the factors involved in the development of a water quality monitoring programme.

TABLE 9.3   Visual effect of oil on surface water (values are approximate)

Appearance of film	Thickness (μm)	Total volume of oil ($l km^{-2}$)
Barely visible under most favourable light conditions	0.04	44
Visible as a silvery sheen on water surface	0.08	88
First trace of colour observed	0.15	176
Bright bands of colour	0.30	351
Colours begin to turn dull	1.00	1168
Colours are much darker	2.00	2337

Reproduced from Ellis (1989) with permission of Macmillan Ltd, London.

On arriving at a sampling site the banks should be examined for at least 30 m in both directions to select the very best position for taking samples and also to ensure that there are no other sources of pollution going into the river. Detailed sketch maps of the location and of the sampling site must be made showing habitat and substrate variation, presence of sewage fungus and macrophytes, bankside condition and vegetation, land use, weather details, location of outfalls, the presence of oil on the water surface (Table 9.3) and any other pertinent information. Good field notes and sketch maps prove invaluable in helping to interpret laboratory data or for refreshing the memory later on. Dated photographs are also useful, especially where prosecutions are possible. Good sampling involves detective work. For example, looking for vehicle tracks down to the banks, animal tracks indicating access by farm animals, locating percolation areas serving isolated houses that are unlikely to be sewered, all factors that can significantly affect local water quality and affect the validity of the sampling site for the specified purpose.

Choosing the correct number and location of sampling sites is critical if the necessary information is to be obtained. For example, if the objective is simply to monitor the changes in quality due to a single point source into a river then it may only be necessary to have a single sampling station downstream after complete mixing has occurred, with perhaps another single station upstream of the point source as a control to eliminate changes caused by factors other than the discharge. In such a situation it is important to select sites as similar as possible (i.e. substrate, depth, velocity, etc.). If the objective is to monitor the whole catchment then a considerable number of sampling sites will be required. These will be selected according to the location of major point sources (e.g. towns, major industries, etc.), confluences of tributaries and diffuse sources associated with land use. For modelling purposes sample sites are used to break the river system up into discrete sectors each containing a major input. The downstream site acts as the upstream control site for the next sector (Section 9.4.2). Sharp (1971) has produced a useful algorithm to help in the selection of sampling sites. Bridges and weirs cause enhanced sedimentation and can significantly alter the biological diversity, and should be avoided as sampling locations. Sites close to roads and bridges can also be affected by local surface run-off (e.g. oil, organic debris and salt). Sample locations, therefore, are a compromise between distance downstream from

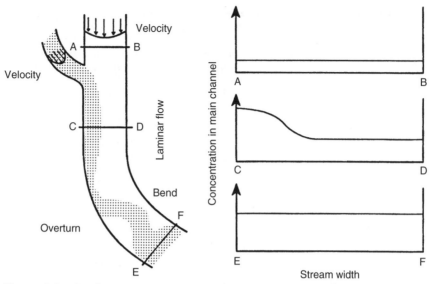

FIGURE 9.2   A tributary entering a river will not mix with the main channel flow due to the maximum velocity, which occurs in the centre of the channel pushing the influent water to the side of the river it has entered. Complete mixing will not occur until laminar flow is disrupted either by a bend in the river (overturn) or a waterfall (turbulent mixing). (Reproduced from Chapman (1996) with permission of UNESCO, Paris.)

inputs, mixing, habitat diversity, substrate, depth, accessibility and safety. Bartram and Ballance (1996) give excellent advice on the selection of sampling sites for both surface and ground waters.

9.2.2 MIXING

Water must be sufficiently well mixed for a single sample of water to be representative of physico-chemical water quality. Water from tributaries or effluent outfalls will have a different density to the water in the main channel. This combined with poor lateral mixing, especially in straight river sections where the flow is laminar, results in the formation of long plumes of unmixed water usually running along one side of the river. In deeper rivers or lakes density stratification may occur. Complete mixing can occur rapidly, but even in small shallow rivers complete mixing may not occur for many kilometres and is normally only achieved when the water turns over at a bend (Fig. 9.2) or the plume is dispersed as it passes over a weir. Effluent plumes can be eliminated by careful siting and design of discharge pipes, or by using weirs or hydraulic jumps immediately below the point of entry. Not establishing complete mixing before sampling below an effluent discharge is one of the commonest sources of error in water quality assessment and will lead to either significantly over- or underestimating its effect.

While rough estimates for complete mixing can be obtained in Table 9.4, in practice complete mixing should be confirmed by taking a series of measurements across the river normally using temperature, conductivity or pH, although this depends on the nature of the input. In deeper rivers (>3 m) where density stratification could also

TABLE 9.4 Approximate distances downstream of influent water entry
for complete mixing to occur in a river or stream

Average width (m)	Mean depth (m)	Estimated distance for complete mixing (km)
5	1	0.08–0.7
	2	0.05–0.3
	3	0.03–0.2
10	1	0.3–2.7
	2	0.2–1.4
	3	0.1–0.9
	4	0.08–0.7
	5	0.07–0.5
20	1	1.3–11.0
	3	0.4–4.0
	5	0.3–2.0
	7	0.2–1.5
50	1	8.0–70.0
	3	3.0–20.0
	5	2.0–14.0
	10	0.8–7.0
	20	0.4–3.0

Reproduced from Bartram and Ballance (1996) with permission
of E. & F. N. Spon, London.

occur then measurements should be taken both near the surface and the bottom of the water column. Uniformity can usually be assumed after mixing so that a single sample for physico-chemical analysis is usually acceptable in all but very large rivers.

9.2.3 SAFETY

The most important factor in field work is safety. Surface waters can often contain pathogens and those taking water or biological samples are particularly at risk from leptospirosis, cryptosporidiosis and, increasingly in Europe, giardiasis (Chapter 12). Those working with surface waters should always wear protective rubber gloves, avoid hand-to-mouth contact (i.e. do not touch your face, smoke, eat or drink without thoroughly washing your hands first with soap and clean water), and all wounds, scratches and skin rashes must be protected from contact with water. Further information on reducing the risk of infection can be found in Section 13.6. Many effluent discharges are chemically dangerous, as are spillages, so there is also a potential risk from chemical injury in some situations.

Apart from pathogen transfer there is also a risk of injury or drowning when sampling. Some general guidelines are given below to help avoid serious injury while carrying out a sampling programme:

(a) Never work alone. One person should always remain on land or in shallow water in order to be in a position to offer assistance.

(b) Always have a first-aid kit and be trained in at least primary first aid.

(c) Always take a mobile phone with you in order to summon assistance.

(d) Never wade out of your depth.

(e) Never use chest waders as these significantly increase the risk of drowning.

(f) Always feel your way in the water using the handle of a hand-net or a pole to avoid deep areas, obstructions and to ensure the substrate is stable. Extreme caution is necessary when the water is discoloured and visibility of the bottom of the river is reduced.

(g) Always ensure waders have a good tread to minimize slipping.

(h) Remain as close to your vehicle and road access as possible.

(i) Always use a boat with an experienced trained crew for sampling deep waters (>1 m) or employ an alternative sampling method (e.g. artificial substrates).

(j) Always inform someone responsible of your exact work programme and at what time you will be expected back.

(k) Whenever possible stay on the bank.

(l) All those involved should be strong swimmers.

(m) You should be in good medical condition and fit enough to carry out the tasks required with ease.

(n) Do not work in the dark.

(o) Where the water is fast flowing, relatively deep (>0.6 m), has steep banks or where there is an accumulation of soft sediment then a safety line, or harness and rope, should always be used. Where the river is fast flowing a catenary chain or rope should also be hung across the channel downstream at water level. The use of a skull protector and a self-righting buoyancy aid should also be considered.

(p) Where sampling sites necessitate abseiling or difficult descents to the river only experienced and trained personnel should be used.

(q) If sampling involves working in difficult terrain ensure:
   (i) you are fully trained to work under those conditions;
   (ii) have the necessary survival and rescue skills to deal with emergencies in those conditions.

(r) If sampling involves submersion in the water (e.g. collection of large bivalves) this should only be carried out by a qualified sub-aqua diving team and must not be attempted using a snorkel.

(s) Familiarize yourself with the concepts of risk assessment and minimization, and act on them.

Before engaging in any fieldwork read the very useful notes by Nichols (1980) and Wright (1993), and talk to your employer's safety or occupational hazards officer. Finally ask yourself the questions:

(a) Is there a safer way of obtaining this information?

(b) Is the end result worth the risks involved in acquiring it?

9.2.4 HYDROLOGY

The discharge rate in rivers is the most important hydrological parameter in water quality and is the quantity of water passing through a channel in a given time. It is expressed in cubic metres per second ($m^3 s^{-1}$) or for smaller flows, such as discharges, in litres per second ($1000 l s^{-1} = 1 m^3 s^{-1}$). The discharge rate is calculated using Equation (3.1) where the flow velocity ($m s^{-1}$) is multiplied by the cross-sectional area of the channel (m^2) (Section 3.3). For a simple approximation of discharge rate the velocity should be measured at 60% of the depth from the river surface (Fig. 3.2) to give an approximate mean flow velocity. Flow velocity can only be accurately measured by using a current meter. This consists of a streamline unit with a propeller attached, which is held in the water pointing upstream. As the water turns the propeller a meter attached to the propeller unit measures the number of revolutions per second. Each instrument is individually calibrated and a specific equation is supplied with each propeller to convert the readings into a discharge rate. There are various other more precise methods of measuring discharge rate which include dilution gauging, the use of permanent gauging stations employing vee notch or rectangular weirs, or ultrasonics (where an ultrasonic beam is transmitted at an angle across the channel and reflected back to a transmitter/receiver; discontinuities in the flow reflect and attenuate the sound waves so that the pulses travelling in either direction have slightly different timings due to the velocity of the water). In practice, current metering, dilution gauging and the use of weirs are most commonly employed in rivers for water quality monitoring purposes.

Current metering

As shown in Fig. 3.1 flow velocity varies across the channel of rivers. This can be overcome to provide a more accurate estimate of discharge rate by dividing the river into a number of regular panels across its width (b) (Fig. 9.3). The depth (d) (m) is measured at regular intervals at each panel boundary to give a cross-sectional area for each panel, and the velocity (v) ($m s^{-1}$) is also measured at 60% of the depth at each point. The discharge for each panel (i.e. cross-sectional area × velocity) is then added up to give the total river discharge (Q) using the following equation:

$$Q = \sum \left(\frac{d_n + d_{n+1}}{2} \right) \left(\frac{v_n + v_{n+1}}{2} \right) b \ (m^3 \ s^{-1}) \qquad (9.1)$$

Clearly, the more panels that are used the more precise the value of Q. In practice 6–10 panels are adequate, however in theory none of the panels should be >5% of the total width or contain >10% of the total discharge, making a minimum of 20 panels. Increased accuracy can also be obtained by taking flow velocity at 20 and 80% of the depth and using the average of these two values in Equation (9.1). Measurements must be done at a straight section of river which is free from macrophytes and that has as regular a cross section as possible.

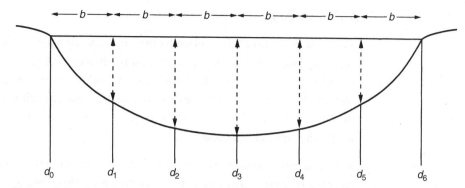

FIGURE 9.3   In order to calculate river discharge rate by current metering river channels are split into separate panels of width (b) with the depth (d) and the flow velocity (v) measured at the boundary of each panel.

EXAMPLE 9.1   Calculate the discharge rate from a river 8 m wide using the data below.

(a) Data collected at river.

Distance from bank (m)	Depth (d) (m)	Flow velocity (v) (m s⁻¹)	Panel width (b) (m)
0	0.0	0.0	1.0
1	0.5	0.4	1.0
2	1.2	0.6	1.0
3	1.8	0.8	1.0
4	2.0	1.0	1.0
5	1.2	0.6	1.0
6	0.8	0.5	1.0
7	0.4	0.3	1.0
8	0.0	0.0	1.0

(b) Discharge rate (Q) is calculated using Equation (9.1):

$[(d_n + d_{n+1})/2]$	$[(v_n + v_{n+1})/2]$	b	q
$[(0.0 + 0.5)/2] = 0.25$	$[(0.0 + 0.4)/2] = 0.20$	1.0	0.050
$[(0.5 + 1.2)/2] = 0.85$	$[(0.4 + 0.6)/2] = 0.50$	1.0	0.425
$[(1.2 + 1.8)/2] = 1.50$	$[(0.6 + 0.8)/2] = 0.70$	1.0	1.050
$[(1.8 + 2.0)/2] = 1.90$	$[(0.8 + 1.0)/2] = 0.90$	1.0	1.710
$[(2.0 + 1.2)/2] = 1.60$	$[(1.0 + 0.6)/2] = 0.80$	1.0	1.280
$[(1.2 + 0.8)/2] = 1.00$	$[(0.6 + 0.5)/2] = 0.55$	1.0	0.550
$[(0.8 + 0.4)/2] = 0.60$	$[(0.5 + 0.3)/2] = 0.40$	1.0	0.240
$[(0.4 + 0.0)/2] = 0.20$	$[(0.3 + 0.0)/2] = 0.15$	1.0	0.030
		Total	5.335

Total discharge rate for river $(Q) = 5.34 \, \text{m}^3\text{s}^{-1}$.

Dilution gauging

Where a river has an irregular channel or is too difficult to measure using current metering due to depth, flow or accessibility, then discharge rate can be calculated by dilution gauging. A tracer (e.g. dye or salt) is added to the river at a known concentration. After complete mixing downstream the concentration of the tracer is measured and from its dilution the discharge rate can be calculated. Two methods are used, constant and slug injection.

With constant injection, a tracer of known concentration (C_1) $(mg\,l^{-1})$ is added to the river at a constant rate (q_1) $(m^3 s^{-1})$. The concentration downstream will increase until it reaches a constant rate (C_2) so long as the tracer continues to be added at a constant rate upstream. Under steady-state conditions the total mass of tracer added equals the total mass of tracer passing the downstream sampling point. Therefore, the input concentration multiplied by the input rate equals the diluted concentration multiplied by the discharge rate

$$C_1 \times q_1 = C_2 \times q_2 \tag{9.2}$$

Therefore the discharge rate (Q) equals q_2:

$$q_2 = \frac{C_1 \times q_1}{C_2} \tag{9.3}$$

As the discharge of the river has been increased by the addition of the tracer, Equations (8.2) and (8.3) are more correctly written as:

$$C_1 \times q_1 = C_2(q_2 + q_1) \tag{9.4}$$

$$q_2 = q_1 \frac{C_1 - C_2}{C_2} \tag{9.5}$$

EXAMPLE 9.2 Calculate the discharge rate (Q) of a river using sodium chloride as a simple tracer. Its recovery downstream is measured in the field as conductivity rather than using chloride analysis.

(a) Establish the relationship between NaCl concentration and conductivity in the laboratory. In practice the relationship will be approximately:

$$NaCl\ (mg\,l^{-1}) = Conductivity\ (\mu S\,cm^{-1}) \times 0.46 \tag{9.6}$$

(b) Determine the background conductivity levels in the river at the sampling station downstream of tracer addition.

(c) Add NaCl tracer by continuous injection over unit time. Twenty litres of $200\,\mathrm{g\,l}^{-1}$ solution of NaCl (i.e. $200\,000\,\mathrm{mg\,l}^{-1}$) is added over 6 min giving a q_1 value of $0.056\,\mathrm{l\,s}^{-1}$.

(d) Continuously measure the downstream conductivity concentration below the mixing zone. The conductivity will rise, reach a plateau and, after tracer injection stops, will slowly return to normal background level. Background conductivity is $250\,\mu\mathrm{S\,cm}^{-1}$ and the plateau conductivity is $425\,\mu\mathrm{S\,cm}^{-1}$, so the increase in conductivity is $175\,\mu\mathrm{S\,cm}^{-1}$. Using Equation (9.6), this corresponds to a NaCl concentration of $175 \times 0.47 = 82.3\,\mathrm{mg\,l}^{-1}$.

$$q_2 = q_1 \frac{C_1 - C_2}{C_2}$$

$$q_2 = 0.056\frac{200\,000 - 82.3}{82.3} = 1361\,\mathrm{s}^{-1}$$

Therefore, the discharge rate of the river (Q) is $0.136\,\mathrm{m}^3\mathrm{s}^{-1}$.

Slug injection involves a known volume (V) of tracer being added instantaneously into the river and then downstream, after complete mixing, the increase and decrease in tracer concentration is recorded and plotted as a curve where the Y-axis is tracer concentration $(\mathrm{mg\,l}^{-1})$ and the X-axis is time (s). By calculating the area under the curve the discharge rate (Q) can be calculated using the following equation:

$$Q = \frac{\text{Input concentration } (C_1) \times \text{Input volume } (V)}{\text{Area under the curve}} \tag{9.7}$$

EXAMPLE 9.3   Using 5 l of a $200\,\mathrm{g\,l}^{-1}$ solution of NaCl the area under the curve downstream of the injection point was found to be $7350\,\mathrm{mg\,l}^{-1}$. This is measured over the total period, in seconds, when the concentration is above background. So using Equation (9.7):

$$Q = \frac{200\,000\ (\mathrm{mg\,l}^{-1}) \times 5.0\ (\mathrm{l})}{7350\ (\mathrm{mg\,l}^{-1})(\mathrm{s})} = 1361\,\mathrm{s}^{-1}$$

Therefore, the discharge rate of the river (Q) is $0.136\,\mathrm{m}^3\mathrm{s}^{-1}$.

Weirs

Permanent gauging stations use weirs that are carefully constructed to control the flow over them, permitting the discharge rate to be very accurately calculated from the height of the water directly upstream of the weir. Large rivers employ wide rectangular weirs to ensure there is no impedance to flow. For smaller channels, such as streams

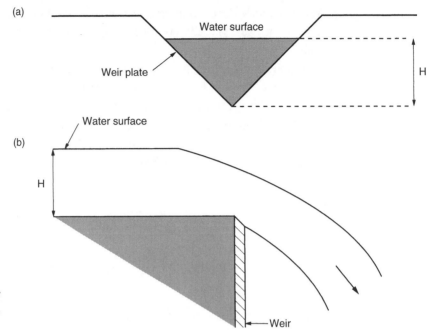

FIGURE 9.4
Sectional drawings
of a 90° vee notch
weir, (a) cross sec-
tion and (b) side
profile through weir,
where *H* is the verti-
cal distance (head)
between the point of
the weir and the true
surface of the water.

and discharges, vee notch weirs are used. Measurements are recorded either continuously using a float recorder with the float inside a vertical pipe partially submerged below the water surface and attached to a recording device, or manually from a measuring staff to show the height of the water. Weirs must meet two basic criteria:

1. the water must flow over a sharp edge;
2. the water must be discharged into the air (i.e. air must pass freely under the jet of water).

The mostly widely employed weir is the 90° vee notch weir which can record a wide variation in discharge rate (Q) and is cheap to construct (Fig. 9.4). If the two conditions above are satisfied then Q is related to the height (H) of the water behind the weir as:

$$Q = 1.34 \, H^{2.48} \; (\text{m}^3 \, \text{s}^{-1}) \tag{9.8}$$

H is the vertical distance (head) from the point of the weir to the free water surface in metres. Specific equations are required for different angled vee notch weirs.

9.2.5 CHEMICAL SAMPLING

The collection of water samples from surface waters is normally done by hand. It is only necessary to have sterile bottles if microbial parameters are to be measured (Chapter 12), otherwise bottles should be acid washed and thoroughly rinsed. Bottles

TABLE 9.5   Summary of the key water quality analyses

Pollutant or parameter	Measurement technique	Response time
pH	Electrometric	10 s
BOD	Dilution/incubation	5 days
COD	Dichromate oxidation	2 h
Metals	ASS	–
Organometallics	GC–AAS	–
Nitrate	Colorimetric	5 min
Nitrate	UV spectrophotometric	1 min
Formaldehyde	Photometric	6 min
Phenols	GC	30 min

Reproduced from Harrison (1992) with permission of Royal Society of Chemistry, Cambridge.

must be clearly marked with the time and date of sampling, site location and the name of the person taking the sample. They should be opened only when required, rinsed and then, holding the bottle firmly near its base, it should be submerged approximately 20 cm below the surface with the open mouth facing slightly downwards towards the current. Gradually turn the bottle upright to fully fill it and replace the stopper. There is a range of automatic samplers that can take separate water samples either at regular intervals or when a certain control parameter (e.g. dissolved oxygen, pH, conductivity, etc.) exceeds certain pre-set limits. Composite samplers may also be used. Samples are taken regularly over a set period and amalgamated in one large sample container. Such samplers can also be flow-related taking samples proportional to the discharge rate. Automatic samplers are widely used to monitor effluent discharges in order to check compliance with consent conditions.

Once collected, samples should be stored for transport back to the laboratory in an insulated plastic box containing ice packs (i.e. picnic or cool box) to maintain the temperature as close to 4°C as possible. Bottles used for microbial analysis should be sealed in sterile plastic bags to reduce the chance of any contamination during transport. If there is a delay >2 h between collection and analysis, samples for both microbial and BOD analysis need to be cooled rapidly after collection by prefilling the insulated transport box with cold water in which the ice packs are placed. Otherwise they should be kept as cool as possible and in the dark. Samples for chemical analysis should be carried out within 24 h of collection, except for BOD, which must be done as soon as possible. Samples may require specific storage and preservation conditions depending on the analysis (Bartram and Ballance, 1996). Physico-chemical analysis of samples in the laboratory must always be carried out using standard methods (e.g. APHA, 1989 or International Standards Organization Series). There is a wide range of instrumental analytical methods available making analysis very much quicker. Some analyses require elaborate pre-treatment such as acid digestion for metal analysis by atomic absorption spectrophotometry. Typical times for the analysis of the common water quality parameters are given in Table 9.5.

9.2.6 BIOLOGICAL SAMPLING

Friedrich *et al.* (1996) identified six main categories of biological sampling methods (see Table 9.22):

1. *Ecological.* These methods are based on the response of aquatic organisms to pollution and its effects on their habitat. Methods are based on community structure using indicator organisms that are sensitive to pollutants, and includes biotic and diversity indices (Section 9.3) using macro-invertebrates or indices using other groups, such as macrophytes or plankton.

2. *Microbial.* Primarily the use of bacteriological indicators of faecal pollution (e.g. *Escherichia coli* and other coliforms). Differentiation between animal and human faecal pollution. Also the typing of bacteria to identify sources of organic pollution (Section 12.2).

3. *Physiological and biochemical.* Response of organisms to variations in water quality. Methods include the growth rate of algae and bacteria, oxygen production potential of water using algae, etc. (Section 15.3).

4. *Bioassays and toxicity tests.* Rapid determination of acute toxicity of water-using test organisms (Section 15.3).

5. *Chemical analysis of biota.* Organisms known to be able to bioaccumulate contaminants are exposed to, or collected from, suspect water and the level of contamination accumulated in their tissues or organs measured.

6. *Histological and morphological.* Examination of organisms for morphological indications of environmental stress (e.g. lesions, tumours, etc.).

The main UK standard methods, which are similar to the equivalent ISO methods, are summarized in Tables 9.6 and 12.4, while US standard biological methods are given in APHA (1989). The majority of biological monitoring of surface waters worldwide is ecologically based using macro-invertebrates. While there are a number of different methods of sampling benthic macro-invertebrates the commonest methods are outlined below (Table 9.7). The Water Framework Directive (WFD) lists standard methods for monitoring the quality parameters (elements) listed. These are given in Annex V of the Directive and summarized in Table 9.8. Biological communities tend to have a clumped rather than random distribution, determined largely by substrate and habitat niche requirements. Therefore, quantitative sampling is often difficult. If used in conjunction with a biological index then specific hydrological features or substrate types may be specified. Routine surveillance requires only a semi-quantitative approach while ecological studies tend to require a more quantitative approach examining all the various habitats at a particular site. Different statistical methods require different quality of data. Nearly all state agencies employ a standard hand-net procedure for river sampling including the UK Environment Agency which employs the Biological Monitoring Working Party (BMWP) score index and the Irish Environmental Protection Agency which uses the Q score index (Section 9.3).

TABLE 9.6   There are a wide range of standard methods published by the Standing Committee of Analysts (SCA) in the series *Methods for the Examination of Waters and Associated Materials* (published by Her Majesty's Stationery Office, London). There are specific methods for all the important ions, compounds, parameters and methods used in water analysis. Below are listed the important biological methods used in water quality analysis. See also Table 12.4

Date of publication	Methods for the examination of waters and associated materials
	Series title
1979	Hand-net sampling of aquatic benthic macro-invertebrates 1978
1982	Acute toxicity testing with aquatic organisms 1981
1982	Quantitative samplers for benthic macro-invertebrates in shallow flowing waters 1980
1983	The bacteriological examination of drinking water supplies 1982
1983	Methods for the isolation and identification of *Salmonellae* (other than *Salmonella typhi*) for water and associated materials 1982
1984	Methods of biological sampling of benthic macro-invertebrates in deep rivers 1983
1984	Sampling of non-planktonic algae (benthic algae and periphyton) 1982
1985	Methods of biological sampling: sampling macro-invertebrates in water supply systems 1983
1985	Methods of biological sampling: a colonization sampler for collecting macro-invertebrate indicators of water quality in lowland rivers 1983
1986	The direct determination of biomass of aquatic macrophytes and measurement of underwater light 1985
1987	Methods for the use of aquatic macrophytes for assessing water quality 1985–86
1988	Methods for sampling fish populations in shallow rivers and streams 1983
1991	Enumeration of algae, estimate of cell volume and use in bioassays 1990
1994	A review and methods for the use of epilithic diatoms for detecting and monitoring changes in river water quality 1993
1994	The microbiology of water 1994: Part 1. Drinking water
1995	Methods for the isolation and identification of human enteric viruses from waters and associated materials 1995

Kick sampling using a hand-net is the standard approach employed by most European countries (Newman *et al.*, 1992). The hand-net must be of an approved standard design (ISO, 1985). It must have a 900 μm mesh net with a 275 mm bag depth and a 230 × 255 mm frame opening, and a handle ⩾1.5 m. The hand-net is ideal for sampling shallow streams and rivers where there is a gravel substrate. The hand-net is placed firmly into the surface of the substrate while held facing upstream. An approximate standard area of substrate in front of the net is kicked and/or hand washed for a standard time, usually 2–3 min. At best the method is semi-quantitative. Where the substrate is too large to be effectively sampled by the kick technique then the substrate must be hand washed immediately in front of the net. Hand washing is less effective than kick

TABLE 9.7   Main sampling options for freshwater biota

Biota main requirement/constraint	Sampling option
Periphyton	
Non-quantitative/selective	Scraping
Quantitative	Artificial substrate
	Periphyton sampler
Phyto- and zooplankton	Plankton net, bottle
Benthic invertebrates	
Shallow water/gravel substrate	Hand-net
Deep water/fine substrate	Grab
Deep water/fine sediment/surface dwellers only	Dredge
Shallow water/large substrate	Hand wash
Moderate depth/any substrate	Artificial substrate
Drift invertebrates	Drift net
Fish	
Deep waters/selective	Net
Moderate waters	Electrofishing
Macrophytes	
Shallow water (wading)	Collection by hand
Deep water (from a boat)	Collection by hand or grab

TABLE 9.8   Recommended ISO methods for use with the WFD

Macro-invertebrate sampling
ISO 5667-3:1995 Water quality – Sampling. Part 3: Guidance on the preservation and handling of sampling

EN 27828:1994 Water quality – Methods for biological sampling. Guidance on hand-net sampling of benthic macro-invertebrates

EN 28265:1994 Water quality – Methods for biological sampling. Guidance on the design and use of quantitative samplers for benthic macro-invertebrates on stony substrata in shallow waters

EN ISO 9391:1995 Water quality – Sampling in deep waters for macro-invertebrates. Guidance on the use of colonization, qualitative and quantitative samplers

EN ISO 8689-1:1999 Biological classification of rivers. Part I: Guidance on the interpretation of biological quality data from surveys of benthic macro-invertebrates in running waters

EN ISO 8689-2:1999 Biological classification of rivers. Part I: Guidance on the presentation of biological quality data from surveys of benthic macro-invertebrates in running waters

Detection and enumeration of faecal indicator bacteria in water
ISO 6461-1:1986 Detection and enumeration of the spores of sulphite-reducing anaerobes (clostridia). Part 1: Method by enrichment in a liquid medium

ISO 6461-2:1986 Detection and enumeration of the spores of sulphite-reducing anaerobes (clostridia). Part 2: Method by membrane filtration

ISO 7704:1985 Evaluation of membrane filters used for microbiological analyses

TABLE 9.8   (*Continued*)

ISO 7899-1:1984 Detection and enumeration of faecal streptococci. Part 1: Method by enrichment in a liquid medium

ISO 7899-2:1984 Detection and enumeration of faecal streptococci. Part 2: Method by membrane filtration

ISO 9308-1:1990 Detection and enumeration of coliform organisms, thermotolerant coliform organisms and presumptive *E. coli*. Part 1: Membrane filtration method

ISO 9308-2:1990 Detection and enumeration of coliform organisms, thermotolerant coliform organisms and presumptive *E. coli*. Part 2: Multiple tube (most probable number) method

FIGURE 9.5
A Surber sampler is a standard hand-net with a quadrate attached to the front of the net to allow quantitative sampling. (Reproduced from Hellawell (1978) with permission of the Water Research Centre plc, Medmenham.)

sampling so a longer sampling period is required (i.e. up to 10 min). Kick sampling should be used in deeper water or if the water is polluted. In lakes, as there is no current to wash the disturbed animals into the net the person collecting the sample must use wide sweeps through the water after disturbing the substrate with their feet. The Environment Agency procedure is given in Section 9.5.

The actual period of sampling and number of replicate samples will vary according to the sampling objectives. As a general rule all available habitats at each site should be sampled in proportion to their occurrence. It is good practice to repeat sampling at several places across the river channel to include different microhabitats within the riffle and then combine them to create a single representative sample. However, as every site is different, use whatever technique seems appropriate to the prevailing conditions, although it is advisable never to exceed a total collection period of 3 min as this will result in an enormous sorting and identification task as well as unnecessary depletion of organisms at the site.

If quantitative samples are required then a Surber sampler is used (Fig. 9.5). This employs a 300 mm square quadrate attached to a standard hand-net and sampling of the substrate is restricted to this area only. Side nets are attached from the quadrate to the net to prevent any loss of animals. Improved designs use cylinder or box samplers to reduce loss of animals washed from the substrate (Fig. 9.6), which allows them to be used in deeper water. The substrate is vigorously disturbed within the cylinder

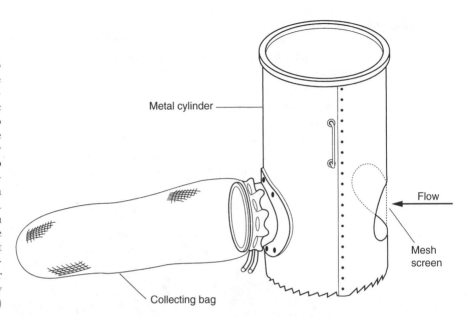

FIGURE 9.6
Surber samplers are
generally con-
structed from plastic
or metal cylinders to
allow them to be
used in slightly
deeper water and to
reduce the possibil-
ity of losing fauna
during collection.
(Reproduced from
Department of the
Environment
(1983a) with per-
mission of Her
Majesty's Stationery
Office, London.)

Metal cylinder

Flow

Mesh
screen

Collecting bag

using a pole and the water passing through the mesh screen carries the displaced biota into the collecting bag.

In deeper waters, either grabs or dredges must be used (Fig. 9.7) (ISO, 1991). They are difficult to use effectively as the operator is unable to see what is going on. They are used for deep water with fine substrates, normally lakes or lowland rivers, so they must be used from a boat. Comparative studies under experimental conditions have shown them to be inefficient and often unreliable in rivers, being severely restricted by the size of the substrate (Table 9.9).

Where the substrate is not suitable or the river conditions are unsuitable for either pond net or Surber sampler, then artificial substrates are probably the best sampling option. A wide variety of materials have been used. The selected substrate material is placed into a mesh bag and placed on the riverbed and the animals allowed to colonize it over a minimum period of a month. The bags containing the substrate are carefully recovered from the river and the animals removed and processed just as any other sample. While there is a standard sampler design which uses plastic wastewater treatment filter media (standard aufwuchs unit) (Fig. 9.8) it is usually best to use graded natural substrate collected from the actual river. The artificial substrates are either weighted down using concrete blocks and tied to the bank, or fixed in position using a steel rod. Careful positioning of substrates in the river is very important:

(a) to avoid detection from the bank and possible disturbance or damage;
(b) to avoid injury to other river users (e.g. anglers, canoeists, etc.);
(c) to ensure that they are not affected by changes in water depth (i.e. that they do not become exposed during low flows or alternatively cannot be recovered during high flows).

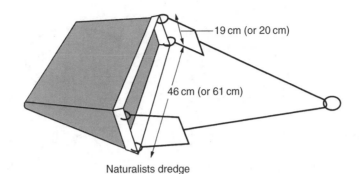

19 cm (or 20 cm)

46 cm (or 61 cm)

Naturalists dredge

FIGURE 9.7
Examples of dredges
and grabs used for
finer sediments.
(Reproduced from
Department of the
Environment
(1983a) with per-
mission of Her
Majesty's Stationery
Office, London.)

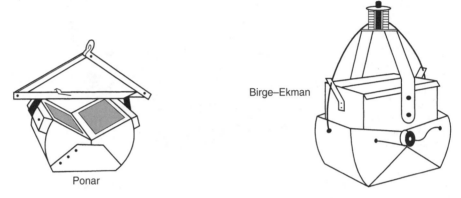

Birge–Ekman

Ponar

TABLE 9.9 Efficiencies of grab samplers in recovering benthic macro-invertebrates at the surface and 30 mm below the surface of a fine and coarse gravel substrate

Grab	% efficiency			
	Substrate 2–4 mm		Substrate 8–16 mm	
	Surface	30 mm below surface	Surface	30 mm below surface
Ponar	100	70	88	50
Weighted ponar	100	70	100	50
Van–Veen	87	56	50	50
Birge–Ekman	73	37	30	7
Allan	51	36	25	7
Friedinger	59	7	30	7
Dietz–La Fond	22	26	25	20

Reproduced from Elliot and Drake (1981) with permission of Blackwell Science Ltd, Oxford.

Drift nets are placed in flowing water to collect animals moving actively or passively drifting downstream. Collections of drift are usually very selective as certain animals (e.g. insect larvae and nymphs) are more likely to drift than heavy species (e.g. molluscs and heavy cased caddis). So, the proportion of species in benthic samples will not be the same as those in drift. However, drift samples are useful for estimating the recovery/recolonization of damaged river sections and also give an indication of the health of the river upstream. In general, <0.5% of macro-invertebrates passively move downstream

FIGURE 9.8
The UK standard colonization unit used as an artificial substrate for macro-invertebrates in rivers sites not suitable for sampling with either a hand-net or Surber sampler. (Reproduced from Department of the Environment (1983b) with permission of Her Majesty's Stationery Office, London.)

each day. Normal drift involves small movements of <2 m per single movement although some species may move up to 50 m each time. The major factor causing drift is food availability, and drifting allows rapid colonization of new channels or areas denuded of species due to very high flows. Generally, the effect of drift is counterbalanced by the active movement of species back upstream while on the surface and by adult insects (e.g. mayflies and stoneflies) flying back upstream to lay eggs.

The pattern of distribution of macro-invertebrates is such that very large numbers of samples are required for a reliable estimate of population density. In fact, accurate measurements of population density are not really possible as a large number of samples are required even to recover all the species or families present (Fig. 9.9). Many samplers and sampling techniques only recover biota from the top surface of substrate. However, estimates indicate that up to 80% of all the organisms can be located >75 mm below the surface and most burrowing forms are only found below the surface. Therefore, grab samplers are particularly poor, as they do not penetrate very deep into the substrate. So, in practice, only a small proportion of the population is recovered by normal sampling.

A preliminary survey should be carried out if possible to determine the number of replicates required to give a certain degree of precision. For most ecological studies the level of precision used is 20% (i.e. $D = 0.2$). The number of replicates (n) can be estimated using the following equation:

$$n = \left(\frac{s}{Dx} \right)^2 \tag{9.9}$$

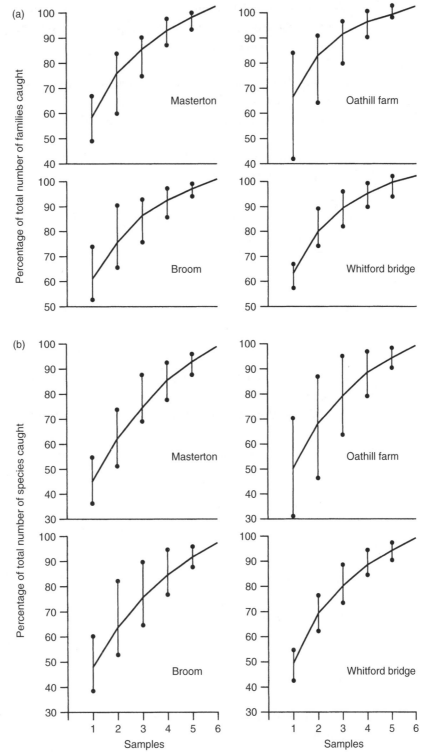

FIGURE 9.9
Taxon accretion for different river sites. Mean, maximum and minimum values derived from all possible combinations of samples for (a) family accretion and (b) species accretion. (Reproduced from Furse *et al.* (1981) with permission of Elsevier Science Ltd, Oxford.)

where D is the index of precision, x the mean and s the standard deviation. The results of the example below show the problem of obtaining precise quantitative results of freshwater macro-invertebrates.

EXAMPLE 9.4   Calculate the optimum number of replicates (n) required to give a degree of precision of 20% for each site of a polluted river using (a) number of species per sample and (b) fauna abundance in riffle samples.

(a) Use Equation (9.9) to calculate n from the number of species collected in five replicates from six sites.

Site	Number of species per replicate					Mean	Standard deviation	n
	1	2	3	4	5			
1	13	11	11	9	12	11.2	1.48	0.4
2	11	13	8	6	5	8.6	3.36	3.8
3	1	1	1	1	2	1.2	0.45	3.5
4	2	2	2	3	4	2.6	0.89	2.9
5	8	4	6	7	7	6.4	1.52	1.4
6	7	1	6	7	3	4.8	2.68	7.8

The optimum number of replicates for 20% precision of species diversity varies between 1 and 8.

(b) Using the total number of individuals collected (a) is repeated.

Site	Faunal abundance per replicate					Mean	Standard deviation	n
	1	2	3	4	5			
1	542	222	287	122	435	321.6	167.6	6.8
2	306	540	104	396	23	273.8	211.3	14.9
3	7	2	16	5	11	8.2	5.5	11.0
4	13	10	13	13	19	13.6	3.3	1.5
5	89	13	41	29	24	39.2	29.6	14.3
6	414	3	33	772	21	248.6	339.1	46.5

The optimum number of replicates for 20% precision of total faunal density varies between 2 and 47.

9.3 BIOLOGICAL INDICES

Raw biological data exist as lists of species' (or taxa) names and abundance. This is difficult for non-biologists to interpret in order to compare the water quality either between sites (spatially) or at sites over time (temporally). In practice, it is impossible to use all the data collected over a long time scale involving numerous stations and involving numerous species and parameters. One of the best ways to reduce or condense the bulk of material collected from river surveys and represent it clearly and concisely is to use a comparative index. Indices utilize all the data to produce a single value that can then be used for rapid and accurate analysis. While a loss of information is inevitable, this must be balanced against the overall gain in comprehension, especially for the non-scientist. In fact, the use of indices often gives new insight on data. Indices have been developed for use with algae, macrophytes and micro-organisms (Hellawell, 1986). The indices considered here deal with macro-invertebrates.

Biological indices are essentially of two types.

1. *Pollution indices* are observations of the responses of indicator species or higher taxa to pollution. So, environmental quality is assessed by observing the presence (or absence) of characteristic species or communities of known tolerances and preferences. The fundamental problem with these indices is that presence or absence may be due to factors other than organic pollution on which the indices are based. For example, presence, and therefore a better rating, may be due to favourable conditions such as high turbulence. Conversely absence, and a lower rating, could be due to natural factors such as water hardness or low turbulence, or non-organic pollution such as increased temperature or toxic contamination.
2. *Diversity indices* depend up on the theoretical concepts of animal or plant communities. Therefore, measuring the extent that the observed community structure differs from the assumed model assesses the degree of change in water quality. These indices are especially useful for environmental impact assessment (EIA) as they are not limited by specific pollutants but measure the overall effect of all pollutants both known and unknown. However, as no clue is given in the index as to the source of pollution this has to be assessed by other methods.

In practice it is best to use indices comparatively, not as absolute values, as each lotic or lentic system has unique physico-chemical conditions that affect the fauna. Therefore, indices should be used over long periods at one site or to examine the spatial variation along a river. Where practicable pollution and diversity indices should be used together.

9.3.1 POLLUTION INDICES

Early descriptive classification

The saprobian system was developed by Kolwitz and Marsson in 1908 who related community structure to quality status (Hellawell, 1986). The saprobic system is purely descriptive relating species to certain levels of organic pollution. Saprobity is used to summarize the degree of pollution or level of organic matter discharged to a lotic system. Four classifications are used (Fig. 9.10). Oligosaprobic is used to describe clean and healthy river sections, polysaprobic is the zone of extreme pollution closest to the discharge point, mesosaprobic is the zone of recovery (oxidation) being divided into

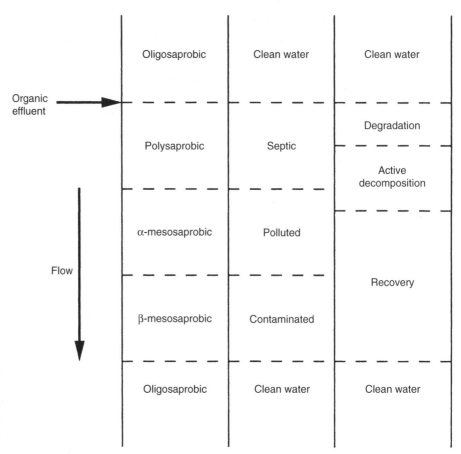

FIGURE 9.10
Various pollutional
classifications used
with lotic systems
compared with
saprobity.

the alpha (α)-mesosaprobic zone of heavy pollution and the beta (β)-mesosaprobic
zone of moderate pollution. Complete recovery is characterized by the return to
oligosaprobic conditions. This descriptive system is still extremely valuable as a lot of
information can be obtained by simply looking for key indicators. The key characteristics of each zone are described next.

Polysaprobic zone (P) – extremely heavily polluted zone
Chemical
(a) Rapid putrefaction process occurs at this stage, therefore anaerobic conditions
are expected.
(b) The higher degradation stages of the proteins are present partly as peptones and
peptides but degradation extends to amino acids, characterized by the presence
of albumens, polypeptides and carbohydrates.
(c) Ammonia, H_2S and CO_2 are produced as end products of degradation.

Physical
(a) Dirty grey colour.
(b) Faecal to mouldy smell.

(c) Very turbid due to enormous quantities of bacteria and colloids.

(d) Normally bottom of watercourse is composed of black digesting sludge and the reverse sides of stones are coloured black by a coat of iron sulphide.

Biological

(a) Most autotrophic organisms are missing.

(b) Community comprised of a few groups, but individuals are very numerous.

(c) Bacteria abundant $>10^6 \, \mathrm{ml}^{-1}$

(d) *E. coli* is abundant if organic waste is faecal in origin, not if industrial.

(e) Mainly blue-green algae and flagellate protozoa and amoeba.

(f) Few invertebrates, only those that have haemoglobin (e.g. *Tubifex, Chironomus riparius*) or those that have organs for the intake of air (e.g. *Eristalis* spp.).

(g) All fish eliminated.

α-mesosaprobic zone (α-M) – oxidation zone of heavy pollution

Chemical

(a) High concentration of breakdown products, amino acids and their degradation products, mainly fatty acids.

(b) More oxygen present, oxidation occurring and no anaerobic muds.

(c) Oxygen normally <50% saturation.

Physical

(a) Dark grey colour.

(b) Smells mouldy or other unpleasant smell (Table 13.4) due to residues of protein and carbohydrate fermentation.

Biological

(a) Sewage fungus zone forming cotton-wool-like coating of streambed.

(b) Bacteria $<10^6 \, \mathrm{ml}^{-1}$.

(c) Few algae.

(d) Mainly filamentous bacteria/fungi.

(e) Protozoa common, both flagellates (e.g. *Bodo*) and ciliates (e.g. *Paramecium* and *Colpidium*).

(f) Few invertebrates, similar to polysaprobic zone.

(g) No fish.

β-mesosaprobic zone (β-M) – oxidation zone of moderate pollution

Chemical

(a) Reduction processes are completed with the last residues of protein degradation, such as amino acids, fatty acids and ammonia found in low concentrations.

(b) Area of ammonical compounds.

(c) Aerobic waters with oxygen never <50%. Diurnal variations in dissolved oxygen governed by photosynthesis. Degradation no longer affecting oxygen concentrations significantly.

Physical
(a) Water transparent or slightly turbid.
(b) Uncoloured and odour free.

Biological
(a) Bacteria always $<10^5 \, ml^{-1}$.
(b) More tolerant microfauna. Many more algal spp. but dominated by filamentous algae (e.g. *Cladophora* – blanket weed).
(c) Protozoa dominated by ciliates especially stalked species (i.e. *Peritrichia*).
(d) Invertebrates still restricted but greater diversity than in previous zone. Dominated by *Asellus* with Mollusca and Hirudinae (leeches) common.
(e) Coarse fish becoming more abundant, especially those able to tolerate low dissolved oxygen concentrations (e.g. bream and chub).

Oligosaprobic zone (O) – zone of complete oxidation
Chemical
(a) All waste products broken down to stable inorganic or organic compounds.
(b) Dissolved oxygen normally saturated (100%), although if algae is still abundant then there may be diurnal variation.

Physical
(a) Clear, no odour, no colour.

Biological
(a) Bacteria $<100 \, ml^{-1}$.
(b) Wide range of animals and plants, including insect larvae and salmonids.
(c) Macrophytes and mosses common.
(d) Periphyton dominated by diatoms and a few green and blue-green algae.

Saprobity index

This is a numerical development of the saprobien classification. Based on the degree of saprobity (s) and the relative frequency of each species (h) a saprobien index (I_s) can be estimated:

$$I_s = \frac{\sum sh}{\sum h} \tag{9.10}$$

This is the saprobity index of Pantle and Buck (Hellawell, 1978), where s is the degree of saprobity where each species is rated according to its saprobity (i.e. oligosaprobic 1, β-mesosaprobic 2, α-mesosaprobic 3 and polysaprobic 4); h is the relative abundance (i.e. 1 is rare, 3 is frequent and 5 is abundant). The resulting index I_s gives the degree of pollution:

I_s	Pollution rating
1.0–1.5	Very slight
1.5–2.5	Moderate
2.5–3.5	Strong
3.5–4.5	Very severe

An example of this index is given in Table 9.10. The saprobic index requires good taxonomic expertise and a knowledge of the saprobity of individual species (Table 9.11). An interesting example of the use of saprobic indices using algae is given by Thorpe and Williams (1980).

Trent biotic index

This pollution or biotic index is based on two effects of organic pollution:

1. a reduction in community diversity;
2. the progressive loss of certain key macro-invertebrate groups from the clean-water fauna in response to organic pollution.

Water quality is split into 11 different index classes ranging from X for the highest quality to O for septic conditions (Table 9.12). Except for these two extremes all the other classes can be obtained from different combinations of community diversity and indicators. For example, index class VII can arise from seven different sets of circumstances.

This index has a number of disadvantages. It takes no account of quantitative data. So the presence of a single good indicator species that may have drifted into the area may result in water quality being overestimated. Neither does it relate results to community biomass by standard sampling methods, indeed it is desirable to sample as many microhabitats as possible. The index remains popular due to its simplicity and the restriction to easily identified groups. With practice collection, sorting and processing can all be carried out in the field. The main problem with the index is that it is insensitive at sites where there is a high diversity. This has been overcome by extending the index from 10 to 15 by increasing the number of categories for the total number of groups present (Table 9.13).

EXAMPLE 9.5 The score is calculated by first deciding the number of groups present using part 2 of Table 9.12. Using the species list in Table 9.10 the number of groups present, as defined by the Trent biotic index (TBI), is calculated as:

(a) *Dendrocoelum lacteum* (platyhelminth – flatworm) (1 group).
(b) *Tubifex tubifex* (oligochaeta – worm) (1 group).

TABLE 9.10   Example of the saprobic index of Pantle and Buck using a simple database

Species	Saprobic rating	s	Abundance (No.)	h	sh
Asellus aquaticus	α	3	80	5	15
Baetis rhodani	β	2	18	3	6
Chironomus riparius	p	4	12	3	12
Dendrocoelum lacteum	β	2	5	2	4
Ephemerella ignita	β	2	1	1	2
Erpobdella octoculata	α	3	2	1	3
Gammarus pulex	o	1	12	3	3
Lymnaea stagnalis	β	2	27	4	8
Metriocnemus hygropetricus	β	2	38	4	8
Tubifex tubifex	p	4	8	2	8
				$\Sigma h = 28$	$\Sigma sh = 69$

Therefore, index $(I_s) = \Sigma sh/\Sigma h = 69/28 = 2.46$
This indicates that the water is showing signs of moderate organic pollution

TABLE 9.11   Saprobic rating of a selection of commonly occurring European macro-invertebrates

Species largely insensitive to severe organic pollution (polysaprobic)

Chironomidae
Camptochironomus tentans
Chironomus plumosus
Chironomus riparius

Oligochaeta
Limnodrilus claparedeanus
Limnodrilus hoffmeisteri
Rhyacodrilus coccineus
Tubifex tubifex

Species tolerant of severe organic pollution (α-mesosaprobic)

Chironomidae
Anatopyria plumipes
Apsectrotanypus trifascipennis
Psectrotanypus varius
Prodiamesa olivacea

Megaloptera
Sialis fulginosa
Sialis lutaria

Trichoptera
Anabolia laevis
Hydropsyche angustipennis
Hydropsyche ornatula
Hydropsyche pellucidula

Coleoptera
Haliplus lineatocollis

Odonata
Coenagrion pulchellum

Ephemeroptera
Baetis vernus
Caenis horaria
Heptagenia longicauda
Siphlonurus lacustris

Oligochaeta
Limnodrilus udekemianus
Lumbriculus variegatus
Nais barbata
Nais communis
Nais elinguis
Peloscolex ferox
Psammoryctes barbatus

Hirudinea
Erpobdella octoculata
Glossiphonia heteroclita
Haemopsis sanguisuga
Helobdella stagnalis
Hirudo medicinalis

Crustacea
Asellus aquaticus

Platyhelminthes
Dugesia tigrina
Planaria torva

TABLE 9.11   (*Continued*)

Species tolerant of moderate organic enrichment (β-mesosaprobic)

Chironomidae
Ablabesmyria monilis
Corynoneura celtica
Corynoneura scutellata
Cricotopus bicinctus
Cricotopus triannulatus
Macropelopia nebulosa
Metriocnemus fuscipes
Metriocnemus hygropetricus
Microtendipes cloris
Polypedilum laetum

Ephemeroptera
Baetis rhodani
Brachycercus harisella
Caenis moesta
Cloeon dipterum
Cloeon rufulum
Ephemera danica
Ephemerella ignita
Heptagenia sulphurea
Leptophlebia marginata

Plecoptera
Isoperla grammatica

Heteroptera
Brychius elevatus
Nepa rubra
Notonecta glauca
Odonata
Ischnura elegans

Trichoptera
Anabolia nervosa
Athripsodes annulicornis
Athripsodes cinereus
Crunoecia irrorata
Geora pilosa
Hydroptila tineoides
Limnephilus sericeus
Limnephilus sparsus
Polycentropus flavomaculatus
Potamophylax stellatus
Psychomia pusilla

Oligochaeta
Aelosoma niveum
Aelosoma quaternarium
Aelosoma tenebrarum
Amphichaeta leydigii
Branchiura sowerbyi
Chaetogaster cristallinus
Chaetogaster palustris
Criodrilus lacuum
Euclyodrilus hammoniensis
Nais bretscheri
Nais variabilis
Psammoryctes abicolus
Stylaria lacustris
Stylodrilus heringianus

Hirudinea
Glossiphonia complanata
Hemiclepsis marginata
Theromyzon tessulatum

Mollusca
Anodonta cygnaea
Bithynia tentaculata
Lymnaea stagnalis
Physa fontinalis
Planorbis planorbis
Radix auricularia
Sphaerium corneum
Theodoxus fluviatilis
Unio pictorum

Platyhelminthes
Dendrocoelum lacteum
Dugesia lugubris
Polycelis nigra
Polycelis tenuis

Coleoptera
Dryops ernesti
Dryops luridus
Haliplus fluviatilis
Helichus substriatus
Hydrobius fuscipes
Potamonectes assimilis

Species largely intolerant of organic pollution (oligosaprobic)

Chironomidae
Brillia longifurca
Brillia modesta
Cladotanytarsus mancus
Eukiefferiella clavescens
Eukiefferiella longicalcar

Oligochaeta
Haplotaxis gordioides
Nais pseudobtusa

Mollusca
Ancylus fluviatilis
Bythinella austriaca

(*Continued*)

TABLE 9.11 (*Continued*)

Heterotrissocladius marcidus	*Pisidium supinum*
Microcricotopus bicolor	*Pisidium personatum*
Micropsectra atrofasciata	*Sphaerium solidum*
Microtendipes britteni	
Parametriocnemus stylatus	Platyhelminthes
Psectrocladius dilatatus	*Crenobia alpina*
Pseudodiamesa branickii	*Polycelis felina*
Plecoptera	Trichoptera
Amphinemura sulcicollis	*Agapetus fuscipes*
Brachyptera risi	*Agraylea multipunctata*
Capnia bifrons	*Apatania fimbriata*
Chloroperla torrentium	*Brachycentrus montanus*
Dinocras cephalotes	*Hydropsyche siltali*
Leuctra nigra	*Ithytrichia lamellaris*
Leuctra moselyi	*Lasiocephala basalis*
Nemoura cambrica	*Leptocerus tineiformis*
Nemurella picteti	*Philopotamus montanus*
Perla bipunctata	*Ptilocolepus granulatus*
Perlodes dispar	*Rhyacophila philopotamo ides*
Protonemura meyeri	*Rhyacophila tristis*
Taeniopteryx nebulosa	*Silo nigricornis*
	Silo pallipes
Ephemeroptera	*Trichostegia minor*
Ameletus inopinatus	
Baetis alpinus	Crustacea
Ecdyonurus dispar	*Gammarus pulex*
Ecdyonurus venosus	
Rhithrogena semicolorata	
Paraleptophlebia submarginata	

Reproduced from Hellawell (1986) with permission of Kluwer Academic Publishers, Dordrecht, The Netherlands.

(c) *Erpobdella octoculata* (hirudinea – leech) (1 group).

(d) *Limnaea stagnalis* (mollusca – snail) (1 group).

(e) *Asellus aquaticus* (crustacea – isopod), *Gammarus pulex* (crustacea – amphipod) (2 groups).

(f) *Ephemerella ignita* (ephemeroptera – mayfly) (1 group).

(g) *Baetis rhodani* (ephemeroptera – mayfly) (1 group).

(h) *Metriocnemus hygropetricus* (chironomidae – chironomid) (1 group).

(i) *Chironomus riparius* (chironomidae – chironomid) (1 group).

Therefore there are 10 groups present so column 5 is used. The most pollution-tolerant species is identified next. In this case there are no Plecoptera, but two ephemeropterans are present, but as *Baetis rhodani*, which is tolerant to organic pollution, is counted with the Trichoptera, only *Ephemerella ignita* is counted. Therefore, the score is on line 4. The biotic score is found where line 4 and column 5 coincide which is score *VI* in the above example indicating a water quality of doubtful–fairly clean.

TABLE 9.12   The Trent biotic index

Part 1: Classification of biological samples

Key indicator groups	Diversity of fauna	Total number of group (see Part 2) present					Line no.
		0–1	2–5	6–10	11–15	16+	
Column no.: 1	2	3	4	5	6	7	
		Biotic index					
Plecoptera	More than one species	–	VII	VIII	IX	X	1
nymphs present	One species only	–	VI	VII	VIII	IX	2
Ephemeroptera	More than one species[a]	–	VI	VII	VIII	IX	3
nymphs present	One species only[a]	–	V	VI	VII	VIII	4
Trichoptera	More than one species[b]	–	V	VI	VII	VIII	5
larvae present	One species only[b]	IV	IV	V	VI	VII	6
Gammarus present	All above species absent	III	IV	V	VI	VII	7
Asellus present	All above species absent	II	III	IV	V	VI	8
Tubificid worms and/or red chironomid larvae present	All above species absent	I	II	III	IV	–	9
All above types absent	Some organisms such as *Eristalis tenax* not requiring dissolved oxygen may be present	0	I	II	–	–	10

[a]*Baetis rhodani* excluded.
[b]*Baetis rhodani* (Ephemeroptera) is counted in this section for the purpose of classification.

Part 2: Groups
The term 'Group' here denotes the limit of identification which can be reached without resorting to lengthy techniques. Groups are as follows:

 1 Each species of Platyhelminthes (flatworm)
 2 Annelida (worms) excluding *Nais*
 3 *Nais* (worms)
 4 Each species of Hirudinea (leeches)
 5 Each species of Mollusca (snails)
 6 Each species of Crustacea (Hog-louse, shrimps)
 7 Each species of Plecoptera (stonefly)
 8 Each species of Ephemeroptera (mayfly) excluding *Baetis rhodani*
 9 *Baetis rhodani* (mayfly)
10 Each family of Trichoptera (caddisfly)
11 Each species of Neuroptera larvae (alderfly)
12 Family Chironomidae (midge larvae) except *Chironomus thummi* (=*riparius*)
13 *Chironomus thummi* (blood worms)
14 Family Simulidae (blackfly larvae)
15 Each species of other fly larvae
16 Each species of Coleoptera (beetles and beetle larvae)
17 Each species of Hydracarina (water mites)

TABLE 9.13 The extended version of the TBI

Biogeographical region: Midlands, England		Total number of groups present									
		0–1	2–5	6–10	11–15	16–20	21–25	26–30	30–35	36–40	41–45
		Biotic indices									
Plecoptera nymphs present	More than one species	–	7	8	9	10	11	12	13	14	15
	One species only	–	6	7	8	9	10	11	12	13	14
Ephemeroptera nymphs present	More than one species[a]	–	6	7	8	9	10	11	12	13	14
	One species only[a]	–	5	6	7	8	9	10	11	12	13
Trichoptera larvae present	More than one species[b]	–	5	6	7	8	9	10	11	12	13
	One species only[b]	4	4	5	6	7	8	9	10	11	12
Gammarus present	All above species absent	3	4	5	6	7	8	9	10	11	12
Asellus present	All above species absent	2	3	4	5	6	7	8	9	10	11
Tubificid worms and/or red chironomid larvae present	All above species absent	1	2	3	4	5	6	7	8	9	10
All above types absent	Some organisms such as Eristalis tenax not requiring dissolved oxygen may be present	0	1	2	–	–	–	–	–	–	–

[a]Baetis rhodani excluded.
[b]Baetis rhodani (Ephemeroptera) is counted in this section for the purpose of classification.

TABLE 9.14   The Chandler biotic score index

Groups present in sample	Abundance in standard sample				
	Present (1–2)	Few (3–10)	Common (11–50)	Abundant (51–100)	Very abundant (100+)
	Points scored				
Each species of *Planaria alpina*, Taenopterygidae, Perlidae, Perlodidae, Isoperlidae, Chloroperlidae	90	94	98	99	100
Each species of Leuctridae, Capniidae, Nemouridae (excluding *Amphinemura*)	84	89	94	97	98
Each species of Ephemeroptera (excluding *Baetis*)	79	84	90	94	97
Each species of cased caddis, Megaloptera	75	80	86	91	94
Each species of *Ancylus*	70	75	82	87	91
Each species of *Rhyacophila* (Trichoptera)	65	70	77	83	88
Genera of *Dicranota, Limnophora*	60	65	72	78	84
Genus of *Simulium*	56	61	67	73	75
Genera of Coleoptera, Nematoda	51	55	61	66	72
Genera of *Amphinemura* (Plecoptera)	47	50	54	58	63
Genera of *Baetis* (Ephemeroptera)	44	46	48	50	52
Genera of *Gammarus*	40	40	40	40	40
Each species of uncased caddis (excluding *Rhyacophila*)	38	36	35	33	31
Each species of Tricladida (excluding *P. alpina*)	35	33	31	29	25
Genera of Hydracarina	32	30	28	25	21
Each species of Mollusca (excluding *Ancylus*)	30	28	25	22	18
Each species of Chironomids (excluding *C. riparius*)	28	25	21	18	15
Each species of *Glossiphonia*	26	23	20	16	13
Each species of *Asellus*	25	22	18	14	10
Each species of Leech (excluding *Glossiphonia, Haemopsis*)	24	20	16	12	8
Each species of *Haemopsis*	23	19	15	10	7
Each species of *Tubifex* spp.	22	18	13	12	9
Each species of Chironomus riparius	21	17	12	7	4
Each species of *Nais* spp.	20	16	10	6	2
Each species of air breathing species	19	15	9	5	1
No animal life			0		

Chandler biotic score index

This index overcomes the omission of relative abundance of indicator groups in the TBI by recognizing five levels of abundance and weighting the score of each indicator accordingly. For example, each abundant sensitive species obtains a very high score while each abundant tolerant species obtains a very low score. The individual score for each species is added to give the total score. The lower the total score the greater the degree of pollution (Table 9.14). If no macro-invertebrates were present the score would be 0, while the score for a very clean river is rarely in excess of 1500. Like the TBI, the more difficult groups to identify such as the chironomidae (midge flies) and

TABLE 9.15   Worked example of Chandler score index using the data from Table 9.10

Species	No.	Abundance	Score
Asellus aquaticus	80	Abundant	14
Baetis rhodani	18	Common	48
Chironomus riparius	12	Common	12
Dendrocoelum lacteum	5	Few	33
Ephemerella ignita	1	Present	79
Erpobdella octoculata	2	Present	24
Gammarus pulex	12	Common	40
Lymnaea stagnalis	27	Common	25
Metriocnemus hygropetricus	38	Common	21
Tubifex	8	Few	18
		Biotic score	314

annelida (worms) have both been included as a single group. A worked example is given in Table 9.15.

The main disadvantage to this index is that it requires extensive species identification and the relative abundance of each species to be assessed. This makes it far more time consuming than the TBI. Like all indices the Chandler score index does not provide an absolute value but rather a comparative value of water quality. This is because diversity varies between rivers due to factors independent of organic pollution (e.g. hardness, reaeration, etc.). Other factors, such as sample size and technique, can also affect the recovery of species. This problem can be overcome by dividing the total score by the number of individual scores to give an average score out of 100. This gives a value independent of diversity which can then be compared more widely.

BMWP biotic indices

Biotic indices have been developed independently in a number of European countries, including the UK and Ireland. All are developments from the TBI. In the UK, the BMWP biotic score or the BMWP biotic index was developed between 1975 and 1980. Scores are given for the presence of each key family group, which are then added together (Table 9.16) (Hawkes, 1998). Individual scores range from 1 (organic pollution tolerant) to 10 (intolerant) (Figs 4.4 and 4.5). Total scores range from 0 to 250. While there is an increase in taxonomic demand compared to the TBI, the BMWP index recognizes the need to split up particular groups. For example, the gravel/sand-cased caddis (e.g. *Stenophylax*) is a very clean indicator and so scores 10, while the net-spinning caddis (e.g. *Hydropsyche instabilis*) is pollution tolerant and scores only 5. So, the TBI suffers from the insensitivity of broad groupings. Chironomids, for example, are represented in many different functional feeding groups and are, therefore, tolerant to a wide range of river conditions and pollution (Section 4.1). However, the chironomids, like the oligochaetes, are a bad example as due to identification difficulties they are generally excluded from indices.

TABLE 9.16   The BMWP biotic score

	Families	*Score*
Mayflies	Siphlonuridae, heptageniidae, Leptophlebiidae, Ephemerellidae, Potamanthidae, Ephemeridae, Ecdyonuridae	10
Stoneflies	Taeniopterygidae, Leuctridae, Capniidae, Perlodidae, Perlidae, Chloroperlidae	
River bug	Aphelocheiridae	
Caddisflies	Phryganeidae, Molannidae, Beraeidae, Odontoceridae, Leptoceridae, Goeridae, Lepidostomatidae, Brachycentridae, Sericostomatidae	
Crayfish	Astacidae	8
Dragonflies	Lestidae, Agriidae, Gomphidae, Cordulegasteridae, Aeshnidea, Corduliidae, Libellulidae	
Caddisflies	Psychomyidae, Philopotamiidae	
Mayflies	Caenidae	7
Stoneflies	Nemouridae	
Caddisflies	Rhyacophilidae, Polycentropidae, Limnephilidae	
Snails	Neritidae, Viviparidae, Ancylidae	6
Caddisflies	Hydroptilidae	
Mussels	Unionidae	
Shrimps	Corophiidae, Gammaridae	
Dragonflies	Platycnemididae, Coenagrionidae	
Water bugs	Mesoveliidae, Hydrometridae, Gerridae, Nepidae, Naucoridae, Notonectidae, Pleidae, Corixidae	5
Water beetles	Haliplidae, Hygrobiidae, Dytiscidae, Gyrinidae, hydrophilidae, Clambidae, Helodidae, Dryopidae, Elminthidae, Chrysomelidae, Curculionidae	
Caddisflies	Hydropsychidae	
Craneflies	Tipulidae	
Blackflies	Simuliidae	
Flatworms	Planariidae, Dendrocoelidae	
Mayflies	Baetidae	4
Alderflies	Sialidae	
Leeches	Piscicolidae	
Snails	Valvatidae, Hydrobiidae, Lymnaeidae, Physidae, Planorbidae	3
Cockles	Sphaeriidae	
Leeches	Glossiphoniidae, Hirudidae, Erpobdellidae	
Hog-louse	Asellidae	
Midges	Chironomidae	2
Worms	Oligochaeta (whole class)	1
Hoverflies	Syrphidae	1

As with the Chandler index, and other biotic score systems, a major drawback with the BMWP is that the index does not give a clear total score out of 10, but a variable total score dependent on diversity. This can be overcome to some extent by using the average score per taxon (ASPT) where n is the number of individual scores:

$$\text{ASPT} = \frac{\sum \text{scores}}{n} \qquad (9.11)$$

TABLE 9.17 Rating values for BMWP score and ASPT as it relates to water quality in habitat-rich rif-fle areas; the OQR provides an index value that can be directly related to water usage

BMWP	ASPT	Rating	Water quality	OQR	Index
151+	6.0+	7	Excellent quality	6.0+	A++
121–150	5.5–5.9	6	Excellent quality	5.5	A+
91–120	5.1–5.4	5	Excellent quality	5	A
61–90	4.6–5.0	4	Good quality	4–4.5	C, B
31–60	3.6–4.5	3	Moderate quality	3–3.5	E, D
15–30	2.6–3.5	2	Poor quality	2–2.5	G, F
0–14	0–2.5	1	Very poor quality	1–1.5	I, H

The ASPT is less influenced by seasonal variation than total score, therefore providing consistent estimates of quality regardless of life cycles. Originally two scores were proposed for the BMWP score, one for eroding sub-strata and a second for depositing sub-strata. The ASPT system also eliminates the need for different scoring dependent on substrate characteristics. Also the ASPT is less influenced by sample size than the total score as a larger sample size results in a higher BMWP score. The ASPT largely overcomes problems of variability in factors such as sampling technique, sample size, efficiency of sample processing and substrate (Hawkes, 1998).

Extence et al. (1987) have devised a method to relate the BMWP score and the ASPT to the quality required for specific water uses. The BMWP and ASPT scores are each given a rating according to Table 9.17. An average value (i.e. overall quality rating (OQR)) is calculated using the equation below:

$$OQR = \frac{BMWP\ rating + ASPT\ rating}{2} \tag{9.12}$$

The water quality index (WQI) is then calculated from the OQR using Table 9.17 where an OQR of 3.0 is equivalent to index E, 2.5 to F, 2.0 to G and 1.0 to I. To support a salmonid fishery a minimum index of B is required, for a coarse fishery D, the lowest index for drinking water using minimum treatment is C and D where impoundment is used prior to water treatment. Water with an index of G is only acceptable as a low-amenity resource. A different table is required for poor riffle and pool areas (Extence et al., 1987).

The nature of biological communities is determined by a wide range of entirely natural factors associated with catchment geology, geographical location and climate. So, a good biotic score in one region may be considered to be poor in another, making the use of such indices impracticable for setting national or regional ecological standards. This is overcome to a certain extent by the use of a predictive model called RIVPACS (river invertebrate prediction and classification system). Using eight main environmental variables (i.e. distance from source, substrate type, altitude, discharge category, depth, width, latitude and longitude) and a further six optional variables (including slope, alkalinity, and temperature range and mean), prediction of the macro-invertebrate composition of a river site, not influenced by anthropogenic activities, can be made for any particular season or seasons, excluding winter. In this way the actual macro-invertebrate community can be compared to that predicted from an unpolluted site. The ratio of the observed to predicted community status using BMWP or ASPT is expressed as a quotient (i.e. an environmental quality index or EQI) (see Table 9.28) (Wright *et al.*, 1993). The biological assessment procedure used for UK rivers is fully explained in Section 9.5.

Irish biotic score index

In Ireland, a system developed by the Environmental Research Unit places groups of animals into five broad classes A–E, of which group A is the most sensitive, B is less sensitive, C is relatively tolerant, D tolerant and E comprises of the most tolerant forms (Table 9.18) (McGarrigle *et al.*, 1992).

Using combinations of the relative abundance of these groups in either eroding or depositing sites then water quality is expressed as a Q value where Q1 is bad and Q5 is excellent water quality (Table 9.19).

Biotic index (Q value)	Water quality
5 (diversity high)	Good
4 (diversity slightly reduced)	Fair
3 (diversity significantly reduced)	Doubtful
2 (diversity low)	Poor
1 (diversity very low)	Bad

Intermediate values are possible (e.g. Q4–5, Q3–4, Q2–3, Q1–2), so that there are nine Q groupings overall. From this four categories of water quality are calculated:

Q value	Category of river quality
Q5, Q4–5, Q4	Unpolluted
Q3–4	Slightly polluted
Q3, Q2–3	Moderately polluted
Q2, Q1–2, Q1	Seriously polluted

Toxic influences are indicated by use of the suffix 0 (e.g. Q1/0). The relationship between the quality rating Q and water quality parameters is shown in Table 9.20.

The *Gammarus*: *Asellus* ratio

This simple index of organic pollution was devised for non-biologists by Whitehurst (1991). All that is required is the ability to distinguish these very familiar organisms from other river biota, a simple collecting technique and recording method. *Gammarus*

TABLE 9.18 The Irish quality rating (Q) is based on the relative proportions of characteristic groups of species indicating specific quality groups

Indicator groups: key taxa

Group A Sensitive forms	Group B Less sensitive forms	Group C Relatively tolerant forms	Group D Tolerant forms	Group E Most tolerant forms
Plecoptera (*excluding* *Leuctra*, Nemouridae), Heptageniidae, Siphlonuridae	*Leuctra,* Nemouridae, Baetidae (excluding *B. rhodani*), Leptophlebiidae, Ephemerellidae, Ephemeridae, Cased Trichoptera (excluding Limnephilidae, Hydroptilidae, Glossosomatidae), Odonata (excluding Coenagriidae), *Aphelocheirus, Rheotanytarsus*	Tricladida, Ancylidae, Neritidae, Astacidae, *Gammarus,* *Baetis rhodani,* Caenidae, Limnephilidae, Hydroptilidae, Glossosomatidae, Uncased Trichoptera, Coleoptera, Coenagriidae, Sialidae, Tipulidae, Simuliidae, Hemiptera (excluding *Aphelocheirus*) Hydracarina	Hirudinea, Mollusca (excluding Ancylidae, Neritidae), *Asellus,* Chronomidae (excluding *Chironomus* and *Rheotanytarsus*)	Tubificidae, *Chironomus*

Reproduced from Newman *et al.* (1992) with permission of the European Commission, Luxembourg.

TABLE 9.19 Calculation of the Irish quality rating (Q) based on the relative proportion of each group described in Table 9.18

Quality rating (Q)	Relative abundance of indicator groups				
	A	B	C	D	E
Eroding sites					
Q5	++++	+++	++	+−	+−
Q4	++	++++	+++	++	+−
Q3	−	+−	++++	+++	++
Q2	−	−	+−	++++	+++
Q1	−	−	−	+−	++++
Depositing sites					
Q5	+−	++++	+++	++	+−
Q4	−	++	++++	++	+−
Q3	−	+−	++	+++	+++
Q2	−	−	+−	+++	+++
Q1	−	−	−	−	++++

++++: well represented or dominant; +++: may be common; ++: may be present in small numbers; +−: sparse or absent; −: usually absent.

TABLE 9.20 Relationship between Irish quality rating (Q) and expected water and biological quality criteria

Class	Class A		Class B		Class C	
Q ratings / Water quality	5 / Good	4 / Fair	3–4 / Doubtful to fair	3 / Doubtful	2 / Poor	1 / Bad
Pollution status	Unpolluted	Unpolluted	Slight Pollution	Moderate pollution	Heavy pollution	Gross pollution
Biodegradable organic wastes	Absent	Absent	Absent	In advanced stages of mineralization	Heavy load	Very heavy load
BOD	Normal, i.e. <3 mg/l	Normal, i.e. <3 mg/l	Close to or normal	May be high at times	High	Very high
Dissolved oxygen	Typically ranges from 80 to 120% of saturation	May fluctuate above and below 80–120%	Fluctuates widely	Fluctuates very widely. Potential fish kills	Low during the day. May be zero at night	Very low or zero at all times
Bottom siltation	None	None	May be light	May be considerable	Heavy	Heavy and commonly anaerobic
'Sewage fungus'	Absent	Absent	Absent	May be small amounts	Usually abundant	Usually abundant
Algae	Diverse communities not excessive in development	Moderate to abundant developments	Abundant	Abundant. May completely blanket riverbed	May be abundant	Ranges from none to abundant
Macrophytes	Usually diverse communities. Development	May be well developed	Usually abundant	Abundant. May completely overgrow river if blanket algae allows	Tolerant forms only. May be abundant	Only the most tolerant types
Macro-invertebrates (from fast areas)	Usually diverse communities. Sensitive species may be common	Some reduction in diversity, density increases	Sensitive forms absent or scarce. Total numbers may be very high	Sensitive forms absent. Diversity falls. Tolerant species common	Tolerant forms only	Only the most tolerant types or none
Potential beneficial uses	High-quality waters suitable for supply and all other fisheries. High amenity value	Waters of somewhat less high quality than Q5 but usable for substantially the same purposes	Usually good game fisheries but fish at risk due to possible fluctuations in dissolved oxygen. Suitable for supply. Moderate to high amenity value	Coarse fisheries. Not likely to support a healthy game fishery. Suitable for supply after advanced treatment	Fish absent or only sporadically present. May be used for low-grade industrial abstraction. Low amenity value	Fish absent. Likely to produce nuisance smells. Very low or zero amenity value
Condition	Satisfactory	Satisfactory	Borderline	Unsatisfactory	Unsatisfactory	Unsatisfactory

pulex normally inhabits well-oxygenated riffle areas of rivers, living between stones, and intolerant of organic pollution. In contrast *Asellus aquaticus* is not normally part of the riffle community, but is associated with depositing silty substrates. It will, however, invade riffle areas during organic enrichment where it often becomes the dominant organism. Therefore, large numbers of *Asellus* in riffle areas indicate organic pollution, with the ratio of *Gammarus* to *Asellus* able to successfully highlight areas suffering from organic enrichment. There is a strong negative correlation between BOD, NH_4, PO_4 and the ratio, the ratio being highest when these are all low. The main problem with this index is that distribution of crustaceans is severely restricted in acid (soft) waters.

9.3.2 DIVERSITY INDICES

Pollution alters community structure by reducing species diversity and increasing the size of populations within that community. So, unpolluted clean water has a high diversity with most species present having a low abundance, while polluted dirty water has a low diversity with most species having a large or high abundance. Environmental stress frequently reduces community diversity. Therefore changes in a diversity index, or any other parameter of community structure, can be used to assess the degree of environmental stress, assuming that the change in index value is related to the intensity of pollution.

Indices of community diversity

There are a number of different indices that relate the number of species to the total number of individuals. For example,

Menhinick's index:

$$I = \frac{S}{\sqrt{N}} \tag{9.13}$$

Margalef's index:

$$I = \frac{S - 1}{\log_e N} \tag{9.14}$$

where I is the diversity index, S is the number of species present and N is the total number of individuals (Hellawell, 1978). These indices are not independent of sample size so careful comparative sampling must be carried out. It is theoretically possible for identical scores to be obtained from quite different communities (Table 9.21). In cases of moderate pollution, diversity may not be reduced significantly although there may be a change in species composition, with more tolerant species replacing less tolerant ones.

TABLE 9.21 Comparison between different communities with identical diversity index values using the Margalef and Menhinick indices

Community	Number of individuals of species					Total number of individuals N	Number of species S	Index value Margalef I	Menhinick I
	n_1	n_2	n_3	n_4	n_5				
A	20	20	20	20	20	100	5	0.87	0.5
B	40	30	15	10	5	100	5	0.87	0.5
C	96	1	1	1	1	100	5	0.87	0.5

Reproduced from Hellawell (1978) with permission of the Water Research Centre plc, Medmenham.

Therefore, diversity indices can miss subtle changes in water quality. This is overcome to a certain extent by the sequential comparison index (SCI).

The most widely used diversity index is the Shannon–Wiener (i.e. the Shannon index H'):

$$H' = -\sum_{i=1}^{s} \frac{n_i}{n} \log_e \left(\frac{n_i}{n} \right) \tag{9.15}$$

where n_i is the number of individuals of species i and n the total number of individuals in the sample and s is the number of species in the sample. The calculation for each species is added together to give the index H'. A value of $H' > 3$ indicates clean water while values <1 indicate severe pollution with values rarely exceeding 4.5. Intermediate values are characteristic of moderate pollution (Hellawell, 1986).

EXAMPLE 9.6 Using the data in Table 9.15 calculate the Shannon–Wiener diversity index using Equation (9.15).

Species	n_i	n_i/n	$\log_e (n_i/n)$	$-\Sigma(n_i/n) \log_e (n_i/n)$
Asellus aquaticus	80	0.394	−0.931	0.367
Baetis rhodani	18	0.089	−2.423	0.216
Chironomus riparius	12	0.059	−2.828	0.167
Dendrocoelum lacteum	5	0.025	−3.704	0.093
Ephemerella ignita	1	0.0049	−5.313	0.026
Erpobdella octoculata	2	0.0099	−4.620	0.046
Gammarus pulex	12	0.059	−2.828	0.167
Lymnaea stagnalis	27	0.133	−2.017	0.268
Metriocnemus hygropetricus	38	0.187	−1.676	0.313
Tubifex tubifex	8	0.039	−3.234	0.126
	$n = 203$			$H' = 1.789$

A H' value of 1.79 is doubtful to fair quality.

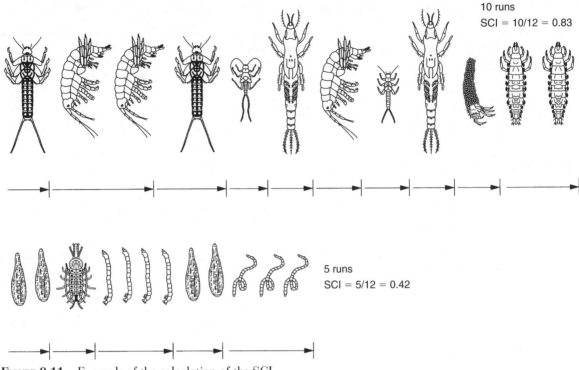

FIGURE 9.11 Example of the calculation of the SCI.

Sequential comparison index

This simple diversity index requires no taxonomic knowledge only the ability to differentiate between different specimens by shape, size and colour. It estimates relative differences in biological diversity by distinguishing runs of individual specimens in a sample row of individuals of known number. The collected individuals are randomly aligned against the straight edge of a sampling tray and then each current specimen is compared with the previous one. When a difference is found then this is classed as the end of a run and so on (Fig. 9.11). The diversity index (DI) is calculated as a score (DI_1)

$$DI_1 = \frac{\text{Number of runs}}{\text{Number of specimens}} \qquad (9.16)$$

The main difficulty of this technique is obtaining a random sequence of macroinvertebrates from the sample. A minimum of 250 individuals should be used to calculate a score. Smaller samples can be used and the animals remixed randomly between counts to obtain the minimum sample size, although a minimum of 50 individuals should be used for each count. The closer the score is to 1.0 the more diverse the sample, while the closer to zero the less diverse and so the more polluted the sample (Hellawell, 1986).

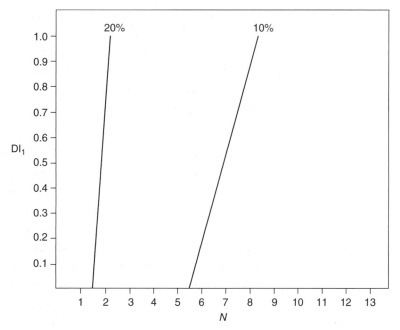

FIGURE 9.12
The number of replicates (N) of DI_1 required to give diversity index within 10 and 20% of the true value with 95% confidence limits in the calculation of the SCI. (Reproduced from Hellawell (1978) with permission of the Water Research Centre plc, Medmenham.)

In order to obtain a mean score within 20% or 10% of the true value with 95% confidence limits, the DI_1 should be repeated between two and eight times, respectively, to give a mean score (DI_2) (Fig. 9.12). The DI_2 score is a comparative value within fixed limits, however some workers have multiplied by the total number of species to give a third score (DI_t), although this needs taxonomic expertise:

$$DI_t = DI_2 \times \text{Number of taxa} \tag{9.17}$$

9.3.3 USING BIOLOGICAL INDICES

Each index provides slightly different information and the selection of indices should be based on the following factors:

(a) Value factor: amount, type and suitability of information obtained.
(b) Cost factor: actual cost of staff and investment in training including taxonomic expertise – this should include the time to do the collection of samples, sorting animals and doing the necessary identification and counts.
(c) Accuracy factor: the precision, repeatability and reproducibility of the index (i.e. the chance of making a mistake, and the effect of such a mistake on the overall value and interpretation of the data; the ability to obtain the same value using different and/or identical methods and personnel for collection, sorting, identification and counting).

Overall the TBI and Chandler score index are both very useful for water management purposes, but there are very large differences between the cost factors. For example, the TBI can be done in the field and an assessment made within 30 min, while the Chandler score index has to be calculated in the laboratory involving full sorting, counting and identification which could take up to 6 h per sample, depending on experience. Some indices are more appropriate for research, others for routine monitoring, others for EIA. However, once a full sort has been made and all the specimens identified and counted, then all the indices can be used with the exception of the SCI, for which the original mixed sample is required.

Biological indices have a number of disadvantages. For example, factors other than pollution can affect the streambed community, such as substrate, alkalinity, and the natural variation in allochthonous inputs. Industrial pollution may cause different conditions to those caused by sewage so that pollution (biotic) indices are not applicable. Accurate identification is difficult and time-consuming, especially of certain groups. Absence of some species may reflect normal life cycle and this may distort seasonal scores. Natural events, such as severe flooding, can remove macrophytes, organic material and even substrate. Quantitative sampling of macro-invertebrates is difficult and it is important to choose similar substrates when sampling. Pollution (biotic) indices only respond to organic pollution while diversity indices are used to measure non-organic or unknown pollutants. Diversity indices have been criticized as being potentially unreliable primarily due to season, size of sample and level of identification (Pinder *et al.*, 1987). However, other factors also been identified as affecting diversity indices, such as sampling method, depth of sampling and duration of sampling. The assumption is made that in more diverse communities a lower probability exists that individuals chosen at random will belong to the same species. However in practice species often have a clumped, rather than random distribution due to microhabitat, breeding or behaviour. Abundance in particular is also readily influenced by extraneous factors such as seasonality, within-site variation in substrate characteristics such as organic content, etc. For that reason the use of species diversity (S) only may be more indicative of environmental change.

However, because of their relatively long life cycles macro-invertebrates are good monitors of the long-term purity of water. Periodic sampling may, therefore, be more representative than chemical monitoring. Certain groups can give misleading signals. For example, the snail *Physa* is tolerant of organic pollution but very intolerant of Cu^{2+}. The *Plecoptera*, on the other hand, are intolerant of organic pollution but fairly tolerant of heavy metal pollution. So, care must be taken when using individual groups as indicators. Almost all the biotic indices have been developed exclusively for lotic systems, and often those species that are indicative of clean water in rivers do not generally live in lentic systems (Section 4.4). Alternative indices using chironomids or oligochaetes that dominate fine sediments have been developed for assessing water quality in lakes (Premazzi and Chiaudani, 1992). Ephemeroptera, trichoptera and Plecoptera are all sensitive to most types of pollutants just as Diptera and tubificid worms are tolerant. Therefore, these groups are often used as indicators (Rosenberg and Resh, 1993).

The move towards multivariate analysis and the use of predictive models, such as RIVPACs and AUSRIVAS (Section 9.3.4), requires a very high level of taxonomic expertise, being able to identify a wide range of organisms to species level with a high level of accuracy. This is a reversal of the trend for simpler and more rapid biological assessment techniques of the past 40 years. Other organisms apart from macro-invertebrates are also used for assessing water quality and include bacteria, algae, macrophytes and fish. Macrophytes in particular have been very successful indicators of river water quality being broadly classified into four groups representing a down-stream continuum from upland oligosaprobic to lowland eutrophic (Holmes *et al.*, 1998). Algae has also been used to assess eutrophication in rivers and lakes (Kelly and Whitton, 1998). For example, the Algal Abundance Index (AAI) is based on a semi-quantitative scale that ranks algal abundance or cover as abundant, common, rare or absent. The index is calculated as:

$$AAI = \frac{(\text{No. of abundant records} \times 2) + \text{No. of common records}}{\text{No. of site visits}} \times 10 \qquad (9.18)$$

Values <20 indicate oligotrophic conditions while >70 indicates hypereutrophication (Marsden *et al.*, 1997). Kelly (1998) has proposed a more sophisticated index using diatoms called the Trophic Diatom Index (TDI).

Today a wide range of biochemical, serological and cellular tests have been used to provide data on biomass, metabolism, etc. in an attempt to monitor pollution in fresh-waters but these all require elaborate techniques, instrumentation and are time con-suming (Table 9.22).

9.3.4 MULTIVARIATE ANALYSIS

Indices reduce complex data into single values for ease of analysis and interpretation. However, full data sets allow for greater understanding of the effects and responses of the biota to changes in water quality. Allowing the relationship between communities, individual species and water quality variables to be explored, thus relating community modification to environmental change.

There are a number of approaches to multivariate analysis with cluster analysis and ordination principally used to interpret biological and physico-chemical data (Norris and Georges, 1993; Waite, 2000). Two-way indicators species analysis (TWINSPAN) and detrended correspondence analysis (DECORANA) have become the two most widely used multivariate methods for analysing benthic community data. Sites can be arranged into a hierarchy based on their taxonomic composition using TWINSPAN. This allows sites to be clustered identifying species that demonstrate the greatest difference between sites. Ordination is carried out by DECORANA, which identifies sites with similar taxonomic composition. It then relates the ordinated groups to specific measured physico-chemical variables. An excellent example of their

TABLE 9.22 Principal biological approaches to water quality assessment, their uses, advantages and disadvantages

	Ecological methods		Microbiological methods	Physiological and biochemical methods	Bioassays and toxicity tests	Chemical analysis of biota	Histological and morphological studies
	Indicator species[a]	Community studies[b]					
Principal organisms used	Invertebrates, plants, algae	Invertebrates	Bacteria	Invertebrates, algae, fish	Invertebrates, fish	Fish, shellfish, plants	Fish, invertebrates
Major assessment uses	Basic surveys, impact surveys, trend monitoring	Impact surveys, trend monitoring	Operational surveillance impact surveys	Early warning monitoring, impact surveys	Operational surveillance, early warning monitoring impact surveys	Impact surveys, trend monitoring	Impact surveys, early warning monitoring, basic surveys
Appropriate pollution sources or effects	Organic matter pollution, nutrient enrichment, acidification	Organic matter pollution, toxic wastes, nutrient enrichment	Human health risks (domestic and animal faecal waste), organic matter pollution	Organic matter pollution, nutrient enrichment, toxic wastes	Toxic wastes, pesticide pollution, organic matter pollution	Toxic wastes, pesticide pollution, human health risks (toxic contaminants)	Toxic wastes, organic matter pollution, pesticide pollution
Advantages	Simple to perform. Relatively cheap. No special equipment or facilities needed	Simple to perform. Relatively cheap. No special equipment or facilities needed. Minimal biological expertise required	Relevant to human health. Simple to perform. Relatively cheap. Very little special equipment required	Usually very sensitive. From simple to complex methods available. Cheap or expensive options. Some methods allow continuous monitoring	Most methods simple to perform. No special equipment or facilities needed. Fast results. Relatively cheap. Some continuous monitoring possible	Relevant to human health. Requires less advanced equipment than for the chemical analysis of water samples	Some methods very sensitive. From simple to complex methods available. Cheap or expensive options
Disadvantages	Localized use. Knowledge of taxonomy required. Susceptible to natural changes in aquatic environment	Relevance of some methods to aquatic systems not always tested. Susceptible to natural changes in aquatic environment	Organisms easily transported, therefore may give false positive results away from source	Specialized knowledge and techniques required for some methods	Laboratory based tests not always indicative of field conditions	Analytical equipment and well-trained personnel necessary. Expensive	Specialized knowledge required. Some special equipment needed for certain methods

[a]For example, biotic indices; [b]For example, diversity or similarity indices.
Reproduced from Chapman (1996) with permission of UNESCO, Paris.

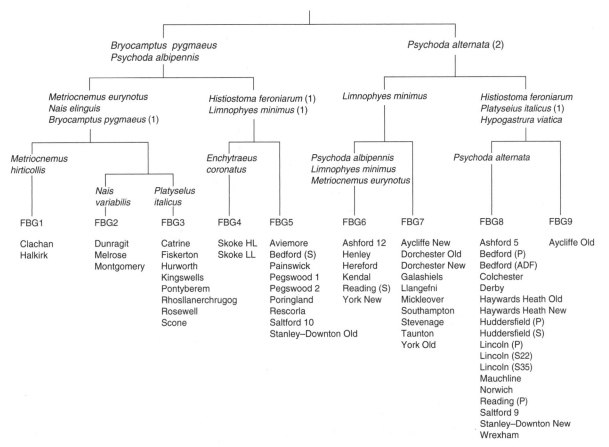

FIGURE 9.13 TWINSPAN classification of the percolating filter beds based on macro-invertebrate species. The indicator species for various divisions are shown. The parentheses next to an indicator species denotes abundance threshold: $1 = >100\,l^{-1}$ medium; $2 = >1000\,l^{-1}$ medium. (Reproduced from Learner and Chawner (1998) with permission of the British Ecological Society, London.)

use has been in the understanding of the role of macro-invertebrates in percolating filters (Section 16.2) (Learner and Chawner, 1998). Using TWINSPAN, detailed species and abundance lists from 59 separate filter beds were grouped into nine species associations (Fig. 9.13). DECORANA ordination was then used to determine the abiotic variables affecting faunal composition (Fig. 9.14). The study showed that the abiotic variables most likely to effect faunal composition were organic loading, film accumulation, air temperature, size of media and age of the bed. For the first time it showed that as all these variables except air temperature are operator controlled, that the invertebrate community structure could be modified or even managed. The software for these two programmes is widely available and has been recently adapted by the Centre for Ecology and Hydrology in England for use on normal PCs.

The enormous databases that have been collected over decades by national monitoring programmes have allowed multivariate analysis to produce more accurate ways of assessing water quality and environmental change through biological assessment. Using environmental data it is now possible to predict the probability of particular species

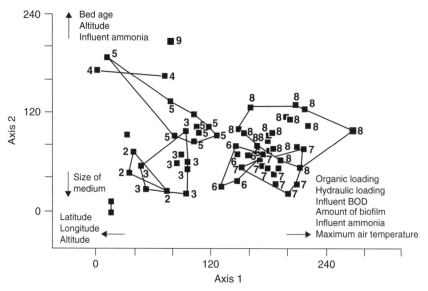

FIGURE 9.14 DECORANA ordination of the percolating filter beds. The filter-bed group to which each bed belongs, as classified by TWINSPAN (Fig. 9.13), is indicated by the numerals. The environmental variables that correlated with the axes are also shown; the arrows indicate the direction of increasing magnitude. (Reproduced from Learner and Chawner (1998) with permission of the British Ecological Society, London.)

making up the community structure at a particular site. Deviation from the predicted assemblage of species is used as the basis of assessing water quality or impact. This is the basis of RIVPACS (Section 9.3.1) which is used as the basis for biological general water quality assessment (GQA) by both the Environment Agency in England and Wales and the Scottish Environmental Protection Agency in Scotland (Section 9.5). The latest version, RIVPACS III+, provides site-specific predictions of the macro-invertebrate fauna in non-impacted situations. It does this for individual seasons (i.e. spring, summer and autumn) or for pairs or all three seasons combined. Prediction is based on up to 12 environmental variables (Section 9.3.1). The software produces predicted species lists, as well as BMWP or ASPT scores, which are then compared with field results, collected using strict field and laboratory protocols to provide a site assessment. The predictive software is based on 35 classification groups that have been derived from the macro-invertebrate fauna recorded at 614 high-quality (pristine) reference sites distributed throughout the British Isles. In total, 637 species of macro-invertebrate were recorded at these sites. Similar systems have been developed around the world; one of the most interactive is AUSRIVAS, which is used in Australia.

9.4 CHEMICAL DATA

9.4.1 CHEMICAL INDICES

Many WQI have been developed to analyse chemical data, and while many require complex computation there are several that are simple to use (House and Ellis, 1987).

One of the most well known was developed for use in Scotland (Scottish Development Department, 1976). Key variables are selected and given different weightings to produce a water quality rating table (Table 9.23). From this a rating or score is obtained for each chemical value:

$$\text{WQI} = \frac{\left(\displaystyle\sum_{i=1}^{n} \text{Water quality ratings}\right)^2}{100} \tag{9.19}$$

where n is the number of water quality parameters rated in Table 9.24.

The weightings ensure that certain parameters affect the overall score more or less depending on their significance for the water use for which the index has been devised. An example of the calculation of the WQI is given in Table 9.24.

The advantage of a WQI is that it is more sensitive than the broad classifications used in some biotic indices. Also by using a number of different variables a clear overall picture of quality is obtained than if simply one or two variables were used for assessment. They are particularly useful where the river is essentially clean and some idea of the effect of remedial measures is required. Such indices can be linked very successfully to models and the setting of consent conditions. WQI can be devised for a wide variety of different water pollution conditions or water uses by using key indicator parameters. For example, for organic pollution particular emphasis would be on BOD, NH_4 and dissolved oxygen, for eutrophication the key parameters would be total phosphorus, total nitrogen, temperature and dissolved oxygen. An example of a WQI to measure the impact of acid mine drainage on surface waters is described by Gray (1996).

9.4.2 MASS BALANCE AND MODELLING

A mass balance approach is used to obtain maximum benefit from a river's natural self-purification capacity. The capacity of a river to assimilate waste (i.e. the waste assimilation capacity (WAC)) provides a link between water quality standards and effluent emission standards (fixed by consent). As the river alters both chemically and physically along its length then the WAC must be calculated at particular points along the river. This can be done using simple mathematical models (for example Streeter–Phelps) or computer simulation models (for example Qual II, etc.) or by using mass balance (McGarity, 1977). Calculation of the effects of organic discharges on the oxygen concentration of rivers is discussed in Section 5.3.

The WAC is calculated as:

$$\text{WAC} = (C_{\text{max}} - C_{\text{back}}) \times F_{95} \times 86.4 \text{ kg BOD day}^{-1} \tag{9.20}$$

where C_{max} is the maximum permissible BOD (mg l^{-1}) which in Ireland is 3 mg l^{-1} for salmonid and 6 mg l^{-1} for cyprinid waters, C_{back} is the background (upstream) BOD

TABLE 9.23   Weighted water quality rating table for the variables used in the calculation of the WQI

Weighted quality rating	DO (% sat)	BOD_5 (mg l⁻¹)	Ammonia (mg l⁻¹)	E. coli (no./100 ml)	pH	Total oxidized nitrogen (mg l⁻¹)	Ortho-phosphate (mg l⁻¹)	Sus. solids (mg l⁻¹)	Conductivity (µS cm⁻¹)	Temperature (°C)
18	93–109									
17	88–92									
16	85–87									
15	81–84	0–0.9								
14	78–80	1.0–1.9								
13	75–77	2.0–2.4								
12	72–74	2.5–2.9	0–0.09	0–249						
11	69–71	3.0–3.4	0.10–0.14	250–999						
10	66–68	3.5–3.9	0.15–0.19	1000–3999						
9	63–65	4.0–4.4	0.20–0.24	4000–7999	6.5–7.9					
8	59–62	4.5–4.9	0.25–0.29	8000–14 999	6.0–6.4 8.0–8.4	0–0.49	0–0.029			
7	55–58	5.0–5.4	0.30–0.39	15 000–24 999	5.8–5.9 8.5–8.7	0.50–1.49	0.030–0.059	0–9		
6	50–54	5.5–6.1	0.40–0.49	25 000–44 999	5.6–5.7 8.8–8.9	1.50–2.49	0.060–0.099	10–14	50–189	
5	45–49	6.2–6.9	0.50–0.59	45 000–79 999	5.4–5.5 9.0–9.1	2.50–3.49	0.100–0.129	15–19	0–49; 190–239	0–17.4
4	40–44	7.0–7.9	0.60–0.99	80 000–139 999	5.2–5.3 9.2–9.4	3.50–4.49	0.130–0.179	20–29	240–289	17.5–19.4
3	35–39	8.0–8.9	1.00–1.99	140 000–249 999	5.0–5.1 9.5–9.9	4.50–5.49	0.180–0.219	30–44	290–379	19.5–21.4
2	25–34	9.0–9.9	2.00–2.99	250 000–429 999	4.5–4.9 10.0–10.4	5.50–6.99	0.220–0.279	45–64	380–539	21.5–22.9
1	10–24	10.0–14.9	4.00–9.99	430 000–749 999	3.5–4.4 10.5–11.4	7.00–9.99	0.280–0.369	65–119	540–839	23.0–24.9
0	0–9	15+	10+	750 000+	0–3.4 11.5–14	10+	0.370+	120+	840+	25+

DO: dissolved oxygen; BOD_5: 5-day BOD; Sus. solids: suspended solids.
Reproduced from Nutall (1983) with permission of Enterprise Ireland, Dublin.

TABLE 9.24   The calculation of the WQI using the rating table (Table 9.23); in this example the conductivity value is missing so that the new total is 94 which is used to adjust the calculation.

	Variable								
	DO	BOD$_5$	NH$_4$-N	E. coli	pH	TON	Ortho-P	Sus. solids	Temper-ature
Observed value	52	3.6	0.9	17 000	7.2	1.22	0.28	13	14
Water quality rating	6	10	4		7 9	7	1	6	5

Sum (Σ) of water quality ratings of the nine variables = 55.

Correction factor for the missing conductivity rating = 1/0.94.
Equation used to calculate WQI = (Σ water quality ratings)^2/100.

Therefore in this example WQI = $(55 \times (1/0.94))^2/100 = 34.2$.

BOD$_5$: 5-day BOD; DO: dissolved oxygen; Sus. solids: suspended solids; TON: total oxidized nitrogen.

(mg l^{-1}), F_{95} is the 95th percentile flow (m^{3}s^{-1}), and 86.4 the factor to convert units to load per day. Therefore, the WAC dictates the possible development of the river in terms of organic loading and secondly, how much treatment dischargers must provide to their waste water in terms of BOD removal prior to discharge.

EXAMPLE 9.7   Calculate the available waste assimilative capacity of a salmonid river with a BOD of 1.8 mg l^{-1} and a 95th percentile flow of 0.75 m^{3}s^{-1}.

Using Equation (9.20):

$$WAC = (3.0 - 1.8) \times 0.75 \times 86.4 = 77.8 \text{ kg BOD day}^{-1}$$

The concentration of a pollutant downstream of a discharge can be calculated using the mass balance equation where the river and effluent discharge rates, the concentration of the pollutant upstream in the river and in the effluent are all known:

$$T = \frac{FC + fc}{F + f} \tag{9.21}$$

where F is river flow upstream (m^{3}s^{-1}), C is the concentration of pollutant upstream (mg l^{-1}), f is the effluent flow (m^{3}s^{-1}), c is the effluent concentration (mg l^{-1}) and T is the concentration of pollutant downstream (mg l^{-1}).

Thus the permitted (consent) concentration of pollutant that can be safely discharged can be calculated as:

$$c = \frac{T(F + f) - CF}{f} \tag{9.22}$$

where T is now the water quality standard set for that stretch of river and c the concentration permitted to achieve it.

An alternative formula for the calculation of T is:

$$T = \frac{P_{eff} + (P_{us} \times \text{Dilution factor})}{\text{Dilution factor} + 1} \tag{9.23}$$

where P_{eff} is the effluent pollutant concentration, P_{us} is the upstream pollutant concentration (mg l^{-1}) and dilution factor is F/f.

EXAMPLE 9.8 Calculate the downstream BOD concentration (T) of a pollutant after an effluent discharge using mass balance. The effluent discharge rate (f) is $0.05 \, \text{m}^3\text{s}^{-1}$ and its BOD (c) 200 mg l^{-1}. The receiving water discharge rate (F) is $12.5 \, \text{m}^3\text{s}^{-1}$ with a mean BOD (C) of 2 mg l^{-1}.

(a) Using the standard mass balance Equation (9.21):

$$T = \frac{(12.5 \times 2) + (0.05 \times 200)}{12.5 + 0.05} = \frac{25 + 10}{12.55} = 2.8 \, \text{mg l}^{-1}$$

(b) Alternatively Equation (9.23) can be used:

$$T = \frac{200 + (2 \times 250)}{250 + 1} = 2.8 \, \text{mg l}^{-1}$$

The toxicity of ammonia in water depends on its form (i.e. ionized or unionized), which is controlled by the pH and temperature of the water (Fig. 6.11). Equilibrium reactions in water control the form that ammonia takes:

$$NH_3 + H_2O \leftrightarrow NH_3 \cdot H_2O \tag{9.24}$$

$$NH_3 \cdot H_2O \leftrightarrow NH_4^+ + OH^- \tag{9.25}$$

Unionized (free or gaseous forms) ammonia (NH_3) is most harmful to freshwater life and to fish in particular. As the pH increases, becoming more alkaline, the greater the proportion of the total ammonia ($NH_3 + NH_4^+$) transformed from the less toxic ionized form (NH_4^+) to the unionized form. Critical concentrations of total ammonia that should not be exceeded in freshwaters at specific temperatures and pH are given in Table 9.25. As a significant shift in pH can be caused by algal activity (Fig. 6.10), total ammonia levels should be very low in rivers susceptible to eutrophication (Table 9.26).

TABLE 9.25 Concentrations $(mg\,l^{-1})$ of total ammonia in freshwater that contain an unionized ammonia concentration of $0.02\,mg\,NH_3\,l^{-1}$ (the threshold concentration of $0.02\,mg\,NH_3\,l^{-1}$ is the long-term toxic effect level for fish; lethal levels are about 10 times greater)

Temperature (°C)	pH							
	6.0	6.5	7.0	7.5	8.0	8.5	9.0[a]	9.5[a]
5	160	51	16	5.1	1.6	0.53	0.18	0.071
10	110	34	11	3.4	1.1	0.36	0.13	0.054
15	73	23	7.3	2.3	0.75	0.25	0.09	0.043
20	50	16	5.1	1.6	0.52	0.18	0.07	0.036
25	35	11	3.5	1.1	0.37	0.13	0.055	0.031
30	25	7.9	2.5	0.8	0.27	0.10	0.045	0.028

[a] Criteria may be unduly low if there is low free carbon dioxide in the water.

TABLE 9.26 The G and I values for ionized (NH_4^+) and unionized (NH_3) ammonia $(mg\,l^{-1})$ set by the European Community (EC) Freshwater Fish Directive (78/659/EEC)

Fishery	G (guide)		I (mandatory)	
	NH_4^+	NH_3	NH_4^+	NH_3
Salmonid	≤0.04	≤0.005	≤1.0	≤0.025
Cyprinid	≤0.2	≤0.005	≤1.0	≤0.025

Reproduced with permission of the European Commission, Luxembourg.

While there has been much commercial interest in the development of computer simulation models to examine contaminant behaviour in groundwater, estuaries and coastal waters, there are few that deal with surface freshwaters. For that reason modelling has centred around four models produced by the US Environmental Protection Agency. These models are widely available and are used both for regulatory purposes as well as research.

Enhanced Stream Water Quality Model (QUAL2E) is designed for well-mixed lakes and dendritic riverine systems. It can use up to 15 water quality parameters and can simulate benthic and carbonaceous oxygen demand, atmospheric reaeration and their effects on dissolved oxygen dynamics, nutrient cycles and algal production. Its main use is in the determination of total maximum daily loads (TMDLs) to surface waters for catchment management. It can be used to model the effects of both point and diffuse sources of pollution. *Water Quality Analysis Simulation Program (WASP5)* is able to simulate and predict the transport and fate of a wide range of contaminants in surface waters. *Storm Water Management Model (SWMM)* is the most widely used storm water run-off simulation and predictive model as it also allows water quality data to be used as well as hydraulic data. It simulates the movement of water and contaminants from surface run-off via the storm water and sewer network to the receiving water. Akan and Houghtalan (2003) give examples of its application and compare it to the Hydraulic Engineering Center Hydraulic Modelling System (HEC-HMS) produced by the US Army Corps of Engineers. The final Environmental

Protection Agency (EPA) model is a watershed hydrology model (*Hydrological Simulation Program FORTRAN (HSPF)*). This integrates stream water quality with hydrological data providing a powerful predictive tool. The model requires extensive real-time climatic data to predict the effects of run-off on receiving waters and sediment quality.

In the UK, regulatory authorities and water companies primarily use the model PC-QUASAR to manage river water quality. Originally developed by the Centre for Ecology and Hydrology the software can simulate and predict river flow and physico-chemical quality (i.e. ammonia, BOD, dissolved oxygen, *E. coli*, pH, nitrate, temperature, and a conservative pollutant or tracer). It is also widely used to calculate consent conditions in order to meet river water quality standards. By providing river flow and water quality estimates at each reach boundary over a period of time, proposed changes in the river's use, flow or quality can be assessed.

All these models are widely available and now run on Windows. Although the software is largely available as free downloads from the Internet, they do need access to a manual and technical support if they are to be successfully employed. As the programmes are constantly being refined and updated it is advisable to purchase supported software.

9.5 THE UK GENERAL QUALITY ASSESSMENT CLASSIFICATION SCHEME

The Environment Agency has been developing a new water quality assessment procedure since the WFD was proposed. The new GQA classification scheme for rivers uses six GQA grades of quality, A to F, representing very good to bad quality, respectively. Assessment is based on four different GQA methods (i.e. chemical, biological, nutrient and aesthetic) (Environment Agency, 1998).

Chemical GQA
This is an assessment against chemical standards (i.e. dissolved oxygen, BOD and ammonia), expressed as percentiles, that should not be exceeded (or fall below in the case of dissolved oxygen) (Table 9.27). Sites are sampled 12 times a year and the data collated over a period of 3 years to obtain the required precision of results.

Biological GQA
Eighty-three groups of macro-invertebrates are used while assessing biological quality using the BMWP system (Section 9.3.1). Two values are used, the number of taxa and the average score per group (taxon), giving a value out of 10 (i.e. 10 being the highest quality and low values indicating poor quality). These numbers are then compared to those expected to occur at a similar site but under pristine conditions using the mathematical model RIVPACS (Section 9.3.1). The ratio of the actual value from

TABLE 9.27   Standards for the Environment Agency chemical GQA of water quality

GQA grade	Description	Dissolved oxygen (% saturation)	BOD (mg l^{-1})	Ammonia (mg N l^{-1})
		10th percentile	90th percentile	90th percentile
A	Very good	80	2.5	0.25
B	Good	70	4	0.6
C	Fairly good	60	6	1.3
D	Fair	50	8	2.5
E	Poor	20	15	9.0
F	Bad	<20	>15	>9.0

Reproduced from Environment Agency (1998) with permission of the Environment Agency, Bristol.

TABLE 9.28   Link between the biological GQA and EQI

Grade	EQI	
	Taxa	ASPT
A	0.85	1.00
B	0.70	0.90
C	0.55	0.77
D	0.45	0.65
E	0.30	0.50
F	<0.3	<0.50

Reproduced from Environment Agency (1998) with permission of the Environment Agency, Bristol.

sampling compared to the predicted values for both the number of taxa and the average score per group are known as the ecological quality indices (EQI), so an EQI of 1.0 indicates pristine conditions. Using Table 9.28, GQA grades of water quality are allocated with the lowest grade assigned if different (Table 9.29). Two samples are collected annually one in the spring (March–May) and the other in the autumn (September–November). The method used is a 3-min kick sample using a hand-net in shallow water, or three–five trawls with a medium naturalists' dredge in deeper water supplemented by a 1-min sweep with a hand-net (Section 9.2.6). Both are followed by 1-min visual search for animals on the water surface or attached to rocks, vegetation or other material.

Nutrient GQA

Nutrients are assessed on the basis on average values of orthophosphate and nitrate concentration in rivers. The selection of phosphate and nitrate classes is difficult due to wide natural variation of P due to geology and soils, as well as inputs from wastewaters and surface run-off from agricultural land (Table 9.30). Total reactive phosphorus and total oxidized nitrogen are measured using unfiltered samples. The average is based on monthly samples calculated over a 3-year period. Thus the average value used for 2004 includes data from 2002 and 2003, some 36 samples in all.

TABLE 9.29   GQA scheme for different grades of water quality based on biology

Grade	Outline description
A: very good	Biology similar to (or better than) that expected for an average and unpolluted river of this size type and location. High diversity of groups, usually with several species in each. Rare to find dominance of any one group
B: good	Biology falls a little short of that expected for an unpolluted river. Small reduction in the number of groups that are sensitive to pollution. Moderate increase in the number of individuals in the groups that tolerate pollution
C: fairly good	Biology worse than expected for an unpolluted river. Many sensitive groups absent or number of individuals reduced. Marked rise in numbers of individuals in groups that tolerate pollution
D: fair	Sensitive groups scarce and contain only small numbers of individuals. A range of pollution tolerant groups present, some with high numbers of individuals
E: poor	Biology restricted to pollution-tolerant species with some groups dominant in terms of the numbers of individuals. Sensitive groups rare or absent
F: bad	Biology limited to a small number of very tolerant groups, such as worms, midge larvae, leeches and water hog-louse present in very high numbers. In the worst case, there may be no life present

Reproduced from Environment Agency (1998) with permission of the Environment Agency, Bristol.

TABLE 9.30   Standards for the Environment Agency nutrient GQA of water quality based on orthophosphate and nitrate

Grade	Grade limit		Description
	Average ($mg\,P\,l^{-1}$)	Average ($mg\,NO_3\,l^{-1}$)	
1	<0.02	<5	Very low
2	>0.02 to 0.06	>5 to 10	Low
3	>0.06 to 0.1	>10 to 20	Moderately low
4	>0.1 to 0.2	>20 to 30	Moderate
5	>0.2 to 1.0	>30 to 40	High
6	>1.0	>40	Very high

Reproduced with permission of the Environment Agency, Bristol.

Aesthetic GQA

The visual appearance of rivers can be a limiting factor for many uses even though the chemical and biological quality may be adequate. Aesthetic quality is assessed by examining both riverbanks (50-m long section, 5 m wide). The type and number of items of litter are counted and the presence of oil, foam, sewage fungus and orche noted, and the colour and odour of the water examined. Using Table 9.31 a score (class) is allocated for each category. The total score then gives a GQA grade of quality from 1 to 4 (Table 9.32). Note that a different GQA grade of quality is used for aesthetic water quality compared to the biological and chemical GQA schemes due to its subjectivity.

TABLE 9.31   Scoring system for the aesthetic GQA of water quality based on different indicator categories

Litter (number of items)

Type of litter	Class 1	Class 2	Class 3	Class 4
Gross	None	2	6	>6
General	5	39	74	>74
Sewage	None	5	19	>19
Faeces	None	3	12	>12

Oil, scum, foam, sewage fungus ochre (percentage cover)

Class 1	Class 2	Class 3	Class 4
0	5	25	>25

Colour

Intensity	Blue/green	Red/orange	Brown/yellow/straw
Colourless	Class 1	Class 1	Class 1
Very pale	Class 1	Class 2	Class 1
Pale	Class 3	Class 3	Class 2
Dark	Class 4	Class 4	Class 3

Odour

Definitions

Group I	Tolerated or less indicative of water quality. Musty, earthy, woody
Group II	Indicators of poor quality. Farmy, disinfectant, gas, chlorine
Group III	Indicators of very poor quality. Sewage, polish or cleaning fluids, ammonia, oily smells, bad eggs (sulphide)

Classification

Intensity of odour	Group I	Group II	Group III
None	Class 1	Class 1	Class 1
Faint	Class 1	Class 2	Class 3
Obvious	Class 2	Class 3	Class 4
Strong	Class 3	Class 4	Class 4

Reproduced from Environment Agency (1998) with permission of the Environment Agency, Bristol.

The Environment Agency has made significant progress in quantifying the statistical variability and precision of their data, especially the biological GQA data. This has resulted in a unique multi-factor assessment procedure of water quality with a realistic approach to precision, resulting in a much greater potential for the use of such data in river management. Table 9.33 gives an overview of river water quality in England and Wales.

River Habitat Survey

An important catchment management tool in the UK is the River Habitat Survey, which is designed to conserve and restore wildlife habitats, including floodplains (Environment Agency, 1998). It is an important part of River Basin Characterization exercise being carried out under the WFD, providing a detailed assessment of habitat quality indicating where conservation is required and where restoration would be of most benefit. Developed by the Environment Agency the survey is a standardize system for assessing rivers primarily on the basis of their physical structure. Using

TABLE 9.32   Aesthetic GQA classification scheme

Parameter	Allocation of points for each class			
	Class 1	Class 2	Class 3	Class 4
Sewage, litter	0	4	8	13
Odour	0	4	8	12
Oil	0	2	4	8
Foam	0	2	4	8
Colour	0	2	4	8
Sewage, fungus	0	2	4	8
Faeces	0	2	4	6
Scum	0	1	3	5
Gross litter	0	0	1	3
General litter	0	0	1	3
Ochre	0	0	0	1

The points allocated for each parameter are summed to give the total score. The grade is then assigned as:

Parameter	Allocation of points for each class	Total score
Grade 1	Good	1, 2 or 3
Grade 2	Fair	4, 5, 6 or 7
Grade 3	Poor	8, 9, 10 or 11
Grade 4	Bad	>11

Reproduced from Environment Agency (1998) with permission of the Environment Agency, Bristol.

TABLE 9.33   River water quality in England and Wales based on the Environment Agency GQA schemes

	River length (%) in each quality grade						Total km
	A	B	C	D	E	F	
Chemical GQA							
1988–90	17.7	30.1	22.8	14.4	12.7	2.3	34 161
1993–95	26.8	32.7	21.3	10.2	8.1	0.9	40 277
1994–96	27.1	31.5	21.2	10.4	8.8	1.0	40 804
Biological GQA							
1990	24.0	31.6	21.6	9.8	7.3	5.7	30 001
1995	34.6	31.6	18.4	8.1	5.4	1.9	37 555
Nutrient GQA							
1990	8.0	17.7	10.2	13.1	28.0	22.9	23 003
1993–95	14.7	22.6	11.0	13.1	27.3	11.0	34 864

standard forms data is collected from 500 m stretches of river channel, which are broken down into ten 10-m long spot check areas each 50 m apart from which specific data is collected (Table 9.34). Modifications and features missed by the spot checks are identified by a general overview of the 500 m stretch using the sweep-up checklist.

TABLE 9.34   List of features measured at River Habitat Survey check points and during the sweep-up of entire 500 m channel site

Features recorded	At 10 spot-checks	In sweep-up
Predominant valley form		✓
Predominant channel substrate	✓	
Predominant bank material	✓	
Flow type(s) and associated features	✓	✓
Channel and bank modifications	✓	✓
Bankface and banktop vegetation structure	✓	
Channel vegetation types	✓	✓
Bank profile (unmodified and modified)		✓
Bankside trees and associated features		✓
Channel habitat features	✓	✓
Artificial features	✓	✓
Features of special interest		✓
Land use	✓	✓

Reproduced from Environment Agency (1998) with permission of the Environment Agency, Bristol.

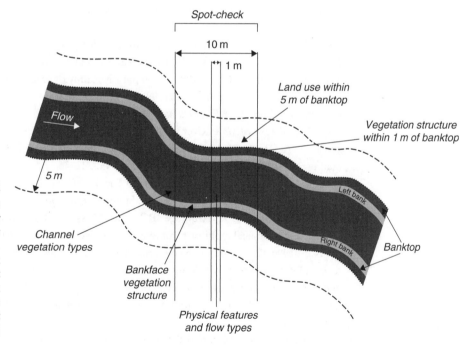

FIGURE 9.15 The key features measured at River Habitat Survey check points. (Reproduced from Environment Agency (1998) with permission of the Environment Agency, Bristol.)

Initially, map based information is gathered including map reference, altitude, slope, geology, height of source and distance from source, etc. This is followed by a detailed field survey which records key features of the channel both within the stream and the banks, and also land use within 5 m of the banktop (Fig. 9.15). The data is then used to generate two scores: a habitat quality assessment (HQA) score and a habitat modification score (HMS) which is based on the type and extent artificial features have been used at the site.

REFERENCES

Akan, A. O. and Houghtalan, R. J., 2003 *Urban Hydrology Hydraulics, and Storm Water Quality: Engineering Applications for Computer Modelling*, John Wiley and Sons, Inc., New York.

APHA, 1989 *Standard Methods for the Examination of Water and Waste Water*, 17th edn, American Public Health Association, Washington, DC.

Bartram, J. and Ballance, R., 1996 *Water Quality Monitoring: A Practical Guide to the Design and Implementation of Freshwater Quality Studies and Monitoring Programmes*, E. & F. N. Spon, London.

Caulcutt, R. and Boddy, R., 1983 *Statistics for Analytical Chemists*, Chapman and Hall, London.

Chapman, D. (ed.), 1996 *Water Quality Assessments*, E. & F. N. Spon, London.

Demayo, A. and Steel, A., 1996 Data handling and presentation. In: Chapman, D., ed., *Water Quality Assessments*, E. & F. N. Spon, London, pp 511–612.

Department of the Environment, 1983a *Methods of Biological Sampling of Benthic Macro-invertebrates in Deep Rivers. Methods for the Examination of Waters and Associated Materials*, Her Majesty's Stationery Office, London.

Department of the Environment, 1983b *Methods of Biological Sampling: A Colonization Sampler for Collecting Macro-invertebrates, Indicators of Water Quality in Lowland Rivers, Methods for the Examination of Waters and Associated Materials*, Her Majesty's Stationery Office, London.

Elliot, J. M., 1977 *Some Methods for the Statistical Analysis of Benthic Invertebrates*. Scientific Publication No. 25, Freshwater Biological Association, Windermere.

Elliot, J. M. and Drake, C. M., 1981 A comparative study of seven grabs for sampling macro-invertebrates, *Freshwater Biology*, **11**, 99–120.

Environment Agency, 1998 *River Habitat Quality: The Physical Character of Rivers and Streams in the UK and the Isle of Man*. River Habitat Survey Report No. 2. Environment Agency, Bristol.

Extence, C. A., Bates, A. J., Forbes, W. J. and Barham, P. J., 1987 Biologically based water quality management, *Environmental Pollution*, **45**, 221–36.

Friedrich, G., Chapman, D. and Beim, A., 1996 The use of biological material. In: Chapman, D., ed., *Water Quality Assessments*, E. & F. N. Spon, London, pp 174–242.

Furse, M. T., Wright, J. F., Armitage, P. D. and Moss, D., 1981 An appraisal of pond-net samples for biological monitoring of lotic macro-invertebrates, *Water Research*, **6**, 79–89.

Gray, N. F., 1996 The use of an objective index for the assessment of the contamination of surface and ground water by acid mine drainage, *Journal of the Chartered Institution of Water and Environmental Management*, **10**, 332–40.

Harrison, R. M. (ed.), 1992 *Understanding Our Environment*, Royal Society of Chemistry, Cambridge.

Hawkes, H. A., 1998 Origin and development of the Biological Monitoring Working Party score system, *Water Research*, **32**, 964–8.

Hellawell, J. M., 1978 *Biological Surveillance of Rivers*, Water Research Centre, Stevenage.

Hellawell, J. M., 1986 *Biological Indicators of Freshwater Pollution and Environmental Management*, Elsevier Applied Science, London.

Holmes, N. T. H., Boon, P. J. and Rowell, T. A., 1998 A revised classification for British rivers based on their aquatic plant communities. *Aquatic Conservation*, **8**, 555–78.

House, M. A. and Ellis, J. B., 1987 The development of water quality indices for operational management, *Water Science and Technology*, **19**(9), 145–54.

ISO, 1985 *Water Quality – Methods of Biological Sampling – Guidance on Hand-net Sampling of Aquatic Benthic Macro-invertebrates*, International Standard 7828, International Organization for Standardization, Geneva.

ISO, 1990 *Water Quality Sampling*, International Standard 5667, International Organization for Standardization, Geneva.

ISO, 1991 *Water Quality Sampling in Deep Water for Macro-invertebrates. Guidance on the Use of Colonization, Qualitative and Quantitative Samplers*, International Standard 9391, International Organization for Standardization, Geneva.

Kelly, M. G., 1998 The use of the Trophic Diatom Index: a new index for monitoring eutrophication in rivers. *Journal of Applied Phycology*, **7**, 433–44.

Kelly, M. G. and Whitton, B. A., 1998 Biological monitoring of eutrophication in rivers. *Hydrobiologia*, **384**, 55–67.

Learner, M. A. and Chawner, H. A., 1998 Macro-invertebrate associations in sewage filter-beds and their relationship to operational practice. *Journal of Applied Ecology*, **35**, 720–47.

Marsden, M. W., Smith, M. R. and Sargent, R. J., 1997 Trophic status of rivers in the Forth catchment, Scotland. *Aquatic Conservation*, **7**, 211–21.

McGarrigle, M. L., Lucey, J. and Clabby, K. C., 1992 Biological assessment of river water quality in Ireland. In: Newman, P. J., Piavaux, M. A. and Sweeting, R. A., eds, *River Water Quality: Ecological Assessment and Control*, Office for the Official Publications of the European Communities, Luxembourg, pp 371–97.

Newman, P. J., Piavaux, M. A. and Sweeting, R. A., 1992 *River Water Quality: Ecological Assessment and Control*, Office for the Official Publications of the European Communities, Luxembourg.

Nichols, D., 1980 *Safety in Biological Fieldwork*, Chartered Institute of Biology, London.

Norris, R. H. and Georges, A., 1993 Analysis and interpretation of benthic macroinvertebrate surveys. In: Roseberg, D. M. and Resh, V. H., eds., *Freshwater Biomonitoring and Benthic Macroinvertebrates*, Chapman and Hall, New York, pp 234–86.

Nutall, D., 1983 The use of an objective index as an aid to water quality management in Ireland, *Irish Journal of Environmental Science*, **2**, 19–31.

Pinder, L. C. V., Ladle, M., Gledhill, T., Bass, J. A. B. and Mathews, A., 1987 Biological surveillance of water quality. 1. A comparison of macroinvertebrate surveillance methods in relation to assessment of water quality in a chalk stream. *Archive Fur Hydrobiologie*, **109**, 207–26.

Premazzi, G. and Chiaudani, G., 1992 Current approaches to assess water quality in lakes. In: Newman, P. J., Piavaux, M. A. and Sweeting, R. A., eds, *River Water Quality: Ecological Assessment and Control*, Office for the Official Publications of the European Communities, Luxembourg, pp 249–308.

Rosenberg, D. M. and Resh, V. H. (eds), 1993 *Freshwater Biomonitoring and Benthic Macroinvertebrates*, Chapman and Hall, New York.

Scottish Development Department, 1976 *Development of a Water Quality Index*, Report AR3, SDD, Edinburgh.

Sharp, W. E., 1971 A topographical optimum water sampling plan for rivers and streams. *Water Resources Research*, **7**, 1641–6.

Thorpe, M. W. and Williams, I. L., 1980 *A Guide to the Use of Algae for the Biological Surveillance of Rivers*, Technical Memorandum 20, Water Data Unit, Reading.

Waite, S., 2000 *Statistical ecology in practice*, Prentice Hall, Harlow.

Whitehurst, I. T., 1991 The *Gammarus: Asellus* ratio as an index of organic pollution, *Water Research*, **25**, 333–9.

Wright, B., 1993 *Guidance Note for the Management of Lone Working*, Natural Environment Research Council, Swindon.

Wright, J. F., Furse, M. T. and Armitage, P. D., 1993 RIVPACS – a technique for evaluating the biological quality of rivers in the UK, *European Water Pollution Control*, **3**(4), 15–25.

FURTHER READING

Croft, P. S., 1986 A key to the major groups of British freshwater invertebrates. *Field Studies*, **6**, 531–79.

Fitter, R. and Manuel, R., 1986 *Field Guide to Freshwater Life of Britain and North-west Europe*, Collins, London.

Hellawell, J. M., 1977 Change in natural and managed ecosystems: detection, measurement and assessment, *Proceedings of the Royal Society of London, B*, **197**, 31–57.

Rosenberg, D. M. and Resh, V. H. (eds), 1993 *Freshwater Biomonitoring and Benthic Macro-invertebrates*, Chapman and Hall, New York.

Thorne, C. R., 1998 *Stream Reconnaissance Handbook: Geomorphical Investigation Analysis of River Channels*, John Wiley and Sons, Chichester.

Weber, C. I. (ed.), 1973 *Biological Field and Laboratory Methods for Measuring the Quality of Surface Waters and Effluents*, USEPA 679/4/73/001, US Environmental Protection Agency, Washington.

Wong, P. T. S. and Dixon, D. G., 1995 Bioassessment of water quality, *Environmental Toxicology and Water Quality*, **10**, 9–17.

Part III Drinking Water Treatment and Supply

Chapter 10 Water Treatment and Distribution

INTRODUCTION TO TREATMENT

All water used for supply originates from the atmosphere as precipitation (rain, snow and hail). This collects either above ground in rivers, natural lakes, man-made impounding reservoirs or below ground in aquifers (Chapter 2). Water rapidly absorbs both natural and man-made substances generally making the water unsuitable for drinking prior to some form of treatment.

The objective of water treatment is to produce an adequate and continuous supply of water that is chemically, bacteriologically and aesthetically pleasing. More specifically, water treatment must produce water that is:

(a) palatable (i.e. no unpleasant taste);
(b) safe (i.e. does not contain pathogens or chemicals harmful to the consumer);
(c) clear (i.e. free from suspended solids and turbidity);
(d) colourless and odourless (i.e. aesthetic to drink);
(e) reasonably soft (i.e. allows consumers to wash clothes, dishes, themselves, without use of excessive quantities of detergents or soap);
(f) non-corrosive (i.e. to protect pipework and prevent leaching of metals from tanks or pipes);
(g) low-organic content (i.e. high-organic content results in unwanted biological growth in pipes and storage tanks that often affects quality).

The two European Directives that currently govern the quality of raw water used for drinking water abstraction, the Surface Water (75/440/EEC) and the GroundWater (80/68/EEC) Directives are to be repealed in 2007 and replaced with specific provisions in the Water Framework Directive (2000/60/EC) (Section 8.1). The original Drinking Water Directive (80/778/EEC) was revised in 1998 (98/83/EEC) setting new standards for water supplied to consumers (Tables 8.6 and 8.7). Water treatment plants must be able to produce a finished product of consistently high-quality regardless of

TABLE 10.1   Main water treatment unit processes

Treatment category	Unit process
Intake	
Pre-treatment	Coarse screening
	Pumping
	Storage
	Fine screening
	Equalization
	Neutralization
	Aeration
	Chemical pre-treatment
Primary treatment	Coagulation
	Flocculation
	Sedimentation
Secondary treatment	Rapid sand filtration
	Slow sand filtration
Disinfection	
Advanced treatment	Adsorption
	Activated carbon
	Fe and Mn removal
	Membrane processes
Fluoridation	
Distribution	

how great the demand might be. Like wastewater treatment, water treatment consists of a range of unit processes, usually used in series (Table 10.1). The cleaner the raw water, then the fewer treatment steps are required and hence the overall cost of water is less. The most expensive operations in conventional treatment are sedimentation and filtration, while water softening can also be very expensive. Groundwater is generally much cleaner than surface waters and so does not require the same degree of treatment, apart from aeration and disinfection before supply (Fig. 10.1). Naturally occurring substances that may need to be reduced or removed in groundwaters include iron, hardness (if $>300\,\text{mg}\,l^{-1}$ as $CaCO_3$) and carbon dioxide (CO_2). Substances originating from humans are becoming increasingly common in groundwaters, and those requiring treatment include nitrates, pathogens and trace organics, such as pesticides. Surface water requires more complex treatment due to poorer quality (Figs 10.2 and 10.3), although the quality of surface waters can be very high (e.g. upland reservoirs). The selection of water resources for supply purposes depends not only on the nature of the raw water and the ability of the resource to meet consumer demand throughout the year, but also on the cost of treating the water. Each of the major unit processes used for water treatment are examined below, although other water treatment processes used for corrosion control or the removal of organic compounds and trace elements are discussed in Chapter 20. Known collectively as advanced water treatment processes, they include softening, ion-exchange, adsorption, membrane filtration including reverse osmosis and chemical oxidation.

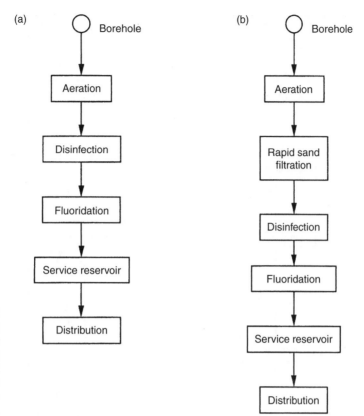

FIGURE 10.1
Sequence
of treatment for
groundwater sources
of (a) excellent
quality and (b) with
moderate concentra-
tions of iron.

10.1 UNIT PROCESSES

The selection of treatment processes used depends on the raw water quality and its seasonal variability (Stevenson, 1998). Process selection is far more complex than in wastewater treatment (Chapter 14), although some idea of the key unit processes used for treating particular problems is given in Table 10.2. Pre-treatment may be required if the water is of poor quality and includes screening (coarse and fine), storage, equalization (often provided by storage), neutralization, aeration and chemical pre-treatment. Normally most waters can be treated solely using conventional unit processes without the need for pre-treatment except for screening to remove fish, natural debris and litter, etc.

10.1.1 PRELIMINARY SCREENING

Large-scale works are rarely close to a suitable source of water (e.g. upland reservoirs), therefore raw water must be conveyed to the treatment plant either by pipe or open conduit/channel. The raw water is screened through a set of coarse screens (100 mm spacing) to remove gross solids, such as litter and branches, before being conveyed to the plant. Prior to treatment it is screened again through fine screens or,

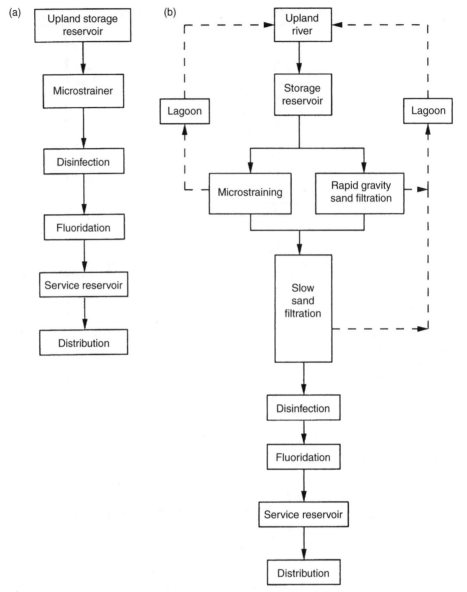

FIGURE 10.2
Typical sequence of unit treatment processes for water abstracted from (a) an upland storage reservoir and (b) an upland river, both of good quality.

if considerable fine solids or algae are present, then micro-straining may be used (i.e. a circular drum-type screen made from fine stainless steel mesh with 25 000 apertures cm^{-2}) before the next stage.

10.1.2 STORAGE

Storage is used primarily for water abstracted from lowland rivers (e.g. the River Thames serving, London) to improve water quality before treatment, as well as to ensure adequate supplies at periods of peak demand (i.e. equalization) (Table 10.3). There are a number of physico-chemical processes at work during storage that affects

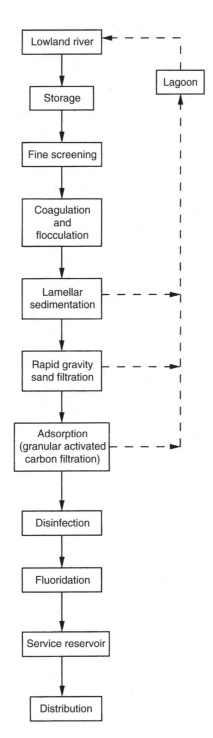

FIGURE 10.3
Typical sequence of
unit treatment
processes for a low-
land river supply of
moderate to poor
quality.

water quality. These include sedimentation and ultraviolet (UV) radiation. Long-term storage encourages flocculation and settlement and it is recommended for all waters with a suspended solids concentration $>50\,\mathrm{mg}\,\mathrm{l}^{-1}$. Filtration can only deal effectively with suspended solids concentration of $<10\,\mathrm{mg}\,\mathrm{l}^{-1}$. Variations in quality

TABLE 10.2   Selection of unit processes for the removal of specific parameters

Parameter	Water treatment process options
Algae	Powdered activated carbon adsorption, microscreens, rapid filtration
Colour	Activated carbon adsorption, coagulation, flocculation, filtration
Floating matter	Coarse screens
Hardness	Coagulation, filtration, lime softening
Coliforms	
>100 per $100\,ml^{-1}$	Pre-chlorination, coagulation, filtration post-chlorination
>20 per $100\,ml^{-1}$	Coagulation, filtration post-chlorination
<20 per $100\,ml^{-1}$	Post-chlorination
Hydrogen sulphide	Aeration
Fe and Mn	Pre-chlorination, aeration, coagulation, filtration
Odour and taste	Aeration, activated carbon adsorption
Suspended solids	Fine screens, microscreens
Trace organics	Activated carbon adsorption
Turbidity	Coagulation, sedimentation, post-chlorination

TABLE 10.3   The improvement brought about by storage in the quality of water abstracted from the River Great Ouse at Grafham, prior to full treatment

	Before storage	After storage
Colour (Hazen)	30	5
Turbidity (NTU)	10	1.5
Ammoniacal N $(mg\,l^{-1})$	0.3	0.06
Biochemical oxygen demand $(mg\,l^{-1})$	4.5	2.5
Total hardness $(mg\,l^{-1})$	430	280
Presumptive coliforms $(100\,ml^{-1})$	6500	20
Escherichia coli most probable number $(100\,ml^{-1})$	1700	10
Colony counts (ml^{-1})		
3 days at 20°C	50 000	580
2 days at 37°C	15 000	140

Reproduced from Saxton (1970) with permission of the Institution of Water and Environmental Management, London.

caused by floods or surface run-off can be largely eliminated by storage. Exposure to natural UV radiation destroys some pathogens, colour is bleached by sunlight, while tastes and odours are oxidized by UV. Hardness is also reduced by liberation of CO_2 by algae that converts the bicarbonates into carbonates which precipitate out of solution (Section 3.4).

Problems with storage include atmospheric pollution and fallout, pollution from birds and algal development (if stored for >10 days). The land requirement is high, making storage expensive in terms of capital cost. If storage reservoirs are deep they can become thermally stratified like a natural lake (Section 2.1.2) (Fig. 2.2).

10.1.3 AERATION

Water from groundwater sources, from the bottom of stratified lakes or reservoirs, or polluted rivers will contain little or no dissolved oxygen. If anaerobic water is passed through the treatment system it can damage filters and adversely affect coagulation. Therefore water needs to be mechanically aerated before treatment, which is most easily achieved by using either a cascade or fountain system. Apart from providing oxygen for purification and improving overall quality, aeration also reduces the corrosiveness of water by driving off CO_2 and raising the pH. However, aeration alone cannot reduce the corrosive properties of acid water, neutralization using lime may also be needed. Aeration also removes Fe and Mn from solution, which affects taste of water and stains clothes. Iron is only soluble in water at pH <6.5 and in the absence of oxygen, therefore aeration converts soluble iron into an insoluble hydroxide form which can be removed by unit processes. Aeration improves the taste of water by also stripping out of solution hydrogen sulphide and volatile organic compounds.

10.1.4 CHEMICAL PRE-TREATMENT

Normally chemical treatment is used as an advanced treatment system after conventional treatment has been completed (Section 20.5). Two pre-treatment processes are generally employed. Pre-chlorination is used when there is a high-coliform count in the raw water due to surface water run-off from agriculture, upstream sewage treatment plant failure or during floods. The process is most successful with low-turbidity waters and, as chlorine oxidizes and precipitates iron and manganese, it is most efficient when these metals are present in low concentrations. Doses are much higher than for post-chlorination, rising to $5 \, \text{mg} \, \text{l}^{-1}$. Activated carbon is used to remove algae, reduce colour, odour and organic compounds. As with pre-chlorination it may only be used intermittently. Powered activated carbon (PAC) is added to the water as a slurry just before coagulation or sand filtration at doses up to $20 \, \text{mg} \, \text{l}^{-1}$. The sand filter retains the PAC where it adsorbs the problem compounds. The PAC needs to be in contact with the water as long as possible so adding it prior to coagulation ensures the largest possible contact time (Section 20.6). As an alternative to pre-chlorination, pre-ozonization is also used both on its own and in conjunction with granular activated carbon (GAC).

10.1.5 COAGULATION

After fine screening most of the remaining suspended solids will be colloidal in nature, such as clays, metal oxides, proteins and micro-organisms. These particles are extremely small (<10 μm) and have negligible settling velocities (Table 10.4), although the exact settling velocity depends on the specific gravity (SG) of the particles. For example, a 1 mm sand particle (SG 2.65) has a settling velocity of $100 \, \text{mm} \, \text{s}^{-1}$ compared to $0.7 \, \text{mm} \, \text{s}^{-1}$ for an alum floc (SG 1.05) of the same size. All

TABLE 10.4   Settling velocity of particles as a function of size

Particle size (µm)	Settling velocity (m h⁻¹)
1000	600
100	2
10	0.3
1	0.003
0.1	0.00001
0.01	0.0000002

Reproduced from Tebbutt (1979) with permission of Butterworth-Heinemann Ltd, Oxford.

TABLE 10.5   Properties of variety of coagulants

Type of water	Alum	Ferric salts	Polymer
Type 1: high turbidity, high alkalinity (easiest to coagulate)	Effective over pH range 5–7. No need to add alkalinity or use coagulant aid	Effective over pH range 6–7. No need to add alkalinity or use coagulant aid	Cationic polymers usually very effective. Anionic and non-ionic may also work
Type 2: high turbidity, low alkalinity	Effective over pH range 5–7. May need to add alkalinity to control pH. Coagulant aid not needed	Effective over pH range 6–7. May need to add alkalinity to control pH. Coagulant aid not needed	Cationic polymers usually very effective. Anionic and non-ionic may also work
Type 3: low turbidity, high alkalinity	Relatively high dose needed to form sufficient floc. pH near to 7. Coagulant aid may help	Relatively high dose needed to form sufficient floc. Coagulant aid may help	Will not work well alone due to low turbidity. Adding a clay to increase turbidity may be effective
Type 4: low turbidity, low alkalinity (most difficult to coagulate)	Relatively high dose needed to form sufficient floc. pH near to 7. Alkalinity or clay need to be dosed to produce type 2 or 3 water	Relatively high dose needed to form sufficient floc. pH near to 7. Alkalinity or clay need to be dosed to produce type 2 or 3 water	Will not work well alone due to low turbidity. Adding a clay to increase turbidity may be effective

Low turbidity <10 NTU; high turbidity >100 NTU; low alkalinity <50 mg $CaCO_3$; high alkalinity >250 mg $CaCO_3$.

tend to be negatively charged, so colloidal particles generally repel each other preventing aggregation and settlement. Coagulants are used to induce particle agglomeration (i.e. coagulation). Selection of a suitable coagulant depends on the nature of the particles, especially their affinity to adsorb water, and their electrical charge. Particles are either hydrophobic (i.e. do not adsorb water) or hydrophilic (i.e. do adsorb water), and are generally negatively charged making them stable (i.e. remain in suspension). Altering their charge destabilizes the particles so that they agglomerate and settle. A wide variety of coagulants are available, the most common are alum ($AlSO_4$), aluminium hydroxide, ferric chloride and ferric sulphate (Table 10.5). The mechanisms of coagulation are complex including adsorption, neutralization of charges and entrainment with the coagulant matrix. The coagulant is added to the

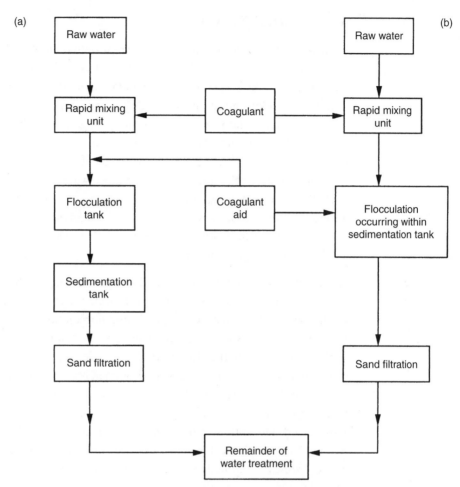

(a)

(b)

Raw water

Raw water

Rapid mixing unit

Coagulant

Rapid mixing unit

Flocculation tank

Coagulant aid

Flocculation occurring within sedimentation tank

Sedimentation tank

Sand filtration

Sand filtration

Remainder of water treatment

FIGURE 10.4 Different alternatives to the addition of coagulant and coagulant aids (polyelectrolytes), (a) is widely used in the USA and (b) is common practice in the UK.

water in a special vessel (coagulant rapid mixing unit) and rapidly mixed for 20–60 s to produce a microfloc. Mixing is achieved by hydraulic jump, jet injection or rapid mixing using a propeller or paddle. The mixture is then gently agitated in the flocculation tank using a mixing rotor in a separate tank for 20–60 min to achieve flocculation where the microfloc grows to form a floc that will readily settle. Owing to the longer retention times this tank is 50–60 times larger than the coagulant rapid mixing unit. If the mixing rotor is too fast then the microflocs can be broken up causing coagulation to fail. Often coagulants alone are not adequate to promote satisfactory floc formation. Coagulant aids, which are polyelectrolytes that include a wide range of synthetic organic polymers, such as polyacrylamide, polyethylene oxide and polyacrylic acid, are added to the water after coagulation using a small rapid mixing tank before flocculant mixing (Fig. 10.4). Lime addition and pH correction may also be necessary for optimum coagulation and prevention of discharge of coagulant with the finished water. The flocs are then removed by sedimentation. Coagulation is discussed further in Section 20.2.

10.1.6 SEDIMENTATION

Sedimentation in water treatment is different to that employed with wastewater due to the nature of the solids to be removed. For that reason it is commonly referred to as clarification. The suspended and colloidal material in surface water that results in unwanted turbidity and colour in drinking water is removed by sedimentation once it has been converted into a settleable form by coagulation. This produces a flocculant suspension that is removed by a combination of sedimentation and filtration. Where dense discrete particles are present in surface waters due to flooding then plain sedimentation similar to that used for primary wastewater treatment is required (Section 14.2). This type of sedimentation is used for type I settling where the particles settle out individually without any flocculation or coagulation between particles (Section 20.3). In practice where surface waters are subject to high-silt loads during high flows and flooding, storage reservoirs are used.

Upward flow settlement tanks, called sludge blanket clarifiers, provide enhanced flocculation as well as floc separation. Inflowing water enters at the base of the tank which results in turbulent mixing promoting flocculation. The water moves upwards through a zone of suspended sludge expanding or fluidizing it to form a thick layer or blanket. This affords the floc particles even further opportunity for contact and aggregation with other particles. A well-defined interface between the clarified water and the sludge ensures that the supernatant water overflows continuously to the decanting channels free of flocs. The density of the sludge blanket is controlled by the sludge bleed. The upflow blanket clarifier is typical of type III or hindered settlement (Section 20.3). Settlement is hindered by the upflow velocity of the water resulting in different zones, with the densest part of the sludge blanket at the top with the highest settling velocities and the weakest at the bottom with the lowest settling velocities.

Older and smaller treatment works employ a hopper-bottom-shaped tank (Fig. 10.5). Newer and larger plants use flat-bottom tanks where the water enters at the base of the tank in a downwards direction, to create maximum turbulence, via a system of perforated pipes that ensures even distribution and loading. Water can be pulsed, rather than loaded continuously, to enhance flocculation and prevent solid accumulation on the tank floor. Clarified water overflows into the decanting channels at the water surface (Fig. 10.6).

Sedimentation tanks employing lamellar designs are now common. Bundles of inclined submerged tubes, parallel plates, honeycomb structures or corrugated asbestos sheeting are included into the settling zone to enhance flocculation and separation. Parallel compartments are formed by the plates or tubes that act as independent settlement chambers. The lamellar media is set at a critical angle that encourages the floc to settle onto the bottom surface of each passage forming a layer of sludge which then slides down the passage into a collecting chamber from which the sludge is removed. The media must have narrow vertical gaps although the width is irrelevant. The efficiency of separation is greatly enhanced as the baffles prevent

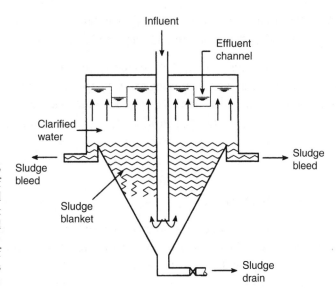

FIGURE 10.5 Schematic layout of a hopper-bottomed sludge-blanket clarifier. (Reproduced from Casey (1997) with permission of John Wiley and Sons Ltd, Chichester.)

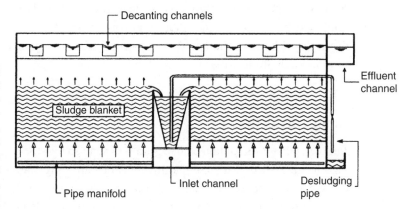

FIGURE 10.6 Schematic layout of a flat-bottomed sludge-blanket clarifier. (Reproduced from Casey (1997) with permission of John Wiley and Sons Ltd, Chichester.)

the onset of convection currents making the process less susceptible to temperature than normal sludge blanket clarifiers. Shorter hydraulic retention times are possible permitting increased loadings by up to 25%.

Hopper-bottomed tank designs set up weak circulation patterns that significantly reduce the upflow rate that can be employed to 1.0–1.5 m h^{-1}. In contrast flat-bottom units provide a true upflow pattern allowing higher upflow rates of between 2 and 4 m h^{-1}. The upflow velocity (V_s) can be calculated using the following equation:

$$V_s = \frac{Q}{A} \tag{10.1}$$

where Q is the inflow rate and A the plan area of the tank. If the upflow velocity exceeds the settling velocity then flocs will be discharged with the clarified water. In the UK it is common practice to use a single tank for both flocculation and settlement, while in the USA separate tanks are used (Fig. 10.4).

Settlement is affected by temperature due to the drag coefficient imposed by the water. This decreases with increasing water temperature thereby increasing the rate of settlement. So as similar particles can take up to twice as long to settle in the winter than in the summer, the design of settlement tanks must take this into account (Binne *et al.*, 2002).

10.1.7 FILTRATION

At this stage only fine solids and soluble material are left in the water, which has a typical suspended solids concentration of $<10\,\text{mg}\,\text{l}^{-1}$. Even micro-strainers are not able to retain solids $<60\,\mu\text{m}$, so filtration is required to remove this residual material. The process of filtration is the most widely used water treatment process. Water is passed through a porous bed of inert medium, usually silica or quartz sand, from which small particulate matter is strained. Filters are classified as either slow or rapid, and can be operated either by gravity or under pressure, where the water is forced through the medium under pressure, significantly increasing the flow-through rate (Solt and Shirley, 1991).

Rapid gravity filters

Rapid gravity filters contain coarse grades of sand so that the gaps between the grains (i.e. the interstices) are comparatively large allowing the water to pass rapidly through to the underdrains removing only the suspended solids. These filters are usually open-top units operated by gravity (Fig. 10.7), although pressure filters are also used. Filters are made up of a layer of coarse sand 0.5–1.0 mm in diameter and 600 mm deep on top of a 300 mm deep bed of graded gravel of sizes ranging from 12 to 25 mm, giving an overall depth of 0.9–1.2 m. Sufficient depth above the media must be allowed for an adequate depth of water to provide enough hydraulic head for the water to pass rapidly through the medium. The process employs depth filtration where all the filter medium is involved in the filtration process so that the quality of the water improves with depth. Treatment is primarily by physical removal with particles and order of magnitude smaller than the interstices in the sand being retained. Apart from straining particles out of suspension, other removal processes include chemical and physical adsorption, sedimentation and adhesion (Metcalf and Eddy, 1991). For this reason the medium rapidly becomes blocked with retained solids and needs frequent cleaning. Filtration rates vary between 5 and $10\,\text{m}\,\text{h}^{-1}$, giving loading rates of 5–$6\,\text{m}^3\,\text{m}^2\,\text{h}^{-1}$ (Table 10.6). Rapid gravity filters need to be backwashed every 20–60 h depending on loading and suspended solid concentration of raw water. The filter will normally require backwashing as the head loss across the filter

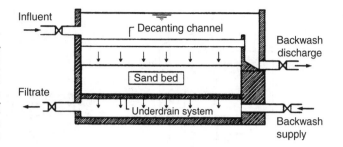
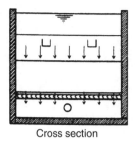

FIGURE 10.7
Schematic layout of a rapid gravity sand filter. (Reproduced from Casey (1997) with permission of John Wiley and Sons Ltd, Chichester.)

TABLE 10.6   Comparison of typical design parameters of slow and rapid gravity sand filters

Parameter	Slow sand filter	Rapid sand filter
Rate of filtration $(m\,h^{-1})$	0.08–0.15	5–10
Loading rate $(m^3\,m^2\,h^{-1})$	< 0.1	5–6
Depth of sand (m)	0.6–1.0	0.6–1.0
Overall depth of media (m)	1.0–2.0	0.9–1.2
Diameter of sand (mm)	0.15–0.3	0.5–1.0
Mixed media	Not possible	Possible
Duration of operation between cleaning (days)	60–100	1–3
Area of particle removal	Surface of filter	Entire depth
Biological action	Yes	No
Pre-treatment required	None	Coagulation flocculation and sedimentation

approaches 2.5 m. This is done by first scouring with compressed air and then backwashing with clean water. The sludge produced from these filters requires further treatment and disposal. The backwash water should not be recirculated back to the start of the treatment process but discarded due to possible high concentrations of pathogens. After backwashing there is a short period when the filter produces poor-quality filtered water (e.g. a turbidity of 0.5–1.0 nephelometric turbidity unit, NTU). This is due to material disturbed during backwashing but not being fully flushed out. So as the sand resettles within the filter, once water is again passed through, there is a period of 30–60 min when the interstices within the sand are slowly closing up to their minimum diameter. It is during this short period at start up that pathogens and other particles can find their way through the filter and contaminate the filtered water. So initially all filtered water should be recycled back to the inlet until the turbidity returns to normal (e.g. 0.1 NTU). When the filter media is backwashed the media is fluidized, regrading the media according to its density. This allows different types of media to be used in the filter in discrete layers to improve performance (e.g. anthracite). Pressure units are employed at small treatment plants where they eliminate the need for a separate pumping stage (Fig. 10.8). While high-loading rates are possible a major disadvantage to operators is that the medium cannot be visually

(a)

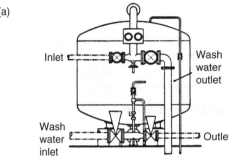

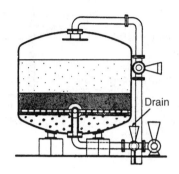

FIGURE 10.8
Schematic layout of (a) a vertical pressure water filter and (b) a horizontal pressure filter. (Reproduced from Solt and Shirley (1991) with permission of Ashgate Publishing Ltd, Aldershot.)

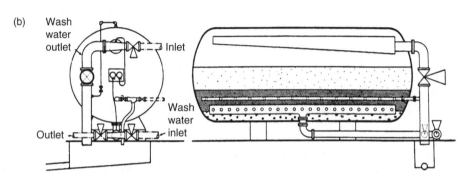

inspected. Rapid sand filters are usually employed after chemical coagulation and sedimentation, and are also used in wastewater treatment for tertiary treatment (Solt and Shirley, 1991).

Slow sand filters

In contrast slow sand filters employ a much finer sand of 0.15–0.3 mm in diameter. Slow sand filters are normally rectangular in plan with a layer of the fine sand up to 1 m in depth laid on a graded layer of coarse sand and gravel to prevent the fine sand blocking the system of underdrains beneath (Fig. 10.9). On the surface of the sand a gelatinous layer develops which is rich in micro-organisms (e.g. bacteria, protozoa, algae) called the schmutzdecke. It is this layer that is largely responsible for the treatment of the water by both physical removal of particles and by the biological treatment of dissolved organic matter and nutrients. The top 2 mm is an autotrophic layer, a mixture of algae and nitrifying bacteria, removing N and P and releasing oxygen. Below this, heterotrophic bacteria dominate where residual organic matter is removed. The heterotrophs extend up to 300 mm within the sand layer. Raw water is allowed to flow onto the surface filter to a depth that will provide sufficient hydraulic head to drive it through the filter to give a hydraulic retention time of about 2 h. Design details are given in Table 10.4. The quality of water is excellent (<1 NTU). However, as the rate of filtration is very slow (<0.1 m^3 m^2 h^{-1}) large areas of filters are required (Example

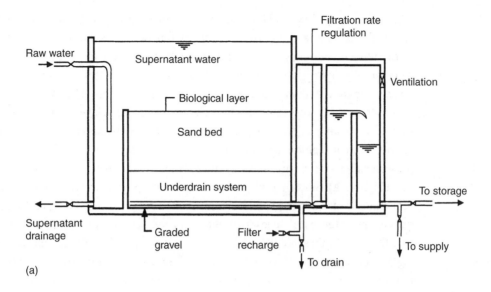

(a)

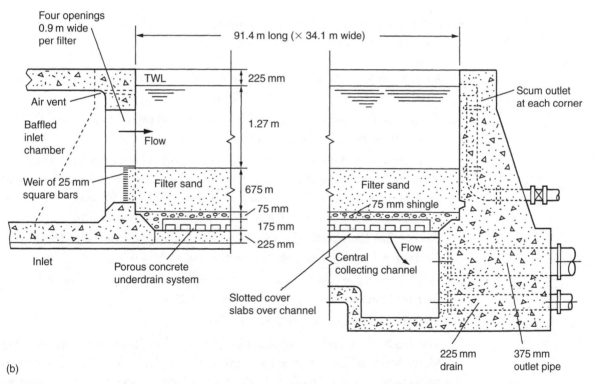

(b)

FIGURE 10.9 Schematic layout of a slow sand filter (a) showing constructional detail of typical filter (b). (Reproduced from (a) Casey (1997) with permission of John Wiley and Sons, Chichester and (b) Twort *et al.* (1994) with permission of Arnold, London.)

10.1). The cost of slow sand filtration is much higher in capital terms compared to rapid gravity filters. They are also expensive to operate because solids retained at the surface of the filter increasingly impede drainage and so the top 10–15 mm of sand must be mechanically skimmed off every 2–4 months after the filter has been drained. This takes some time to complete requiring extra standby filters. The fine sand layer must also be topped up every few years to maintain an adequate layer of fine sand. This layer should not be allowed to reduce to <0.5 m in depth. Slow sand filtration can be used on its own to treat high-quality surface and groundwaters, although if there are turbidity problems then pre-filtration using either a rapid sand filter or micro-strainer is necessary (Fig. 10.2a). Slow sand filters do not work with coagulants, so there are significant reductions in the sludge produced compared with rapid sand filters, and require high-quality raw waters with the low turbidities normally associated with reservoirs and lakes (<10 NTU). Rapid sand filters have been used as roughing filters to increase the loading rate of slow sand filters and to enable them to be used for treating poorer-quality raw waters. A layer of GAC can also be incorporated between the media layers to improve performance and to remove trace organics.

EXAMPLE 10.1 Compare the tank dimensions of a rapid gravity sand filter and a slow sand filter treating $12\,000\,\text{m}^3$ of water per day.

(a) Design a slow sand filter to treat $1200\,\text{m}^3\,\text{day}^{-1}$ at a filtration rate of $0.15\,\text{m}\,\text{h}^{-1}$. Required tank area $= (1200/24) \times (1/0.15) = 333\,\text{m}^2$.

Tanks should not normally exceed $200\,\text{m}^2$ in area, so two tanks are required 17 m long $\times$ 10 m wide giving a total area of $340\,\text{m}^2$.

The depth of the tank is calculated by allowing 0.5 m for the underdrains, 1.5 m for the height of the hydraulic head (this should be the same as the depth of sand used but no less than 1.0 m), and 1.5 m for the depth of sand used (1.0 m of fine sand on 0.5 m coarse sand), giving an overall depth of 3.5 m.

Overall tank dimensions are 3.5 m depth $\times$ 17 m long $\times$ 10 m wide. This gives a total volume for two tanks of these dimensions of $1190\,\text{m}^3$.

(b) *Design a rapid gravity filter to treat $1200\,\text{m}^3\,\text{day}^{-1}$ at filtration rate of $8\,\text{m}\,\text{h}^{-1}$.* Required tank area $= (1200/24) \times (1/8) = 6.25\,\text{m}^2$.

Tank surface dimensions are 2.5 m $\times$ 2.5 m.

The depth of the tank is calculated by allowing 0.5 m for the underdrains, 1.5 m for the height of the hydraulic head (this should be the same as the depth of sand used but no less than 1.0 m), and 1.0 m for the depth of sand used, giving an overall depth of 0.3 m.

Overall tank dimensions are 3.0 m depth $\times$ 2.5 m long $\times$ 2.5 m wide. This gives a total volume for the tank of $18.8\,\text{m}^3$.

(c) *Comparison of systems*: The rapid gravity filter requires a tank 53 times smaller in area than a slow sand filter to treat the same volume of water each day. While the quality of water will be better from slow sand filtration, the capital investment is much greater. In both cases standby filters are required to maintain treatment capacity during either backwashing or scrapping, which significantly influences cost. In practice there should never be less than four slow sand filters at a treatment plant.

Since the middle of the century rapid gravity filters have been used in preference to slow sand filters, mainly due to cost, land requirement and flexibility of operation. The increasing problem of pathogenic protozoan cysts (i.e. *Crytosporidium* and *Giardia*), which are unaffected by chlorination and can break through rapid sand filters resulting in outbreaks of disease in the community (Chapter 12), has renewed interest in slow sand filtration. Sludges from both rapid and slow sand filters must be disposed of carefully as they are highly contaminated with pathogenic microorganisms removed from the raw water. The theory of filtration is considered in detail by the American Water Works Association (1990).

10.1.8 DISINFECTION

Water is not sterilized by disinfection. Rather the risk of infection is reduced to an acceptable level by controlling the numbers of bacteria and viruses by using a disinfection method (Section 12.3.3). Three disinfection options are available:

1. Ozone has powerful oxidation properties but has no residual action to protect the water during distribution. It also oxidizes colour, taste and odour in the water. A dose of 1 ppm destroys all bacteria within 10 min, but is far more expensive than chlorine and has to be manufactured on site.
2. UV radiation is widely used for small plants and individual houses. Like ozone it has no residual action and so must be used close to the point of use to prevent recontamination by biofilm development within the distribution network (Section 11.3) or by pollution due to damaged pipework.
3. Chlorination is not quite as powerful as ozone it has a lasting residual effect making it an effective disinfectant of water. It is also relatively easy to handle and cost effective.

Chlorination is the most widely used disinfection process worldwide. Chlorine reacts with water to form hypochlorous acid (HOCl) and hydrochloric acid (HCl). The former is a weak acid which readily dissociates to give hypochlorite ions (OCl^-) and a chemical equilibrium develops between the associated and unassociated forms, both of which are disinfectants. Hypochlorite solution is often used as a chlorine source:

$$Cl_2 + H_2O \leftrightarrow HCl + HOCl$$
$$\updownarrow \qquad \updownarrow$$
$$H^+ + Cl^- \quad H^+ + OCl^- \qquad (10.2)$$

The degree of dissociation is pH dependent being suppressed as the pH falls. So if the pH is >9.0 then 100% of the Cl_2 is in the chlorite form, falling to 50% at pH 7.5. Where the pH is <5.0 then 100% of the Cl_2 is in the hypochlorous acid form. Hypochlorous acid is about 80 times more effective than the hypochlorite ion so disinfection is more effective under acid conditions. The dosing rate depends on the rate of flow and the residual required, normally 0.2–0.5 mg l^{-1} after 30 min. Chlorine has a strong taste and odour in drinking water with average taste thresholds of 0.16 and 0.45 mg l^{-1} at pH 7 and 9, respectively (Section 11.2).

Hypochlorous acid and hypochlorite are free chlorine residuals. Chlorine reacts rapidly with any reducing agents and organic matter present in water. Therefore residual chlorine may not last long. Ammonia also reacts with Cl_2 to form chloramines some of which retain a disinfection potential and are known as combined chlorine residuals. Monochloramines (NH_2Cl), dichloramines ($NHCl_2$) and trichloramines (NCl_3) can all be formed depending on the relative concentration of chlorine and ammonia and the pH:

$$NH_4 + HOCl \rightarrow NH_2Cl + H_2O + H^+ \tag{10.3}$$

$$NH_2Cl + HOCl \rightarrow NHCl_2 + H_2O \tag{10.4}$$

$$NHCl_2 + HOCl \rightarrow NCl_3 + H_2O \tag{10.5}$$

The mono- and di-forms have less effective disinfecting properties compared to the free residuals but are much longer lasting. Combined residuals require up to 100 times longer contact time than free residuals for the same degree of treatment. Ammonia may be added deliberately to form combined residuals to prevent chlorine from reacting with phenols, which can result in taste problems, or with trace organics that are toxic. In general, the cleaner the water the longer the residual effect.

Breakpoint chlorination is a technique used to ensure that sufficient free chlorine remains in the water to offer sufficient disinfection. The free chlorine residual is controlled by altering the chlorine to ammonia weight ratio. The breakpoint chlorination curve (Fig. 10.10) explains how the process works. Iron and Mn exert a chlorine demand (a–b) so no residual is produced until this demand is satisfied. Where the ratio of chlorine to ammonia is <5:1 at pH 7 or 8, then monochloramine is produced (b–c). At higher ratios some dichloramine is produced, but as the ratio approaches 7.6:1 the chloramines are oxidized by the excess chlorine to nitrogen gas resulting in a rapid loss of combined residual chlorine in water. Above a ratio of 7.6:1 all the chlorine residual is present as free chlorine residuals.

The disadvantages associated with the chlorination of domestic water supplies such as the formation of trihalomethanes (THMs) are discussed in Section 11.2.

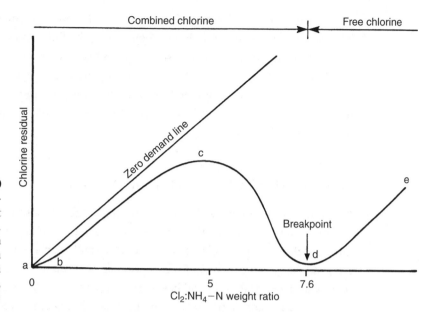

FIGURE 10.10
Breakpoint chlorina-
tion curve. See text
for full explanation.
(Reproduced from
Gray (1994) with
permission of John
Wiley and Sons Ltd,
Chichester.)

10.1.9 FLUORIDATION

Fluoride is a naturally occurring element in many groundwaters, although is generally rare in surface waters. Where fluoride is considered to be beneficial in the prevention of dental caries water has the fluoride concentration artificially raised to
1 mg l^{-1} (Section 11.2). Fluoride must be added after coagulation, lime softening and
activated carbon treatment, as it can be lost during any of these unit processes.
Filtration will not reduce the fluoride concentration provided it is fully dissolved
before reaching the filter. As chlorination has no effect on fluoride, then fluoridation
is normally carried out on the finished water either before or after chlorination.

Fluorine in its pure elemental state is very reactive and so is always used as a compound. Those most widely used for the fluoridation of supplies are ammonium
fluosilicate ($(NH_4)_2SiF_6$) which comes as white crystals, fluorspar or calcium fluoride
(CaF_2) a powder; fluosilicic acid (H_2SF_6) a highly corrosive liquid which must be kept
in rubber lined drums or tanks; sodium fluoride (NaF) which is either a blue or white
powder; and sodium silicofluoride (Na_2SiF_6) which is also either a white or blue powder. All these compounds dissociate (break up) in water to yield fluoride ions (F^-),
with the release of the fluoride almost complete, and silicate (SiO_3). Silicofluorides
(SiF_6) are generally used in water treatment in the British Isles, with sodium silicofluoride the cheapest and most convenient to use. The reaction that takes place when
it is added to water is shown below:

$$SiF_6 + 3H_2O \leftrightarrow 6F^- + 6H^+ + SiO_3 \tag{10.6}$$

The compound is emptied into hoppers and then slowly discharged into a tank by a
screw feed where it is mixed with water to get it into solution. After pH correction the

fluoride solution is added to the finished water. Fluoride is removed by precipitation with calcium and excess aluminium coagulant in the finished water, with the precipitate forming in the distribution system. Care is taken to ensure that the dosage is kept at the correct level. This is done by monitoring the weight of chemical used each day and the volume of water supplied. Also fluoride concentration is tested continuously using an ion-specific electrode and meter to ensure the final concentration is as near to $1.0 \, \text{mg} \, \text{l}^{-1}$ as possible.

10.1.10 ADVANCED WATER TREATMENT

Conventional treatment will normally produce safe wholesome water as defined in 'Introduction to treatment'. However, increasing drinking water standards (e.g. nitrate, chlorinated organics, chlorine-resistant pathogens, taste and odour problems, etc.) and the growing demand for exceptionally high-quality water by many industries (e.g. pharmaceutical, food and beverage, computer manufacture, etc.) has resulted in many physico-chemical post-treatment processes being employed at water treatment plants. These include chemical removal by softening (Section 20.5), adsorption (Section 20.6), ion-exchange (Section 20.7), and membrane filtration including nanofiltration and reverse osmosis (Section 2.8). In order to meet the new pesticide standard of $0.1 \, \mu\text{g} \, \text{l}^{-1}$ and remove traces of pharmaceutical drugs from drinking water, GAC filtration (Section 20.6.2) and ozonization are becoming increasingly common. Membrane filtration and reverse osmosis are equally widespread.

10.1.11 SLUDGE PRODUCTION AND DISPOSAL

Sludges arising from water treatment are very different from those generated from wastewater treatment, although water treatment sludges are often richer in certain pathogens, especially cysts and oocysts. As these sludges are microbially hazardous, careful final disposal is vital. The bulk of the sludge can be either organic matter derived from the removal of suspended solids and colour, or chemical in nature from the use of coagulants and lime. Sludge arises from a variety of different stages in water treatment. The origins of solids are:

(a) *Suspended solids in raw water*. Unlike wastewater treatment the suspended solids concentration of raw water is not normally measured, turbidity being the normal measure of solids concentration. The suspended solids concentration $(\text{mg} \, \text{l}^{-1})$ can be estimated as twice the turbidity value as expressed in NTU.

(b) *Colour removal*. This is particularly a problem where upland sources, which are fed from peaty soils, are used for supply. The sludge production $(\text{mg} \, \text{l}^{-1})$ is estimated to be 20% of the colour removed in degrees Hazen.

(c) *Dissolved chemicals that precipitate during treatment*. These include dissolved iron and manganese, and also the reduction of hardness by softening. If iron is precipitated as ferric hydroxide then $1.9 \, \text{mg} \, \text{l}^{-1}$ of dry solid (DS) is produced for every $\text{mg} \, \text{l}^{-1}$ of iron removed. Lime and soda water softening produce significant quantities of sludge, which can be calculated in mg DS per litre of water softened as:

$$(\text{lime dose } (\text{mg} \, \text{l}^{-1}) + 1.26) \times (\text{decrease in hardness}$$
$$(\text{mg} \, \text{CaCO}_3 \, \text{l}^{-1}) - 0.26) \times \text{sodium carbonate dose } (\text{mg} \, \text{l}^{-1})$$

(d) *Coagulation chemicals.* When coagulants precipitate as hydroxides then the sludge production for each $\text{mg} \, \text{l}^{-1}$ of metal or coagulant used is Al $2.9 \, \text{mg} \, \text{l}^{-1}$, Al_2O_3 $1.5 \, \text{mg} \, \text{l}^{-1}$, Fe $1.9 \, \text{mg} \, \text{l}^{-1}$ and FeO_3 $1.35 \, \text{mg} \, \text{l}^{-1}$.

(e) *Other chemicals.* These include chemicals added during treatment such as powdered activated carbon or bentonite, with the weight added being the weight added to the sludge.

(f) *Biological growth.* This is only significant when slow sand filtration is used. It is not possible to estimate the amount generated, but it is very small.

In 1998 130 756 tonnes DS of water treatment derived sludge was produced in the UK from supplying $16\,500 \, \text{Ml} \, \text{d}^{-1}$ of drinking water, with sludge from alum coagulation comprising 44.8% of the total, iron coagulation 32.9%, softening 17%, natural material 4.5% and other sources, such as solids from GAC adsorption backwashing, 0.8%. The commonest routes for sludge disposal are landfill and discharge to sewer (Table 10.7). Novel disposal methods include: spreading onto agricultural land (natural sludge) 4.4%, (iron sludge) 26.5%, (softening sludge) 16.8%, (alum sludge) 21.7%; land reclamation (iron sludge) 21.2%; soil conditioning (natural sludge) 3.6%; incorporation into construction materials, such as bricks and synthetic aggregates (alum and iron sludges) 5.8% (Simpson *et al.*, 2002).

Solids concentration is expressed as either weight (*w*) of DS per unit volume (*v*), that is DS (*w/v*), or as a proportion of sludge weight that is DS (*w/w*). For water treatment sludges the solids concentration varies from 1 to $50 \, \text{kg} \, \text{m}^3$, with most $5 \, \text{kg} \, \text{m}^3$ with a SG close to 1. Sludge concentration is also expressed as a percentage using *w/v*. Clarifier

TABLE 10.7 Disposal routes for water treatment sludges in the UK during 1997

Disposal route	Mass	
	tonnes DS	*%*
Landfill (offsite)	61 779	51.3
Landfill (onsite)	7660	6.4
Foul sewer	29 599	24.6
Transported to sewage treatment plant	5052	4.2
Lagoon	2423	2.0
Discharge to surface water	2706	2.2
Novel methods	11 128	9.2

sludges have solids concentration of 0.1–0.8% DS with production rates of 1.8% and 2.2% of volume of water treated for ferric and alum sludges, respectively. Production rates in excess of 3.5% of the volume treated indicates operational problems. In contrast filter wash water generates a weak sludge with 0.03% DS with a production rate of 3% of volume treated. However, as the level of treatment improves, so the amount of sludge produced increases.

Water treatment sludges are normally thickened to 5–10% DS before dewatering and subsequent disposal. Dewatering is normally done with a filter press producing a strong cake with high-solids content. The technology employed is very similar to that used for wastewater sludges (Section 21.2).

10.2 WATER DISTRIBUTION

After treatment the finished water has to be conveyed to consumers. This is achieved using a network of pipes known as water mains. There are two broad categories of water main. Trunk mains are the largest and used for transporting large volumes of water from the source to the treatment plant, and from there to a service reservoir or tower. There are no branch or service pipe connections to trunk mains with the water often at a very high pressure. Distribution mains are the network of pipes that bring the water from the service reservoir to the consumer's property. These are smaller and vary in size. The network is highly branched, to which connections to individual houses are made. Distribution mains form loop systems, which equalize the pressure, and ensures the water is used rapidly and kept mixed. Spurs or dead ends are used for cul-de-sacs. These have lower pressure at the end of the pipework resulting in a build-up of material, such as debris, animals and fibres at the end. The water tends to be old which causes further quality problems. While distribution network design and operational practice varies from country to country, the practice employed in the UK that is described below is very similar to that in other countries.

The distribution network comprises of pipes of various sizes ranging down from 18 to 6 in. New systems use 100 or 150 mm (4–6 inches) rigid pipes, down to 63 or 90 mm which are flexible. Materials used include iron, asbestos cement (no longer used) uPVC (unplasticized polyvinyl chloride) or MDPE (medium density polyethylene). All underground pipes are now colour coded to prevent accidents. Water is blue, gas is yellow, electricity is black and telephone is grey.

Demand for water, like electricity, varies diurnally therefore it is very expensive to provide a mains system from the treatment plant to carry maximum flow. This is overcome by the provision of service reservoirs and water towers. These also provide a reserve capacity if the treatment plant or trunk main fails. Service reservoirs ensure peak demands are met in full while smaller pumps and trunk mains are used which are capable of coping with average daily flow rather than peak flow, which is 50–80% greater.

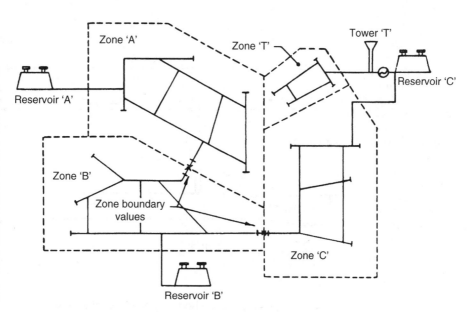

FIGURE 10.11
Schematic diagram
showing that water
distribution net-
works are broken
down into opera-
tional zones each
supplied by a
service reservoir.
(Reproduced from
Latham (1990)
with permission of
the Chartered
Institution of Water
and Environmental
Management,
London.)

The size of service reservoirs depend on the population served, but they must provide storage for at least 24–36 h. The larger ones are constructed in concrete, while smaller ones are made from steel or brick. Such tanks must be watertight and sealed to ensure quality is protected. Each service reservoir serves a water supply zone, which can supply a maximum of 50 000 people, and must ensure an adequate hydraulic head to produce sufficient pressure within mains to push the water up to household storage tanks (Fig. 10.11). A minimum hydraulic head of 30 m is required for fire fighting. In practice the maximum head permitted is 70 m, as above this the loss of water via leaks is very high. Such high pressure also causes excessive wear on household equipment (e.g. taps, stop valves, washing machine valves, ball valves, etc.) and may even prevent some stop valves from shutting off (e.g. dish washers and washing machines). Noise in household plumbing, such as knocking and banging, increases with pressure.

As water is normally fed to the consumer by gravity in flat areas, water towers are required to ensure adequate hydraulic pressure. These tend to serve smaller water supply areas and are expensive to operate due to pumping, even though this is usu- ally done at night during low electricity demand periods. Flats and office blocks require their own pumps to supply a sufficient head. The installation, operation and maintenance of such pumps are the responsibility of the owner.

Hydrants are required primarily for fire fighting but are also used to flush debris out of the mains. The location of a hydrant depends on the requirements of the water company and the fire brigade. In the UK water companies have a legal requirement to locate hydrants where the fire brigade require them, although the cost is borne by the local authority and not the water company. The hydrants are located underground to protect them from frost and damage, and are tested regularly by the fire brigade.

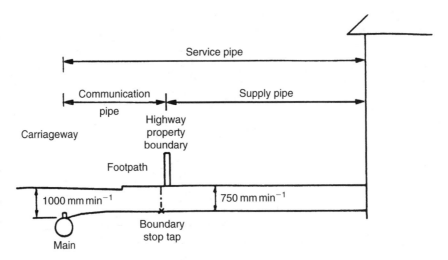

FIGURE 10.12
Arrangement of
service pipes show-
ing minimum depths
to which pipes
should be buried.
(Reproduced from
Latham (1990)
with permission of
the Chartered
Institution of Water
and Environmental
Management,
London.)

The location of hydrants is shown by a yellow plaque with a large black letter H. Two numbers are shown on the plaque above and below the horizontal bar of the letter H. The top number refers to the diameter of the main, which is usually 100 mm for a spur system and 75 mm for loop main. While the lower number is the distance of the hydrant from the plaque in metres which is located directly at right angles to the plaque. The hydrant cover is also painted yellow.

Unlike the other utilities (i.e. gas, telephone and electricity) that take responsibility to connect their supply within the house, this is not the case with water. The household plumbing is connected to the mains by means of a service pipe. For a single dwelling this pipe is <25 mm diameter (usually 13 mm), while if it serves several dwellings then a larger diameter pipe is required (>25 mm). The service pipe is separated by a boundary stop tap into the communication and supply pipes (Fig. 10.12). The communication pipe goes from the main to boundary stop tap and is owned and maintained by the water company. The supply pipe, which is owned and maintained by the owner of the property, runs from the boundary stop tap into the house where there is another stop tap inside the building. Communication pipes are usually short which is not always the case with the supply pipe. This may be very long and cross under roads, across private property and, if old, may even run under buildings making them very difficult to locate and repair. In the UK prior to 1974 Local Authorities were responsible for housing and water supply and they often employed joint supply pipes serving several houses that often passed through the cellars of the houses served (Fig. 10.13). Major problems have arisen as water usage has increased resulting in the end house connected having low pressure. As the supply pipe is jointly owned this leads to serious disputes about paying for maintenance and repair. Joint supply pipes are rarely used now, and when they cause problems are replaced with individual supply pipes.

Supply pipes can be of mild steel, iron, copper, lead, polyethylene or uPVC. Lead was commonly used in older houses because it was flexible but this can result in leaching

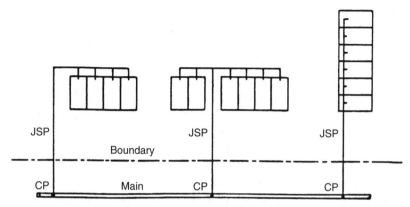

Figure 10.13 Examples of typical joint supply arrangements where CP is the communication pipe and JSP the joint supply pipe. (Reproduced from Latham (1990) with permission of the Chartered Institution of Water and Environmental Management, London.)

of lead into the water. Water companies are required to replace communication pipes made out of lead with polyethylene. Grants are available to householders to replace similar supply pipes. In practice this is done at the same time and can be achieved by pulling the new service pipe through the old one. In older houses (<1965) the electricity supply was generally earthed via the service pipe. So if replaced the electricity supply must be re-earthed as plastic is a non-conductor of electricity.

REFERENCES

American Water Works Association, 1990 *Water Quality and Treatment: A Handbook of Community Water Supplies*, McGraw-Hill, New York.

Binne, C., Kimber, M. and Smethurst, G., 2002 *Basic Water Treatment*, 3rd edn, Thomas Telford, London.

Casey, T. J., 1997 *Unit Treatment Processes in Water and Waste-water Engineering*, John Wiley and Sons, Chichester.

Gray, N. F., 1994 *Drinking Water Quality: Problems and Solutions*, John Wiley and Sons, Chichester.

Latham, B., 1990 *Water Distribution*, Chartered Institution of Water and Environmental Management, London.

Metcalf and Eddy, 1991 *Waste-water Engineering: Treatment, Disposal and Reuse*, McGraw-Hill, New York.

Simpson, A., Burgess, P. and Coleman, S. J., 2002 The management of potable water sludge: present situation in the UK. *Journal of the Chartered Institution of Water and Environmental Management*, **16**, 260–3.

Solt, G. S. and Shirely, C. B., 1991 *An Engineer's Guide to Water Treatment*, Avebury Technical, Aldershot.

Stevenson, D. G., 1998 *Water Treatment Unit Processes*, Imperial College Press, London.

Tebbutt, T. H. Y., 1979 *Principles of Water Quality Control*, Pergamon Press, London.

Twort, A. C., Crowley, F. C., Lawd, F. M. and Ratnayaka, P. P., 1994 *Water Supply*, 4th edn, Arnold, London.

FURTHER READING

Bryant, E. A., Fulton, G. P. and Budd, G. C., 1992 *Disinfection Alternatives for Safe Drinking Water*, Van Nostrand Reinhold, New York.

Camel, V. and Bermond, A., 1998 The use of ozone and associated oxidation processes in drinking water treatment. *Water Research*, **32**, 3208–22.

Chilton, R. (ed.), 1977 *Water Pipeline Systems: Leakage Management, Network Optimization, and Pipeline Rehabilitation Technology*, BHR Group Conference Series 23, Mechanical Engineering Publications, London.

Kawamura, S., 2000 *Integrated Design and Operation of Water Treatment Facilities*, John Wiley and Sons, New York.

Latham, B., 1990 *Water Distribution*, Chartered Institution of Water and Environmental Management, London.

Libra, J. A., Gottschalk, C. H. and Saupe, A., 2000 *Ozonation of Water and Wastewater*, John Wiley and Sons, Chichester.

Open University, 1975 *Water Distribution, Drainage, Discharge and Disposal*, PT272-7, Open University Press, Milton Keynes.

Vigneswaran, S. and Visvanathan, C., 1995 *Water Treatment Processes: Simple Options*, CRC Press, Boca Raton, Florida.

Chapter 11 Drinking Water Contamination

Introduction

Water problems can arise at four different points in the water supply cycle: at the resource, the treatment stage, during distribution to the consumer's house, and within the household plumbing system (Table 11.1). The problem of pathogens is considered separately in Chapter 12. A problem with quality is defined as a failure of a water supply to meet the minimum standards laid down in the EC Drinking Water Directive (Section 8.1).

11.1 Problems Arising from Resource

While the natural quality of drinking water depends primarily on the geology and soils of the catchment, other factors such as land use and disposal of pollutants are also important. In general, impermeable rocks such as granite are associated with turbid, soft, slightly acidic and naturally coloured waters. Groundwater resources associated with hard rock geology are localized and small, so supplies come mainly from surface waters such as rivers and impounding reservoirs. In contrast permeable rocks such as chalk, limestone and sandstone produce clear, hard waters (high Ca^{2+} and Mg^{2+}) which are slightly alkaline. Permeable rocks also form large underground resources making water available from both surface and groundwater sources (Table 11.2). Even after treatment the main physico-chemical characteristics of the supply remain.

Wherever people work and live there will be an increase in toxic substances, non-toxic salts and pathogens entering the water cycle. Industrial contamination, although more localized, is often more serious. The extensive nature of agriculture makes it the most serious threat to drinking water quality, mainly due to the diffuse nature of such pollution, making it difficult to control. Any material or chemicals that find their way into the resource may need to be removed before supply, their removal generally being technically difficult and expensive. The main contaminants in water prior to treatment are listed in Table 11.3. As a general rule the cleaner the raw water, the cheaper the finished water is to produce, and the safer it is to drink. So where the concentration of these compounds is excessively high then the resource may be rejected due to the cost of adequate treatment.

TABLE 11.1   Main sources of drinking water contamination

Resource	Natural geology
	Land use
	Pollution
Treatment	Unit process efficiency
	Chemicals added to clear water
	Chemicals added for protection of consumer
Distribution	Material of pipes, coating
	Organisms
	Contamination
Home plumbing	Materials of pipework/tank
	Contamination
	Poor installation

TABLE 11.2   General water quality characteristics of four water resource types commonly used for supply purposes

	Chalk borehole	Lowland river	Upland river or reservoir	
			Peaty hard rock catchment	Non-peaty hard rock catchment
pH	Slightly alkaline	Neutral or slightly alkaline	Acid	Neutral or slightly acidic
Hardness	Very hard	Moderate to very hard	Soft	Soft
Ammonia	Low	High	Low	Low
Pollution	Normally absent	Present	Absent	Normally absent
Microbial quality	Good	Very poor	Moderately good	Moderately good
Colour	None	Low	Brown	Slight
Taste/odour	None	Bad at times	None	None

TABLE 11.3   Sources of drinking water contamination

Resource based	Arising from water treatment	Arising from distribution system	Arising from home plumbing system
Nitrate	Aluminium	Sediment	Lead
Pesticides	Discolouration	Discolouration	Copper
Industrial solvents	Chlorine	Asbestos	Zinc
Odour and taste	Odour and taste	Odour and taste	Odour and taste
Iron	Iron	Iron	Fibres
Manganese	Trihalomethanes	PAH	Corrosion
Hardness	Fluoride	Animals/biofilm	Pathogens
Pathogens	Pathogens	Pathogens	
Algal toxins	Nitrite		
Radon and radioactivity			

11.1.1 TASTE AND ODOUR

Water should be palatable rather than taste and odour free, but as people have very different abilities to detect tastes and odours at low concentrations, this is often difficult to achieve. Offensive odours and tastes account for the majority of consumer complaints about water quality. Odour and taste problems can be a very transitory phenomenon, so it is unlikely that the standardized monitoring involving taking just four or five samples a year will detect all but the most permanent odour and taste problems.

It is extremely difficult to chemically analyse tastes and odours, mainly because of the very low concentrations at which many of these chemicals can be detected by humans. In addition, offensive odours can be caused by a mixture of several different chemicals. The identification of odours and tastes is normally carried out subjectively by trained technicians or by a panel of testers. Acceptable levels of odour and taste in drinking water are determined by calculating odour and taste threshold levels. There are two widely used measures of taste and odour. Both are based on how much a sample must be diluted with odour- and taste-free water to give the least perceptible concentration. Threshold number (TN) is most widely used:

$$TN = \frac{A + B}{A} \qquad (11.1)$$

where A is the volume of the original sample and B is the volume of dilutent. So $A + B$ is the total volume of sample after dilution to achieve no perceptible odour. So a sample with a TN of 1 has no odour or taste.

Dilution number is used in the EU Drinking Water Directive:

$$DN = \frac{B}{A} \qquad (11.2)$$

The two are related as $DN = TN - 1$. So a sample with a DN of 3 has a TN of 4.

Standards for taste and odour are generally the same. In the US, the maximum contaminant level (MCL) for odour is a TN of 3 (no temperature specified), while the WHO only requires that the taste and odour shall be acceptable. The EU originally set a guide DN value of 0 and two MAC values of 2 at 12°C and 3 at 25°C in the first Drinking Water Directive (80/778/EEC). The problem was that the Directive did not specify how to interpret the standard and no methodology was specified. In contrast the new Drinking Water Directive (98/83/EC) does not set a numerical value, rather it states that taste must be acceptable to consumers with no abnormal changes. To help with monitoring there are now several standard methods for assessing taste and odour in the UK (Table 11.4) and in the US (ASTM, 1984; APHA, 1998).

TABLE 11.4   Key methods for the assessment of taste and odour in drinking water

The assessment of taste, odour and related aesthetic problems in drinking waters 1998. Blue Book No. 171. Standing Committee of Analysts, Environment Agency, London.

The Microbiology of Drinking Water 2004. Part 11: Taste, odour and related aesthetic problems. Blue Book No. 196. Standing Committee of Analysts, Environment Agency, London.

The Microbiology of Drinking Water 2004. Part 12: Methods for the Isolation and Enumeration of Micro-organisms Associated with Taste, Odour and Related Aesthetic Problems. Blue Book No. 197. Standing Committee of Analysts, Environment Agency, London.

TABLE 11.5   The chemicals causing specific groups of odours and their possible source

Odour	*Odour producing compound*	*Possible source*
Earthy/musty	Geosmin 2-Methylisoborneol 2-Isopropyl-3-methoxy pyrazine Mucidine 2,3,6-Trichloroanisole	Actinomycetes; blue-green algae
Medicinal or chlorphenolic	2-Chlorphenol 2,4-Dichlorphenol 2,6-Dichlorphenol	Chlorination products of phenols
Oily	Napthalene Toluene	Hydrocarbons from road run-off; bituminous linings in water mains
Fishy, cooked vegetables or rotten cabbage	Dimethyltrisulphide Dimethidisulphide Methyl mercaptan	Breakdown of algae and other vegetation
Fruity and fragrant	Aldehydes	Ozonation by-products

Reproduced from Gray (1994) with permission of John Wiley and Sons Ltd, Chichester.

Odour problems are categorized according to the origin of the substance causing the problem. Substances can be present in the raw water, be added or created during water treatment, arise within the distribution system or arise in the domestic plumbing system. Of course, the quality of the raw water will often contribute to the production of odours during treatment and distribution.

It is not always straightforward to identify the source of an odour or taste in drinking water (Table 11.5). There are seven common causes:

1. *Decaying vegetation*: algae produce fishy, grassy and musty odours as it decays, and certain species can cause serious organoleptic problems when alive.
2. *Moulds and actinomycetes*: produce musty, earthy or mouldy odours and tastes; they tend to be found where water is left standing in pipework and also when the water is warm – they are frequently found in the plumbing systems of large buildings

such as offices and flats, but are also associated with water-logged soil and unlined boreholes.

3. *Iron and sulphur bacteria*: both bacteria produce deposits which release offensive odours as they decompose.

4. *Iron, manganese, copper and zinc*: the products of metallic corrosion all impart a rather bitter taste to the water.

5. *Sodium chloride*: excessive amounts of sodium chloride will make the water taste initially flat or dull, then progressively salty or brackish.

6. *Industrial wastes*: many wastes and by-products produced by industry can impart a strong medical or chemical taste or odour to the water – phenolic compounds which form chlorphenols on chlorination are a particular problem.

7. *Chlorination*: chlorine by itself does not produce a pronounced odour or taste unless the water is overdosed during disinfection; chlorine will react with a wide variety of compounds to produce chlorinated products many of which impart a chlorinous taste to the water.

Musty or earthy odours are commonly reported in drinking water. They are generally caused by actinomycetes and to a lesser extent by blue-green algae and fungi. The main organic compounds produced which cause such odours are geosmin and 2-methylisoborneol, and consumer complaints will follow if concentrations of either of these exceed $8–10\,\mathrm{ng}\,\mathrm{l}^{-1}$ (Table 11.5). Dimethyl sulphide is also produced by blue-green algae and although it has a higher odour threshold than geosmin, it produces a grassy odour which intensifies into a very unpleasant septic/piggy odour with increasing concentration of algal biomass.

11.1.2 IRON AND MANGANESE

Both iron and manganese are extremely common metals and are found in large amounts in soil and rocks, although normally in an insoluble form. However, due to a number of complex reactions which occur naturally in the ground, soluble forms of the metals can be formed which can then contaminate any water passing through. Therefore, excess iron and manganese are a common phenomenon of groundwaters, especially those found in soft groundwater areas.

The solubility of iron depends largely on pH and redox potential. While the normal level of soluble iron in groundwaters is $<1\,\mathrm{mg}\,\mathrm{l}^{-1}$ even slight changes in water chemistry can lead to significant increases in solubility of the metal. The natural conditions in the majority of groundwaters in the British Isles are on the boundary between soluble (Fe^{2+}) and insoluble iron ($Fe(OH)_3$) (Fig. 11.1). Any reduction in pH or redox potential due to dissolved CO_2, presence of humic acids from peaty soils or contamination by organic matter causing deoxygenation will all increase the soluble iron content of the water. Boreholes allow oxygen to enter aquifers, especially confined aquifers (Fig. 2.4), and this, combined with the removal of CO_2 from solution, causes

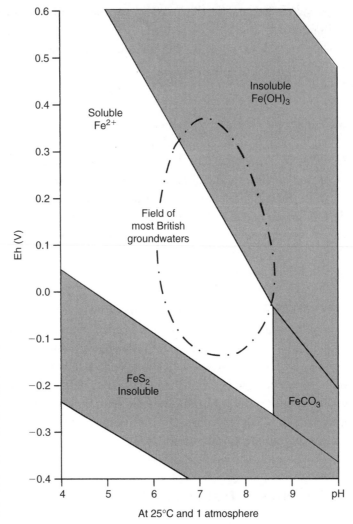

FIGURE 11.1
Phase-stability dia-
gram for the solubil-
ity of iron in
groundwater. Normal
range for UK
groundwaters is
shown by the
dotted line.
(Reproduced from
Clark (1988) with
permission of John
Wiley and Sons Ltd,
Chichester.)

the iron to precipitate out as ferric hydroxide. This is accompanied by the growth of iron bacteria that form slimy filamentous colonies that coat the wall of the boreholes and clog screens. Eventually the deposited iron can reduce the supply of water by reducing the flow into the borehole and by damaging the pump.

Discolouration of water by iron is inconvenient but is only an aesthetic problem. It becomes more irritating when the iron causes staining of laundry and discolouration of vegetables, such as potatoes and parsnips, during cooking. But more important is that iron has quite a low taste threshold for such a common element, giving the water a strong unpleasant taste, which ruins most beverages, made with the water. High concentrations of iron in the water can react with the tannin in tea causing it to turn odd colours, usually inky black. Iron is so ubiquitous that some will find its way into nearly all supplies. The taste threshold is about $0.3\,\mathrm{mg\,l^{-1}}$ ($300\,\mathrm{\mu g\,l^{-1}}$) although it varies considerably between individuals. The EU Drinking Water Directive sets a maximum concentration value of $0.2\,\mathrm{mg\,l}^{-1}$, although ideally this should not exceed $0.05\,\mathrm{mg\,l}^{-1}$ (Table 8.7). The WHO (1993) guideline was set at the higher concentration of $0.3\,\mathrm{mg\,l}^{-1}$, but in the latest review (WHO, 2004) a guide value has not been set as iron is not of concern to health at the concentrations normally observed in drinking water with water rejected on grounds of taste or appearance well below the health-based value is reached.

Manganese is identical to iron in all respects and indeed causes all the same taste and staining problems. Staining is more severe than iron, as is the unacceptable taste manganese imparts to the water. For that reason a much stricter drinking water limit has been imposed at $0.05\,\mathrm{mg\,l}^{-1}$ ($50\,\mathrm{\mu g\,l}^{-1}$), the concentration at which laundry and sanitary ware becomes discoloured grey/back. Although the WHO (2004) has a health-related guide value of $0.4\,\mathrm{mg\,l}^{-1}$, a much lower value must be set in practice to avoid consumer complaints. While the major source of manganese in the diet is food, all the manganese in water is readily bioavailable. However, unless the concentration in drinking water is excessively high, it is unlikely that it will significantly contribute to the daily intake of the metal. Toxicity is not a factor with manganese as it will be rejected on the grounds of taste or appearance well below the health-based value is reached. Excessive manganese intake does cause adverse physiological effects, particularly neurological, but not at the concentrations normally associated with water in Europe.

11.1.3 NITRATE

Nitrate is a common component in food, and the percentage intake from drinking water is linked directly to its concentration. For example, at concentrations of 10, 50 and $100\,\mathrm{mg\,l}^{-1}$ as NO_3 the total daily intake from drinking water is 20%, 55% and 71% on average of the total daily intake, with the rest coming from food. The current EU limit for nitrate is $50\,\mathrm{mg\,l}^{-1}$, which is equivalent to $11.3\,\mathrm{mg\,l}^{-1}$ as nitrogen (Table 8.6). In the US, there is a mandatory MCL value of $10.0\,\mathrm{mg\,l}^{-1}$ as nitrogen which is equivalent to $44\,\mathrm{mg\,l}^{-1}$ as nitrate (Section 8.1). A new mandatory nitrite standard of $0.5\,\mathrm{mg\,l}^{-1}$ has been set in the UK, with a combined nitrate/nitrite parameter also used which is calculated as:

$$\frac{\text{Nitrate}}{50} + \frac{\text{Nitrite}}{3} \tag{11.3}$$

which must not exceed 1.0 at the consumer's tap.

The main concern of nitrate in potable supplies is methaemoglobinaemina (IM), or blue baby syndrome in infants. The nitrate combines with haemoglobin in red blood cells to form methaemoglobin, which is unable to carry oxygen, and so reduces the uptake of oxygen from the lungs. Normal methaemoglobin levels in blood are between 0.5% and 2.0% while levels in excess of 10% results in a blue tinge to the skin of infants. Death is caused when 45–65% of the haemoglobin has been converted. Infants <3 months of age have different respiratory pigments that combine more rapidly with nitrate than haemoglobin making them more susceptible to the syndrome. IM is primarily associated with bacteriologically contaminated water and is limited to those on undisinfected well or spring supplies. With 98% of Europeans on mains supplies, in practice, infantile IM is now almost non-existent. It can, however, be a rare complication in gastroenteritis regardless of nitrate concentration or intake. While no direct evidence of nitrate causing the condition has so far been found, very high nitrate levels in water ($>100\,\text{mg}\,\text{l}^{-1}$ as NO_3) must be considered as potentially hazardous especially to young children, children with gastroenteritis and pregnant women. Most nitrate is excreted in the urine largely unchanged, while some bacterial reduction to nitrite occurs in the stomach. Up to 25% of nitrate is recirculated in saliva during which most is converted to nitrite by bacteria. In the intestines nitrite reacts in an acidic environment to form N-nitroso compounds with amines and amides, which may be associated with gastric and bladder cancer (ECETOC, 1988).

Nitrate finds its way into water resources either by leaching into aquifers or as run-off into surface waters from agricultural land. There has been a steady build up of nitrate in both ground and surface waters since the 1940s due to a combination of increasing fertilizer usage per unit area and more grassland being ploughed up for crop production. Nitrate concentrations in rivers fluctuate more rapidly than groundwaters showing a seasonal pattern with highest concentrations corresponding with periods of run-off. This is particularly evident in a wet autumn after soil nitrogen levels have built up over a dry summer. While levels in many rivers have now stabilized due to better agricultural practice, the levels in many groundwaters continue to rise due to the time delay as pollutants move slowly through the aquifer. This also has important implications for rivers fed by groundwater. As with other pollutants, the concentration of nitrate in groundwater depends largely on aquifer type. Even if farmers stopped using nitrogen fertilizers completely, most ground waters would continue to have very high nitrate concentrations for a minimum of 20–30 years. So to ease the pressure in the short term, it will be necessary to remove nitrates from water, which is both expensive and technically difficult. There is growing interest in the possibility of farmers being required to pay for advanced water treatment by utilizing the 'polluter pays' principle to charge for nitrate leached from land. However, the dispersed nature of nitrate leaching makes the implementation of such charges very difficult. The Water Framework Directive provides an integrated river basin management system that allows much greater control of the pollution of all water resources from land-based sources (Section 7.3). The Nitrate Directive specifies a number of measures for the protection of surface and ground waters from agricultural nitrate pollution (Section 8.4). The approach adopted by most countries is a mixture of treatment and resource protection strategies. The approach selected is dictated purely by economics and curiously enough prevention of contamination may be more expensive than removing the nitrate at the water treatment plant. Ways of removing nitrates from water supplies include:

1. *Replacement*: expensive as new water mains and a suitable alternative supply is required. In practice, this option is normally restricted to small and usually isolated contaminated groundwater resources only.

2. *Blending*: controlled reduction of nitrate to an acceptable concentration by diluting nitrate-rich water with low nitrate water. Currently widely practised throughout Europe, blending becomes increasingly more expensive if the nitrate concentration continues to rise. Not only is a suitable alternative supply required, but also facilities to mix water at the correct proportions. While groundwater nitrate concentrations do not fluctuate widely from month to month, surface water nitrate levels can display large seasonal variations with particularly high concentrations occurring after heavy rainfall following severe drought periods. Most blending operations involve diluting contaminated groundwater with river water, so such seasonal variations can cause serious problems to water supply companies.

3. *Storage*: some removal of nitrate can be achieved by storing water for long periods of time in large reservoirs. The nitrate is reduced to nitrogen gas by bacteria under the low oxygen conditions that exist in the sediments of the reservoir, a process known as denitrification.

4. *Treatment*: the nitrate can be removed by ion exchange or microbial denitrification. Both are expensive and continuous operations. Water purification at this advanced stage is technically complex and difficult to operate continuously at a high efficiency level. Also malfunction of nitrate reduction treatment plants can cause further water contamination. Ion-exchange systems use a resin that replaces nitrate ions with chloride ions (Section 20.7). While the system is efficient it does produce a concentrated brine effluent that requires safe disposal. Nitrate removal brings about other changes in drinking water composition, and the long-term implications on the health of consumers' drinking water treated by ion exchange is unknown. This is the process in which most water companies are investing and is now widely in use throughout the UK.

5. *Selective replacement*: rather than treating the entire water supply many water supply companies use a system of identifying those most at risk within the community from high nitrates, and supplying an alternative supply such as bottled water or home treatment units. This is by far the cheapest approach for the water industry, although is dependent on legal constraints (i.e. quality standards).

Farmers are encouraged to reduce nitrate contamination by cutting down on fertilizer use, modifying cropping systems and simple control measures (Table 7.6). In terms of good agricultural practice this includes the following:

1. Fallow periods should be avoided. The soil should be kept covered for as long as possible, especially during winter, by the early sowing of winter cereals, intercropping or using straw mulch.

2. Avoid increased sowing of legumes unless the subsequent crop can utilize the nitrogen released by mineralization of nitrogen-rich residues.

3. Grassland should not be ploughed.

4. Tillage should be minimized and avoided in the autumn. Direct drilling should be used whenever possible.

5. Slopes should, whenever possible, be cultivated transversely to minimize run-off.
6. Manure should not be spread in autumn or winter. It should be spread evenly and the amount should not exceed crop requirements.
7. Amount of nitrogen fertilizer and manure should be applied at times and in quantities required by plants to ensure maximum uptake. The amount of available nitrogen in the soil should also be considered. Computer models are available to help the farmer calculate very accurately the amount of fertilizer required for a particular crop grown on a specific soil type; alternatively, consultants should be used to advise on fertilizer application rates. With the increasing problem of eutrophication in surface waters, and the large residual phosphorus concentration present in many soils, nutrient management plans should consider both nitrogen and phosphorus.

11.1.4 ORGANIC COMPOUNDS

With the development of gas chromatography coupled with mass spectrometry it has become possible to identify hundreds of organic compounds in drinking water. Although some are naturally occurring compounds, the majority are synthetic such as pesticides and industrial solvents. Others are synthesized when organic compounds react with chlorine during disinfection (i.e. trihalomethanes (THMs)) (Section 11.2.1). Nearly all organic compounds found in drinking water are toxic and most are also thought to be carcinogenic, even at very low concentrations ($<1\,\mu g l^{-1}$).

Pesticides are not only associated with large lowland arable farms but from other farm-based activities, such as sheep dipping in upland areas. There is a wide range of other users from railway companies to local authorities, forestry groups to the average gardener to name but a few. This is reflected in the high concentrations of essentially non-agricultural herbicides found in drinking water, in particular simazine and atrazine (Table 11.6).

Only five industrial solvents are frequently found in drinking water (Table 11.7), although pollution by these solvents is very widespread. Contamination is largely restricted to large industrial cities such as Milan where in 1982 60% of the 621 wells tested contained chlorinated solvents in excess of $50\,\mu g l^{-1}$. In the Netherlands a study of 232 boreholes showed 67% contained trichloroethylene (TCE), 60% trichloromethane (TCM) and 19% perchloroethylene (PCE). In the US, the problem is also widespread. In New Jersey, 20% of the 315 wells tested contained industrial solvents with TCE levels in excess of $100\,\mu g l^{-1}$ in some wells (Loch et al., 1989). The situation in the UK is reviewed by Rivett et al. (1990).

Chlorinated organics have higher densities but lower viscosities than water, while aromatic hydrocarbons such as benzene, toluene and fuel oils have lower densities but are more viscous. In groundwaters they behave quite differently. The former solvents tend to accumulate in the base of the aquifer with their direction of movement

TABLE 11.6   Commonest pesticides in European drinking waters are all
herbicides except dimethoate which is an insecticide

Frequently occurring
Atrazine, chlorloluron, isoproturon, MCPA, mecoprop, simazine
Commonly occurring
2,4,D, dicamba, dichlorprop, dimethoate, linuron, 2,4,5,T

TABLE 11.7   Major industrial solvents in groundwaters

Solvent	*Uses*	*Notes*
TCE: trichloroethylene or trichloroethene	Solvent cleaning, dry cleaning	Being replaced by TCA
PCE: perchloroethylene or tetrachloroethene	Metal cleaning	Being replaced by TCA
TCA: methylchloroform or 1,1,1-trichloroethane	Metal/plastic cleaning, general solvent	Less toxic (both acute and chronic) than TCE and PCE
DCM: dichloromethane or methylene chloride	Paint striping, metal cleaning, pharmaceuticals, aerosol propellants, acetate film	
TCM: trichloromethane or chloroform	Fluorocarbon synthesis, pharmaceuticals	

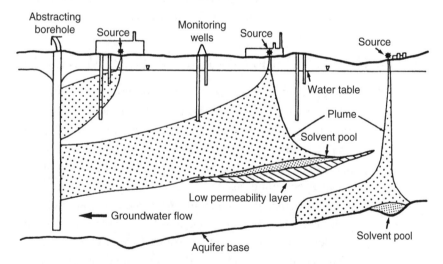

FIGURE 11.2   Behaviour of chlorinated solvents in groundwaters and the interactions
with a pumped extraction borehole and unpumped monitoring wells. (Reproduced
from Rivett *et al.* (1990) by permission of the Chartered Institution of Water and
Environmental Management, London.)

governed by the slope and form of the base of the aquifer (Fig. 11.2). Some of the solvent dissolves into the water forming distinct plumes governed by the direction and
velocity of the groundwater. The latter hydrocarbons tend to float on top of the
groundwater forming pancakes of material. Migration follows the flow of the groundwater but remains at the surface of the water table. If the water table rises, the

aromatic hydrocarbons rise with it and are largely retained within the porous unsaturated layer once the water table falls. In this way the movement of these compounds is restricted resulting in localized pollution while the chlorinated organics can affect large areas of the aquifer. Further details on groundwater contaminants and their treatment are given in Mather *et al.* (1998) and Nyer (1992).

11.1.5 THE FUTURE

In Europe, only a third or less of supplies come from surface waters, and these are subject to extensive treatment including extensive storage and bankside infiltration. In the UK, 75% of all supplies come from surface waters that generally receive only conventional treatment. Therefore, it is inevitable that the supplies will be more prone to problems such as colour, iron and aluminium from upland catchments, and taste, odour and THMs in lowland abstractions after treatment. With dwindling water resources and ever-increasing demands, it is inevitable that water will have to be reused over and over again, and that companies will have to exploit water from resources of poor quality. The most effective way to control these problems is to prevent contamination of raw water supplies in the first place. In Europe, this is being achieved by the introduction of aquifer protection zones, nitrate-sensitive areas and in tighter controls of disposal of hazardous and toxic wastes through the Water Framework Directive and the proposed new Groundwater Directive (Section 8.4).

11.2 WATER QUALITY PROBLEMS ARISING FROM WATER TREATMENT

The principle of water treatment is to make raw water safe and palatable to drink. During treatment a wide variety of chemicals are used. Very occasionally excessive amounts of these chemicals are added due to poor operation or accidents. However, small amounts of these chemicals are normally discharged with the finished water due to the nature of the processes themselves. These include aluminium, iron and organic compounds used as coagulants, such as polyacrylamide. These are needed to remove particulate matter and with good operational management the carryover should be negligible. These chemicals cause discolouration as well as odour and taste problems. The new Drinking Water Directive sets limit values for all these compounds (Tables 8.6 and 8.7), with a limit set on acrylamide for the first time in drinking water at $0.1 \, \mu g \, l^{-1}$. This will be difficult for the water industry to achieve as acrylamide monomer is the active ingredient in polyacrylamide flocculants, which are used widely to treat drinking water and to dewater sludges in both water and wastewater treatment. The industry had set a maximum permissible dosing concentration of $0.5 \, mg \, l^{-1}$ of the flocculant, which must not exceed 0.025% free acrylamide monomer in its formulation, resulting in $0.125 \, \mu g \, l^{-1}$ in the finished water assuming all the acrylamide monomer is dissolved. The Drinking Water Inspectorate has set a stricter acrylamide monomer limit for polyacrylamide flocculants at 0.02% ensuring the new drinking water standard can be achieved. The new purity limits in the existing standards for polyacrylamides used for the treatment of water for human consumption (BS:EN 1407:1998 Anionic and non-ionic polyacrylamides and BS:EN 1410:1998 Cationic polyacrylamides) came into force on 25 December 2003. Another factor is that recycled water from sludge treatment will contain all the acrylamide monomer used during the process. This extra loading must also be taken into consideration when calculating flocculant doses. Other chemicals are added deliberately to water, chlorine to protect consumers from pathogens and fluoride to help protect teeth from decay.

11.2.1 CHLORINE AND CHLORINATED ORGANIC COMPOUNDS

Chlorine has a distinctive odour that is responsible for most reported odour and taste problems in drinking waters. Although the WHO has set its health-related guideline value at $5.0 \, mg \, l^{-1}$, in practice concentrations are very much lower than this due to odour and taste problems. Chlorine has an average odour threshold of $0.16 \, mg \, l^{-1}$ at pH 7 and $0.45 \, mg \, l^{-1}$ at pH 9, although the ability to detect chlorine in drinking water varies significantly between individuals. Chloramines, produced when chlorine reacts with ammonia, are more odorous then free chlorine becoming progressively more offensive with increasing number of chlorine atoms. The operator has to strike a balance between the protection of public health by good disinfection and the wholesomeness of water in terms of chlorine taste (Section 10.1).

Phenolic compounds react with chlorine to produce mono- or dichlorphenols, both of which have a strong medicinal odour. So objectionable is the odour that problems can occur even when phenols are present in raw waters at below the current detection limit. Dichlorphenol has an odour threshold of $2 \, \mu g \, l^{-1}$, and as many phenolic compounds occur naturally in lowland rivers such odour problems are an unavoidable seasonal problem.

A wide range of compounds can be formed by chlorine reacting with natural humic material or other organic compounds (Table 11.8). THMs are simple, single carbon compounds which have the general formulae CHX_3 where X may be any halogen atom (e.g. chlorine, bromine, fluorine, iodine or a combination of these). While the risk from THMs is low, some are known to be carcinogenic with evidence showing a link between long-term low-level exposure and rectal, intestinal and bladder cancers. They are only found in waters that have been chlorinated, with those dependent on soft peaty water resources or nutrient-rich lowland rivers for supply particularly at risk. The EU Drinking Water Directive (98/83/EC) has set a limit value for total THMs of $100 \, \mu g \, l^{-1}$ (Tables 8.6). Total THMs is calculated as the sum of the four common THMs in Table 11.9). The World Health Organization (WHO) has set guideline values for four THM compounds (Table 11.9), where the guideline value for THMs is that the sum of the ratio of the concentration of each to its respective guideline value should not exceed 1.

TABLE 11.8   Common disinfection by-products found in drinking water

Chlorine
   Trihalomethanes
      Chloroform (trichloromethane)[a]
      Bromoform (tribromomethane)[a]
      Bromodichloromethane (BDCM)[a]
      Dibromochloromethane (DBCM)[a]
   Other trihalomethanes: chlorinated acetic acids
      Monochloroacetic acid (MCA)[a]
      Dichloroacetic acid (DCA)[a]
      Trichloroacetic acid (TCA)[a]
   Other trihalomethanes: haloacetonitriles
      Dichloroacetonitrile (DCAN)[a]
      Dibromoacetonitrile (DBAN)[a]
      Bromochloroacetonitrile (BCAN)[a]
      Trichloroacetonitrile (TCAN)[a]
   Other
      Chloral hydrate[a]
      3-Chloro-4-(dichloromethyl)-5-hydoxy-2(5H)-furanone (MX)[a]
      Chloropicrin[a]
      Chlorophenols[a]
      Chloropropanones

Ozone
   Bromate[a]
   Formaldehyde[a]
   Acetaldehyde
   Non-chlorinated aldehydes
   Carboxylic acids
   Hydrogen peroxide
Chlorine dioxide
   Chlorite[a]
   Chlorate[a]
Chloramination
   Cyanogen chloride[a]

[a]Included in the 1993 WHO drinking water guidelines.
Reproduced from Gray (1994) with permission of John Wiley and Sons Ltd, Chichester.

TABLE 11.9   WHO standards for the common THMs

Chloroform	$CHCl_3$	$200 \, \mu g \, l^{-1}$
Bromoform	$CHBr_3$	$100 \, \mu g \, l^{-1}$
Dibromochloromethane	$CHBrCl$	$100 \, \mu g \, l^{-1}$
Bromodichloromethane	$CHBrCl_2$	$60 \, \mu g \, l^{-1}$

Reproduced with permission of the World Health Organization, Geneva.

To prevent by-product formation where there is a high concentration of organic matter present in the water that could react with chlorine, either chlorine dioxide or chloramination should be used, or if feasible an alternative disinfection method employed (e.g. ozonization or ultraviolet radiation). However, other by-products can also be formed with alternative disinfectants (Table 11.8). While granular activated carbon (GAC) filtration has been used to remove chlorinated organics, the general approach is to use coagulation to remove dissolved organic matter before disinfection (Sections 10.1 and 20.2). Some interesting treatment solutions are described by Banks (2002).

11.2.2 FLUORIDE

For decades it has been accepted that the addition of fluoride to water supplies reduces the incidence of tooth decay in growing children, with optimum benefit achieved around $1.0 \, \text{mg} \, l^{-1}$. It has also been widely suggested that fluoride is also beneficial to older people in reducing hardening of the arteries and, as fluoride stimulates bone formation, in the treatment of osteoporosis. Today over 250 million people drink artificially fluoridated water. In the US, for example, over half of the water supplies are fluoridated while in the Republic of Ireland all public water supplies are fluoridated where the natural levels are below $1.0 \, \text{mg} \, l^{-1}$. No supplies are fluoridated in Northern Ireland and only about 10% of supplies are currently fluoridated in the UK (Department of the Environment, 1997). In western continental Europe its use has been terminated or never implemented, primarily over health concerns, so that <1% of the population drink artificially fluoridated water. Fluoride is found naturally in waters and high natural levels have to be reduced by mixing with low fluoride water to bring the concentration to $0.9 \, \text{mg} \, l^{-1}$ (Fig. 11.3).

The EU Drinking Water Directive sets limit values for fluoride at $1.5 \, \text{mg} \, l^{-1}$ at 8–12°C, or $0.7 \, \text{mg} \, l^{-1}$ at 25–30°C. This limit value of $1.5 \, \text{mg} \, l^{-1}$ has been adopted in the UK, although in the Republic of Ireland a national limit of $1.0 \, \text{mg} \, l^{-1}$ has been set. The effects of fluoride vary with temperature, so as the temperature increases the concentration permitted in drinking water decreases. Some bottled mineral waters may also contain high levels of fluoride. The US Environmental Protection Agency has set two standards for fluoride: a primary drinking water standard of $4 \, \text{mg} \, l^{-1}$ to protect against skeletal fluorosis, and a non-mandatory secondary drinking water standard of $2.0 \, \text{mg} \, l^{-1}$ to prevent dental fluorosis (Section 8.1), both of which are under review. The WHO have set a guide value of $1.5 \, \text{mg} \, l^{-1}$, although it stresses that the volume of water consumed, and the intake of fluoride from other sources, should be considered when setting national standards.

While it is clear that the early fluoridation campaigns had a significant improvement on the incident on tooth decay, these effects are no longer so obvious within communities (USEPA, 1991). The primary explanation given being the now widespread use of fluoridated toothpaste, fluoride supplements and better dental hygiene in the community. After the age of 12 years calcification of teeth ceases so that fluoride is of no further benefit. Therefore, it does seem more appropriate to prescribe fluoride medications solely for children rather than this mass-medication

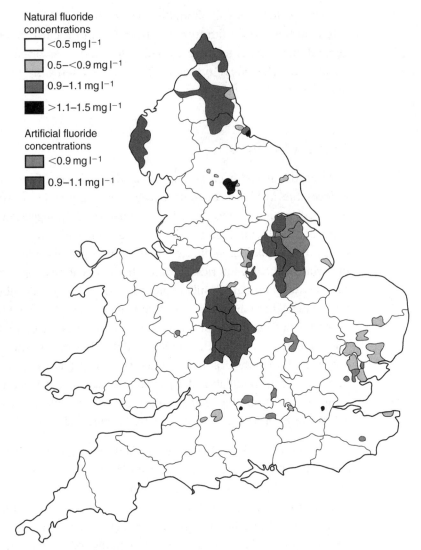

Natural fluoride concentrations
- □ <0.5 mg l^{-1}
- ▨ 0.5–<0.9 mg l^{-1}
- ▨ 0.9–1.1 mg l^{-1}
- ■ >1.1–1.5 mg l^{-1}

Artificial fluoride concentrations
- ▨ <0.9 mg l^{-1}
- ▨ 0.9–1.1 mg l^{-1}

FIGURE 11.3 Natural and artificial fluoride concentrations in water supplies in England and Wales in 1996. (Reproduced from the Department of the Environment (1997) with permission of Her Majesty's Stationery Office.) More detailed information is given in Department of the Environment (1997).

approach. However, this does disadvantage those children who do not have access to fluoride medication.

Excess fluoride causes teeth to become discoloured (fluorosis) and long-term exposure results in permanent grey to black discolouration of the enamel. Children who drink fluoride in excess of 5 mg l^{-1} also develop severe pitting of the enamel. With the increasing exposure of children to fluoride the advantages of water fluoridation may be outweighed by the risk of fluorosis from over-exposure (Dunning, 1986) (Fig. 11.4). Fluoride is cumulative and so any long-term effects of low-dose exposure may well not be realized yet. Leaks and spillages of fluoride can be very serious and even fatal. The general effect of moderately elevated fluoride (30–50 mg l^{-1}) will be mild gastroenteritis and possible skin irritation. For these reasons, many water suppliers have taken the precautionary step of not fluoridating water until more research has been done on the potential long-term health risks.

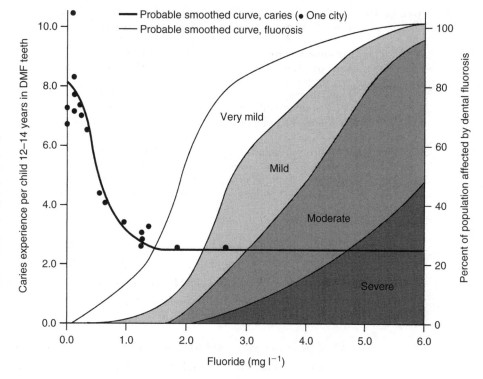

Figure 11.4 Occurrence of dental caries and dental fluorosis in children in relation to the concentration of fluoride in drinking water. (Reproduced from Dunning (1986) with permission of the President and Fellows of Harvard College, and Harvard University Press, Cambridge, Massachusetts.)

11.3 Water Quality Problems Arising from the Distribution System

Water quality problems are commonly associated with the deterioration of the distribution network. Iron pipes can lead to discolouration of water, elevated iron concentrations and sediment problems. Asbestos cement, once widely used for water mains, releases fibres of asbestos into the water. Concentrations of fibres up to $2 \times 10^6 1^{-1}$ are common and, although the USEPA has included asbestos in the National Primary Drinking Water Regulations with an MCL set at 7×10^6 fibres 1^{-1}, there is little evidence of any health risks, which is why no standard has been proposed for Europe. Both iron and asbestos cement distribution pipes are frequently lined with coal tar or bitumen that release polycyclic aromatic hydrocarbons (PAHs) into the water. All distribution mains have resident populations of animals than can cause consumer complaints but the development of biofilms on the walls can lead to pathogen problems (Chapter 12) as well as odour and taste problems.

11.3.1 POLYCYCLIC AROMATIC HYDROCARBONS (PAHs)

Polycyclic (polynuclear) aromatic hydrocarbons also occur in soot, vehicle exhausts and in the combustion products of hydrocarbon fuels. They are generally not very soluble but are readily adsorbed onto particulate matter resulting in high concentrations where suspended solids are present in water. Nearly all PAHs are carcinogenic, although their potency varies, the most hazardous by far being benzo(a)pyrene.

TABLE 11.10 Total concentration of PAHs in drinking water is calculated from the concentration of six reference compounds

Compound names used by WHO	Alternative names used by European Commission
Fluoranthene	Fluoranthene
3,4-Benzofluoranthene	Benzo-3,4-fluoranthene*
11,12-Benzofluoranthene	Benzo-11,12-fluoranthene*
Benzo(a)pyrene	Benzo(a)pyrene
1,12-Benzoperylene	Benzo-1,12-perylene*
Indeno-(1,2,3-cd) pyrene	Indeno-(1,2,3-c-d)-pyrene*

*The reference compounds used in the EU Drinking Water Directive.

PAHs in drinking water are thought to cause gastrointestinal and oesophageal tumours, however, drinking water contributes only a small proportion ($<0.5\%$) of the total adult PAH intake. PAHs are rarely present in the environment on their own and the carcinogenic nature of individual compounds is thought to increase in the presence of other PAH compounds.

While there are over 100 PAH compounds, the WHO have listed six reference compounds that should be routinely analysed. These represent the most widely occurring compounds of this group found in drinking water (Table 11.10). The EU has set a limit value for the sum of the four reference PAH compounds in drinking water at $0.1\,\mu g\,l^{-1}$, and a separate limit value of $0.01\,\mu g\,l^{-1}$ for benzo(a)pyrene. Recent research has confirmed that fluoranthene is not found at concentrations in drinking water that pose a health hazard, and so the WHO (2004) has only proposed a health guideline value in drinking water for benzo(a)pyrene at $0.7\,\mu g\,l^{-1}$.

Groundwater sources and upland reservoirs in the UK contain low levels of PAHs, $<50\,ng\,l^{-1}$, while lowland rivers used for supply contain between 40 and $300\,ng\,l^{-1}$ during normal flows, rising to in excess of $1000\,ng\,l^{-1}$ at high flows. Between 65% and 76% of the PAH in surface water is bound to particulate matter and so can effectively be removed by physical water treatment processes such as sedimentation, flocculation and filtration. The remainder can be either chemically oxidized or removed by activated carbon. Even when PAH levels in raw water are high, due to localized industrial pollution, then water treatment can adequately remove PAH compounds to conform to EU limits. Typical concentrations of PAH in drinking water range from $1\,ng\,l^{-1}$ to $11\,\mu g\,l^{-1}$.

PAHs can arise in drinking water after it leaves the treatment plant. Fluoranthene, which is the most soluble but least hazardous of the PAH compounds, is occasionally found in high concentrations in drinking water samples collected at consumers' taps. This is due to leaching from the coal tar and bitumen linings commonly applied to iron water mains. These linings also contain other PAH compounds which can find their way into the water supply mains. As the lining ages then small particles become eroded from the surface, which are up to 50% PAH by weight. New bitumen lining materials also contain PAHs, although these leach at very much slower rates. Conventional water treatment effectively removes the bulk of the PAH compounds adsorbed onto particulate matter, although the majority of PAH remaining is fluoranthene due to its high solubility. Granular activated carbon can, however, remove 99.9% of the fluoranthene and any other PAHs remaining in the water (Section 20.6). Owing to its relatively high solubility, fluoranthene is the major PAH to accumulate in groundwater. The vast majority of breaches of the limit values for PAHs in drinking water is due to leaching in the distribution system.

11.3.2 ANIMALS AND BIOFILM

Bacteria and fungi are common in water mains. They grow freely in the water, and more importantly form films or slime growths on the inside of the pipe, which makes them far more resistant against attack from residual chlorination. Biofilms can significantly affect the microbial safety of drinking water (Chapter 12), while in an operational sense slime formation is undesirable as it increases the frictional resistance in the pipes, thereby increasing the cost of pumping water through the system (van der Wende *et al.*, 1989). Certain bacteria attack iron pipes increasing the rate of corrosion, while other pipe materials can also be affected. In terms of water quality, microbial slimes can alter the chemical nature of the water through microbial metabolism, reducing dissolved oxygen levels or producing end products such as nitrates and sulphides. Odour and taste problems have been associated with high microbial activity in distribution systems, as have increased concentrations of particulate matter in drinking water (Table 11.5). The microbial slimes are also the major food source for larger organisms that are normally found in the bottom or at the margins of reservoirs, lakes and rivers.

Many upland rivers contain water of an exceptionally high quality having a low density of suspended solids and a good microbial quality. Such supplies may only receive rudimentary treatment, so very small animal species can enter the mains in large numbers. Free swimming (planktonic) species, such as *Daphnia* (water flea), are unable to colonize the mains, while many of the naturally occurring bottom sediment-dwelling species (benthic), such as *Cyclops, Asloma* and *Nais*, easily adapt to live in the mains and even form breeding populations. Those species which give rise to most complaints are not necessarily those species which enter the mains in the greatest numbers from the resource, rather it is those which are able to successfully colonize and reproduce within the distribution system. Most treated water contains particulate organic or plant material in the form of algae, although the majority of organic matter present will be in dissolved form (>85%). The majority of the animals present are either feeding off the particulate organic matter or the microbial slime, which itself is feeding off the dissolved organic matter. These animals are eaten by carnivores, thus forming very basic food chains. The density or number of animals in the mains will increase if there is an increase in organic matter or algae entering the system from the source, while the accumulation of organic material including dead animals provides a rich source of food and an ideal habitat for many animals. For this reason animals are often associated with discoloured water and often cause discolouration themselves. Also, as organic particles tend to accumulate towards the end of spur mains, in cul-de-sacs for example, then the problem of discolouration, sediment and animals tend to increase towards the end of the pipework.

The source of water is important in promoting animal growth. Groundwaters contain less dissolved and particulate organic matter than surface waters, as it is removed by filtration as the water moves through the aquifer. Also, far fewer animals naturally occur in such waters to replenish or inoculate the mains. The lack of light in aquifers prevents algal development, which is an important source of organic matter in many other sources. It is the eutrophic lowland reservoirs or rivers that have the greatest potential for supporting large animal populations, but these particular sources normally receive considerable and often complex treatment prior to entering the mains. In practice, it is upland sources, which receive more basic treatment only, that give rise to major consumer complaints of animals in drinking water. The greater the level of organic matter in supplies the greater the problems from excessive slime growths in distribution systems, often made worse by not having animals present to graze and so control slime development.

Animals occur in private supplies mainly because the water is being taken either directly from a stream or spring, or from a wide borehole or old type of well. Where wells are used, covering them to exclude the light will ensure that no algae will develop and, of course, ensure no frogs can enter to spawn. In streams where the water is taken from a small chamber built into the streambed then it is inevitable that some animals will be sucked into the pipe when the water is being used in the house. In both cases it is advisable to fit a series of mesh screens over the inlet, starting with a broad grade and ending up with a narrower grade to ensure both animals and silt are excluded. Mesh screens on the inlet pipes of wide shallow wells are also advisable. These screens will, of course, require periodic cleaning, but they will ensure that the larger species are kept out of the supply. Periodic examination of the water storage tank should also be made for animals and, if present, it should be drained and carefully cleaned out ensuring all the silt and any surface slime growths in the tank are also removed using a weak solution of bleach.

All water mains contain animals and to date some 150 different species have been identified in British water distribution networks. Most of these species are aquatic and enter the mains from the treatment plant, either during construction or maintenance work, or during a temporary breakdown. The major sources of complaints are due to truly aquatic species that are able to colonize the mains and successfully breed. The largest animals regularly reported in British water mains are the water louse (*Asellus*) and the water shrimp (*Gammarus*) (Section 4.3). Once the animal has been identified then the water supply company may take the following action depending on the species. The isolated appearance of a single organism, although potentially distressing, should not be taken as a serious indication of a problem. Often animals, such as *Asellus* or mussels, are present in very large numbers in water mains without ever appearing at the consumer's tap and without causing any reduction in water quality. Conversely, the creature that falls into the sink from the tap may have been the only one. Incidentally, animals usually appear at the kitchen tap, or less frequently end up in the storage tank. If animals appear in taps not directly connected to the rising main, then the water storage tank should be examined immediately. Whenever possible invertebrates should be eliminated before they pass into the distribution system. Once in the distribution network a number of control options are available. Chemical controls include disinfectants (chlorine and chloramines) and pesticides, such as pyrthrins and copper sulphate. Physical controls include hydrant flushing, cleaning and relining pipes, and elimination of dead ends and areas of low flow. In practice, flushing is used in the first instance and if the problem persists then more drastic action will be required.

11.4 WATER QUALITY PROBLEMS ARISING IN HOME PLUMBING SITUATIONS

The major problems arising within the house are elevated concentrations of lead, copper and zinc, odour and taste, fibres from loft insulation, corrosion and pathogens. However, most home-based problems are associated with corrosion of pipework and storage tanks leading to contamination of water by lead, copper, zinc or iron.

11.4.1 CORROSION

Corrosion usually involves electrochemistry. The components of an electrochemical cell are: an anode and a cathode which are sites on the household pipework, either different metals coupled together or the same metal; the external circuit which is connection between the anode and cathode (i.e. the pipe); the internal circuit which is the conducting solution (i.e. the water). When a cell is formed, oxidation occurs at the cathode which is negatively charged. Metal dissolution occurs at the anode releasing metal into solution (Fig. 11.5). The difference in potential is a major factor in corrosion. Potential differences on a pipe of the same material is caused by minute differences in surfaces caused during manufacture or by stress caused during installation. Any imperfections result in differences in potential, but major potential differences, resulting in rapid corrosion, are found when different metals are coupled together. This is known as galvanic corrosion. The galvanic series is the relative tendency of metals to corrode and can be used to select suitable metals to couple together. In decreasing order of corrosivity the metals are manganese, zinc, aluminium, steel and iron, lead, tin, brass, copper and bronze. The further apart two metals are on this scale, the greater the potential difference between them if coupled and so the greater the rate of corrosion (e.g. iron pipes will corrode if connected by brass fittings). The commonest problem is where a piece of copper pipe is used to replace a section of lead pipe. The copper pipe becomes a cathode and the lead pipe an anode, resulting in

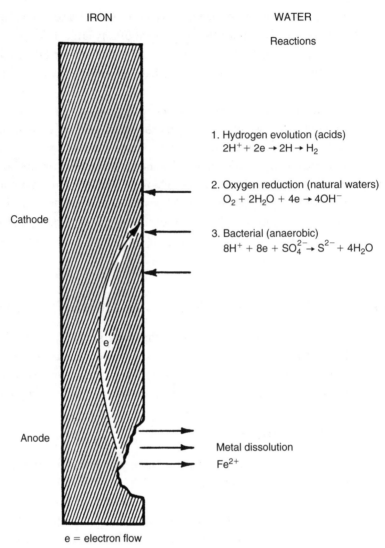

IRON

WATER

Reactions

1. Hydrogen evolution (acids)
$2H^+ + 2e \rightarrow 2H \rightarrow H_2$

2. Oxygen reduction (natural waters)
$O_2 + 2H_2O + 4e \rightarrow 4OH^-$

Cathode

3. Bacterial (anaerobic)
$8H^+ + 8e + SO_4^{2-} \rightarrow S^{2-} + 4H_2O$

e

Anode

Metal dissolution
Fe^{2+}

e = electron flow

FIGURE 11.5
Main pathways leading to the corrosion of metal surfaces in household plumbing. (Reproduced from Ainsworth *et al.* (1981) with permission of WRc plc, Medmenham.)

the lead pipe corroding away, releasing metal into solution. So in household plumbing it is best to use all the same metal (e.g. copper pipes and fittings) or metals very close on the electrochemical scale (e.g. copper pipes and brass/bronze fittings).

Galvanized steel tanks (steel coated with zinc) are particularly susceptible to corrosion as any crack or blemish on the surface coating of zinc will form a cell. The zinc becomes the anode and the iron the cathode, and so zinc is leached into the water. Corrosion of galvanized steel is increased significantly if coupled with copper tubing (upstream) and the water is cuprosolvent. The leached copper is deposited in the tank causing rapid corrosion. A common problem is where water is returned back to a galvanized tank by copper overflow pipes (e.g. overflow from a copper hot water cylinder).

The rate of corrosion rapidly increases at higher temperatures and when the water is acidic. Therefore, corrosion is more pronounced in soft water areas and in the hot water circuit of the household plumbing, with copper most at risk. Corrosion can be controlled by installing sacrificial anodes. These are made out of different metals depending on the cathode requiring protection. For example, magnesium alloy fitted onto a steel rod is used to protect galvanized steel tanks, while aluminium rods are used in copper cylinders. These corrode instead of existing pipework or tanks. Acidic water is normally neutralized during treatment to prevent corrosion to consumers' plumbing, while in the UK Water By-laws are used to protect supplies from poor plumbing (White and Mays, 1989). By-law 60 does not permit dissimilar metals to be connected unless effective measures are taken to prevent deterioration (i.e. installation of a sacrificial anode) or where deterioration is unlikely to occur through galvanic action. Metals can be mixed so long as the sequence in downstream order of use is galvanized steel, uncoated iron, lead and copper. Plastic pipes are now widely used instead of copper and they can also be safely used with any other material.

11.4.2 LEAD

Lead is rarely found in water resources in concentrations in excess of $10 \, \mu g l^{-1}$. However, in many areas where soft acidic water resources are used elevated lead levels in drinking waters are common. This is because most houses built before 1964 will have some lead piping and fittings, while houses constructed before 1939 have extensive lead pipework. Many millions of houses have lead service pipes connecting the house to the water main (Section 10.2). Lead was widely used for this purpose because being so malleable it allowed slight movement of the pipework without fracture, rather like the modern plastic pipes used today. There is currently a major campaign throughout the EU to replace lead service pipes.

Plumbosolvency (i.e. the ability of the water to dissolve lead) is strongly affected by the pH and alkalinity of the water. Below pH 8.0 the rate of leaching increases rapidly so that at pH 6.5, which is common in many areas receiving soft acidic upland waters, lead is rapidly dissolved from pipes. Other factors affecting plumbosolvency are water temperature, contact time between the lead and water, and the surface area of lead exposed to the water. Lead can also be a short-term problem in modern buildings. Tin–lead-soldered joints are widely used to connect copper pipework. This source of lead does not normally result in a serious increase in lead uptake as it is generally only the internal pipework that is affected and not the cold-water tap in the kitchen which is connected directly to the rising main using a plastic pipe. In 1986, the US Congress banned the use of solder containing lead in potable water supply and plumbing systems. Only lead-free materials can now be used. Solder is considered to be lead free if it contains no more than 0.2% lead, while pipes and fittings must not exceed 8.0% lead. Water coolers (drinking fountains) are commonly used in schools in the US, and were found to have lead-lined water reservoirs and lead-soldered parts. This was leading to elevated lead levels in the water. Legislation, in the form of the Lead Contamination Control Act 1988, amended the Safe Drinking Water Act to protect children from this exposure by banning the sale or manufacture of coolers that are not completely lead free.

Lead is odourless, tasteless and colourless when in solution, so that even relatively high concentrations of lead in drinking water are undetectable by the consumer unless chemically analysed. Lead is not removed by boiling and in fact is concentrated, with elevated levels found in kettles that are not fully emptied before refilling. The body readily absorbs lead, with young children, infants and foetuses, which are rapidly growing, absorbing lead five times faster than adults. Lead attacks the nervous system resulting in behavioural abnormalities and mental retardation in the young and unborn. Adults with specific disorders who require increased intake of water or who need renal dialysis are also at risk. Foetuses are particularly vulnerable and there is strong evidence to link stillbirths with the exposure of pregnant women to high lead levels in drinking water.

Until the introduction of the new Drinking Water Directive into national legislation in 2000, the standard for lead was $50 \, \mu g l^{-1}$. This will be reduced to the WHO guideline of $10 \, \mu g l^{-1}$ within 15 years in a number of incremented reductions by specified target dates (Section 8.1). The USEPA set the current MCL for lead at $50 \, \mu g l^{-1}$ in 1975 with no MCLG value specified. In August 1988, it proposed an MCLG value of zero and an MCL value of just $5 \, \mu g l^{-1}$ for water leaving the treatment plant and entering the distribution system. If the lead level exceeds certain threshold values, so-called 'no-action levels', then certain remedial measures must be taken by the supplier. For example, if the average lead level in the system exceeds $10 \, \mu g l^{-1}$ or the pH is <8.0 in more than 5% of the samples collected at consumers' taps, then corrosion control is required. If 5% of these samples exceed $20 \, \mu g l^{-1}$ then a public education programme must be instigated to alert consumers to the problem and to what preventative action should be taken.

Among the remedial measures that can be taken to reduce lead levels in drinking water are:

(a) to reduce the level of lead in the resource, although this is rarely a problem;
(b) to alert the consumer of the possibility of high lead levels and to identify those at risk and what action should be taken to reduce exposure;
(c) to reduce the corrsosivity of water by neutralization;
(d) by lining lead service pipes; or
(e) by replacing lead service pipes (Gray, 1994).

Lead phosphates are more insoluble than lead carbonate compounds and in practice the addition of small quantities of orthophosphate ($1.5-2.0 \, mg l^{-1}$) into soft waters produces a substantial reduction in lead solubility. Phosphated waters have extremely flat solubility curves over a wide range of pH in soft waters compared to those with no orthophosphate present (Cardew, 2002).

11.4.3 COPPER

Copper is normally present in natural waters at low concentrations, except in metal-liferous areas. Its presence in drinking water is almost always due to the attack of copper plumbing. Where the water is cuprosolvent the metal is leached from the copper pipework into solution giving rise to blue coloured water, staining of sanitary ware and occasionally taste problems. Concentrations in drinking water are generally low with an astringent taste caused by copper at concentrations in excess of $2\,mg\,l^{-1}$ ($2000\,\mu g\,l^{-1}$). Above this concentration the consumer will reject the water due to an astringent taste and so copper rarely causes health problems. Water should not be drunk from the hot water taps, as the water which is heated in the copper cylinder will have been in contact with copper for some time at elevated temperatures allowing maximum corrosion to occur and resulting in levels high enough to cause nausea. Cuprosolvency occurs with soft acidic waters or with private supplies from boreholes where free CO_2 in the water produces a low pH. Apart from pH, other important factors in cuprosolvency are hardness, temperature, age of pipe and oxygen. The inside of the pipe is protected by a layer of oxides, but water with a pH of <6.5 and a low hardness of $<60\,mg\,l^{-1}$ as $CaCO_3$ will result in corrosion. Hot water accelerates the corrosion rate, although oxygen must be present. If absent then corrosion will be negligible regardless of the temperature. Corrosion of pipes can be severe causing bursts in pipework, usually the hot water pipes, the central heating system or the hot water cylinder. Under such conditions some form of corrosion control is recommended. Raising the pH of the water to 8.0–8.5 overcomes most corrosion problems. Temporary elevated copper concentrations are a particular problem associated with new plumbing. Unless the water is acidic, then the problem should quickly resolve itself as the pipework becomes coated with a layer of oxides.

The Drinking Water Directive sets its limit value at $2000\,\mu g\,l^{-1}$ while the National Primary Drinking Water Regulations in the US have a threshold value of $1\,mg\,l^{-1}$ for copper. In accordance with the 1986 amendment to the Safe Drinking Water Act, the USEPA have proposed guide (MCLG) and mandatory (MCL) levels of $1.3\,mg\,l^{-1}$ entering the distribution system. Corrosion control is required if copper levels at the kitchen tap exceed $1.3\,mg\,l^{-1}$ in 5% of the samples tested or, as with lead, if the pH is <8.0 in more than 5% of the samples tested. In their 2004 guidelines, the WHO have set a health-based guide value of $2\,mg\,l^{-1}$, but at concentrations above $1\,mg\,l^{-1}$ staining of laundry and sanitary ware occurs. Copper begins to impart a blue colour to water at concentrations above 4–$5\,mg\,l^{-1}$. Such water should not be used for consumption purposes.

REFERENCES

Ainsworth, R. G. *et al.*, 1981 *A Guide to Solving Water Quality Problems in Distribution Systems*, Technical Report 167, Water Research Centre, Medmenham.

APHA, 1998 *Standard Methods for the Examination of Waters and Wastewaters*, 20th edn, American Public Health Association, Washington, DC.

ASTM, 1984 Standard test method for odour in water, D1292-80. *Annual Book of ASTM Standards, Section 11: Water and Environmental Technology*, **11.01**, 218.

Banks, J., 2002 Low-cost solutions for trihalomethane compliance, *Journal of the Institution of Water and Environmental Management*, **16**, 264–9.

Cardew, P. T., 2002 Plumbosolvency control in soft waters, *Journal of the Institution of Water and Environmental Management*, **16**, 199–206.

Clark, L., 1988 *The Field Guide to Water Wells and Boreholes*, John Wiley and Sons, Chichester.

Department of the Environment, 1980 *Odour and Taste in Raw and Potable Waters 1980, Methods for the Examination of Waters and Associated Materials*, HMSO, London.

Department of the Environment, 1997 *Digest of Environmental Protection and Water Statistics No. 19*, HMSO, London.

Dunning, J. M., 1986 *Principles of Dental Public Health*, 4th edn, Harvard University Press, Massachusetts.

ECETOC, 1988 *Nitrates and Drinking Water*, ECETOC Technical Report 27, European Chemical Industry Ecology and Toxicology Centre, Brussels.

Gray, N. F., 1994 *Drinking Water Quality: Problems and Solutions*, John Wiley and Sons, Chichester.

Loch, J. P. G., van Dijk-Looyaard, A. and Zoeteman, B. C. J., 1989 Organics in ground water. In: Wheeler, A., Richardson, M.L. and Bridges, J., eds, *Watershed 89, the Future of Water Quality in Europe*, Proceedings of the International Association of Water Pollution Research and Control, 17–20th April 1989, Pergamon Press, London, pp 39–55.

Mather, J., Banks, D., Dumpleton, S. and Feror, M. (eds), 1998 *Groundwater Contaminants and Their Migration*, Geological Society Special Publication No. 128, The Geological Society, London.

Nyer, E. K. 1992 *Groundwater Treatment Technology*, 2nd edn, Van Nostrand Reinhold, New York.

Rivett, M. O., Lerner, D. N. and Lloyd, J. W., 1990 Chlorinated solvents in UK aquifers, *Water and Environmental Management*, **4**, 242.

USEPA, 1991 *ARARs Q's and A's Compliance With New SDWA National Primary Drinking Water Regulations for Organic and Inorganic Chemicals (Phase II)*, US Environmental Protection Agency, Washington, DC.

Van der Wende, Characklies, W. G. and Smith, D. B., 1989 Biofilms and bacterial drinking water quality, *Water Research*, **23**, 1313.

White, S. F. and Mays, G. D., 1989 *Water Supply Bylaws Guide*, 2nd edn, Water Research Centre and Ellis Horwood, Chichester.

FURTHER READING

Gilbert, C. E. and Calabrese, E. J., 1992 *Regulating Drinking Water Quality*, Lewis Publishers, Boca Raton, Florida.

Mallevialle, J. and Suffet, I. H., 1987 *Identification and Treatment of Tastes and Odours in Drinking Water*. American Water Works Association Research Foundation, Denver.

Pontius, F. W. (ed.), 1990 *Water Quality and Treatment: A Handbook Of Community Water Supplies*, McGraw-Hill, New York.

WHO, 2004 *Guidelines for Drinking Water Quality: 1. Recommendations*, 3rd edn, World Health Organization, Geneva.

Chapter 12   Pathogens and Their Removal

INTRODUCTION

Microbial contamination is the most critical risk factor in drinking water quality with the potential for widespread waterborne disease. Illness derived from chemical contamination of drinking water supplies is negligible when compared to the number due to microbial pathogens (Galbraith *et al.*, 1987; Herwaldt *et al.*, 1992).

CLASSIFICATION OF PATHOGENS

The term pathogenic is applied to those organisms that either produce or are involved in the production of a disease. Three different groups of pathogenic microorganisms can be transmitted via drinking water to humans; these are bacteria, viruses and protozoans. All are transmitted by the faecal–oral route, so largely arise either directly or indirectly by contamination of water resources and supplies by sewage or animal faeces. It is theoretically possible, although unlikely, that other pathogenic organisms, such as nematodes (roundworm or hookworm) and cestodes (tapeworm), may also be transmitted via drinking water.

Diseases closely associated with water are classified according to their mode of transmission and the form of infection into four different categories: waterborne, water-washed, water-based and water-related diseases.

1. *Waterborne diseases*. These diseases occur where a pathogen is transmitted by ingestion of contaminated water. The classical waterborne diseases are mainly low-infective dose infections, such as cholera and typhoid, with all the other diseases high-infective dose infections that include infectious hepatitis and bacillary dysentery. All waterborne diseases can also be transmitted by other routes, which permits faecal material to be ingested.
2. *Water-washed diseases*. These include faecal–orally spread disease or disease spread from one person to another facilitated by a lack of an adequate supply of

water for washing. The incidence of all these diseases will fall if adequate supplies of washing water, regardless of microbial quality, are provided. These are diseases of mainly tropical areas, and include infections of the intestinal tract, the skin and eyes. Most of the intestinal infections are diarrhoeal diseases responsible for the high mortality rates among infants in hot climates. The infections of the skin and mucous membranes are non-faecal in origin, and include bacterial skin sepsis, scabies and cutaneous fungal infections (such as ringworm). Diseases spread by fleas, ticks and lice are also included in this category, such as epidemic typhus, rickettsial typhus and louse-borne fever.

3. *Water-based infections*. These diseases are caused by pathogenic organisms which spend part of their life cycle in aquatic organisms. All these diseases are caused by parasitic worms, with the severity of the infection depending on the number of worms infesting the host. The two commonest water-based diseases are schistosomiasis due to the trematode *Schistosoma* spp. and guineaworm, which is the nematode *Dracunculus medimensis*. *Schistosoma* worms use aquatic snails as intermediate hosts and are estimated as infecting as many as 200 million people, while the guineaworm uses the small crustacean *Cyclops* spp. as its intermediate host.

4. *Water-related diseases*. These are caused by pathogens carried by insects that act as mechanical vectors and which live near water, all these diseases are very severe and the control of the insect vectors is extremely difficult. The most important water-related diseases include two viral diseases: yellow fever transmitted by the mosquito *Aedes* spp. and dengue carried by the mosquito *Aedes aegypti* which breed in water. Gambian sleeping sickness, trypanosomiasis, is caused by a protozoan transmitted by the riverine tsetse fly (*Glossina* sp.) which bites near water, while malaria is caused by another protozoan (*Plasmodium* sp.) which is transmitted by the mosquito *Anopheles* sp. that breeds in water (Bradley, 1993). Climate change will have a significant effect on the distribution of these diseases, especially those in Category 4 (Section 23.2).

STRATEGIES FOR CONTROLLING PATHOGEN TRANSFER

A barrier approach is the key strategy for controlling the health risks posed by microbes in drinking waters. This involves the treatment of wastewaters to remove pathogens as well as the treatment of raw waters, which includes disinfection (Fig. 12.1). Monitoring water supplies for the presence of specific pathogens is difficult and largely impractical, so a more indirect approach is adopted where water is examined for indicator bacteria whose presence in water implies some degree of contamination. The use of indicator organisms, in particular the coliform group, as a means of controlling the possible presence of pathogens has been paramount in the approach to assessing water quality adopted by the World Health Organization (WHO), US

Environmental Protection Agency (USEPA) and the European Union (EU) (Section 8.1). This approach is based on the assumption that there is a quantifiable relationship between indicator density and the potential health risks involved. A water quality guideline is established (Fig. 12.2) which is a suggested upper limit for the density of an indicator organism above which there is an unacceptable risk (Cabelli, 1978). An

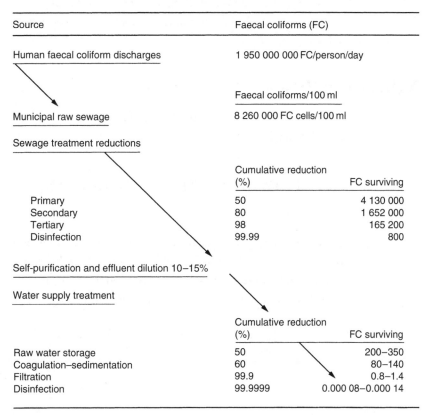

Source	Faecal coliforms (FC)
Human faecal coliform discharges	1 950 000 000 FC/person/day
	Faecal coliforms/100 ml
Municipal raw sewage	8 260 000 FC cells/100 ml

Sewage treatment reductions

	Cumulative reduction (%)	FC surviving
Primary	50	4 130 000
Secondary	80	1 652 000
Tertiary	98	165 200
Disinfection	99.99	800

Self-purification and effluent dilution 10–15%

Water supply treatment

	Cumulative reduction (%)	FC surviving
Raw water storage	50	200–350
Coagulation–sedimentation	60	80–140
Filtration	99.9	0.8–1.4
Disinfection	99.9999	0.000 08–0.000 14

FIGURE 12.1 The use of barriers is vital in the control of pathogens in water supplies. (Reproduced from Geldreich (1991) with permission of John Wiley and Sons Inc., New York.)

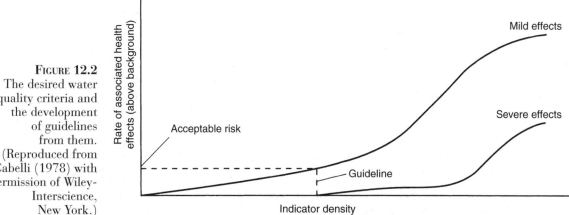

FIGURE 12.2 The desired water quality criteria and the development of guidelines from them. (Reproduced from Cabelli (1978) with permission of Wiley-Interscience, New York.)

example is given for determining the risk of infection from bathing waters in Section 8.2 (Fig. 8.1). Ideally water destined for human consumption should be free from micro-organisms; however, in practice this is an unattainable goal.

The extent to which strategies, such as the barrier approach and the establishment of allowable limits for bacteria in water, have been successful in maintaining water quality is seen by the dramatic decline in epidemic and endemic waterborne bacterial diseases, such as typhoid fever and cholera, in the more developed regions of the world (Sobsey *et al.*, 1993).

12.1 WATERBORNE PATHOGENS

The major waterborne disease-causing pathogens are summarized in Table 12.1 and are considered below.

12.1.1 PRIMARY BACTERIAL PATHOGENS

Salmonella

The various serotypes that make up the genus *Salmonella* are the most important group of bacteria affecting the public health of both humans and animals in Western Europe; wild, domestic and farm animals often acting as reservoirs of human salmonellosis. Water resources can become contaminated by raw or treated wastewater, as

TABLE 12.1 Bacterial, viral and protozoan diseases generally transmitted by contaminated drinking water.

Agent	Disease	Incubation time
Bacteria		
Shigella spp.	Shigellosis	1–7 days
Salmonella spp.		
S. typhimurium	Salmonellosis	6–72 h
S. typhi	Typhoid fever	1–3 days
Enterotoxigenic *Escherichia coli*	Diarrhoea	12–72 h
Campylobacter spp.	Gastroenteritis	1–7 days
Vibrio cholerae	Gastroenteritis	1–3 days
Viruses		
Hepatitis A	Hepatitis	15–45 days
Norwalk-like agent	Gastroenteritis	1–7 days
Virus-like particles <27 nm	Gastroenteritis	1–7 days
Rotavirus	Gastroenteritis	1–2 days
Protozoa		
Giardia lamblia	Giardiasis	7–10 days
Entamoeba histolytica	Ameobiasis	2–4 weeks
Cryptosporidium	Cryptosporidiosis	5–10 days

Reproduced from Singh and McFeters (1992) with permission of John Wiley and Sons Inc., New York.

well as by effluents from abattoirs and animal-processing plants (Gray, 2004). It is commonly present in raw waters but only occasionally isolated from finished waters, chlorination being highly effective at controlling the bacteria. Typical symptoms of salmonellosis are acute gastroenteritis with diarrhoea, and can also be associated with abdominal cramps, fever, nausea, vomiting, headache and, in severe cases, even collapse and possible death. The most serious diseases associated with specific species are typhoid fever (*Salmonella typhi, S. typhi*) and paratyphoid (*S. paratyphi* and *S. schottmuelleri*). Salmonellosis carries a significant mortality amongst those with acquired immuno deficiency syndrome (AIDS) and poses significant problems in its management.

Shigella

Shigella causes bacterial dysentery or shigellosis and is one of the most frequently diagnosed cause of diarrhoea in the US. Shigellosis is a problem of both developed and developing countries, with the eastern Mediterranean countries considered as an endemic region for the disease. The species of this bacterial genus are rather similar in their epidemiology to that of *Salmonella* except they rarely infect animals and do not survive quite so well in the environment. When the disease is present as an epidemic it appears to be spread mainly by person-to-person contact, especially between children, shigellosis being a typical institutional disease occurring in over-crowded conditions. There has been a significant increase in the number of outbreaks arising from poor quality drinking water contaminated by sewage. Of the large number of species (>40), only *S. dysenteriae, Shigella sonnei, S. flexneri* and *S. boydii* are able to cause gastrointestinal disease. *S. sonnei* and *S. flexneri* account for >90% of isolates, although it is *S. dysenteriae* type 1 which causes the most severe symptoms due to the production of the shiga toxin.

Cholera

While the disease in now extremely rare in the developed world, major waterborne outbreaks occur in developing countries, war zones and disaster areas. It is still endemic in many areas of the world especially those which do not have adequate sanitation and in particular situations where the water supplies are continuously contaminated by sewage. However, over the past 15–20 years the incidence and spread of the disease has been causing concern which has been linked to the increasing mobility of travellers and the speed of travel. An infected person or symptomless carrier of the disease excretes up to 10^{13} bacteria daily, enough to theoretically infect 10^7 people! Up to 10^6–10^7 organisms are required to cause the illness; hence cholera is not normally spread by person-to-person contact. It is transmitted primarily by drinking contaminated water, but also by eating food handled by a carrier, or which has been washed with contaminated water, and is regularly isolated from surface waters in the UK and the US. It is an intestinal disease with characteristic symptoms: that is, sudden diarrhoea with copious watery faeces, vomiting, suppression of urine, rapid dehydration, lowered temperature and blood pressure, and complete collapse. Without immediate medical therapy the disease has a >60% mortality rate, the patient dying within a few hours of first showing the symptoms, although with suitable treatment the mortality rate can be reduced to <1%. *Vibrio cholerae* are natural inhabitants of brackish and saline waters, and are rapidly inactivated under unfavourable conditions, such as high-acidity or high-organic-matter content of the water, although in cool unpolluted waters the bacterium will survive for up to 2 weeks. Survival is even greater in estuarine and coastal waters.

12.1.2 OPPORTUNISTIC BACTERIAL PATHOGENS

Opportunistic bacteria are usually found as part of the normal heterotrophic bacterial flora of aquatic systems and may also exist as part of the normal body microflora (Table 12.2) (Reasoner, 1992). These organisms are normally not a threat to healthy individuals but under certain circumstances they can lead to infection in certain segments of the community, in particular newborn babies, the elderly and the immuno-compromised. It is thought that numerous hospital-acquired infections are attributable to such organisms (Du Zuane, 1990). Some of the organisms listed in Table 12.2 are also considered as primary pathogens, meaning that they are also

TABLE 12.2 Typical opportunistic bacterial pathogens that are isolated from drinking water

Acinetobacter spp.
Achromobacter xylosoxidans
Aeromonas hydrophila
Bacillus spp.
Campylobacter spp.[a]
Citrobacter spp.
Enterobacter aerogenes
Enterobacter agglomerans
Enterobacter cloacae
Flavobacterium meningusepticum
Hafnia alvei
Klebsiella pneumoniae[a]
Legionella pneumopila[a]
Moraxella spp.
Mycobacterium spp.
Pseudomonas spp. (non-aeruginosa)[a]
Serratia fonticola
S. liquefaciens
S. marcescens
Staphylococcus spp.[a]
Vibrio fluvialis[a]

[a] Indicates that the organism may be a primary pathogen.
Reproduced from Reasoner (1992) with permission of
US Environmental Protection Agency, Washington, DC.

capable of being primary disease-causing agents rather than secondary invaders. Of these organisms, those of particular concern at present include *Campylobacter* spp., enterotoxigenic *Escherichia coli* (*E. coli*), *Mycobacteria* spp., *Legionella pneumophila* and aeromonads.

Campylobacter

Campylobacter is a major cause of gastroenteritis, being more common than *Salmonella*. In the US, the annual incidence of this organism is between 30 and 60 per 100 000 of the population (Skirrow and Blaser, 1992). In developing countries, outbreaks of *Campylobacter* enteritis are a major cause of morbidity and mortality in the first 2 years of life. While *Campylobacter* enteritis is essentially a foodborne disease, with the most important reservoirs of the bacterium being meat, in particular poultry, and unpasturized milk, waterborne transmission has been implicated in several large outbreaks. Waterborne transmission of *Campylobacter* occurs in untreated, contaminated waters, in situations where faulty disinfection has occurred or where waters have been contaminated by sewage or animal wastes (Tauxe, 1992). Household pets, farm animals and birds are all known to be the carriers of the disease. There is a definite seasonal variation in numbers of *Campylobacters* in river water, with greatest numbers occurring in the autumn and winter. *Campylobacter jejuni* (*C. jejuni*) serotypes are common in human infections and especially common downstream of sewage effluent sites, confirming sewage effluents as important sources of *C. jejuni* in the aquatic environment. Gulls are known carriers and can contaminate water supply reservoirs while they roost. Dog faeces in particular are rich in the bacterium, making contamination of surface waters used for abstraction by surface (urban) run-off an important source of contamination. *Campylobacter* can remain viable for extended periods in streams and groundwaters. Survival of the bacterium decreases with increasing temperature but with survival at 4°C in excess of 12 months.

Escherichia coli 0157:H7

E. coli 0157:H7 causes haemorrhagic colitis, haemolytic–uraemic syndrome and is a major cause of kidney disease in children. Like *C. jejuni*, this organism is generally associated with food, in particular beef and milk, but in recent years has been implicated in a number of waterborne outbreaks. The number of organisms required to initiate infection is thought to be <100.

Mycobacteria

Mycobacteria spp. are now recognized as being opportunistic pathogens of considerable significance that cause a wide range of diseases in humans, including pulmonary disease, cervical lymphadenopathy as well as localized and soft tissue infections. It is well established that *Mycobacteria* are commonplace in all types of aquatic environments, including estuaries, ocean water, groundwater, surface water and distribution systems (Jenkins, 1991). The majority of waterborne mycobacterial outbreaks are attributable to treatment deficiencies, such as inadequate or interrupted chlorination; but other factors may also influence the growth of this organism in water supplies, such as pitting and encrustations found inside old water pipes which protect bacteria from exposure to free chlorine. *Mycobacteria* can also colonize areas where water is moving slowly, as in water distribution systems in large buildings, such as blocks of flats, offices and hospitals, thus continuously seeding the system (Du Moulin and Stottmeier, 1986). Disease associated with these bacteria is steadily rising, particularly among patients with AIDS.

Aeromonads

Aeromonas spp. have been implicated as the causative agent in a number of waterborne outbreaks and are considered to be an important, and often fatal, cause of non-gastrointestinal illness in immuno-compromised individuals. Aeromonads have been isolated from both chlorinated and unchlorinated drinking water supplies occurring in greatest numbers during the summer months. They have also been isolated in waters containing no *E. coli* and few total coliforms (Schbert, 1991).

Legionella pneumophila

The bacterium *Legionella pneumophila* has been associated with domestic water systems, especially hot water stored at between 20°C and 50°C. Heat exchangers, condensers in air conditioning units, cooling towers and shower heads have all been found to be havens for the bacteria leading to human infection. Infection is by inhalation of aerosols from contaminated appliances. The bacteria survive and grow within phagocytic cells, multiplying in the lungs causing bronchopneumonia and tissue damage. The bacterium is widespread in natural waters so any water supply can become contaminated. *Legionella* are commonly found in hospital water systems, and hospital-acquired (nosocomial) Legionnaires' disease is now a major health problem (Joseph *et al.*, 1994).

It appears that a long retention time, and the presence of key nutrients, such as iron, provides ideal conditions for the bacterium to develop. Therefore iron storage cisterns subject to corrosion are particularly susceptible. In addition, water pipes that are installed alongside hot water pipes or other sources of heat may also permit the bacteria to develop. Prevention of *Legionella* infection is normally achieved by either hyperchlorination or thermal eradication of infected pipework, although these methods are not always successful or with a permanent effect. The only effective preventative method has been shown to be the use of ultraviolet (UV) sterilizers as close to the point of use as possible.

TABLE 12.3   Some human enteric viruses and the diseases they cause

Virus group	Serotypes	Some diseases caused
Enteroviruses		
Polioviruses	3	Paralysis, aseptic meningitis
Coxsackievirus		
A	23	Herpangia, aseptic meningitis, respiratory illness, paralysis, fever
B	6	Pleurodynia, aseptic meningitis, pericarditis, congenital heart disease anomalies, nephritis, fever
Echovirus	34	Respiratory infection, aseptic meningitis, diarrhoea, pericarditis, myocarditis, fever and rash
Enteroviruses (68–71)	4	Meningitis, respiratory illness
Hepatitis A virus (HAV)		Infectious hepatitis
Reoviruses	3	Respiratory disease
Rotaviruses	4	Gastroenteritis
Adenoviruses	41	Respiratory disease, acute conjunctivitis, gastroenteritis
Norwalk agent (calcivirus)	1	Gastroenteritis
Astroviruses	5	Gastroenteritis

Reproduced from Bitton (1994) with permission of Wiley-Liss Inc., New York.

12.1.3 VIRUSES

There are over 120 distinct known types of human pathogenic viruses. Of most concern for drinking water are those which cause gastrointestinal illness (enteric viruses) which includes enteroviruses, rotaviruses, astroviruses, calciviruses, Hepatitis A virus, Norwalk virus and other 'small round' viruses (West, 1991). However, the health risks presented by these viruses are not just restricted to gastroenteritis (Table 12.3).

Viral contamination arises when sewage containing pathogenic viruses contaminates surface and groundwaters that are subsequently used as sources of drinking waters. Most viruses are able to remain viable for several weeks in water at low temperatures, so long as there is some organic matter present. Of most concern in Britain and the US is viral hepatitis. There are three subgroups, Hepatitis A which is transmitted by water, Hepatitis B which is spread by personal contact or inoculation and which is endemic in certain countries such as Greece, and Hepatitis C which is a non-A- or B-type hepatitis virus. Hepatitis A is spread by faecal contamination of food, drinking water and areas, which are used for bathing and swimming. Epidemics have been linked to all these sources, and it appears that swimming pools and coastal areas used for bathing which receive large quantities of sewage are particular sources of infection. There is no treatment for Hepatitis A, with the only effective protection being good personal hygiene, and the proper protection and treatment of drinking water

(Section 13.6). Symptoms develop 15–45 days after exposure, and include nausea, vomiting, muscle ache and jaundice. Hepatitis A virus accounts for 87% of all viral waterborne disease outbreaks in the US (Craun, 1986).

Two viruses which have caused recent outbreaks of illness due to drinking water contamination are Norwalk virus and rotavirus (Cubitt, 1991). Norwalk virus results in severe diarrhoea and vomiting. It is of particular worry to the water industry in that it appears not to be affected by normal chlorination levels. Also it seems that infection by the virus only gives rise to short-term immunity while life-long immunity is conferred by most other enteric viruses. Rotavirus is a major contributor to child diarrhoea syndrome. This causes the death of some 6 million children in developing countries each year. This is not a serious problem in developed countries due to better hygiene, nutrition and health care. Outbreaks do occur occasionally in hospitals, and although associated with child diarrhoea, it can be much more serious if contracted by an adult.

Viruses are usually excreted in numbers several orders of magnitude lower than that of coliforms (APHA, 1992). Because they only multiply within living susceptible cells, their numbers cannot increase once excreted. Once in a cell-free state, their survival and infectivity in the aquatic environment depends on a variety of biotic (i.e. type of virus, bacterial and algal activity, and predation by protozoa) and abiotic factors (i.e. temperature, suspended matter, pH, salinity, UV light penetration, organic compounds, adsorption to suspended matter and aggregation). Temperature is considered to be the most important factor influencing viral destruction outside the host cell, being rapidly inactivated once exposed to temperatures in excess of 50°C. Suspended solids provide a certain degree of protection for viruses. Adsorption onto organic matter can prevent inactivation by UV light. Once adsorbed the viruses can settle from suspension and survive for long periods in sediments to become resuspended if the water becomes turbulent.

In practice viruses generally pass unaffected through wastewater treatment plants and so will be found in surface waters receiving both treated and untreated sewage (Gray, 2004). Therefore effective water treatment disinfection is critical to prevent viruses entering potable water supplies. Enteric viruses have been isolated from drinking waters that have been treated by chlorination or other processes, such as ozonation and chemical coagulation. Such drinking waters contain chlorine levels originally thought to be virucidal ($0.1–0.3 \, \text{mg} \, \text{l}^{-1}$). Initially the ability to survive chlorination was thought to be due to a lack of contact time with chlorine. It is now well established that some enteric viruses are more resistant to chlorination than coliforms. Exposure to over $2 \, \text{mg} \, \text{l}^{-1}$ of chlorine for 30 min is required to inactivate the infectivity of the Hepatitis A virus, while exposure to as much as $5–6 \, \text{mg} \, \text{l}^{-1}$ chlorine for 30 min may be required to destroy the infectivity of the Norwalk virus. It has been suggested that such resistance may be due to the protective effect of viral aggregation (West, 1991).

12.1.4 PROTOZOA

Protozoan pathogens of humans are almost exclusively confined to tropical and sub-tropical areas, which is why the increased occurrence of *Cryptosporidium* and *Giardia* cysts in temperate areas is causing so much concern. However, with the increase in travel, carriers of all diseases are now found worldwide, and cysts of all the major protozoan pathogens occur in European sewages from time to time. Two other protozoan parasites which occur occasionally in the UK are *Entamoeba histolytica* (*E. histolytica*), which causes amoebic dysentery, and *Naegteria fowleri*, which causes the fatal disease amoebic meningoencephalitis.

Giardia lamblia

Giardia lamblia is a flagellated protozoan that is a significant cause of gastroenteritis ranging from mild to severe and disabling disease with a worldwide distribution being significantly more common in children than in adults. Cysts are passed from the body in the stools and are ovoid, 14–16 μm long, 6–12 μm wide and are quadri-nucleate. *Giardia* cysts are relatively resistant to environmental conditions and are capable of survival once excreted for long periods, especially in winter.

Transmission of *Giardia* cysts may be by faecal contamination of either hands, food or water supplies. The disease is a zoonosis and cysts from a human source can be infective to about 40 different species of animal including pets and vice versa. Cysts are therefore widely distributed in the environment entering waterways through sewage or storm water discharges or via the droppings of infected animals. Giardiasis is now established to be one of the most common causes of waterborne diseases in the developed world. Most *Giardia* outbreaks occur in waters where chlorination is the only form of water treatment. This resistance to disinfection levels typically used in water treatment indicates the need for additional treatment barriers. In recognition of this, recent amendments to the US Safe Drinking Water Act now requires that all surface waters intended for human consumption must undergo filtration to specifically remove cysts and sufficient disinfection to destroy *Giardia* and prevent disease transmission. To date no such provisions exist in European legis-lation. There is no way of preventing infection except by adequate water treatment and resource protection. Current disinfection practices are generally inadequate as the sole barrier to prevent outbreaks. Boiling water for 20 min will kill cysts, while the use of 1 μm pore cartridge filters to treat drinking water at the point of use are also effective.

Cysts are generally present in very low densities and so samples must be concentrated first. This is achieved by using ultra-filtration cassettes or finely wound polypropylene cartridges (APHA, 1992). As cyst numbers cannot be amplified by *in vitro* cultivation, they are generally detected by immuno-fluorescence with poly- or monoclonal anti-bodies or by direct phase contrast microscopy (Table 12.4).

TABLE 12.4 The standard methods associated with the sampling, isolation and enumeration of pathogens in drinking water, recreational waters and sewage sludge in the UK. These are available online from the Environment Agency

Drinking water

The Microbiology of Drinking Water 2002 – Part 1: Water quality and public health. MDW2002(1). Standing Committee of Analysts, Environment Agency, London.

The Microbiology of Drinking Water 2002 – Part 2: Practices and procedures for sampling. MDW2002(2). Standing Committee of Analysts, Environment Agency, London.

The Microbiology of Drinking Water 2002 – Part 3: Practices and procedures for laboratories. MDW2002(3). Standing Committee of Analysts, Environment Agency, London.

The Microbiology of Drinking Water 2002 – Part 4: Practices and procedures for laboratories. MDW2002(3). Standing Committee of Analysts, Environment Agency, London.

The Microbiology of Drinking Water 2002 – Part 5: Isolation and enumeration of enterococci by membrane filtration. MDW2002(5). Standing Committee of Analysts, Environment Agency, London.

The Microbiology of Drinking Water 2004 – Part 6: Methods for the isolation and enumeration of sulphite-reducing clostridia and Clostridium perfringens *by membrane filtration.* Blue Book No. 192. Standing Committee of Analysts, Environment Agency, London.

The Microbiology of Drinking Water 2002 – Part 7: The enumeration of heterotrophic bacteria by pour and spread plate techniques. MDW2002(7). Standing Committee of Analysts, Environment Agency, London.

The Microbiology of Drinking Water 2002 – Part: Methods for the isolation and enumeration of Aeromonas and Pseudomonas aeruginosa by membrane filtration. MDW2002(8). Standing Committee of Analysts, Environment Agency, London.

The Microbiology of Drinking Water 2004 – Part 9: Methods for the isolation and enumeration of Salmonella and Shigella by selective enrichment, membrane filtration and multiple tube most probable number techniques. Blue Book No. 194. Standing Committee of Analysts, Environment Agency, London.

The Microbiology of Drinking Water 2002 – Part 10: Methods for the isolation of Yersinia, Vibrio and Campylobacter by selective enrichment. MDW2002(10). Standing Committee of Analysts, Environment Agency, London.

The Microbiology of Drinking Water 2004 – Part 12: Methods for the isolation and enumeration of micro-organisms associated with taste, odour and related aesthetic problems. Blue Book No. 197. Standing Committee of Analysts, Environment Agency, London.

Recreational water

The Microbiology of recreational and environmental waters 2000. Blue Book No. 175. Standing Committee of Analysts, Environment Agency, London.

Sewage sludge

The Microbiology of Sewage Sludge 2003 – Part 1: An overview of the treatment and use in agriculture of sewage sludge in relation to its impact on the environment and public health. Blue Book No. 188. Standing Committee of Analysts, Environment Agency, London.

The Microbiology of Sewage Sludge 2003 – Part 2: Practices and procedures for sampling and sample preparation. Blue Book No. 189. Standing Committee of Analysts, Environment Agency, London.

The Microbiology of Sewage Sludge 2003 – Part 3: Methods for the isolation and enumeration of Escherichia coli, *including verocytotoxigenic* Escherichia coli. Blue Book No. 190. Standing Committee of Analysts, Environment Agency, London.

The Microbiology of Sewage Sludge 2003 – Part 3: Methods for the isolation and enumeration of Salmonellae. Blue Book No. 191. Standing Committee of Analysts, Environment Agency, London.

The Microbiology of Sewage Sludge 2004 – Part 4: Methods for the detection, isolation and enumeration of Salmonellae. Blue Book No. 195. Standing Committee of Analysts, Environment Agency, London.

Cryptosporidium parvum

Cryptosporidium parvum is a coccidian protozoan parasite and while the first case of human infection was not recorded until 1976, cryptosporidiosis is acquired by ingesting viable oocysts that are ovoid, between 4 and 6 μm in diameter, and generally occur in low numbers in water. Clinical symptoms of cryptosporidiosis include an influenza-like illness, diarrhoea, malaise, abdominal pain, anorexia, nausea, flatulence, malabsorption, vomiting, mild fever and weight loss. Generally this disease is not fatal among healthy individuals. However, in young malnourished children, it can cause severe dehydration and sometimes death (Smith, 1992). In the immuno-compromised, including those with AIDS or those receiving immuno-suppressive drugs, and also those with severe malnutrition, cryptosporidiosis can become a life-threatening condition causing profuse intractable diarrhoea with severe dehydration, malabsorption and wasting. Sometimes the disease spreads to other organs. These symptoms can persist unabated until the patient eventually dies. In the UK, cryptosporidiosis is currently the fourth most commonly identified cause of diarrhoea in which a parasitic, bacterial or viral cause was established (Department of the Environment and Department of Health, 1990) and is particularly difficult to treat. Little is known of the exact infectious dose size, but a single oocyst may be enough to cause infection, although outbreaks of cryptosporidiosis are usually associated with gross contamination. The organism is not host specific and is capable of infecting many species of mammal, bird and reptile. Oocysts from humans are infective for numerous mammals.

Studies of water resources in the UK and the US have found that oocysts commonly occurred in all types of surface water (lakes, reservoirs, streams and rivers) including pristine waters with densities ranging from 0.006 to 2.5 oocysts l^{-1} (Department of the Environment and Department of Health, 1990; Le Chevallier *et al.*, 1991). Significantly higher numbers of oocysts are found in water resources receiving untreated or treated wastewaters, while oocysts tend to occur much less frequently in groundwaters. Oocysts can survive for up to 18 months depending on the temperature. Most of the oocysts found in both surface and groundwaters are derived from agricultural sources. Cattle and infected humans can excrete up to 10^{10} oocysts during the course of infection, so that cattle slurry, wastewater from marts and sewage should all be considered potential sources of the pathogen. The most important outbreak of cryptosporidia in the US in recent years occurred in April 1993 in Milawakee. The water distribution system serving 800 000 people was contaminated by raw water from a river swollen by spring run-off. In all 370 000 people became ill, 4400 were admitted to hospital and approximately 40 died (Jones, 1994).

It has been established that chlorination at levels used in water treatment is ineffective against oocysts while ozone can achieve 90% inactivation at 1 ppm of ozone for 5 min. The critical factor in water treatment is the recycling of the backwash water from rapid sand filtration. This water can contain up to 10 000 oocysts l^{-1} with a very high chance of resultant break through if recycled. In practice well-operated treatment processes with proper filtration and disposal of filter backwash water should be capable of achieving 99% reduction in oocyst concentration (Section 10.1.7). Their detection in water samples relies on filtration of large volumes of water (100–500 l) to remove oocysts and examination of the concentrate by microscopy. Most methods available for oocyst detection are adaptations of those used for *Giardia* detection (Table 12.4).

12.1.5 CONCLUSION

Human enteric viruses and protozoan parasites possess certain traits, which aid waterborne transmission and which have contributed to their increase in recent years (West, 1991). These include:

(a) an ability to be excreted in faeces in large numbers during illness;
(b) failure of conventional sewage treatment to remove them;
(c) they can survive as an environmentally robust form or they demonstrate resilience to inactivation whilst in an aquatic environment;
(d) they are largely resistant to common disinfectants used in drinking water treatment;
(e) most importantly, they only require low numbers to elicit infection in hosts consuming or exposed to water.

These factors are compounded by the difficulty of isolating and accurately detecting these pathogens in both treated and untreated effluents, as well as in both surface and groundwater resources. For both viruses and protozoa, large sample volumes must be examined in order to detect small numbers of organisms. Propagation and identification may take several days. In addition, by the time the outbreak is eventually recognized, it is usually long after the initial contamination event and that water is no longer available for examination. This is particularly the case with protozoan pathogens when a minor operational error during backwashing sand filters may result in the breakthrough of cysts and oocysts into the treated water.

12.2 MICROBIAL ASSESSMENT OF DRINKING WATER QUALITY

In order to prevent pathogen transfer to consumers via drinking water, microbial analysis is required. Chemical analysis can only be used for the assessment of water treatment efficiency and to monitor compliance to legal standards. Biological examination of water is used to detect the presence of algae and animals that may affect treatment or water quality, and to identify possible defects in the distribution network.

12.2.1 MONITORING PATHOGENS

Routine monitoring of pathogenic micro-organisms in drinking water is extremely difficult because:

(a) pathogens are outnumbered by the normal commensal bacterial flora in both human and animal intestines;
(b) large volumes of water ($\leqslant 2$l) must be filtered to ensure pathogens are recovered;
(c) isolation of pathogens require specific and complicated tests often using special equipment;
(d) positive identification may require further biochemical, serological or other tests.

It is therefore impracticable to examine all water samples on a routine basis for the presence or absence (P-A) of all pathogens. In order to routinely examine water supplies a rapid and preferably a single test is required. The theory being it is more effective to examine a water supply frequently using a simple general test, as most cases of contamination of water supplies occur infrequently, than only occasionally by a series of more complicated tests. This has led to the development of the use of indicator organisms to determine the likelihood of contamination by faeces.

The use of indicator organisms is a widely accepted practice, with legal standards based on such organisms. The main criteria for selection of an indicator organism are:

(a) they should be a member of the normal intestinal flora of healthy people;
(b) they should be exclusively intestinal in habit and therefore exclusively faecal in origin if found outside the intestine;
(c) ideally they should only be found in humans;
(d) they should be present when faecal pathogens are present and only when faecal pathogens are expected to be present;
(e) they should be present in greater numbers than the pathogen they are intended to indicate;
(f) they should be unable to grow outside the intestine with a die-off rate slightly less than the pathogenic organisms;
(g) they should be resistant to natural environmental conditions and to water and wastewater treatment processes in a manner equal to or greater than the pathogens of interest;
(h) they should be easy to isolate, identify and enumerate;
(i) they should be non-pathogenic.

While no organism can meet all these conditions consistently, in temperate regions *E. coli* fulfils most of these requirements with other coliform organisms, faecal streptococci and *Clostridium perfringens* also widely used. The EU Drinking Water Directive (Section 8.1) specifies numerical standards for *E. coli* and enterococci (faecal streptococci) in drinking water; *E. coli*, enterococci, *Pseudomonas aeruginosa* and total viable counts of heterotrophic bacteria at 22°C and 37°C in bottled waters (Table 8.6). Routine (indicator) monitoring is restricted to coliforms, *Clostridium perfringens* and heterotrophic bacteria at 22°C only (Table 8.7).

These three groups are able to survive for different periods of time in the aquatic environment. Faecal streptococci die fairly quickly outside the host and their presence is an indication of recent pollution. *E. coli* (faecal coliforms) can survive for several weeks under ideal conditions and are far more easily detected than the other indicator bacteria. Because of this it is the most widely used test although the others are often used to confirm faecal contamination if *E. coli* is not detected. Sulphate-reducing clostridia (*Clostridium perfringens*) can exist indefinitely in water. When *E. coli* and faecal streptococci are absent, its presence indicates remote or intermittent pollution. It is especially useful for testing lakes and reservoirs, although the spores do eventually settle out of suspension. The spores are more resistant to industrial pollutants than the other indicators and it is especially useful in waters receiving both domestic and industrial wastewaters. It is assumed that these indicator organisms do not grow outside the host and, in general, this is true. However, in tropical regions *E. coli* in particular is known to multiply in warm waters and there is increasing evidence that *E. coli* is able to reproduce in enriched waters generally, thus indicating an elevated health risk. Therefore, great care must be taken in the interpretation of results from tropical areas, so the use of bacteriological standards designed for temperate climates are inappropriate for those areas. The most widely used identification and enumeration techniques for indicator micro-organisms in drinking waters are summarized below. However, these methods are constantly being refined and updated with the latest UK standard methods listed in Table 12.4.

12.2.2 HETEROTROPHIC PLATE COUNT BACTERIA

Heterotrophic plate counts (HPCs) represent the aerobic and facultatively anaerobic bacteria that derive their carbon and energy from organic compounds (Table 12.5). Certain HPC organisms are considered to be opportunistic pathogens (Table 12.2) and have been implemented in gastrointestinal illness (Section 12.1.2).

Heterotrophic bacteria are commonly isolated from raw waters and are widespread in soil and vegetation, and can survive for long periods in water and rapidly multiply, especially at summer temperatures. There is also concern that these organisms can rapidly multiple in bottled waters, especially if not stored properly once opened (Gray, 1994). The EU Drinking Water Directive requires that there is no significant increase from background levels of HPC bacteria in either tap or bottled waters. While HPC bacteria are not a direct indicator of faecal contamination, they do indicate variation in water quality and the potential for pathogen survival and regrowth.

TABLE 12.5   Example of the HPC bacteria isolated from distribution and raw waters

Organism	Distribution water		Raw water	
	Total	Percentage of total	Total	Percentage of total
Actinomycete	37	10.7	0	0
Arthrobacter spp.	8	2.3	2	1.3
Bacillus spp.	17	4.9	1	0.6
Corynebacterium spp.	31	8.9	3	1.9
Micrococcus luteus	12	3.5	5	3.2
Staphylococcus aureus	2	0.6	0	0
S. epidermidis	18	5.2	8	5.1
Acinetobacter spp.	19	5.5	17	10.8
Alcaligenes spp.	13	3.7	1	0.6
F. meningosepticum	7	2.0	0	0
Group IVe	4	1.2	0	0
Group MS	9	2.6	2	1.3
Group M4	8	2.3	2	1.3
Moraxella spp.	1	0.3	1	0.6
Pseudomonas alcaligenes	24	6.9	4	2.5
P. cepacia	4	1.2	0	0
P. fluorescens	2	0.6	0	0
P. mallei	5	1.4	0	0
P. maltophilia	4	1.2	9	5.7
Pseudomonas spp.	10	2.9	0	0
Aeromonas spp.	33	9.5	25	15.9
Citrobacter freundii	6	1.7	8	5.1
Enterobacter agglomerans	4	1.2	18	11.5
Escherichia coli	1	0.3	0	0
Yersinia enterocolitica	3	0.9	10	6.4
Group llk biotype l	0	0	1	0.6
Hafnia alvei	0	0	9	5.7
Enterobacter aerogenes	0	0	1	0.6
Enterobacter cloacae	0	0	1	0.6
Klebsiella pneumoniae	0	0	0	0
Serratia liquefaciens	0	0	1	0.6
Unidentified	65	18.7	28	17.8
Total	347	100	157	99.7

Reproduced from Bitton (1994) with permission of Wiley-Liss Inc., New York.

The HPC organisms typically found in raw waters and within water supply distribution systems are listed in Table 12.5. HPCs are carried out normally using the spread plate method using yeast extract agar (YEA) and incubated at 22°C for 72 h and 37°C for 24 h, respectively. Results are expressed as colony forming units (cfu) per ml. Counts at 37°C are especially useful as they can provide rapid information of possible contamination of water supplies (Department of the Environment, 1994).

HPCs have long been employed to evaluate water quality (Table 12.6) although less importance is currently placed on HPCs for assessing the potability of drinking water. It is considered that their value lies mainly in indicating the efficiency of various water

TABLE 12.6   Sanitary quality of water based on HPC

Quality	Number of bacteria/ml
Excessively pure	<10
Very pure	10–100
Pure	100–1000
Mediocre	1000–10 000
Impure	10 000–100 000
Very impure	>100 000

treatment processes including disinfection, as well as the cleanliness and integrity of the distribution system. They are more useful in assessing the quality of bottled waters, which may be stored for long periods before being sold for consumption. The US National Primary Drinking Water Regulations now include maximum contaminant levels (MCLs) of no >500 cfu ml^{-1} for HPCs (USEPA, 1990) although this is primarily to reduce possible interference with the detection of coliforms.

12.2.3 ENUMERATION TECHNIQUES

The exact methods employed to enumerate indicator bacteria are specified by the legal standards used. Two techniques are principally used, membrane filtration and the multiple tube methods. The EU Drinking Water Directive specifies that *E. coli*, enterococci (faecal streptococci), *Pseudomonas aeruginosa* and *Clostridium perfringens* must all be isolated using the membrane filtration method (Table 12.4), although the multiple tube method is still widely used for clostridia because of the need to incubate under anaerobic conditions.

Coliforms do not only occur in faeces, they are normal inhabitants of water and soil. The presence of coliforms in a water sample does not necessarily indicate faecal contamination, although in practice it must be assumed that they are of faecal origin unless proved otherwise. The total coliform count measures all the coliforms present in the sample. However, only *E. coli* is exclusively faecal in origin with numbers in excess of $10^8 g^{-1}$ of fresh faeces. So it is important to confirm *E. coli* is present. Routine coliform testing comprises of two tests giving the total coliform count and faecal coliform (*E. coli*) count.

Membrane filtration

The membrane filtration method is now widely used for all coliform testing. Known volumes of water are passed through a sterile membrane filter with a pore size of just 0.45 μm (Fig. 12.3). This retains all the bacteria present. The membrane filter is placed onto a special growth medium, which allows the individual bacteria to grow into colonies. Special media, which only allow specific bacteria to grow, are used.

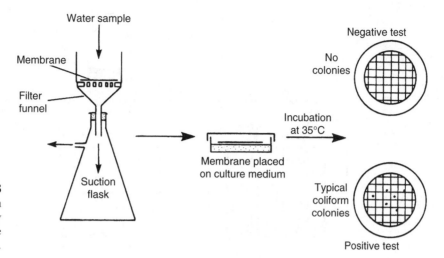

FIGURE 12.3
Major steps in
coliform testing by
the membrane
filtration technique.

For total coliforms the membrane filter is placed onto membrane-enriched teepol medium, which contains the detergent teepol to inhibit non-intestinal bacteria from growing. A different medium (m-ENDO) is normally used in the US (APHA, 1992). The membrane filter on the culture medium is incubated. During this incubation period, the nutrients diffuse from the culture medium through the membrane and the coliform bacteria are able to multiply and form recognizable colonies that can then be counted. Membranes are incubated at 30°C for 4 h followed by 14 h at 37°C for total coliforms or 14 h at 44°C for *E. coli*, the colonies of which are a distinctive yellow colour. Faecal streptococci are determined using the same technique except that different culture media and incubation conditions are required (i.e. membrane *Enterococcus* agar medium and incubated for 4 h at 37°C followed by 44 h at 45°C with colonies appearing either red or maroon in colour) (Table 12.4).

Multiple tube

The principle of the test is that various volumes of sample water are inoculated into a series of tubes containing a liquid medium that is selective for coliform bacteria (Fig. 12.4). From the pattern of positive and negative growth responses in the dilution series, a statistical estimate of the number of coliforms and subsequently *E. coli* can be made. This is called the most probable number (MPN) estimate, which is calculated by reference to probability tables. The MPN is expressed as the number of cells per 100 ml of sample. The technique is carried out in two discrete stages. The first estimates the number of coliforms present on the assumption that all the tubes, which show acid and gas productions contain coliform organisms. Because this assumption is made, this is known widely as the presumptive coliform count. The second stage tests for the presence of *E. coli*. The first stage is completed over 48 h, while the second takes a further 24 h.

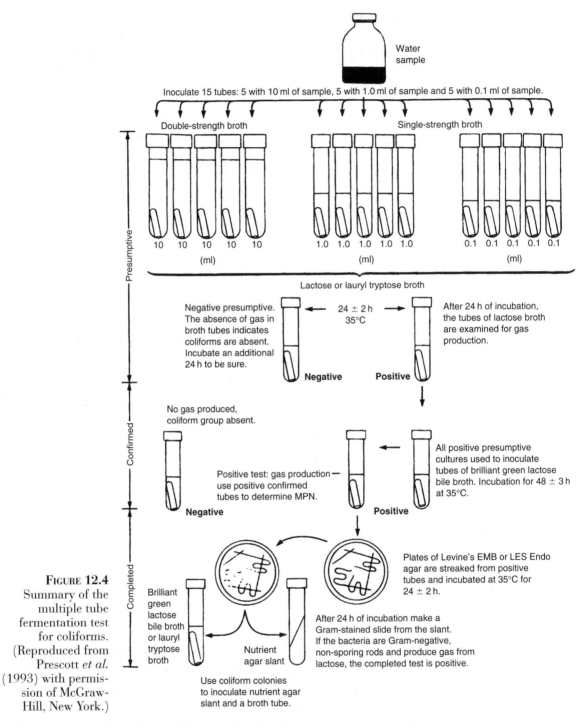

FIGURE 12.4 Summary of the multiple tube fermentation test for coliforms. (Reproduced from Prescott *et al.* (1993) with permission of McGraw-Hill, New York.)

The membrane filtration technique is considered to have many advantages over the multiple tube method for water testing. These include:

(a) presumptive coliform counts are available in a shorter time (18 h);

(b) it is a simpler test with less steps;

(c) there is considerable saving in the laboratory in the amounts of culture media, labour and glassware required;
(d) larger volumes of sample may be processed;
(e) it is possible to carry out filtration in the field;
(f) false negative results due to the development of aerobic and anaerobic spore bearing organisms are unlikely to occur.

Details of all microbial methods are given in *Standard Methods* for the US (APHA, 1992) and for the UK in *Report on Public Health and Medical Subjects No. 71* (Department of the Environment, 1994), although the latter is now regularly updated online by the Environment Agency (Table 12.4).

12.2.4 EMERGING TECHNOLOGIES

There are many emerging new technologies for microbial water testing, primarily to detect coliforms. These are focused on complex biochemical techniques, such as hybridization and polymerase chain reaction (PCR), gene probe technology and monoclonal antibody methods, which allow single bacterial cells to be detected (Gleeson and Gray, 1997). Enzyme detection methods are now widely in use, especially in field situations. Total coliform detection is based on the presence of β-galactosidase, an enzyme that catalyses the breakdown of lactose into galactose and glucose; while *E. coli* is based on the detection of β-glucuronidase activity. These tests are known as ONPG and MUG methods, respectively, after the substrates used in the tests and are now accepted standard monitoring methods in the US. These ONPG–MUG tests can be used to give an MPN value or simply indicate the presence or absence (P-A) of coliforms or *E. coli*. These tests will become increasingly important as there is a general swing away from standards based on microbial density to those simply based on P-A of coliforms in a sample. The best-known commercial ONPG–MUG preparations are currently Colilert® (Access Analytical, Branford, CT), Coliquick® (Hach Co., Loveland, CO) and Colisure® (Millipore Co., Bedford, MA). The ingredients for these new tests come in powder form (in test tubes for the quantitative MPN method and in containers for P-A analysis). A measured amount of water is added to each tube or container and the powder dissolves into a colourless solution. The tubes are placed in an incubator for 24 h at 35°C. The solution in tubes with total coliforms will be yellow which is then exposed to a hand-held fluorescent light. If the tube contains *E. coli* the solution will fluoresce brightly. The specificity of this method eliminates the need for confirmatory and completed tests.

The coliform test is still widely considered the most reliable indicator for potable water. However, in recent years there has been growing dissatisfaction with the use of coliforms as indicator organisms. Recent years have seen increasing reports of waterborne outbreaks largely as a result of protozoan and viral agents in waters considered safe to drink under current legislation, which relies largely on the coliform test. The major deficiencies identified with the use of coliforms as indicators for drinking water quality assessment are:

(a) the regrowth of coliforms in aquatic environments;
(b) the regrowth of coliforms in distribution networks;
(c) suppression by high background bacterial growth;
(d) they are not directly indicative of a health threat;
(e) a lack of correlation between coliforms and pathogen numbers;
(f) no relationship between either protozoan or viral numbers;
(g) the occurrence of false positive and false negative results.

This has been reviewed by Gleeson and Gray (1997).

12.2.5 STANDARDS

EU legislation

In December 1998 the newly revised Drinking Water Directive was published (Section 8.1). As well as complying with new fixed parameters, water supplied under the Directive will have to be free of pathogenic micro-organisms and parasites in numbers constituting a danger to public health (Table 8.6). Although as before there are no guide values given for specific viruses, protozoans or bacterial pathogens. The New Directive gives separate maximum permissible concentrations for the microbial parameters for both tap and bottled waters, except natural mineral waters that have their own Directive. For tap water maximum permissible concentrations are given for *E. coli* (0 per 100 ml) and enterococci (0 per 100 ml). For bottled waters the maximum permissible concentrations are stricter than previously with *E. coli* (0 per 250 ml) and enterococci (0 per 50 ml). *Pseudomonas aeruginosa* has also been included for the first time with a maximum permissible concentration of 0 per 250 ml. Maximum colony counts at 22°C and 37°C are 100 and 20 ml^{-1}, respectively. The microbial parameters are listed in a new section, Part A, of the Directive. Part B contains chemical parameters and Part C indicator parameters. Included in Part C are total coliforms with a maximum permissible concentration of 0 per 100 ml for tap waters and 0 per 250 ml for bottled waters; and also included are total bacterial counts (22°C) which must not show any abnormal change with a maximum permissible value of 0 per ml in tap waters and sulphite-reducing clostridia at 0 per 100 ml (Table 8.7). Specific and detailed notes are given on analysis, including the composition of all recommended media. The recommended analyses in the proposed Directive are given in Table 12.7.

US legislation

The USEPA set a MCL based on the P-A concept for coliforms. This revised MCL came into force on 31 December 1990. The recommended sampling frequency of water supplies is dependent on the population served. For systems requiring >40

TABLE 12.7   The recommended microbial methods in the EU Drinking Water Directive
(98/83/EEC)

Total coliforms
Membrane filtration followed by incubation on membrane lauryl broth for 4 h at 30°C
followed by 14 h at 37°C. All yellow colonies are counted regardless of size.

E. coli
Membrane filtration followed by incubation on membrane lauryl broth for 4 h at 30°C
followed by 14 h at 44°C. All yellow colonies are counted regardless of size.

Faecal streptococci
Membrane filtration followed by incubation on membrane *Enterococcus* agar for 48 h at
37°C. All pink, red or maroon colonies which are smooth and convex are counted.

Sulphite-reducing clostridia
Maintain the sample at 75°C for 10 min prior to membrane filtration. Incubate on
tryptose–sulphite–cycloserine agar at 37°C under anaerobic conditions. Count all black
colonies after 24 and 48 h incubation.

Pseudomonas aeruginosa
Membrane filtration followed by incubation in a closed container at 37°C on modified
Kings A broth for 48 h. Count all colonies which contain green, blue or reddish-brown
pigment and those that fluoresce.

Total bacteria counts
Incubation in a YEA for 72 h at 22°C and for 24 h at 37°C. All colonies to be counted.

Reproduced with permission of the European Commission, Luxembourg.

samples per month, <5% of samples must be total coliform positive. For systems
where the frequency of analysis is <40 samples per month, then no >1 sample per
month may be total coliform positive. If a sample is found to be total coliform positive then repeat samples must be taken within 24 h. Repeat samples must be taken at
the same tap and also at adjacent taps within five service connections (Section 10.2),
both up- and downstream of the original sample point. If these repeated samples are
also total coliform positive, then the samples must be immediately tested for faecal
coliforms or *E. coli*. If these prove positive then the public must be informed. A full
detail of the revised coliform rule is given by Berger (1992).

A 100 ml sample bottle must be used in analysing total coliforms using one of the following techniques: 10-tube multiple tube fermentation technique, the membrane filtration technique, the P-A coliform test or the minimal-media ONPG–MUG test.
This P-A concept has a number of potential advantages:

(a) Sensitivity is improved because it is more accurate to detect coliform presence
     than to make quantitative determinations.
(b) The concept is not affected by changes in coliform density during storage.
(c) Data manipulation is much improved.

The introduction of new and rapid testing systems will most likely see a similar P-A
coliform test introduced throughout Europe.

WHO recommendations

The WHO (2004) guidelines for drinking water are perhaps the most important standards relating to water quality (Section 8.1). Standards are based solely on the presence of *E. coli* or thermotolerant coliforms, which must not be detectable in any 100 ml sample of water entering or within the distribution system, or in the drinking water delivered to the consumer (Table 8.8). This simple P-A approach makes water safety easier to manage with the presence of either *E. coli* or thermotolerant coliforms indicative that faecal contamination is able to enter the supply system. There are no specific guideline values for other bacteria, or for viruses or parasites. This is because the analytical methods for these organisms are too costly, complex and time consuming for routine laboratory use. Instead, guideline criteria are outlined based on the likely viral content of source waters and the degree of treatment necessary to ensure that even large volumes of water have a negligible risk of containing viruses. It is considered that 'the attainment of the bacteriological criteria and the application of treatment for virulogical reduction should ensure the water presents a negligible health risk' (WHO, 2004).

12.3 REMOVAL OF PATHOGENIC ORGANISMS

Efficient wastewater treatment is critical for the prevention of waterborne disease. Conventional treatment can only remove up to 80–90% of bacterial pathogens with tertiary treatment increasing this to ⩾98%. While wastewater disinfection can increase this to >99.99%, there will still be significant numbers of pathogens present in the final effluent. Dilution and the effects of natural biotic and abiotic factors in surface waters will reduce the density of pathogens further, although all water resources can become contaminated from other sources, such as diffuse agricultural pollution, surface (urban) run-off, and contamination from septic tank systems and landfill leachate. Also, with the emphasis of discharge standards for wastewater treatment plants based largely on environmental criteria, efficiency of pathogen removal is often ignored and usually poor. This places an ever-increasing burden on the final barrier, water treatment. With increasing concerns over the transfer of antibiotic resistance between bacteria in biological wastewater treatment systems and the increasing occurrence of chlorine-resistant pathogens, microbial discharge standards for sewage treatment plants are likely, making wastewater disinfection inevitable. The effectiveness of water and wastewater treatment in the removal of faecal coliforms is summarized in Fig. 12.1.

12.3.1 ENVIRONMENTAL FACTORS OF SURVIVAL

All pathogens are able to survive for at least a short period of time in natural waters, both fresh and saline, and generally this period is extended at cooler temperatures and if organic pollution is present. In raw sewage 50–75% of the coliforms are associated with particles with settling velocities $>0.05\,cm\,s^{-1}$, so in conventional primary sedimentation at a treatment plant significant removals of enteric bacteria are achievable. Some idea of the effect of time on the survival of pathogenic micro-organisms can be obtained by examining the removal efficiency of storage lagoons which are used at some waterworks to improve water quality prior to treatment and supply. Up to 99.9% reduction of enteric bacteria can be obtained by storage, although this is dependent on temperature, retention time and the level of pollution of the water. For example, a 99.9% reduction in pathogenic viruses by storage requires a retention time of 20 days at 20–27°C but up to 75 days at 4.8°C. Viruses can survive longer than bacteria in natural waters, with some viruses surviving for up to twice as long as certain indicator bacteria (Table 12.8). Temperature is an important factor in the survival of viruses and at low temperatures (4–8°C) survival time will increase.

Both UV radiation and short-wave visible light is lethal to bacteria with the rate of death related to light intensity, clarity of the water and depth. In the dark the death rate of coliforms follows first-order kinetics over the initial period; however, predation by protozoa causes a departure from the log-linear relationship. The death rate is measured as the time for 90% mortality to occur (T_{90}). In the dark at 20°C the T_{90} for coliforms is 49 h while under mid-day sunlight the T_{90} is reduced to 0.3 h. The death rate is also temperature dependent, increasing by a factor of 1.97 for each 10°C rise in temperature, and is proportional to the total radiation received regardless if continuous or intermittent. The amount of radiation required to kill 90% (S_{90}) of coliforms is estimated as $23\,cal.\,cm^{-2}$ ($11.3\,cal.\,cm^{-2}$ for *E. coli*). The S_{90} ratio of faecal streptococci to coliforms varied from 10 to 40 indicating that the die-off rate for faecal streptococci is appreciably slower than for total coliforms. The effect of temperature is far less marked under light conditions and, while variations in salinity have no effect on death rate, it is substantially slower in fresh and brackish waters. The daily solar radiation in southern Britain is between 50 to $660\,cal.\,cm^{-2}$ so there appears to be ample radiation to inactivate all faecal coliforms. Wavelength is also important with about 50% of the lethal effect of solar radiation attributable to wavelengths below 370 nm, 25% to near visible UV (370–400 nm) and 25% to the blue-green

TABLE 12.8 Effects of temperature on the time (t) required for 99.9% inactivation of enteric micro-organisms during storage

Species	Temperature (°C)	Time (t)	Temperature (°C)	Time (t)
Poliovirus	4	27	20	20
Echovirus	4	26	20	16
E. coli	4	10	20	7
S. faecalis	4	17	20	8
Poliovirus type 1, 2, 3	4–8	27–75	20–27	4–20
Coxsackie virus type A2, A9	4–8	12–16	20–27	4–8

Reproduced from Kool (1979) with permission of John Wiley and Sons Ltd, Chichester.

region of the visible spectrum (400–500 nm). The effect of wavelengths >500 nm is negligible. As this die-off rate for coliforms is proportional to the light intensity and is therefore essentially a first-order relationship, it can be expressed as:

$$\frac{dC}{dt} = -kl_o \cdot e^{-az}C \tag{12.1}$$

where C is the coliform concentration at time t at a depth z, and k is a proportionality coefficient, l_o the light intensity just below the water surface and a the effective attenuation coefficient.

12.3.2 WASTEWATER TREATMENT

The removal of pathogenic micro-organisms is brought about by a combination of physical, chemical and biological processes. Physically pathogens are removed by adsorption and settlement, while the overall concentration is reduced by dilution. The chemical nature of the wastewater will determine whether the environmental conditions are suitable for the survival or even the growth of pathogens; however, factors such as hardness, pH, ammonia concentration, temperature and the presence of toxic substances can all increase the mortality rate of the micro-organisms. Biologically, death of pathogens can occur due to a number of reasons including starvation, although predation by other micro-organisms and grazing by macro-invertebrates are important removal mechanisms. During treatment of sewage the microbial flora changes from predominately faecal in character to that found in enriched freshwaters.

It is convenient to look at the wastewater treatment plant as an enclosed system with inputs and outputs. It is a continuous system so the outputs, in the form of sludge and a final effluent, will also be continuous. While a comparison of the number of pathogens in the influent with the final effluent will provide an estimate of overall removal efficiency, it will not give any clues as to the mechanism of removal. Essentially pathogens are either killed within the treatment unit, discharged in the final effluent or concentrated in the sludge which will result in secondary contamination problems if disposed either to agricultural land or into coastal waters. An estimation of the specific death rate $(-\mu)$ of a pathogen can be calculated by accurately measuring all the inputs and outputs of the viable organism (Pike and Carrington, 1979). All the individuals of the population of the pathogen (x), which are assumed to be randomly dispersed within the reactor, have the same chance of dying within a specific time interval $(t - t_o)$. Under steady-state conditions, the causes of death can be assumed to remain constant in terms of concentration (in the case of a toxic substance) or number (of a predator), so the rate of death will be proportional to the number of survivors (x_t) of the original population (x_o). This can be represented by a first-order, exponential 'death' equation:

$$x_t = x_o \cdot e^{-\mu(t-t_o)} \tag{12.2}$$

The problem with using Equation (12.2) in practice is obtaining accurate estimates of the viable micro-organism present in the outputs, especially if the cells have flocculated or are attached to film debris. The type of reactor is also important in the estimation of the death rate. In an ideal plug-flow system (e.g. percolating filter) in which first-order kinetics apply, the fraction of pathogens surviving (x_t/x_o) can be related to the dilution rate or the reciprocal of the retention time (ϕ) according to Equation (12.3):

$$\frac{x_t}{x_o} = \frac{-\mu}{\phi} \tag{12.3}$$

In a continuous-stirred tank reactor (CSTR) the specific death rate is calculated by assuming the rate of change in pathogen concentration within the reactor (dx/dt) equals the input concentration (x_o) minus the output concentration (t), minus those dying within the reactor ($-\mu$). So by assuming that the rate of change is zero within the reactor under steady-state conditions, then:

$$\frac{dx}{dt} = \phi x_o - \phi x_t - \mu x \tag{12.4}$$

Primary sedimentation

Bacteria and viruses are significantly reduced by primary settlement, with settled sludge containing the whole range of pathogens found in raw sewage. Ova (eggs) and cysts of parasites are only significantly removed at this stage in wastewater treatment plants. Settlement efficiency is dependent on the size and density of the ova and cysts, and as their free-falling settling velocities are not much greater than the theoretical upflow velocity, near quiescent conditions are required for optimum removal. The larger, denser ova of *Ascaris lumbricoides* and *Taenia saginata* are more efficiently removed than the smaller cysts of *Entamoeba* spp. The settling velocity of *T. saginata* is about 0.6–0.9 m h^{-1}, although much less if detergents are present, resulting in 68% removal after 2 h and 89% after 3 h settlement.

Activated sludge

The activated sludge process is highly efficient in the removal of pathogenic bacteria and viruses, achieving a 90% removal efficiency or more (Geldreich, 1972). The major removal mechanism of bacteria in the activated sludge process is predation by a variety of amoebae, ciliate protozoa and rotifers. The ciliate protozoa and rotifers feed only on the freely suspended bacteria and not on flocculated forms (Curds *et al.*, 1968) (Fig. 12.5). Amoebae occur in similar numbers to ciliates and have similar yield coefficient biomass and generation times. They also play a significant role in the removal of bacteria by predation and are able to feed on flocculated forms as well as the freely

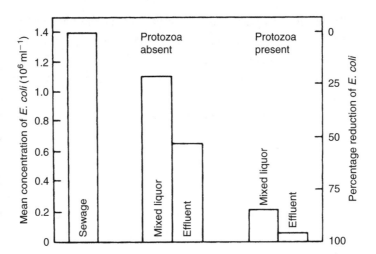

FIGURE 12.5
Removal of *E. coli* in experimental-activated sludge plants operating in the absence or presence of ciliated protozoa. (Reproduced from Curds and Fey (1969) with permission of Elsevier Science Ltd, Oxford.)

suspended bacteria. The estimated specific growth rate of *E. coli* is -7.9 day^{-1} in the presence of protozoa compared with $+0.12$ day^{-1} in the absence of protozoa, suggesting that *E. coli* are capable of very slow growth in sewage when ciliates are absent. In general terms the percentage removal of coliform bacteria is directly related to the specific sludge wastage rate, with 90% removal at a sludge wastage rate of 0.65 day^{-1} which rapidly decreases as the wastage rate increases (Pike and Carrington, 1979).

The prime removal method of viruses is adsorption onto sludge flocs, predation having a negligible effect on concentration in the liquid phase (Section 20.6). Both viruses and bacteria are adsorbed according to the empirical Freundlich adsorption isotherm, where the count of particles adsorbed per unit mass of sludge (Y/m) is proportional to a power (n) of the count of particles (x) in the liquor at equilibrium:

$$\frac{Y}{m} = kx^n \tag{12.5}$$

This adsorption model is for unreactive sites whereas Michaelis–Menten or Monod kinetics are for reactive sites. However, as Pike and Carrington (1979) point out, the successful fitting of an adsorption model to data may not demonstrate that adsorption is the only factor operating since, for example, protozoa attached to the activated sludge flocs and feeding off freely suspended bacteria will quantitatively behave as a continual adsorption site. Upwards of 90% of enteroviruses are removed by the activated sludge process (Geldreich, 1972). Ova and cysts of parasites are able to survive the activated sludge process and are not effectively removed.

Fixed-film reactors

Percolating filters are extremely effective in removing pathogenic bacteria with normal removal efficiencies of >95%. Removal is achieved by similar mechanisms as in the activated sludge process, except that filters are plug-flow systems with a fixed and not a mixed microbial biomass, so that opportunities for contact between pathogens and adsorption sites in the biomass are reduced.

Removal of bacteria is directly related to the bacterial count of the sewage and at low-rate loadings to the surface area of the filter medium. Removal efficiency falls off during winter that suggests that maximum removal of pathogens occurs when the film is most actively growing and under maximum grazing pressure, which is when maximum availability of adsorption sites will occur (Gray, 2004). Ciliate protozoa, rotifers and nematodes can ingest pathogenic bacteria. However, in percolating filters there is a much larger range of macro-invertebrate grazers feeding directly on the film and so indirectly feeding on the pathogens. Once pathogens have been adsorbed onto the film they are essentially 'removed' and their subsequent ingestion by a grazing organism may not be significant. The major limitation of percolating filtration in the removal of pathogens is their physical adsorption from suspension. Rotating biological contactors are also extremely efficient in the removal of pathogenic bacteria, with median removal of *E. coli* normally >99.5%. Percolating filtration is not very effective in the removal of parasite ova and cysts although the nature of the film does allow some retention of ova. Geldreich (1972) in a review on waterborne pathogens quotes removal rates of 18–70% for tapeworm ova and 88–99% for cysts of *E. histolytica*.

Waste stabilization ponds

All stabilization ponds and lagoons are extremely effective in removing pathogenic bacteria, viruses and the other parasites. Removal mechanisms include settlement, predation, inactivation due to solar radiation which is also linked with temperature, increase in pH due to day-time assimilation of carbon dioxide which can reach in excess of pH 9.0, and finally antibacterial toxins produced by algae. It is most appropriate to consider ponds as CSTRs, so the survival of pathogens (x_t/x_0) can be calculated as:

$$\frac{x_t}{x_0} = \frac{1}{[1 - (\mu/\phi)]} \tag{12.6}$$

Ponds are normally in series, so if they have similar dilution rates the survival of pathogens can be calculated using the following equation:

$$\frac{x_n}{x_0} = \frac{1}{[1 - (\mu/\phi)]^n} \tag{12.7}$$

where n is the number of ponds. A problem arises if the number of ponds in series is large (i.e. >5) because then the system behaves more like a plug-flow reactor. In this case the relationship becomes:

$$\frac{x_n}{x_0} = \exp - (\mu/\phi_n) \tag{12.8}$$

where ϕ_n is the dilution rate of the complete system.

Solar radiation is a major removal mechanism of bacteria and viruses in maturation ponds. In the UK at a retention time of 3.5 days the removal rate of *E. coli* was >90% in the summer but fell to 40% in the winter (60% for faecal streptococci) due to the seasonal difference in light intensity (Toms *et al.*, 1975). In the British situation the algal density is rarely sufficient to significantly shift the pH in order to kill the pathogens, in fact the algae can reduce mortality of bacteria and viruses by reducing light intensity by shading. The depth is also an important factor with removal rates reduced with depth.

In tropical and sub-tropical countries the removal of pathogens is as important as biochemical oxygen demand (BOD) removal is in temperate zones, and in many countries more so. Facultative ponds are very efficient in the removal of pathogens (Gloyna, 1971) with removal rates of coliform and streptococci bacteria >99% and viruses inactivated by light so that >90% removals are achieved. The greater efficiency of facultative ponds compared with other systems is due to the much longer retention times, so ponds in series achieve greater removals than single ponds. Major removal mechanisms are the high pH values created by photosynthesis and the higher zooplankton predation rate which is recorded. Anaerobic treatment followed by facultative treatment gives a higher die-off rate of pathogenic bacteria than a facultative pond followed by a maturation pond in series. Pond systems appear extremely effective in the removal of parasites due to the long retention time allowing maximum settlement. Cysts of the protozoans *E. histolytica* and *Giardia lamblia* are almost completely removed, while the helminth parasites, such as *Schistosoma, Ascaris, Enterobius, Ancylostoma* and *Trichuris* are also effectively removed (Gloyna, 1971). Maximum removal occurs in the first pond and as the nematode eggs in particular are highly resistant, extreme care must be taken with the disposal of the raw sludge from the pond if contamination is to be prevented (Section 18.2).

The removal of helminth (nematode) eggs in stabilization ponds is dependent on retention (ϕ). So for ponds in series (including anaerobic, facultative and maturation) the number of eggs in the final effluent (E_f) is calculated using Equation (12.9):

$$E_f = E_i \left[1 - \left(\frac{R_1}{100} \right) \right] \left[1 - \left(\frac{R_2}{100} \right) \right] \left[1 - \left(\frac{R_3}{100} \right) \right] \tag{12.9}$$

where E_i is the number of eggs in the raw sewage (l^{-1}). Values of R are obtained from Table 12.9. If E_f is >1 then an extra maturation pond is required. Maturation pond design is based on the removal of *E. coli* or faecal coliforms (Mara, 1996).

EXAMPLE 12.1 Assuming an egg density of $2000 l^{-1}$ in the raw sewage calculate the final density of eggs in the final effluent after anaerobic lagoon, facultative pond and maturation pond treatment in series. The retention times in the various ponds are 5.0, 3.6 and 8.5 days, respectively.

TABLE 12.9 Estimated values of percentage removal (R) of intestinal nematode eggs in waste stabilization ponds at various retention times (θ) in days

θ	R	θ	R	θ	R
1.0	74.67	4.0	93.38	9.0	99.01
1.2	76.95	4.2	93.66	9.5	99.16
1.4	79.01	4.4	93.40	10	99.29
1.6	80.87	4.6	94.85	10.5	99.39
1.8	82.55	4.8	95.25	11	99.48
2.0	84.08	5.0	95.62	12	99.61
2.2	85.46	5.5	96.42	13	99.70
2.4	87.72	6.0	97.06	14	99.77
2.6	87.85	6.5	97.57	15	99.82
2.8	88.89	7.0	97.99	16	99.86
3.0	89.82	7.5	98.32	17	99.88
3.2	90.68	8.0	98.60	18	99.90
3.4	91.45	8.5	98.82	19	99.92
3.6	92.16			20	99.93
3.8	92.80				

Reproduced from Mara (1996) with permission of John Wiley and Sons Ltd, Chichester.

Using Table 12.9 to calculate R for each pond and Equation (12.9) then:

$$E_f = 2000 \left[1 - \left(\frac{95.62}{100} \right) \right]\left[1 - \left(\frac{92.16}{100} \right) \right]\left[1 - \left(\frac{98.82}{100} \right) \right]$$

$$= 2000(0.0438)(0.0784)(0.0118)$$

$$E_f = 0.081^{-1}$$

Digestion

There are very conflicting results relating to the efficiency of anaerobic digestion and composting to remove pathogens from sewage sludge. Anaerobic digestion certainly reduces the numbers of pathogens considerably but not always completely. While many bacteria, fungi and viruses are rapidly killed by air drying such organisms can survive anaerobic digestion at 20 or 30°C for long periods. For example, *S. typhi* can survive digestion for 12 days at 20°C or 10 days at 30°C, *Ascaris* ova 90 days at 30°C and hookworm ova 64 days at 20°C or 41 days at 30°C. Both the latter ova can also withstand air drying. Complete destruction of pathogens is only possible by heating the sludge to 55°C for 2 h or treating with lime.

Sterilization and disinfection methods

Sterilization and disinfection of final effluents to remove any disease-causing organisms remaining in effluents is not widely practised in Europe, but is common in the

TABLE 12.10 Typical chlorine dosages to achieve disinfection of various wastewaters

Wastewater	Chlorine dose (mg l⁻¹)
Raw sewage	8–15
Settled sewage	5–10
Secondary treated effluent	2–8
Tertiary treated effluent	1–5
Septic tank effluent	12–30

USA. The increasing need to reuse water for supply after wastewater treatment will mean that the introduction of such methods to prevent the spread of diseases via the water supply is inevitable. The two processes are distinct form each other. Sterilization is the destruction of all the organisms in the final effluent regardless of whether they are pathogenic or not, while disinfection is the selective destruction of disease-causing organisms. There are three target groups of organism, the viruses, bacteria and protozoan cysts, and each is more susceptible to a particular disinfection process than the other. The main methods of sterilization and disinfection are either chemical or physical in action. The factors affecting the efficiency of chemical disinfectants are contact time, concentration and type of chemical agent, temperature, types and numbers of organisms to be removed, and the chemical nature of the wastewater. Typical chlorine doses for wastewaters are given in Table 12.10. Viruses require free chlorine to destroy them and at higher doses compared with enteric bacteria. Chemical disinfection has been discussed in Section 10.1. The commonest physical method for removing pathogens is UV radiation, which acts on the cellular nucleic acids destroying bacteria, viruses and any other organisms present. Although expensive in terms of energy, such systems can be highly effective and do not affect receiving water quality. The greatest effect occurs at a wavelength of 265 nm, with low-pressure mercury arc lamps (254 nm) most widely used. The major operational problem is to obtain maximum penetration of the rays to ensure that even turbid effluents are fully sterilized. Numerous systems have been evaluated to obtain maximum exposure of wastewater to the radiation, but the most effective system to date is the use of thin-film irradiation (<5 mm). UV irradiation is particularly useful in preventing contamination of lakes and coastal waters with pathogens, which are popular for bathing. Filtration is also used to remove pathogens from effluents that have received tertiary treatment (Section 20.8).

12.3.3 WATER TREATMENT

Storage

Bacteria and viruses are significantly reduced when water is stored in reservoirs. During the spring and summer, sunlight, increased temperature and biological factors ensure that between 90% and 99.8% reductions of E. coli occur. The percentage reduction is less during the autumn and winter due to the main removal mechanisms being less effective, so expected reductions fall to between 75% and 98%. Lowest reductions occur when reservoirs are mixed to prevent stratification. The greatest decline in *E. coli* and *Salmonella* bacteria occurs over the first week, although the longer the water is stored the greater the overall reduction. Temperature is an important factor. For example, poliovirus was reduced by 99.8% in <15 days at 15–16°C compared with 9 weeks for a similar reduction at 5–6°C. For optimum removal of all micro-organisms of faecal origin then 10 days retention should achieve between 75% and 99% reduction regardless of temperature. Protozoan cysts are not effectively removed by storage because of their small size and density. For example, the settling velocity of *Cryptosporidium* oocysts is $0.5\,\mu m\,s^{-1}$, compared with $5.5\,\mu m\,s^{-1}$ for *Giardia* cysts. So while it would be feasible in a large storage reservoir to remove a percentage of the *Giardia* cysts, if the retention time was in excess of 6 weeks, and mixing and currents were minimal, *Cryptosporidium* oocysts would remain in suspension (Denny, 1991). *Salmonella*, faecal streptococci and *E. coli* are excreted in large numbers by gulls. The presence of gulls, especially in high numbers, on storage reservoirs may pose a serious problem either from direct faecal discharge and/or from rainfall run-off along contaminated banks.

Unit processes

Bacteria and viruses are removed by a number of unit processes in water treatment, especially coagulation, sand filtration and activated carbon filtration (Section 10.1). Coagulation using alum removes about 90% of faecal indicator bacteria, 95–99% of all viruses and about 60–70% of the total plate count bacteria, although these figures vary widely from treatment plant to treatment plant. Other coagulants, such as ferric chloride and ferric sulphate, are not quite as efficient. The use of polyelectrolytes as coagulant aids does not improve removal of viruses. Rapid sand filtration is largely ineffective in removing viruses and bacteria unless the water has been coagulated prior to filtration. In contrast, slow sand filtration is able to remove up to 99.5% of coliforms and 97–99.8% of viral particles from water, although performance is generally worse in the winter (Denny, 1991). Activated carbon can remove viruses, which are adsorbed onto the carbon. This is by electrostatic attraction between positively charged amino groups on the virus and negatively charged carboxyl groups on the surface of the carbon. Removal efficiency is very variable and depends on pH (maximum removal occurs at pH 4.5), the concentration of organic compounds in the water, and the time the filter has been in operation. Removal rates of between 70% and 85% are common. However, these filters can become heavily colonized by heterotrophic micro-organisms. Minute fragments are constantly breaking off the granular-activated carbon and each is heavily coated with micro-organisms. These micro-organisms are not affected to any great extent by disinfection, and so can introduce large numbers of micro-organisms including pathogens into the distribution system (Le Chevallier and McFeters, 1990).

Cysts should be removed effectively by coagulation and the addition of poly-electrolyte coagulant aids should also enhance removal. Using optimum coagulant conditions (as determined by the jar test) then 90–95% removal should be possible. Rapid sand filtration is not an effective barrier for cysts unless the water is coagulated prior to filtration. When used after coagulation then very effective removal of *Giardia* is achieved (99.0–99.9%). The only proven effective method of removing both *Cryptosporidium* and *Giardia* cysts is by slow sand filtration, with 99.98% and 99.99% of cysts removed, respectively (Hibler and Hancock, 1990).

Disinfection

Disinfection is absolutely vital to ensure that any micro-organisms arising from fae-cal contamination of the raw water are destroyed (Section 10.1.8). Chlorination is by far the most effective disinfectant for bacteria and viruses because of the residual dis-infection effect that can last throughout the water's journey through the distribution network to the consumer's tap. The most effective treatment plant design to remove pathogens is rapid and slow sand filtration, followed by chlorination or treatment by pre-chlorination followed by coagulation, sedimentation, rapid sand filtration and post-chlorination. Both of these systems give >99.99% removal of bacterial pathogens including *C. perfringens*. Chlorine and monochloramine have been found to be inef-fective against *Cryptosporidium* oocysts. Ozone at a concentration of about $2\,\mathrm{mg\,l^{-1}}$ is able to achieve a mean reduction in viability of oocysts of between 95% and 96% over a 10-min exposure period.

The disinfection rate determines the level of destruction of pathogens for a given period of contact. The death of micro-organisms is a first-order reaction in respect of time for a given disinfectant and concentration. This is Chick's law and is used to esti-mate the destruction of pathogens by a disinfectant as a function of time.

$$\frac{\mathrm{d}N}{\mathrm{d}t} = -kN \tag{12.10}$$

This can be integrated as:

$$N = N_0 e^{-kt} \tag{12.11}$$

or

$$\ln\frac{N}{N_0} = -kt \tag{12.12}$$

where N is the concentration of viable micro-organisms at time t, N_0 the initial con-centration of viable micro-organisms and k the first-order decay rate. The negative sign indicates that the number of organisms decreases over time. The persistence of pathogens and non-pathogenic micro-organisms in aquatic environments approximate to Chick's law which when plotted as $\ln(N/N_0)$ versus t is a straight line (Fig. 12.6) with the slope being $-k$. If $-\ln(N/N_0)$ is plotted against t instead, then the straight line intercepts 0 and the slope is k. Temperature and salinity particularly affect the decay constant k for any specific micro-organism. This can be adjusted by using Equation (12.13) for temperature or Equation (12.14) for both:

$$k_T = k_{20} \times 1.047^{T-20} \tag{12.13}$$

$$k_T = k_{20} \times (0.006 \times \% \text{ seawater}) \times 1.047^{T-20} \tag{12.14}$$

The rate coefficient k is redefined to incorporate the effect of a specific disinfectant at concentration C:

$$k = k'C^n \tag{12.15}$$

where n is an empirical constant known as the coefficient of dilution. Equation (12.12) can be rewritten as:

$$\ln \frac{N}{N_0} = -k'C^n t \tag{12.16}$$

or

$$\log \frac{N}{N_0} = \frac{-k'C^n t}{2.3} \tag{12.17}$$

So to determine the level of $C^n t$ to achieve the objective (N/N_0):

$$C^n t = \frac{-1}{k'} \ln\left(\frac{N}{N_0}\right) = \frac{-2.3}{k'} \log\left(\frac{N}{N_0}\right) \tag{12.18}$$

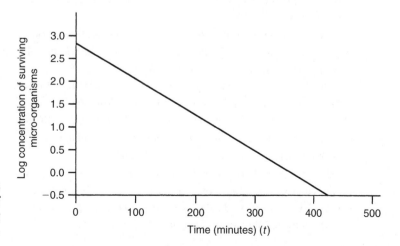

FIGURE 12.6
Destruction of pathogens according to Chick's law.

From Equation (12.18) variations in $\log(N/N_0)$ can be related to variations in $C^n t$ which is the basis of disinfection practices of drinking water. Where $n = 1$ then $\log(N/N_0)$ is proportional to Ct so individual rates for C and t need not be specified. However, where n is not equal to 1 then individual values of C and t are required. Where $n < 1$ reaction rate (t) is the most important factor for pathogen inactivation, while when $n > 1$ disinfection concentration (C) is the major factor (Bryant *et al.*, 1992).

It is a fact that conventional water treatment cannot guarantee the safety of drinking water supplies at all times. Outbreaks of waterborne diseases can and do happen, although very infrequently. With correct operation the chances of pathogenic micro-organisms causing problems to consumers will be reduced even further. The greatest risks come from private supplies that are not treated. The installation of a small UV water sterilizer unit installed under the sink on the rising main directly to the kitchen tap ensures microbially safe water. However, the only effective way to destroy protozoan cysts is to physically remove them. This is achieved by using a 1 μm pore-sized fibre cartridge filter that is placed upstream of the UV sterilizer. The filter will also increase the efficiency of the sterilization by removing any particulate matter that may harbour or shield pathogens from the UV radiation.

Wash water and water-treatment sludge contain all the pathogens removed from the water, and so must be handled as carefully as any other microbiologically hazardous waste. It is important that the sludge is not disposed of in such a way as to recontaminate the raw water source (Section 10.1.11).

References

APHA, 1992 *Standard Methods for the Examination of Water and Waste Water*, 18th edn, American Public Health Association, Washington, DC.

Berger, P. S., 1992 Revised total coliform rule. In: Gilbert, C.E. and Calabrese, E.J., eds, *Regulating Drinking Water*, Lewis, Bocta Raton, FL, pp 161–6.

Bradley, D. J., 1993 Human tropical diseases in a changing environment. In: *Environmental Change and Human Health*, Ciba Foundation Symposium, Vol. 171, John Wiley and Sons, Chichester, pp 146–70.

Britton, G., 1994 *Wastewater Microbiology*, Wiley-Liss, New York.

Bryant, E. A., Fulton, G. P. and Budd, G. C., 1992 *Disinfection Alternatives for Safe Drinking Water*, Van Nostrand Reinhold, New York.

Cabelli, V., 1978 New Standards for enteric bacteria. In: Mitchell, R., ed., *Water Pollution Microbiology*, Vol. 2, Wiley-Interscience, New York, pp 233–73.

Craun, G. F., 1986 *Waterborne Diseases in the United States*, CRC Press, Boca Raton, FL.

Cubitt, D. W., 1991 A review of the epidemiology and diagnosis of waterborne viral infections: II, *Water, Science and Technology*, **24**, 197–203.

Curds, C. R., Cockburn, A. and Van Dyke, J.M., 1968 An experimental study of the role of ciliated protozoa in the activated sludge process, *Water Pollution Control*, **67**, 312–29.

Curds, C. R. and Fey, G. J., 1969 The effect of ciliated protozoa on the fate of *Escherichia coli* in the activated sludge process, *Water Research*, **3**, 853–67.

Denny, S., 1991 *Microbiology Efficiency of Water Treatment*, Report FR0219, Foundation for Water Research, Marlow.

Department of the Environment and Department of Health, 1990 *Cryptosporidium in Water Supplies*, Report of a Group of Experts, HMSO, London.

Department of the Environment, 1994 *The Microbiology of Water 1994: Part 1 Drinking water, Reports on Public Health and Medical Subjects No. 71. Methods for the Examination of Water and Associated Materials*, HMSO, London.

Du Moulin, G. C. and Stottmeier, K. D., 1986 Waterborne *Mycobacteria*: an increasing threat to health, *American Society of Microbiology News*, 525–9.

Du Zuane, J., 1990 *Handbook of Drinking Water Quality Standards and Controls*, Van Nostrand Reinhold, New York.

Galbraith, N. S., Barnett N. J. and Stanwell-Smith, R., 1987 Water and disease after Croydon: a review of waterborne and water associated disease in the United Kingdom (1937–1986), *Journal of the Institution of Water and Environmental Management*, **1**, 7–21.

Geldreich, E. E., 1972 Waterborne pathogens. In: Mitchell, R., ed., *Water Pollution Microbiology*, Wiley-Interscience, New York, pp 207–41.

Geldreich, E. E., 1991 Microbial water quality concerns for water supply use, *Environmental Toxicology and Water Quality*, **6**, 209–23.

Gleeson, C. and Gray, N. F., 1997 *The Coliform Index and Waterborne Disease: Problems of Microbial Drinking Water Assessment*, E. & F.N. Spon, London.

Gloyna, E. F., 1971 *Waste Stabilization Ponds*, WHO Monograph Series, 60, World Health Organization, Geneva.

Gray, N. F., 2004 *Biology of Waste Water Treatment*, 2nd edn, Imperial College Press, London.

Gray, N. F., 1994 *Drinking Water Quality: Problems and Solutions*, John Wiley and Sons, Chichester.

Herwaldt, B. L., Craun, G. F., Stokes, S. L. and Juranek, D. D., 1992 Outbreaks of water disease in the United States: 1989–1990, *Journal of the American Water Works Association*, **84**, 129–35.

Hibler, C. P. and Hancock, C. M., 1990 Waterborne giardiasis. In: McFeters, G. A., ed., *Drinking Water Microbiology*, Springer, New York, pp 271–93.

Jenkins, P. A., 1991 Mycobacterium in the environment. *Journal of Applied Bacteriology Symposium Supplement*, 137–41.

Jones, K., 1994 Inside Science: 73, Waterborne Diseases, *New Scientist*, 1–4.

Joseph, C. A., Watson, J. M., Harrison, T. G. and Bartlett, C. L. R., 1994 Nosocomial Legionnaires' disease in England and Wales 1980–92, *Epidemiology and Infection*, **112**, 329–46.

Kool, H. J., 1979 Treatment processes applied in public water supply for the removal of micro-organisms. In: James, A. and Evison, L., eds, *Biological Indicators of Water Quality*, John Wiley and Sons, Chichester, pp 17.1–17.31.

Le Chevallier, M., Norton, W. D. and Lee, R. G., 1991 Occurrence of *Giardia* and *Cryptosporidium* spp. in surface water supplies, *Applied and Environmental Microbiology*, **57**, 2610–16.

Le Chevallier, M. and McFeters, G. A., 1990 Microbiology of activated carbon. In: McFeters, G. A., ed., *Drinking Water Microbiology*, Springer, New York, pp 104–19.

Mara, D. D., 1996 *Low-Cost Urban Sanitation*, John Wiley and Sons, Chichester.

Pike, E. B. and Carrington, E. G., 1979 The fate of enteric bacteria and pathogens during sewage treatment. In: James, A. and Evison, L., eds, *Biological Indicators of Water Quality*, John Wiley and Sons, Chichester, pp 20.01–32.

Prescott, L. M., Harley, J. P. and Klein, D. A., 1993 *Microbiology*, WCB Publishers, Iowa.

Reasoner, D., 1992 *Pathogens in Drinking Water – Are There Any New Ones?* US Environmental Protection Agency, Washington, DC.

Schbert, R. H., 1991 Aeromonads and their significance as potential pathogens in water, *Journal of Applied Bacteriology Symposium Supplement*, **70**, 131–5.

Singh, A. and McFeters, G. A., 1992 Detection methods for waterborne pathogens. In: Mitchell, R., ed., *Environmental Microbiology*, John Wiley and Sons, New York.

Skirrow, M. B. and Blaser, M. J., 1992 Clinical and epidemiological considerations. In: Nachamkin, I., Blaser, M. J. and Tomkins, L. S., eds, *Campylobacter jejuni, Current Status and Future Trends*, American Society for Microbiology, Washington, pp 3–8.

Smith, H. V., 1992 *Cryptosporidium* and water: a review, *Journal of the Institute of Water Engineers and Scientists*, **6**, 443–51.

Sobsey, M. D., Dufour, A. P., Gerba, C. P., LeChevallier, M. W. and Payment, P., 1993 Using a conceptual framework for assessing risks to human health from microbes in drinking water, *Journal of the American Water Works Association*, **85**, 44–8.

Tauxe, R. V., 1992 Epidemiology of *Campylobacter jejuni* infections in the US and other industrialized nations. In: Nachamkin, I., Blaser, M. J. and Tomkins, L. S., eds, *Campylobacter jejuni, Current Status and Future Trends*, American Society for Microbiology, Washington, DC, pp 9–19.

Toms, I. P., Owens, M., Hall, J. A. and Mindenhall, M. J., 1975 Observations on the performance of polishing lagoons at a large regional works, *Water Pollution Control*, **74**, 383–401.

USEPA, 1990 *Drinking Water Regulations Under the Safe Drinking Water Act*, Fact Sheet, May, US Environmental Protection Agency, Washington, DC.

West, P. A., 1991 Human pathogenic viruses and parasites: emerging pathogens, *Journal of Applied Bacteriology (Symposium Supplement)*, **70**, 1075–145.

WHO, 2004 *Guidelines for Drinking Water Quality, Vol. 1: Recommendations*, 3rd edn, World Health Organization, Geneva.

FURTHER READING

Department of the Environment, 1994 *The Microbiology of Water 1994: Part 1 Drinking Water, Reports on Public Health and Medical Subjects No. 71, Methods for the Examination of Water and Associated Materials*, HMSO, London.

Du Zuane, J., 1990 *Handbook of Drinking Water Quality Standards and Controls*, Van Nostrand Reinhold, New York.

Gameson, A. L. H., 1985 *Investigations of Sewage Discharges to Some British Coastal Waters: 8. Bacterial Mortality, Part 2*, Technical Report: 222, Water Research Centre, Stevenage.

Geldreich, E. E., 1996 *Microbial Quality of Water Supply in Distribution Systems*, Lewis Publishers, Bocta Raton, FL.

Gleeson, C. and Gray, N. F., 1997 *The Coliform Index and Waterborne Disease: Problems of Microbial Drinking Water Assessment*, E. & F.N. Spon, London.

McFeters, G. A. (ed.), 1990 *Drinking Water Microbiology*, Springer, New York.

WHO, 2004 *Guidelines for Drinking Water Quality, Vol. 1: Recommendations*, 3rd edn, World Health Organization, Geneva.

Part IV Wastewater Treatment

Chapter 13 Nature of Wastewater

13.1 COMPOSITION OF WASTEWATER

Wastewater is a complex mixture of natural inorganic and organic material mixed with man-made substances. It contains everything discharged to the sewer, including material washed from roads and roofs, and of course where the sewer is damaged groundwater will also gain entry. It is this complex mixture that ends up at the wastewater treatment plant for purification. In its broadest sense wastewaters can be split into domestic (sanitary) wastewater also known as sewage, industrial (trade) wastewaters and, finally, municipal wastewater which is a mixture of the two. It is increasingly unusual for municipal plants to treat wastewaters from new industries. Most authorities charge for treating industrial wastes (polluter pays principle), so it is generally cheaper for industry to treat its own or at least to pre-treat its wastewaters prior to discharge to the local authority sewer in order to reduce costs.

The strength and composition of sewage changes on an hourly, daily and seasonal basis, with the average strength dependent on per capita water usage, infiltration, surface runoff as well as local habits and diet. The water usage in the US is at least twice as great than in the UK, which is why the sewage in the US is weaker (Table 13.1).

Sewage is 99.9% water with the material that requires to be removed amounting to just 0.1% by volume. This solid material is a mixture of faeces, food particles, grease, oils, soap, salts, metals, detergents, plastic, sand and grit (Fig. 13.1). The organic fraction

TABLE 13.1   Typical raw wastewater composition

Parameter	US (mg l^{-1})	UK (mg l^{-1})
pH	7.0	7.0
BOD	250	350
COD	500	700
Suspended solids	250	400
Ammonia nitrogen	30	40
Nitrate nitrogen	<1	<1
Total phosphorus	10	15

BOD: biochemical oxygen demand; COD: chemical oxygen demand

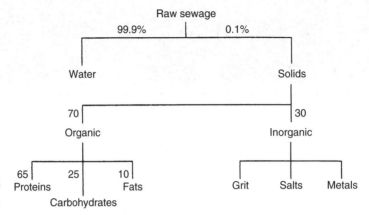

FIGURE 13.1
Composition of
raw sewage.

is composed primarily of proteins, carbohydrates and fats, which reflects the diet of the community served by the treatment system. The aim of wastewater treatment is to convert this solid portion to a manageable sludge (2% dry solids) while leaving only a small portion in the final effluent (0.003% dry solids).

Carbohydrates have a general formulae $(C_6H_{10}O_5)_n$. Complex sugars are broken down into simple sugars which in turn are oxidized into carbon dioxide and water. The commonest sugars in sewage are glucose, sucrose and lactose, which together with the minor carbohydrates present represent a significant portion of the BOD, between 50 and $120 \, mg \, l^{-1}$. Volatile acids are also found in sewage (e.g. acetic, propionic, butyric and valeric), which are responsible for between 6 and $37 \, mg \, l^{-1}$ of the BOD. If sewage is stored it rapidly becomes anaerobic resulting in sugars rapidly being converted to volatile acids. However, as long as sewage is kept moving it will maintain a minimum dissolved oxygen concentration of 1–$2 \, mg \, l^{-1}$. If high concentrations of volatile acids are found in sewage then anaerobosis must be suspected.

Proteins are primarily composed, in decreasing proportion, of carbon, hydrogen, oxygen, nitrogen, sulphur and phosphorus, and many other trace elements. For example, most proteins are comprised of 16% nitrogen. Proteins are broken down to polypeptides, then to individual amino acids, fatty acids, nitrogenous compounds, phosphates and sulphides. Protein is the major source of nitrogen in sewage along with urea, and before the advent of detergents it was also the main source of phosphorus. The per capita production of nutrients in the UK is $5.9 \, g \, N \, day^{-1}$ and $2.0 \, g \, P \, day^{-1}$. The average concentration of total phosphorus in sewage varies from 5 to $20 \, mg \, P \, l^{-1}$ (Section 20.5). Urea can be found in concentrations up to $16 \, mg \, l^{-1}$ in sewage, and is rapidly converted to ammonia under both aerobic and anaerobic conditions. As a general guide the concentration of total nitrogen is directly related to the BOD for any specific domestic raw wastewater. Together carbohydrates and proteins make up 60–80% of the organic carbon in sewage.

Nitrogen (N) is found in a number of different forms in sewage, as organic nitrogen, ammonia and oxidized nitrogen (nitrite and nitrate). Nitrogen is usually measured as

Kjeldhal N, which is the sum of the organic N and ammonia N present. Total nitrogen is the sum of the Kjeldhal N and the oxidized N. In raw and settled sewage there is no oxidized nitrogen present, so the total nitrogen is equivalent to the Kjeldhal N. The range of ammonia N in settled wastewater is between 41 and 53 mg N l^{-1}, and 16–23 mg N l^{-1} as organic nitrogen, which is equivalent to 57–76 mg N l^{-1} as Kjeldhal N.

Fats is a general term and used interchangeably with lipids and grease. It includes the whole range of fats, oils and waxes associated with food. Fats are stable compounds and not readily degraded biologically into fatty acids, which are also found in sewage (e.g. palmitic, stearic and oleic). They can represent a significant proportion of the BOD of sewage (40–100 mg l^{-1}). Soaps are manufactured from fats reacted with either sodium or potassium hydroxide to yield glycerol and the salts of particular fatty acids which is the soap. For example, stearic acid and sodium hydroxide yields glycerol and sodium stearate ($C_{17}H_{35}COONa$).

Not all wastewaters are treatable biologically, and apart from the presence of inhibitory or toxic material, biodegradability depends largely on the C:N:P ratio of the wastewater (Section 15.3). The optimum ratio for biological treatment is 100:5:1. At this ratio all the nutrients are utilized for biological growth and are removed from solution. In raw sewage the normal ratio is much higher at 100:17:5 and for settled sewage 100:19:6. It can be seen that there is an excess of nutrients in sewage in relation to the carbon as measured by the BOD test. This results in the excess nitrogen and phosphorus being discharged in the final effluent as either ammonia or nitrate and phosphate. Food-processing wastewaters are often short of nitrogen and phosphorus, while industrial wastes may be short of organic material (Table 13.2). Where nutrients are limited wastewaters can be mixed with sewage in order to utilize the surplus or a supplement in the form of urea (N) or agricultural grade fertilizer (N and P) can be added. Where supplements are used it may be necessary to dissolve them before adding them to the wastewater.

Between 1960 and 1980 the worldwide per capita consumption of detergents rose from 3.6 to 6.3 kg per annum, and doubling in the US over the same period from 12.8 to 30.1 kg. Since 1965 there has been a ban on the use of hard non-biodegradable alkyl–benzene–sulphonate (ABS)-based detergents that caused extensive foaming during treatment and subsequently in receiving waters. These have been replaced by soft, biodegradable linear–alkylate–sulphonate (LAS) detergents. To increase the washing ability of LAS detergents it has been necessary to incorporate builders, which are mainly polyphosphate compounds. The most widely used is sodium tripolyphosphate which is responsible for 70% of the phosphate load to treatment plants and which leads, along with nitrogen, to eutrophication. The formulation of detergents is complex with the builder making up as much as 30% of the volume of some washing powders (Table 13.3). Alternative builders that do not contain phosphates have been introduced although they are less efficient. Inorganic compounds are usually too bulky and inefficient (e.g. sodium borate), while organic builders have been more widely adopted. The most

TABLE 13.2   Typical industrial wastewater characteristics

Industry	Flow	BOD	Total suspended solids	COD	pH	Nitrogen	Phosphorus
Meat products	Intermittent	High–extremely high	High	High–extremely high	Neutral	Present	Present
Milk handling	Intermittent	Average–high	Low–average	Average–high	Acid–alkaline	Adequate	Present
Cheese products	Intermittent	Extremely high	Average–extremely high	Extremely high	Acid–alkaline	Deficient	Present
Alcoholic beverages	Intermittent	High–extremely high	Low–high	High–extremely high	Alkaline	Deficient	Deficient
Soft drinks	Intermittent	Average–high	Low–high	Average–high	–	Deficient	Present
Textiles	Intermittent–continuous	High	High	High	Alkaline	Deficient	Present
Tanning and finishing	Intermittent	Extremely high	Extremely high	Extremely high	Acid–alkaline	Adequate	Deficient
Metal finishing	Continuous–variable	Low	Average–high	Low	Acidic	Present	Present
Fruit and vegetables	Intermittent	Average–extremely high	Average–extremely high	–	Acid–alkaline	Deficient	Deficient
Paper and allied products	Continuous	Average–extremely high	Low–high	Low–high	Neutral (mech. pulping)	Deficient	Deficient
Pharma-ceutical	Continuous–intermittent	High	Low–high	High	Acid–alkaline	Deficient	Deficient
Plastics and resins	Continuous–variable	Average–high	Low–high	Average–high	Acid–alkaline	–	–

Reproduced with permission of the Environmental Protection Agency, Dublin.

TABLE 13.3   Typical formulation of a household detergent

Chemical function	Example	Proportion (%)
Surfactant	LAS	3–15
Builder	Sodium tripolyphosphate	0–30
Ion exchanger	Zeolite type A	0–25
Antiredeposition agent	Polycarboxylic acids	0–4
Bleaching agent	Sodium perborate	15–35
Bleach stabilizer	Phosphonate	0.2–1.0
Foam booster	Ethanolamide	1–5
Enzyme	Protease	0.3–1.0
Optical brightener	Pyrazolan derivatives	0.1–1.0
Corrosion inhibitor	Sodium silicate	2–7
Fragrance		0.05–0.3

Reproduced from Hunter *et al.* (1988) with permission of Selper Ltd, London.

successful include nitrilotriacetic acid (NTA), polyacrylic acid (PAA) and polycarboxylic acids (PCAs). However, there is growing concern that NTA in particular, but also PAA and PCAs may have unacceptable health and environmental risks associated with their use, and have been banned in some countries. Since December 2002, phosphate-based laundry detergents have been banned in Ireland (Gray, 2004).

Raw sewage is rather a turbid liquid with small but visible particles of organic material that readily settle out of suspension. The colour is normally grey to yellow-brown according to the time of day. Much darker sewage, especially black, indicates anaerobosis. Municipal wastewaters may be affected by dyes and other coloured discharges. Domestic wastewater has a musty odour which is generally non-offensive, however pungent odours can be produced if wastewaters become anaerobic. Some common odours are listed in Table 13.4 and their presence can be helpful in identifying potential operational problems or changes in the influent composition. These odours are generally products of organic degradation, which form volatile, or gaseous compounds that often have very low odour thresholds (i.e. average odour detection limit). For example, methyl mercaptan (CH_3SH) and ethyl mercaptan (C_5H_5SH) have strong decaying cabbage odours with odour thresholds of 0.0011 and 0.00019 ppm, respectively. The temperature of sewage remains several degrees warmer than the air temperature except during the warmest months, because the specific heat of water is much greater than that of air, while the pH is generally neutral.

While the nature of wastewaters have changed significantly in recent years, reflecting the new materials used in the home and in industry, two new components of wastewater have been identified as causing major environmental problems in receiving waters, especially where this water is reused for supply purposes. These are endocrine-disrupting substances, and health care products. Like may micro-pollutants, neither is effectively

TABLE 13.4 Characteristic odour-producing compounds present in wastewaters; they are primarily degradation products of nitrogenous or sulphurous compounds, or other odorous compounds such as those associated with chlorine and phenolic wastes

Compounds	General formulae	Odour produced
Nitrogenous		
Amines	CH_3NH_2, $(CH_3)_3N$	Fishy
Ammonia	NH_3	Ammoniacal pungent
Diamines	$NH_2(CH_2)_4NH_2$, $NH_2(CH_2)_5NH_2$	Rotten flesh
Skatole	$C_8H_5NHCH_3$	Faceal, repulsive
Sulphurous		
Hydrogen sulphide	H_2S	Rotten eggs
Mercaptans	CH_3SH, $CH_3(CH_2)SH$	Strong decayed cabbage
Organic sulphides	$(CH_3)_2S$, CH_3SSCH_3	Rotten cabbage
Sulphur dioxide	SO_2	Pungent, acidic
Other		
Chlorine	Cl_2	Chlorine
Chlorophenol	Cl,C_6H_4OH	Medicinal phenolic

removed during wastewater treatment and so are discharged in the final effluent, or can contaminate groundwaters when sludge is landfilled or spread onto land.

Endocrine-disrupting substances are either natural or synthetic substances that interfere with the normal functioning of the endocrine system, which controls many different physiological processes. Apart from neurobehavioural, growth and development problems, the main concern has been the effect endocrine disrupters have on reproductive processes. Feminization of male fish is increasingly common in lowland rivers in all developed countries. For example, 100% of roach in the River Mersey in 2003 showed feminine characteristics. The main endocrine-disrupting substances found in sewage are the natural female steroid hormones oestrogen and 17β-oestradiol and the synthetic hormone ethinyl oestradiol from the contraceptive pill, which arise from the urine of the female population. Oestrogen mimicking compounds from industrial sources include alkylphenols (APs) and alkylphenol ethoxlates (APEs). However, a much wider group of compounds can also be classed as endocrine-disrupting substances including many pesticides, dioxins and furans and tributyltin (Table 13.5).

TABLE 13.5   Potential endocrine-disrupting substances and level of regulatory control in the UK

Substance	Statutory control
Pesticides	
DDT; 'Drins; Lindane	A, C, D
Dichlorvos; Endosulphan;	B, C, D
Trifluralin; Demeton-*S*-	
Methyl; Dimethoate; Linuron;	
Permethrin	
Herbicides	
Atrazine; Simazine	B, C, D
PCBs	
Polychlorinated biphenyls	C
Dioxins and furans	
Polychlorinated dibenzofuran;	C
Dibenzo-*p*-dioxin congeners	
Antifoul/wood preservative	
Tributyltin	B, C, D
Alkyphenols	
Nonylphenol	None
Nonylphenol ethoxylate	None
Octylphenol	None
Octylphenol ethoxylate	None
Steroids	
Eithinyl oestradiol;	
17β oestradiol; Oestrone	None

A: EC Dangerous Substances Directive (76/464/EEC) List I substance; B: List II substance; C: IIPC prescribed substance, D: Statutory EQSs in place in 2000.

While many of these substances are already covered by the EU Dangerous Substances Directive (Section 8.3), the more common endocrine-disrupting substances are not controlled. The priority is to develop environmental quality standards for these substances and improve analytical monitoring techniques. Control measures are centred on reducing their use by industry and significantly reducing or eliminating them from discharges both from industry and the home to sewerage systems (Environment Agency, 2000).

The release of pharmaceuticals and personal care products (PPCPs) after their ingestion, external application or disposal to sewers or the surface and ground waters has become one of the major environmental issues of the decade. Advances in chemical analysis has shown that drugs from a wide spectrum of therapeutic groups, and their metabolites, as well as personal care products such as fragrances and sunscreens, are present in both surface and ground waters at low but measurable concentrations (ng-μg l^{-1}) (Daughton and Ternes, 1999; Debska et al., 2004). Many of these compounds are also suspected endocrine-disrupting substances. The major route for PPCPs into the aquatic environment is via sewage treatment, where they are widespread. These compounds pass through sewage treatment plants largely unaltered and as convention water treatment is also ineffective in their removal, there is a high probability that such compounds can contaminate drinking water (Fig. 13.2). Advanced treatment processes such as ozonation, chlorination or adsorption using activated carbon are required to remove these trace contaminants. The diversity and quantity of these compounds that are sold annually poses a major risk to all aquatic ecosystems worldwide, with antibiotics, hormones, analgesic and sedative drugs, disinfectants, vitamins and many more all being discharged to surface waters. The most widely reported compounds of pharmaceutical origin in surface waters are

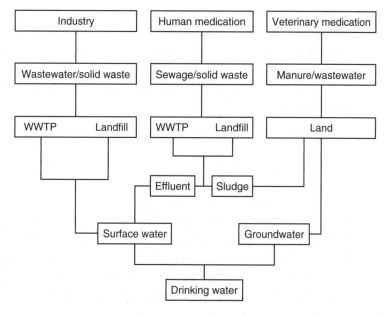

FIGURE 13.2
Pathways of PPCPs into drinking water.

TABLE 13.6 Commonly occurring trace contaminants of pharmaceutical origin in wastewater and lowland rivers

Acetylsalicylic acid	Analgesic/anti-inflammatory
Caffine	Psychomoter stimulants
Carbamazepine	Analgesic/anti-inflammatory
Carboxyibuprofen	Analgesic/anti-inflammatory
Chloramphenicol	Antibiotic
Ciprofloxacin	Antibiotic
Clofibric acid	Lipid-lowering agent
Diazepam	Psychiatric
Diclofenac	Analgesic/anti-inflammatory
Erythromycin	Antibiotic
17β-estradiol	Estrogen
Estrol	Estrogen
Estron	Estrogen
17α-ethinyl estradiol	Estrogen
Hydroxyibuprofen	Analgesic/anti-inflammatory
Ibuprofen	Analgesic/anti-inflammatory
Naproxen	Analgesic/anti-inflammatory
Norflozacin	Antibiotic
Nonylphenol	Estrogen
Primidone	Antiepileptic
Salicylic acid	Multi-purpose
Sulphadizine	Sulphonamide
Sulphomethoxazole	Sulphonamide
Sulphonamides	Sulphonamide
Trimethoprim	Antibacterial

given in Table 13.6. The full extent of the presence of such a complex cocktail of active compounds on aquatic organisms is largely unknown. However, most of these compounds retain their active ingredient after treatment and discharge into the environment. For example, there is increasing evidence that certain pharmaceuticals can inhibit efflux pumps, the membrane-based active-transport system that removes toxicants from inside cells, so making organisms even more vulnerable to toxicity. The worrying issue is that while the effects may be so subtle that they are not recognizable in real time, that they elicit discernible cumulative effects that appear to have no obvious cause.

13.2 SEWAGE COLLECTION

There are two basic types of sewerage systems, combined and separate; sewerage being the network of pipes that collects and transfers the sewage to the wastewater treatment plant. Sewer pipes are usually concrete and laid 1–3 m below ground. Normally laid beneath one side of the road, the precise depth of the sewer pipe depends on the gradient, geology and road surface loading. Pipe diameters vary in the same way as water mains with individual houses served by 0.15 m diameter plastic pipes that feed into pipes that get progressively larger as the number of connections and area served increases. Sewer pipes up to 3 m in diameter are common within towns. The design of sewerage systems is complex and normally done by computer programs that are able to use real-time rain data (Hydraulic Research Laboratory, 1985; White, 1986).

Combined sewerage is common in most European towns and both surface drainage from paved areas and sewage are collected together for subsequent treatment. Most treatment plants are built on the outskirts of towns and so it may take many hours (often in excess of 6–12 h) for wastewater from some parts of the sewerage network to reach the treatment plant depending on the distance the sewage has to travel. During wet weather there is a significant increase in flow to the treatment plant due to surface run-off, this mixture is known as storm water. Treatment plants serving combined sewerage systems are designed to treat up to three times the dry weather flow (3DWF), with storage for up to six times DWF using storm water tanks for about 2 h (Section 14.2.4). Once the storm water tanks are full the excess diluted influent passes directly into the receiving water, which will now be in spate, thus minimizing any impact. During storms the time delay caused by long sewerage distances to sewage treatment plants can often be beneficial allowing time for the sewage to become diluted and for the receiving water to increase its discharge rate to ensure even greater dilution.

Separate sewerage systems overcome the problems of dilution of sewage and fluctuations in the flow to the treatment plant due to climatic conditions. The sewage goes directly to the treatment plant while surface drainage goes to the nearest watercourse. This results in a smaller volume, and less variation in flow, of sewage allowing for smaller treatment plants (2DWF) to be built. The major drawback is that surface drainage reaches receiving waters, which are often only small streams, very quickly before there is time for any significant increase in flow volume to dilute the incoming surface run-off resulting in serious localized pollution. In addition, spillages of toxic or inflammable liquids on roads can often be intercepted at the treatment plants by using the storm water tanks, where they discharge into roadside gully pots connected to combined sewerage systems. When accidents occur where there is a separate system, then the spillage will quickly enter the local watercourse. It is not advisable to use soakaways or percolation areas to dispose of surface run-off as this can lead to groundwater contamination. The treatment of diffuse pollution, including surface run-off, is considered in Section 7.4.

Infiltration of groundwater into damaged sewers is a major problem as it dilutes the sewage and increases the volume of wastewater for treatment. In cities the high leakage rates from water mains is also a major source of infiltration water to sewers. If very badly deteriorated, then up to 80% of the total flow in a sewer may be infiltration water. In the US, the estimated mean infiltration rate has been estimated to be $70 \, \text{m}^3 \, \text{day}^{-1} \, \text{km}^{-1}$ of sewer (Clark et al., 1977). The volume of surface run-off from paved areas can become very large during periods of heavy rainfall. To supply combined sewers large enough

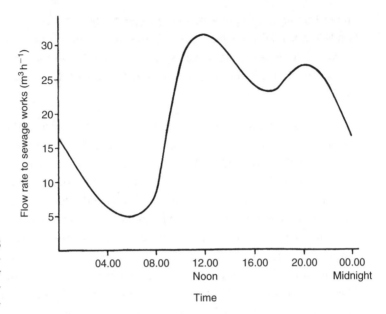

to accommodate both maximum surface run-off and discharged has now become largely unfeasible due to climate change and centralization of treatment systems, requiring storm water overflows to divert portions of the total flow in excess of 6 or 9 DWF to nearby watercourses via combined sewer overflows (CSO). This practice, however, is becoming less common as it can result in significant pollution of receiving waters, even though there will be increased flow in the receiving water to dilute the already diluted wastewater. CSOs also discharge gross solids resulting in toilet paper and other sanitary material getting tangled up in overhanging riverbank vegetation causing both a health problem and aesthetic pollution (Section 9.5). The problem with highly variable surface run-off entering combined sewers and the subsequent need for storm water overflows is the major reason for the development of separate sewerage systems, although the latter are more expensive to construct. The design and construction of sewerage systems is dealt with by White (1986).

The flow rate to treatment plants is extremely variable and, although such flows follow a basic diurnal pattern (hydrograph), each plant tends to have a characteristic flow pattern (Fig. 13.3). Infiltration increases the total volume of sewage without altering the shape of the hydrograph of the diurnal flow. Storm water, however, can alter the shape significantly by hiding peaks and troughs or adding new peaks as the rainfall causes rapid increases in flow. The volume of water supplied is essentially the same as the volume of wastewater discharged to the sewer, resulting in the wastewater hydrograph measured at the treatment plant being largely parallel to the water supply hydrograph, as measured at the water treatment plant but with a lag of many hours. Infiltration and storm water must however be taken into account if one is being used to estimate the other.

Where precise discharge rates are not known simple estimations of flow to sewer can be made by using the following steps:

(a) Domestic contribution per house (Q_d) is calculated using the following equation:

$$Q_d = H \times F \text{ m}^3 \text{ house day}^{-1} \tag{13.1}$$

where H is the habitation factor per house and F is the mean flow per capita per day. So assuming an average habitation factor of 4 and a per capita water usage of 220 l day^{-1} then Q_d is:

$$Q_d = 4 \times 0.22 = 0.88 \text{ m}^3 \text{ house day}^{-1}$$

(b) Industrial contribution (Q_I) is estimated as 0.5–2.0 l s^{-1} ha^{-1} for light industry, 1.5–4.0 l s^{-1} ha^{-1} for medium and 2.0–8.0 l s^{-1} ha^{-1} for heavy industry depending on the degree of recycling and water conservation employed.

(c) The rate of surface run-off depends on numerous factors including intensity of rainfall and the impermeability of the area drained. Impermeability is expressed as a run-off coefficient (C) factor where a watertight surface would have a factor approaching 1 while flat undeveloped land would have a factor of 0.1. The peak rate of run-off (Q) is calculated using the following equation:

$$Q = 0.0278 \, C \, i \, A \text{ m}^3 \text{ s}^{-1} \tag{13.2}$$

where A is the area of the drainage area (ha) and i is the rainfall intensity (cm h^{-1}). Typical values of the run-off coefficient C are given in Table 13.7. The values are only estimates that vary according to rainfall intensity and duration, slope and impermeability. Estimates of percentage cover of each category is best done using a geographical information system (GIS). For accurate estimations of storm water reference should be made to Corbitt (1990).

If wastewaters in sewers become anaerobic then hydrogen sulphide will be produced by the action of sulphate-reducing bacteria. The slightly acidic gas is absorbed into condensation water at the top or crown of the sewer where sulphur bacteria (e.g. *Thiobacillus*), which are able to tolerate low pH levels of 1.0, oxidize the hydrogen sulphide to strong sulphuric acid using atmospheric oxygen:

$$H_2S + O_2 \xrightarrow{\text{Thiobacillus}} H_2SO_4 + \text{energy} \tag{13.3}$$

The strong acid reacts with the lime in the concrete to form calcium sulphate, which weakens the pipe leading to eventual collapse. Crown corrosion is particularly a problem in sewers that are constructed on flat gradients, in warm climates, in sewers receiving heated effluents, with wastewaters with a high sulphur content and in sewers that are inadequately vented. Corrosion is prevented by adequate ventilation to reduce

TABLE 13.7   Range of typical run-off factors (C)

Surface area classification	C factor
Specific values	
Roofs	0.75–0.95
Asphalt and concrete paved areas	0.70–0.95
Macadam road	0.25–0.60
Gravel drive	0.15–0.30
Parks and cemeteries	0.10–0.25
Playgrounds	0.20–0.35
Railway sidings	0.20–0.35
Lawn sandy soil	
Flat (slope < 2%)	0.05–0.10
Average (slope 2–7%)	0.10–0.15
Steep (slope > 7%)	0.15–0.20
Lawns heavy soil	
Flat (slope < 2%)	0.13–0.17
Average (slope 2–7%)	0.18–0.22
Steep (slope > 7%)	0.25–0.35
General values	
Business areas	
City	0.70–0.95
Suburbs	0.50–0.70
Residential	
High density	0.60–0.75
Moderate density	0.40–0.60
Low density	0.30–0.50
Agricultural	
Tillage	0.02–0.05
Permanent grassland	
Flat (slope < 2%)	0.10–0.15
Average (slope 2–7%)	0.15–0.20
Steep (slope > 7%)	0.18–0.30
Undeveloped land	
Flat	0.10–0.20
Sloping	0.20–0.40
Steep rocky slope	0.60–0.80

condensation and expel the hydrogen sulphide, by using non-corrosive materials such as polyvinyl chloride (PVC) or vitrified clay, by preventing deoxygenation (which is a particular problem in rising mains where the wastewater is pumped up hill under pressure without the normal air space above the liquid phase), by the addition of pure oxygen or by using compressed air. Chlorination may be used under exceptional circumstances to inhibit the sulphate-reducing bacteria. The major problem with designing combined sewers is allowing sufficient gradient to maintain a minimum self-cleansing velocity, preventing anaerobic conditions developing at low flows, while at the same time avoiding excessively high velocities at high flows. The minimum pipe velocity in a foul sewer should be >0.75 m s^{-1} while the maximum pipe velocity to prevent scouring or separation of solids from liquids is <3.5 m s^{-1} (White, 1986).

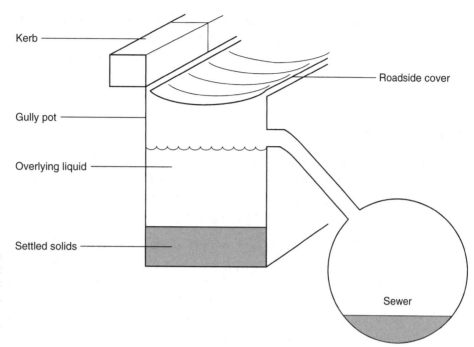

Kerb

Roadside cover

Gully pot

Overlying liquid

Settled solids

Sewer

FIGURE 13.4
A roadside gully pot designed to collect surface water and remove gross solids before allowing it to drain into the sewer.

13.3 SURFACE RUN-OFF

Rainfall washes any material deposited on paved areas, such as roads, into the nearest drain. Roads are drained by the familiar gully pot (Fig. 13.4), and in the storage compartment solid material settles and builds up. This material is a mixture of organic debris, such as leaves, twigs, animal faeces, and of course oil and other compounds from vehicles (Pope, 1980). During dry periods this settled material slowly decays under anaerobic conditions resulting in a highly polluting anaerobic liquid with a BOD often in excess of $7500 \, \mathrm{mg \, l^{-1}}$. When the next storm occurs this stored liquor is displaced from all the gully pots into the sewer resulting in an initial flush of highly contaminated water, which can often cause significant problems at the treatment plant, or severe local water pollution depending on the type of sewerage system employed. The degree of contamination from gully pots depends on:

(a) the intensity and duration of rainfall;
(b) the length of the preceding dry period;
(c) seasonal factors, such as temperature, leaf fall, use of road salt;
(d) the effectiveness in cleaning roadside gullies and pots.

Urban dog populations have been cited as a major source of organic loading to surface water drains. The dog population of Paris is currently (1996) in excess of five million, while the streets of New York are fouled by 68 000 kg of faeces and 405 000 l of urine from pet dogs daily. An average design figure for Northern Europe for dog faeces is

$17\,g\,m^{-2}$ of paved area per year. Dogs are carnivores so their faeces are protein rich resulting in high BOD and N concentrations. Their faeces are also associated with a wide range of pathogens that are transferable to humans (Chapter 12).

13.4 Charging

The analysis of sewage is normally limited to BOD, COD, suspended solids and ammonia. From these parameters the polluting strength of wastewaters can be assessed and charges for treatment levied. Other parameters may be required if the effluent is discharged to a sensitive receiving water or if the sludge is to be used for agriculture.

The most widely used charging system in Europe is the Mogden formulae. It calculates the charge for treatment and disposal of sludge by comparing the strength of the wastewater to normal domestic sewage and then calculating separate charges for collection via the sewerage network per m^3 (R), the primary treatment cost per m^3 (V), the secondary treatment cost per m^3 (B), and the cost of treatment and disposal of sludge per m^3 of wastewater (S). The cost in pence per m^3 of wastewater treated (C) is calculated as:

$$C = R + V + \left(\frac{O_t}{O_s}\right) B + \left(\frac{S_c}{S_s}\right) S \text{ pence m}^3 \qquad (13.4)$$

where O_t is the COD of the wastewater to be charged compared to O_s the average strength of domestic wastewater and S_c is the suspended solid concentration of the wastewater compared to S_s the average concentration in domestic wastewater. Average values for calculating trade charges by water companies in the UK during 1996/97 were R 15.31p, V 13.80p, O_s $542\,mg\,l^{-1}$, B 19.11p, S_s $351\,mg\,l^{-1}$ and S 12.50p. There are significant regional variations in charging formulae and charges (Water Services Association, 1997). Schemes for calculating charges for industrial wastewater treatment and also potable water charges in Europe have been reviewed by the European Commission (1996).

The person or population equivalent (pe) is the amount of BOD or other parameter produced per person per day. It is calculated as:

$$pe = \frac{\text{Average flow (1)} \times \text{Average BOD (mg l}^{-1})}{10^6} \text{ kg BOD day}^{-1} \qquad (13.5)$$

The pe for BOD in the UK varies from 0.045 for residential areas to 0.077 for a large industrial city, giving a mean value of $0.055\,kg\,BOD\,day^{-1}$. In the US, the pe values are 0.052 and 0.063 for domestic wastewater discharged to separate and combined sewers, respectively. Standard pe values in the US and Canada are $0.77\,kg\,BOD\,day^{-1}$

and 0.10 kg suspended solids per day. Equation (13.5) can be modified to express industrial effluents in terms of population equivalents:

$$pe = \frac{\text{Mean flow } (m^3 \text{ day}^{-1}) \times \text{Mean BOD } (mg \, l^{-1})}{0.055 \times 10^3} \quad (13.6)$$

13.5 INDUSTRIAL AND AGRICULTURAL WASTEWATERS

Industrial wastewaters vary considerably in their nature, toxicity and treatability (Table 13.2), and normally require pre-treatment before discharge to sewer (Section 14.3). Nemerow (1978) and Eckenfelder (1989) give excellent reviews of industrial wastewater characteristics and their treatment. Food processing and agricultural wastewaters are very different to industrial wastewaters, being readily degradable and largely free from toxicity (Table 13.8).

Animal wastes are a problem when stocking density exceeds the lands capacity to absorb the waste or where animals are intensively housed. There is very little dilution of animal wastes so they are extremely strong organically and difficult to treat (Table 13.9). In terms of organic production per animal it is convenient to express waste load per animal in comparison to humans. So where a human is 1, a cow is equivalent to 16.4 people, horse 11.2, chicken 0.014, sheep 2.45 and a pig 3.0. It is not possible to treat animal wastes using conventional wastewater treatment technology due to the vast dilution required (Weller and Willetts, 1977). In addition, with some animals that are fed exclusively with artificial feeds, then high concentrations of heavy metals and antibiotics can occur in the waste. For example, pig feed can contain up to 233 and 300 ppm dry weight of copper and zinc, respectively. Little is retained by the animal with 70–80% of the copper and 92–96% of the zinc excreted. The organic load

TABLE 13.8 Comparative strengths of wastewaters from food-processing industries

	BOD $(mg \, l^{-1})$	COD $(mg \, l^{-1})$	PV $(mg \, l^{-1})$	Suspended solids $(mg \, l^{-1})$	pH	Population equivalent per m^3 of waste
Brewery	850	17 000	–	90	4–6	14.2
Cannery						
Citrus	2000	–	–	7000	Acid	33.3
Pea	570	–	–	130	Acid	9.5
Dairy	600–1000	–	150–250	200–400	Acid	10.0–16.7
Distillery	7000	10 000	–	Low	–	116.7
Farm	1000–2000	–	500–1000	1500–3000	7.5–8.5	16.7–33.3
Silage	50 000	–	12 500	Low	Acid	833.3
Potato processing	2000	3500	–	2500	11–13	33.3
Poultry	500–800	600–1050	–	450–800	6.5–9.0	8.3–13.3
Slaughterhouse	1500–2500	–	200–400	800	7	25.0
Sugar beet	450–2000	600–3000	–	800–1500	7–8	7.5–33.3

TABLE 13.9 Average volume, strength and nutrient content of animal wastes

Animal	Volume of waste per adult animal $(m^3 \, day^{-1})$	COD $(mg \, l^{-1})$	BOD$_5$ $(mg \, l^{-1})$	N $(kg \, tonne^{-1})$	P $(kg \, tonne^{-1})$	K $(kg \, tonne^{-1})$	Moisture content (%)
Cow	0.0500	150 000	16 100	11.1	4.5	13.4	87
Pig	0.0045	70 000	30 000	8.9	4.5	4.5	85
Poultry	0.0001	170 000	24 000	38.0	31.3	15.6	32–75[a]

[a] Depending on housing.

from farm animals is three times that generated by the human population in the UK. There are serious problems associated with pathogen transfer between animals and also to humans due to poor disposal and handling of animal wastes. The increase in antibiotic-resistant bacterial pathogens has been linked to the widespread use of antibiotics in animal feeds.

The dairy industry is the largest agricultural sector in Europe. Wastewater originates from a variety of sources. Fluid milk is lost from reception and bottling areas as well as on the farm via milking parlours and storage areas. European farmers are also forced to dump millions of litres of milk towards the end of each year due to over production on their milk quotas. This rich full-cream milk is particularly polluting, with a BOD of up to $100\,000 \, mg \, l^{-1}$, and must be disposed of very carefully and should not be discharged to the sewer in large quantities. Processing plants produce a wide range of wastewater from the manufacture of butter, cheese, evaporated and condensed milk, milk powder, and other milk products. Wastes include milk, separated milk, buttermilk and whey. Washings from butter and cheese manufacture have a BOD of between 1500 and $3000 \, mg \, l^{-1}$. Whey is acidic while the rest is alkaline, although as lactose is rapidly fermented to lactic acid on storage most dairy wastes are acidic. Dairy wastes are rich in lactose (4.5%), protein (3.8%) and fats (3.6%).

Silage liquor is the waste product from ensiling grass. Under anaerobic conditions the cellulose in grass is broken down microbially to produce simple organic acids, which reduces the pH preserving the grass. The waste liquor produced is highly polluting with a pH of 4.5 due to the presence of lactic, acetic, propionic and butyric acids. The wastewater is readily degradable with a high BOD ($50\,000$–$60\,000 \, mg \, l^{-1}$) and nitrogen ($25\,000 \, mg \, N \, l^{-1}$) concentrations. The volume of liquor produced is linked to the moisture content of the grass when ensiled. For example, at dry matter contents of 10–15, 16–20 and >20%, between 360 and 450, 90–225 and <90 l of liquor are produced per tonne of silage, respectively. Silage clamps require leachate collection systems and the liquor which is highly corrosive to normal concrete is usually mixed with slurry and spread on land. It cannot be treated conventionally due to its toxicity to aerobic treatment systems. The introduction of big-bag silage making avoids large point sources of liquor which is dispersed back directly to the land and so has reduced the problem of water pollution.

13.6 WORKING WITH SEWAGE

Although there is a health risk associated with working with sewage it can be minimized by careful work practice. The commonest problems are associated with mild to moderate gastroenteritis, although more serious diseases such as leptospirosis (Weil's disease) and hepatitis have also been associated with sewage (Chapter 12). Asthma and alveolitis can also be caused from aerosols produced at sewage treatment plants due to proteins of living and dead bacteria causing allergic reactions (Gray, 2004).

In the UK, the Health and Safety at Work Act 1974 requires employers to do all that is reasonably possible to protect employees from the risks of illness and injury. More recent legislation including the Control of Substances Hazardous to Health Regulations (COSHH) 1994 and the Management of Health and Safety at Work Regulations 1993 requires employers to carry out risk assessments of the hazards associated with employment, and, in the case of working with sewage, this includes the risk of exposure to pathogenic organisms in sewage. The assessment is used to devise control measures to eliminate or minimize the risks, which includes adequate training and the provision of personal protective equipment. Similar legislation also exists throughout Europe and the US.

Before working with sewage or sewage-related or -contaminated material anyone involved should make sure that they understand the risks to health and the ways in which they can pick up infections. First, they should read the Water Services Association of England and Wales booklet *Guidance on the Health Hazards of Work Involving Exposure to Sewage* published in 1995. Second, they should seek medical advice to ensure that they are not especially at risk and obtain any necessary vaccinations. In Ireland, it is recommended that those working with sewage should be vaccinated for tetanus, polio, and Hepatitis A and B. In the UK, those in contact with sewage are required to carry an occupational health information card bearing the name, address and telephone number of the employers and details of the individual's work. This is particularly important where there is a risk of leptospirosis where the flu-like symptoms are easily overlooked but where early diagnosis is vital.

The British Health and Safety Executive suggest the following good work practices in order to minimize the risk of infection:

(a) use safe systems of work and wear the protective equipment that is provided;
(b) report damaged equipment and get it replaced;
(c) avoid becoming contaminated with sewage;
(d) avoid breathing in sewage-contaminated dust or spray (i.e. aerosols);
(e) do not touch your face or smoke, eat or drink, unless you have washed your hands and face thoroughly with soap and water;
(f) cleanse all exposed wounds, however small, and cover with a sterile waterproof dressing;
(g) change out of contaminated clothing before eating, drinking or smoking;

(h) if you suffer from a skin problem, seek medical advice before working with sewage;

(i) clean contaminated equipment on-site – do not take contaminated clothing home for washing; your employer should deal with this.

Remember that the greatest risk is from direct contamination via hand-to-mouth transfer. There is also a risk of infection through damaged skin, no matter how minor that damage may be (e.g. scratches, rashes, etc.). Avoid working downwind of aeration units where aerosols are formed or near air extraction outlets venting sludge-pressing areas. Wear suitable protective equipment and keep clean. Do not handle any solid material with your hands as this may contain discarded hypodermic needles that may result in puncture wounds. Similar precautions are required when handling collected material in the laboratory. Safety precautions required while taking samples from surface waters are discussed in Section 9.2.3.

Carbon dioxide, methane and hydrogen sulphide are all readily produced when sewage becomes septic. So do not work in confined spaces (e.g. partially emptied sedimentation tanks, deep excavations at sewage works sites) or in enclosed spaces (e.g. manholes or sewers) unless supervised and properly equipped. Always test the air in confined spaces using a hand-held gas detector which should be carried with you if making a general inspection of buildings. A worker exposed to a concentration of hydrogen sulphide of 5000 ppm (which is odourless at this concentration) will collapse immediately and will be dead within 60 s. There is also a high risk of explosion in these situations.

References

Clark, J. W., Viessman, W. and Hammer, M. J., 1977 *Water Supply and Pollution Control*, Harper and Row, New York.

Corbitt, R. A., 1990 *Standard Handbook of Environmental Engineering*, McGraw-Hill, New York.

Daughton, C. G. and Ternes, T. A., 1999 Pharmaceuticals and personal care products in the environment: agents of subtle change? *Environmental Health Perspectives*, **107** (suppl. 6), 907–38.

Debska, J., Kot-Wasik, A. and Namiesnik, J., 2004 Fate and analysis of pharmaceutical residues in the aquatic environment, *Critical Reviews in Analytical Chemistry*, **34**, 51–67.

Eckenfelder, W. W., 1989 *Industrial Water Pollution Control*, McGraw-Hill, New York.

Environment Agency, 2000 *Endocrine Disrupting Substances in the Environment: The Environment Agency's Strategy*, Environment Agency, Bristol.

European Commission, 1996 *Wastewater Charge Schemes in the European Union*, Office for Official Publications of the European Communities, Luxembourg.

Gray, N. F., 2004 *Biology of Waste Water Treatment*, 2nd edn, Imperial College Press, London.

Hunter, M., Motta Marques, M. L. da, Lester, J. and Perry, R., 1988 A review of the behaviour and utilization of polycarboxylic acids as detergent builders, *Environmental Technology Letters*, **9**, 1–22.

Hydraulic Research Laboratory, 1985 *Design and Analysis of Urban Storm Drainage: The Wallingford Procedures V1 and V2*, Hydraulics Research Laboratory, Wallingford.

McKay, G., 1996 Industrial pollutants. In: McKay, G., ed., *Use of Adsorbents for the Removal of Pollutants from Waste Waters*, CRC Press, Boca Raton, FL, pp 31–7.

Nemerow, N. L., 1978 *Industrial Water Pollution*, Addison-Wesley, Massachusetts.

Pope, W., 1980 Impact of man in catchments: (ii) Roads and urbanization. In: Gower, A. M., ed., *Water Quality in Catchment Ecosystems*, John Wiley and Sons, Chichester, pp 73–112.

Water Services Association, 1997 *Waterfacts '96*, Water Services Association, London.

Willer, J. B. and Willetts, S. L., 1977 *Farm Waste Management*, Crosby Lockwood Staples, London.

White, J. B., 1986 *Wastewater Engineering*, 3rd edn, Arnold, London.

FURTHER READING

Butler, D. and Davis, J., 1998 *Urban Drainage*, E. and F. N. Spon, London.

Day, D. L. and Funk, T. L., 1998 Processing manure: physical, chemical and biological treatment. In: Hatfield, J. L. and Stewart, B. A., eds, *Animal Waste Utilization,* Ann Arbor Press, Michigan, pp 243–82.

Gray, N. F., 2004 *Biology of Waste Water Treatment*, 2nd edn, Imperial College Press, London.

Metcalf and Eddy, 1991 *Wastewater Engineering*, 3rd edn, McGraw-Hill, New York.

Redman, M., Winder, L. and Merrington, G., 1998 *Agricultural Pollution*, E. and F. N. Spon, London.

West, P. A. and Locke, R., 1990 Occupational risks from infectious diseases in the water industry, *Journal of the Institution of Water and Environmental Management*, **4**, 520–3.

14.1 REQUIREMENTS OF TREATMENT

The aims of wastewater treatment are:

(a) to convert the waste materials present in wastewaters into stable oxidized end products that can be safely disposed of to inland waters without any adverse ecological effects;
(b) to protect public health;
(c) to ensure wastewater is effectively disposed of on a regular and reliable basis without nuisance or offence;
(d) to recycle and recover the valuable components of wastewater;
(e) to provide an economic method of disposal;
(f) to comply with legal standards and consent conditions placed on dischargers.

A wastewater treatment plant is a combination of separate treatment processes or units (Table 14.1), designed to produce an effluent of specified quality from a wastewater (influent) of known composition and flow rate. The treatment plant is also, usually, required to process the separated solids to a suitable condition for disposal. By a suitable combination of these unit processes it is possible to produce a specified final effluent quality from virtually any type of influent wastewater. While treated wastewaters can be used for groundwater recharge or even be recycled, they are generally discharged to surface waters, primarily rivers. The amount of treatment required depends largely on the water quality objectives of the receiving water and also the dilution capacity available.

Traditional design of treatment plants in the UK and Ireland is based on the Royal Commission standard effluent, which has a biochemical oxygen demand (BOD) of $20 \, \mathrm{mg \, l^{-1}}$ and a suspended solid concentration of $30 \, \mathrm{mg \, l^{-1}}$. This effluent quality causes no deterioration in receiving waters so long as there is at least an eight-fold dilution of clean water available. Most European plant design before 1990 was based on or near this final effluent standard. It is normal to quote final effluent quality as two numeric values, the BOD first and the suspended solid concentration second. So the

TABLE 14.1 Major unit processes in wastewater treatment

Process	Description
Physical unit processes	
Balancing	Where the flow of wastewater produced is variable over time, balancing tanks are used to ensure a constant flow and consistent quality of wastewater is pumped forward for treatment. This reduces both the capacity and cost of treatment
Screening	Screens remove large particles from wastewater. Used early in treatment to protect other treatment processes. Screens can be stationary, vibrating or rotating drums
Sedimentation	Special tanks to separate organic and inorganic solids from liquids
Flotation	Small air bubbles introduced at the base of a tank become attached to suspended particles and float. The particles are then skimmed off the surface as a sludge. Used extensively in dairy, paper, meat packing and paint industries
Hydrocyclone	Removal of dense particles (e.g. sand, grit and glass) from wastewaters is achieved as it enters a conical tank tangentially. As the wastewater spirals through the tank, particles are thrown against the wall by centrifugal forces and fall to the base (point) of the cone from where they can be removed
Filtration	Treated wastewater can be passed through a fine media filter (e.g. sand) in order to further reduce suspended solids concentration ($<20 \, mg \, l^{-1}$). High-performance filters using synthetic fibres to remove particles between 1 and 500 μm from treated wastewater or process streams
Centrifugation	Separation of solids from liquids by rapid rotation of the mixture in a special tapered vessel. Solids are deposited as a thick sludge (20–25% dry solids) either against the inner wall or at the base. Widely used in pharmaceutical, pulp, paper, chemical and food industries, and for dewatering sewage sludge
Reverse osmosis	Under pressure (1500–3000 kPa) water is driven through a semi-permeable membrane with extremely small pores to concentrate ions and other particles in solution and to purify the water. Used to remove and recover contaminants from process waters before discharge to sewer
Ultrafiltration	Similar to reverse osmosis. Particles of 0.005–0.1 μm are removed as they are forced through a micro-porous membrane at pressures up 3000 kPa. Used for removal and recycling of colloidal material including dyes, oils, paints and even proteins from cheese and whey from wastewaters. Able to remove the smallest micro-organisms including viruses and pyrogenic macromolecules
Micro-filtration	Similar to ultrafiltration except used to recover large particles (0.1–5 μm) at lower pressures (100–400 kPa). Widely used in food and drink industry. Micro-porous filters can also be used for the disinfection of process waters and effluents

(Continued)

TABLE 14.1 (*Continued*)

Process	Description
Adsorption	Activated carbon or synthetic resins are used to remove contaminants by adsorption from liquids. Used primarily for the removal of organics from industrial process and wastewaters
Chemical unit processes	
Neutralization	Non-neutral wastewaters are mixed either with an alkali (e.g. NaOH) or an acid (e.g. H_2SO_4) to bring the pH as close to neutral as possible to protect treatment processes. Widely used in chemical, pharmaceutical and tanning industries
Precipitation	Dissolved inorganic components can be removed by adding an acid or alkali, or by changing the temperature, by precipitation as a solid. The precipitate can be removed by sedimentation, flotation or other solids removal process
Ion-exchange	Removal of dissolved inorganic ions by exchange with another ion attached to a resin column. For example, Ca an Mg ions can replace Na ions in a resin thereby reducing the hardness of the water
Oxidation reduction	Inorganic and organic materials in industrial process waters can be made less toxic or less volatile by subtracting or adding electrons between reactants (e.g. aromatic hydrocarbons, cyanides, etc.)
Biological unit processes	
Activated sludge	Liquid wastewater is aerated to allow micro-organisms to utilize organic polluting matter (95% reduction). The microbial biomass and treated effluent are separated by sedimentation with a portion of the biomass (sludge) returned to the aeration tank to seed the incoming wastewater
Biological filtration	Wastewater is distributed over a bed of inert medium on which micro-organisms develop and utilize the organic matter present. Aeration occurs through natural ventilation and the solids are not returned to the filter
Stabilization ponds	Large lagoons where wastewater is stored for long periods to allow a wide range of micro-organisms to breakdown organic matter. Many different types and designs of ponds including aerated, non-aerated and anaerobic ponds. Some designs rely on algae to provide oxygen for bacterial breakdown of organic matter. Sludge is not returned
Anaerobic digestion	Used for high-strength organic effluents (e.g. pharmaceutical, food and drink industries). Wastewater is stored in a sealed tank which excludes oxygen. Anaerobic bacteria breakdown organic matter into methane, carbon dioxide and organic acids. Final effluent still requires further treatment as has a high BOD. Also used for the stabilization of sewage sludge at a concentration of 2–7% solids

TABLE 14.2 Guide to the required dilution for various levels of treatment employed prior to the implementation of the Urban Waste Water Treatment Directive

Ratio of receiving water to effluent	Required standard ($mg\,l^{-1}$)	Treatment level required
≤8	BOD 20 SS 30	At least secondary and possibly tertiary
150–300	BOD 100 SS 60	High-rate biological treatment
300–500	BOD 240 SS 150	Primary sedimentation only
500	Raw sewage	Screening only

Royal Commission standard is 20:30, with plants also constructed to higher or lower standards. Cost is a major factor and the stricter the final effluent standard the more expensive treatment becomes. More relaxed standards apply where there is a large dilution available or where the assimilative capacity of receiving waters is high. In coastal and estuarine waters there is a vast dilution so effluent standards can be relaxed further (Table 14.2).

This dilution-based approach to effluent disposal has been superseded by the introduction of the European Union (EU) Urban Waste Water Treatment Directive (91/271/EEC). Secondary treatment is now mandatory for all inland towns and for those discharging to estuaries with a population equivalent (pe) in excess of 2000, and for towns with a pe >10 000 discharging to coastal waters regardless of receiving water assimilative capacity (Section 8.3) (Fig. 14.1). New minimum treatment standards have been set for treatment plants, or alternatively minimum percentage reductions must be achieved within sensitive catchment areas (Table 14.3). Inland waters generally require tailor-made standards with particular regard to nutrient concentration to prevent eutrophication. Eutrophication is no longer a problem only of lakes as many rivers and coastal waters are showing significant enrichment with subsequent algal blooms (Section 8.2). Effluent discharge licences are now set for a wide range of chemical variables, according to receiving water usage and its existing assimilative capacity. Under the EU Urban Waste Water Treatment Directive, where receiving waters are subject to eutrophication (i.e. waters classified as sensitive), then minimum effluent standards for total nitrogen and phosphorus have been set (Table 14.3). The Directive should not be seen in isolation but rather as a part of the integrated river basin management plan required by the Water Framework Directive (Section 7.3). In the US secondary treatment is also mandatory for wastewater treatment with minimum standards of $<30\,mg\,l^{-1}$ BOD and $<30\,mg\,l^{-1}$ suspended solids as monthly averages.

14.2 BASIC UNIT PROCESSES

Wastewater treatment is a mixture of settlement and either biological or physico-chemical processes (Table 14.1). The action of treatment is one of separation of settleable, suspended and soluble substrate from the water by various sorption

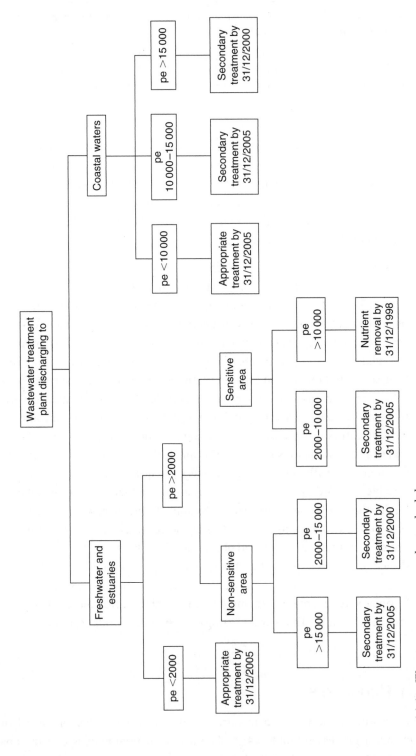

Figure 14.1 Wastewater treatment plant schedule.

TABLE 14.3   Requirements for discharges from wastewater treatment plants under the Urban Waste Water Treatment Directive (91/271/EEC) – values for total phosphorus and nitrogen only apply to discharges >10 000 pe discharging to sensitive waters (i.e. those subject to eutrophication)

Parameter	Minimum concentration	Minimum percentage reduction
BOD$_5$	25 mg O$_2$l^{-1}	70–90
Chemical oxygen demand	125 mg O$_2$l^{-1}	75
Suspended solids	35 mg l^{-1}	90
Total phosphorus	1 mg P l^{-1a}	80
	2 mg P l^{-1b}	80
Total nitrogen	10 mg N l^{-1a}	70–80
	15 mg N l^{-1b}	70–80

[a] 10 000–100 000 pe.
[b] >100 000 pe.

processes to form particles large enough to be removed from suspension by settlement. The concentrated particles form a sludge, which then must be processed separately and disposed of, which can in itself be a major problem. The exact nature of the sludge is dependent on the processes used as well as the nature of the influent wastewater. Successful treatment of wastewaters require a combination of unit processes, the selection of which is dependent upon the nature of the wastewater itself.

In terms of treatment plant design, unit processes are classified into five groups or functions:

1. *Preliminary treatment*. The removal and disintegration of gross solids, the removal of grit and the separation of storm water. Oil and grease are also removed at this stage if present in large amounts (Fig. 14.2). Industrial wastewaters may require pre-treatment before rendered suitable for discharge to sewer or further treatment (Section 14.3).
2. *Primary (sedimentation) treatment*. The first major stage of treatment following preliminary treatment, which usually involves the removal of settleable solids, which are separated as sludge.
3. *Secondary (biological) treatment*. The dissolved and colloidal organic matter is oxidized by micro-organisms.
4. *Tertiary treatment*. Further treatment of a biologically treated effluent to remove remaining BOD, suspended solids, bacteria, specific toxic compounds or nutrients to enable the final effluent to comply with a standard more stringent that can be achieved by secondary treatment alone.
5. *Sludge treatment*. The dewatering, stabilization and disposal of sludge.

Depending on the quality of the final effluent required, not all the stages will be required. For example, preliminary treatment is only generally given to those wastewaters permitted to be discharged to coastal waters via long sea outfalls, to

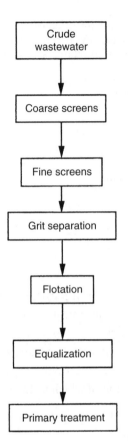

FIGURE **14.2** Summary of preliminary wastewater treatment processes.

ensure that floating debris and gross solids are not washed back onto the beach. Primary treatment is given where effluent is discharged to an estuary, with sludge often dumped at sea from special vessels. The acceptability of the practice of disposing of untreated or primary treated wastewater to coastal waters is becoming increasingly untenable internationally. Secondary treatment is given to all effluent discharged to inland waters. While tertiary treatment is often required if water is abstracted for public supply downstream of the discharge, where eutrophication is a problem, or discharge is to either a lake or groundwater. So the degree of treatment is dictated by the dilution factor. However, with the growing urbanization the maxim 'the solution to pollution is dilution' is no longer appropriate (Table 14.2) and all wastewaters should receive full treatment regardless of the dilution available in the receiving water. Within Europe the Urban Waste Water Treatment Directive reflects this new policy, and secondary and tertiary treatment are becoming increasingly important if the problem pathogens, that are becoming increasingly common in our recreational and supply waters, are to be intercepted (Chapter 12) and eutrophication reduced.

All wastewater treatment plants are based on the generalized layout in Fig. 14.3. The basic unit processes are outlined below. More detailed descriptions of biological and physico-chemical unit processes are given in following chapters.

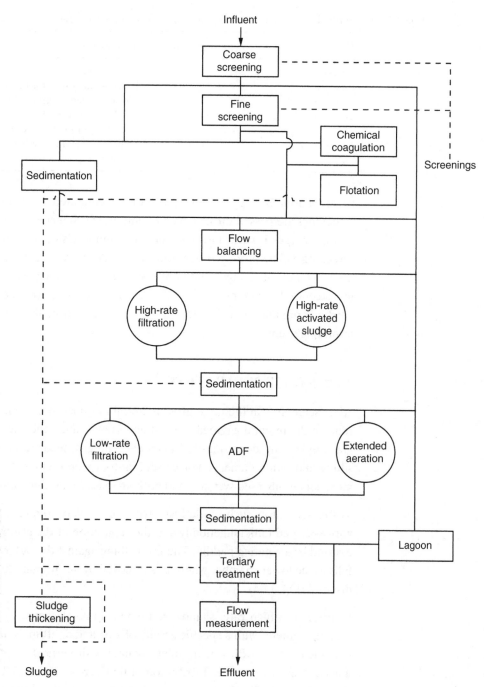

FIGURE 14.3 Summary of the main process options commonly employed at both domestic and industrial wastewater treatment plants. Not all these unit processes may be selected, but the order of their use remains the same.

14.2.1 SCREENS

Inclined parallel rows of steel bars physically remove larger solids that may cause blockage of pipework or damage to pumps. For most domestic plants these are spaced 20–30 mm apart. Floating solids, paper, plastic, wood, etc. are all removed. The removed solids, known as screenings or rakings, are very unhygenic due to faecal matter and

TABLE 14.4 Typical size range of screens used in wastewater treatment

Category	Spacing (mm)	Application
Coarse screens	>10	Removal of large material
Fine screens	1.5–10	Used as substitute for primary sedimentation for extended aeration plants
Very fine screens	0.2–1.5	Used as substitute for primary sedimentation. Used in conjunction with series of larger screens
Micro-screens	0.001–0.3	Effluent polishing. Treatment of inert quarry washings

so are normally bagged and then either incinerated or landfilled. Alternatively, screenings are passed through a comminuter, which macerates it into small particles and discharges it back into the influent channel. Using a 25-mm screen $0.06\,m^3$ of screening will be produced per hour per 10 000 pe treated. These coarse bar screens are used most at sewage treatment plants. However, finer screens made from stainless steel wedge wire mesh are widely used in industrial treatment for removing fine inert and organic solids (Table 14.4). The mesh is normally employed in the form of a rotating drum.

14.2.2 GRIT SEPARATION

After screening all the large gross material has been removed. The next stage is to remove the mineral aggregate (sand and gravel) and other dense material, such as glass, metal and dense plastic fragments. The grit needs to be removed to prevent silting and pump damage. Vortex separators (hydrocyclones) and detritors are the most commonly used systems, and both are based on sedimentation.

Vortex separators such as the Pista grit trap are very compact. Wastewater enters a vortex-shaped tank tangentially and the separation of the grit by centrifugal force is assisted by a rotating paddle. The grit is flung against the outer wall of the tank and falls to the lower chamber where it is removed by air lift pump to a drainage area for disposal (Fig. 14.4).

Detritors are high rate, shallow settlement tanks designed to remove particles of 0.2 mm diameter with a specific gravity of 2.65 and a settling velocity of $0.02\,m\,s^{-1}$. It is a square tank with a wide inlet channel with vertical pivoted deflector baffles angled to ensure an even distribution of flow across the tank. Retention time within the tank is only 30–60 s with the settled grit removed continuously to a peripheral sump by a rotary scraping mechanism (Fig. 14.5). The velocity of the wastewater is maintained at $0.3\,m\,s^{-1}$ to ensure only the dense particles are removed and that all the organic solids remain in suspension. The material removed by the detritor is washed in the classifier (cleansing channel), which is an inclined ramp up which a reciprocating rake pushes the grit against a counter flow of effluent. The grit collects in a disposal unit from where it is landfilled.

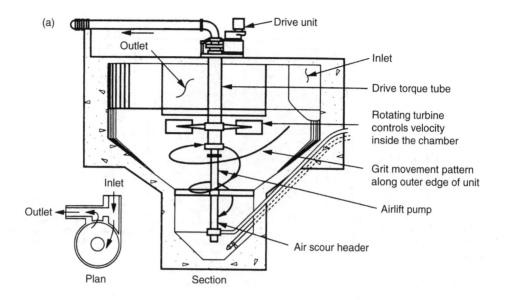

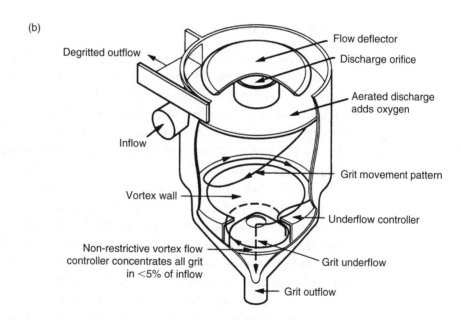

FIGURE 14.4
Grit separation using vortex-type systems: (a) a Pista grit separator manufactured by Smith and Loveless Ltd. and (b) a teacup unit manufactured by Eutek Ltd. (Reproduced from Metcalf and Eddy (1991) with permission of McGraw-Hill, New York.)

14.2.3 OTHER PRELIMINARY PROCESSES

At normal domestic and municipal wastewater treatment plants screening and grit separation is all that is normally necessary. However, if there are excessive amounts of fats, oils or grease in the influent then flotation may be required (Section 20.4) after grit separation. Equalization may subsequently be necessary where there is excessive diurnal variation in either flow rate or raw water quality to the treatment plant (Section 20.1). Removal efficiencies are not applicable to preliminary treatment processes and the strength of raw (crude) wastewater is measured after this level of treatment but before primary sedimentation.

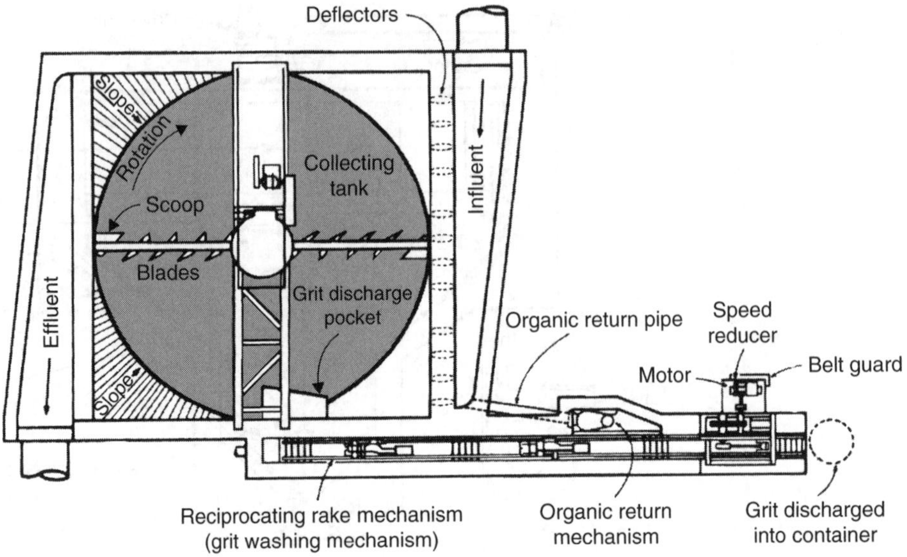

FIGURE 14.5
Outline design of a detritor or square horizontal-flow grit chamber. (Reproduced from Metcalf and Eddy (1991) with permission of McGraw-Hill, New York.)

14.2.4 STORM WATER

The design of treatment plants is based on dry weather flow (DWF). The DWF is the 'average daily flow to the treatment plant during seven consecutive days without rain, following 7 days during which the rainfall did not exceed 0.25 mm in any one day'. When it rains the surface water generally finds its way to the wastewater treatment plant via the sewer. Apart from diluting the wastewater, there are physical constraints at the plant on the volume of wastewater that can be treated. Most treatment plants are designed to fully treat 3DWF, with any excess stored in extra settlement tanks known as storm water tanks. The maximum flow to the treatment plant will be twice this with the excess flow (>6DWF), now highly diluted, discharged directly to the receiving water. So the storm water tanks act as a buffer between the treatment plant and the receiving water. It is normally the first few hours after a storm that the wastewater is at its strongest and, as this is retained in the storm water tanks, it is important that this stored water is treated before final discharge. The excess wastewater is diverted to the storm water tanks using a simple overflow weir after it has passed through the screens and grit separation. The stored wastewater undergoes settlement in the storm water tanks before it is returned to the inlet for full treatment.

The maximum flow of wastewater to the treatment plant can be more accurately estimated as:

$$(PG + I + E) + 1.36P + 2E \, \text{m}^3 \text{day}^{-1} \tag{14.1}$$

where P is the population served; G, the average per capita wastewater production ($\text{m}^3 \text{ca}^{-1} \text{day}^{-1}$) which is normally taken as being between 0.12 and 0.18 $\text{m}^3 \text{ca}^{-1} \text{day}^{-1}$; I, the infiltration rate ($\text{m}^3 \text{day}^{-1}$) and E, the industrial effluent ($\text{m}^3 \text{day}^{-1}$). The DWF here is represented by $(PG + I + E)$ so that the maximum flow to receive full secondary treatment is $(3PG + I + 3E)$ where the infiltration component remains constant (White, 1986).

14.2.5 PRIMARY SEDIMENTATION (PRIMARY TREATMENT)

All the wastewater is passed through a specially constructed sedimentation tank at such a velocity that the fine solids settle out of suspension by gravity. The settleable solids form a sludge in the base of the tank which can be removed as primary sludge. The hydraulic retention time (HRT) is usually a minimum of 2 h at 3DWF. Primary treatment can reduce the BOD by 30–40% and the suspended solids concentration by 40–70%, while up to 50% of faecal coliforms are also removed at this stage. Removal of suspended solids is by a number of physical processes that occur simultaneously within the tank, such as flocculation, adsorption and sedimentation (Casey, 1997). The overflow from the tank is called settled sewage and contains fine suspended (colloidal) and dissolved material only. In order to facilitate servicing, treatment plants require more than one sedimentation tank.

Primary sedimentation tanks clarify sewage by flotation as well as settlement. Any suspended matter which is less dense than water, such as oil, grease or fat will rise to the surface and form a scum. Skimmer blades attached to the scrapper bridge push the scum across the surface of the tank to where it is discharged into a special sump or a mechanically operated skimming device which decants the scum from the surface. While primary sedimentation tanks are designed to remove both floating and settleable material, special flotation tanks may be required before the primary settlement stage if large amounts of floatable material are present in the wastewater (Section 20.4). Removal efficiency depends on the nature of the wastewater and in particular the proportion of soluble organic material present, which can severely affect BOD removal. Some industrial wastes (e.g. food-processing wastewaters) may contain higher proportions of settleable organic matter resulting in BOD removals at the primary sedimentation stage in excess of 60–70%. A retention time of 2–3 h at maximum flow rate for primary sedimentation is adequate to allow almost all settleable solids to settle to the bottom of the tank, but is not long enough for septic conditions to result. However, at flows below DWF, septicity of the sludge can occur due to the extended retention time.

Overflow rates vary between 16 and $32\,m^3m^{-2}day^{-1}$, with $24\,m^3m^{-2}day^{-1}$ the commonly used value for design. The accumulated sludge density is influenced by the overflow rate forming a dense sludge at loadings of $<24\,m^3m^{-2}day^{-1}$ with the scrapers enhancing consolidation, or a dilute sludge at $>32\,m^3m^{-2}day^{-1}$ as the extra hydraulic movement inhibits consolidation (Section 20.3). At higher overflow rates the turbidity of the effluent will increase, especially during periods of maximum flow. Turbid effluents from primary tanks are also caused by excessive weir loadings, which should not exceed $120\,m^3m^{-1}day^{-1}$ for plants receiving $<4000\,m^3day^{-1}$ or

$180 \, m^3 \, m^{-1} \, day^{-1}$ at flows of $>4000 \, m^3 \, day^{-1}$ if solids are not to be carried over the final effluent weir.

There are three main types of sedimentation tanks: upward flow, horizontal flow and radial flow. Upward-flow settlement occurs as sewage moves upwards through the tank causing flocculation to increase the size of the particles (Fig. 14.6). Owing to the nature of these tanks they do not have a scraper, with the sludge which collects at the base being removed by hydrostatic pressure. Such systems are used at small treatment plants and are either conical (45° slope) or pyramidal (60° slope) in cross section with an upward velocity $>1.5 \, m \, h^{-1}$. As these units need to be deep, they are difficult to construct. Horizontal-flow settlement occurs as wastewater moves along a channel with a length to width ratio of 3–4:1 (Fig. 14.7). The floor is sloped (1:100) towards a sump where sludge is collected by a travelling bridge scraper that moves along the entire length of the tank. Such tanks have a minimum depth of 2 m, a surface loading of $30 \, m^3 \, m^{-2} \, day^{-1}$ and a weir loading of $300 \, m^3 \, m^{-1} \, day^{-1}$. The inlet and outlet weirs are usually the full width of the channel to ensure as little turbulence

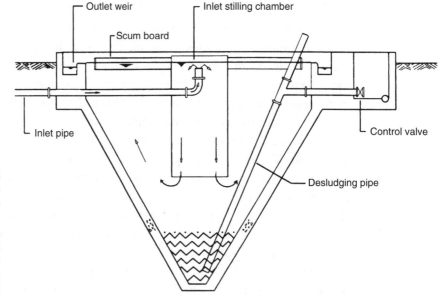

FIGURE 14.6 Upward-flow primary sedimentation tank. (Reproduced from Casey (1997) with permission of John Wiley and Sons Ltd, Chicester.)

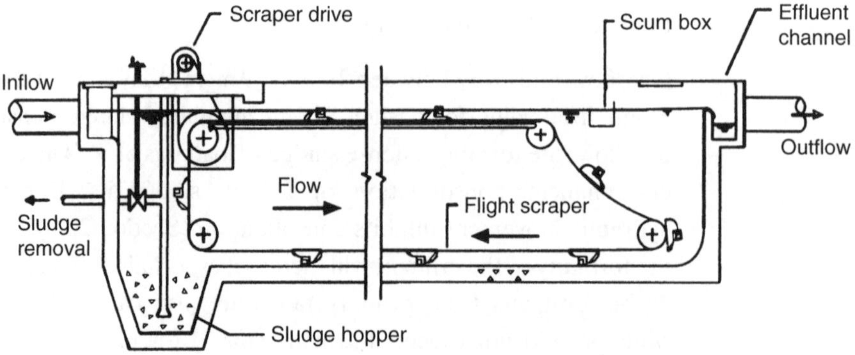

FIGURE 14.7 Horizontal-flow rectangular primary sedimentation tank. (Reproduced from Casey (1997) with permission of John Wiley and Sons Ltd, Chichester.)

within the tank as possible. Such units are very compact often constructed sharing side walls, but are found at large plants only. Radial-flow settlement tanks are circular (5–50 m in diameter) with a central feed for the wastewater. The floor slopes into a central sump where the sludge is gathered by a rotating scraper moving at 1.5–2.5 m min^{-1}. Settlement occurs as wastewater moves up and out, with an optimum upward velocity >1.5 m s^{-1}, minimum side wall depth of 2 m and a surface loading of 30 m^{3}m^{-2}day^{-1} (Fig. 14.8). The overflow weir runs around the periphery of the tank giving low weir loadings of <220 m^{3}m^{-1}day^{-1}. In comparison secondary sedimentation tanks can have a lower retention time (1.5 h) and higher surface loading rates (SLR) (40 m^{3}m^{-2}day^{-1}) than primary sedimentation tanks.

The surface overflow rate (SOR) required for a specific primary total suspended solids (TSSs) removal (between 30% and 70% removal) can be estimated using the following equation:

$$SOR\ (m^3m^{-2}day^{-1}) = 185 - (2.5 \times TSS\ removal\ (\%)) \tag{14.2}$$

Similarly the SOR required for a specific primary standard five-day BOD test (BOD$_5$) removal (between 20% and 40% removal) can be estimated using the Equation (14.3):

$$SOR\ (m^3m^{-2}day^{-1}) = 182 - (4.6 \times BOD_5\ removal\ (\%)) \tag{14.3}$$

There is little difference between the efficiency of horizontal- or radial-flow units for primary settlement and selection is normally based on constructional costs or plant configuration. For example, a circular tank of the same volume and surface area as a rectangular tank usually costs less to build. However, if a group of tanks are required, then rectangular tanks are normally used with common dividing walls which are not only cheaper to construct than individual circular tanks, but also take up considerably less space. Except at very small rural plants, there is always more than one primary

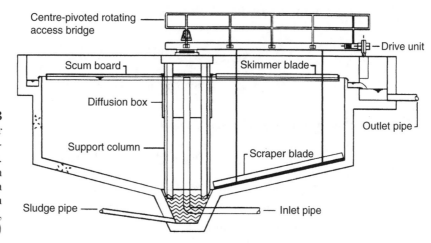

FIGURE 14.8 Radial-flow circular primary sedimentation tank. (Reproduced from Casey (1997) with permission of John Wiley and Sons Ltd, Chichester.)

sedimentation tank to allow for maintenance, with the flow divided equally between the tanks.

Basic design parameters for primary sedimentation tanks can be calculated as shown below.

EXAMPLE **14.1** Design a circular primary sedimentation tank to achieve a 60% reduction of TSSs for a wastewater flow rate (Q) of $6000 \, \mathrm{m^3 \, day^{-1}}$ and with a peak-flow factor of 2.5. Equation (14.2) is used to estimate the SOR which is $35 \, \mathrm{m^3 \, m^{-2} \, day^{-1}}$ which also gives a 32% BOD reduction (Equation (14.3)).

(a) Calculate surface area (A) of tank required:

$$A = \frac{Q}{\mathrm{SOR}} \tag{14.4}$$

$$\frac{6000}{35} = 171.4 \, \mathrm{m^2}$$

(b) Calculate diameter (d) of circular tank using $A = \pi r^2$:

$$d = 14.8 \, \mathrm{m}$$

(c) Calculate tank volume (V) with a side wall depth (d') of 3 m:

$$V = Ad' = 171.4 \times 3 = 514 \, \mathrm{m^3}$$

(d) Calculate HRT in hours:

$$\mathrm{HRT} = \frac{V}{Q} = \frac{514 \times 24}{6000} = 2.06 \, \mathrm{h}$$

(e) Calculate peak-flow SOR ($\mathrm{m^3 \, m^{-2} \, day^{-1}}$):

$$\mathrm{SOR} = \frac{\mathrm{peak \ flow \ factor} \times Q}{A} = \frac{2.5 \times 6000}{171.4}$$
$$= 87.5 \, \mathrm{m^3 \, m^{-2} \, day^{-1}}$$

Primary sludge production is calculated as:

$$W_{\mathrm{ds}} = Q \times \mathrm{TSS} \times F \, \mathrm{kg \, day^{-1}} \tag{14.5}$$

where W_{ds} is the quantity of sludge produced as dry solids ($\mathrm{kg \, day^{-1}}$); Q, the through flow rate ($\mathrm{m^3 \, day^{-1}}$); TSS, the total suspended solids concentration ($\mathrm{mg \, l^{-1}}$) and F, the fraction of solids removed.

EXAMPLE **14.2** A wastewater treatment plant serving a pe of 30 000, a per capita flow rate of 230 l day^{-1}, an influent TSS concentration of 350 mg l^{-1} and with a TSS removal efficiency of 60%. Calculate the flow to the plant, the quantity and volume of sludge produced each day.

(a) Flow to plant:

$$Q = \text{pe} \times \text{per capita flow} = 30\,000 \left(\frac{230}{1000} \right)$$

$$= 6900 \, \text{m}^3 \, \text{day}^{-1}$$

(b) Sludge quantity as dry solids (W_{ds}) (kg day^{-1}) (Equation (14.5)):

$$W_{ds} = Q \times \text{TSS} \times F = 6900 \times 350 \times 0.6 \times \left(\frac{10^3}{10^6} \right)$$

$$= 1449 \, \text{kg day}^{-1}$$

Alternatively W_{ds} can be expressed per unit volume of wastewater treated (kg m^3 day^{-1}):

$$\frac{W_{ds}}{Q} = \frac{1449}{6900} = 0.21 \, \text{kg m}^3 \, \text{day}^{-1}$$

(c) The volume of sludge dry solids produced (V_{ds}) per day is:

$$V_{ds} = \frac{W_{ds}}{p_s} = \frac{1449}{1000} = 1.45 \, \text{m}^3 \, \text{day}^{-1}$$

where p_s is the density of liquid sludge which can be estimated to be the same as water at 1000 kg m^3.

The particle density of sludge is approximately 1700 kg m^{-3} so the exact density of sludge depends on the solid content and is calculated by mass balance. For example, a sludge with a 5% solids content has a density calculated as:

$$p_s v_s = p_w v_w + p_p v_p \tag{14.6}$$

where ($p_s v_s$) is the density and volume of liquid sludge, ($p_w v_w$) is the density and volume of water, and ($p_p v_p$) is the density and volume of sludge particles present:

$$p_s \times \frac{100}{100} = \left(1000 \times \frac{95}{100} \right) + \left(1700 \times \frac{5}{100} \right) \text{kg m}^3$$

$$p_s = 950 + 85 = 1035 \, \text{kg m}^3$$

The volume of liquid sludge produced depends primarily on its water content, although the nature of the solids also has a minor influence. The specific gravity of the dry solids in sludge varies depending on the relative portion of organic (volatile) and mineral (fixed) material present. This is calculated by using Equation (14.7):

$$\frac{W_{ds}}{S_{ds}\,p_w} = \frac{W_f}{S_f\,p_w} = \frac{W_r}{S_v\,p_w} \tag{14.7}$$

where W_{ds} is the weight of dry solids produced, S_{ds} is the specific gravity of solids, p_w the density of water, W_f is the weight of mineral solids, S_f is the specific gravity of the mineral solids, W_v is the weight of organic solids and S_v is the specific gravity of the organic solids.

EXAMPLE 14.3 If a sludge containing 95% water is comprised of solids of which 30% are mineral with a specific gravity of 2.5 while the remainder are organic with a specific gravity of 1.0, then the specific gravity of all the solids is calculated as:

$$\frac{1}{S_{ds}} = \frac{0.30}{2.5} + \frac{0.70}{1} = 0.82$$

$$S_{ds} = \frac{1}{0.82} = 1.22$$

The specific gravity of water is 1.0, so with a 95% water content the specific gravity of the liquid sludge (S_{sl}) is:

$$\frac{1}{S_{sl}} = \frac{0.05}{1.22} + \frac{0.95}{1.0} = 0.99$$

$$S_{sl} = \frac{1}{0.99} = 1.01$$

The volume of liquid sludge (V) produced can be calculated accurately once the specific gravity of the sludge is known using Equation (14.8):

$$V = \frac{W_{ds}}{p_w S_{sl} P_s} \tag{14.8}$$

where P_s is the percentage solids expressed as a decimal.

EXAMPLE 14.4 Using the examples above where the weight of primary sludge produced each day is 1449 kg day^{-1}, the specific gravity of the sludge is 1.01 and the percentage of dry solids is 5%, then the daily volume of primary sludge (V) is:

$$V = \frac{1449}{1000} \times 1.01 \times 0.05 = 28.7 \text{ m}^3 \text{ day}^{-1}$$

Approximations of sludge volume can be made using per capita sludge production rates given in Table 21.1.

14.2.6 BIOLOGICAL (SECONDARY) TREATMENT

After primary treatment the settled wastewater enters a specially designed reactor where it is mixed or exposed to a dense microbial population under aerobic conditions. Micro-organisms, mainly bacteria, convert soluble and colloidal organic matter into new cells (Chapter 15). All that is required is to separate the dense microbial biomass from the purified wastewater. This is done by secondary settlement tanks. The sludge from secondary settlement is very different from primary sludge, being composed mainly of biological cells rather than gross faecal solids (Section 21.1). There are many different biological unit processes (Table 14.1) including fixed film reactors (Chapter 16), activated sludge systems (Chapter 17) and stabilization ponds (Chapter 18).

14.2.7 SECONDARY SEDIMENTATION

Secondary sedimentation tanks separate the microbial biomass produced in the biological treatment unit from the clarified effluent.

The biological growth which is flushed from percolating filters is known as humus and comprises of well-oxidized particles, mainly in the form of dense microbial film, living invertebrates and invertebrate debris, which all readily settle. The volume of sludge produced is much less compared with the activated sludge process. The sludge will form a thin layer at the base of the sedimentation tank, rarely exceeding 0.3 m in depth, if the tank is regularly desludged twice a day. The SOR can be up to 32 m^{3}m^{-2}day^{-1} with a similar weir depth and loading as a primary sedimentation tank.

Settlement of activated sludge is more variable as the flocs are light and more buoyant than solids from percolating filters, which reduces their settling velocities. This is made worse by denitrification at the centre of the tiny sludge flocs which results in small bubbles being produced which buoy up the clusters of particles. This results in a thick sludge blanket that can often take up half of the tank volume at peak flows. Therefore a rapid and continuous method of sludge removal is necessary. This is achieved by not normally having a central sump as in a primary sedimentation tank but a floor slope of 30° to the horizontal, which allows the sludge to be quickly and continuously withdrawn. Another common method of removal is to have numerous uptake tubes attached to the scrapper which forces the sludge up and out by the hydrostatic pressure. Rapid withdraw is achieved across the entire bottom of the tank which ensures that the retention time of solids settling at the periphery of the tank is no greater than at the centre, reducing the age of the floc and the chance of gas production resulting in reduced settleability. Sludge removed by this method encourages a vertical, rather than horizontal, movement in the tank that encourages flocculation and increases sludge density (Section 20.3). Other problems causing reduced settleability of activated sludge effluents are discussed in Section 17.7.

The volume of effluent from a primary sedimentation tank can be considered as being equivalent to the influent, as the volume of sludge withdrawn from the tank is negligible compared with the total volume of wastewater. However, in a secondary settlement tank receiving activated sludge the influent is equivalent to the effluent plus the returned sludge. In this case, the overflow of effluent is used for design purposes and not the influent, which includes recirculation. Secondary and primary sedimentation tanks are similar in design except that special attention is paid to the large volume of flocculant particles in the liquid. Secondary sedimentation tanks are between 3 and 5 m in depth. The capacity of tanks receiving activated sludge effluents is usually based on an HRT of 1.5–2.0 h at maximum flow rate. The maximum rate of flow will be 3DWF plus 1DWF for returned sludge, which will give a DWF retention time of 6–8 h, the same capacity as a primary sedimentation tank. With activated sludge it is critical that sufficient sludge thickening occurs in order to maintain sufficient biomass (mixed liquor suspended solid (MLSS) in $mg l^{-1}$) (Section 17.2.1) in the aeration tank. The sludge-loading rate controls sludge thickening in secondary sedimentation tanks. SOR (Equation (14.3)) should be between 20 and $40 m^3 m^{-2} day^{-1}$. While SLR should be between 50 and $300 kg m^{-2} day^{-1}$:

$$SLR = SOR \times \left(\frac{MLSS}{1000} \right) \times \left(1 + \frac{RAS}{Q} \right) \tag{14.9}$$

where RAS is the returned activated sludge ($m^3 day^{-1}$) and Q the influent flow rate ($m^3 day^{-1}$). Settleability of activated sludge in secondary settlement tanks is measured using the sludge volume index (SVI) (Section 17.2.3), which is linked to the SLR. The higher the SVI the lower the SLR.

14.2.8 TERTIARY TREATMENT

This is used when the standard 20:30 effluent is not sufficient. Main methods are: prolonged settlement in lagoons, irrigation onto grassland or via percolation areas (land treatment), straining through a fine mesh, or filtration through sand or gravel.

Disinfection is used where pathogens are a problem, for example, where a discharge is close to a bathing area. Chlorination of effluents is widely used in the US, while in Europe ozonation or ultraviolet (UV) treatment is used in preference as they cause less environmental damage. Chlorine will oxidize any organic matter remaining in wastewater before it acts as a disinfectant. So residual BOD may have to be removed prior to chlorination. Methods for disinfection are similar to those used for potable water (Section 10.1) but the chlorine dose required is usually 10–15 times more than for potable waters (White, 1992). Non-chemical disinfection using membrane filtration is also used. Nutrient removal is normally incorporated as part of the biological unit process (Section 18.4), but can be added as a tertiary or advanced treatment process, usually in the form of chemical treatment (Section 20.5).

14.2.9 SLUDGE TREATMENT

Sludge acts as a liquid when it contains >90% water, below this it becomes increasingly plastic. Primary sludge is 95–96% water while secondary sludges vary according to the process used. For example, waste activated sludge may be as high as 98.5% water. Therefore there is a need to treat the sludge further to reduce the water content and so reduce the volume of sludge that must be ultimately disposed. Treatment of sludge follows a general sequence:

(a) water content reduced by some form of thickening;
(b) then it is dewatered using chemical addition;
(c) final disposal.

Between stages (a) and (b) sludges may be stabilized to control anaerobic breakdown occurring on storage. This is achieved by anaerobic digestion, aerobic digestion or lime addition (pH >11). Final disposal routes include land spreading (sacrificial land, land reclamation, agricultural use or forestry), landfill, sea disposal (from pipeline or ship but now illegal within the EU), or incineration. This is considered in detail in Chapter 21.

Depending on the wastewater to be treated specialist unit processes may be required to deal with particular problems, such as heavy solids (coarse and fine screening), high grease content (flotation) (Section 20.4) and high organic load (high-rate secondary treatment) (Table 14.1).

14.3 PRE-TREATMENT OF INDUSTRIAL WASTEWATERS

Pre-treatment of industrial wastewaters is often necessary to prevent damage to sewers or the treatment processes employed, as well as to protect the health and safety of staff. This is done at source and is achieved mainly by flow balancing, neutralization, and fat and/or oil separation, although specific recalcitrant, toxic or dangerous pollutants may be recovered and either treated on site or sent off for specialized treatment, normally incineration. Some further treatment options are shown in Fig. 14.9. Waste minimization, spillage management and strict environmental management, coupled with increasing on-site wastewater treatment, have significantly reduced the problems at municipal treatment plants receiving industrial wastewaters. Increasing use of environmental management systems (EMSs) is resulting in industry moving away from end-of-pipe treatment to preventative strategies involving recycling, reuse and product substitution. While the concept of zero discharge is widely embraced, in practice few industries have been successful in eliminating the production of all wastewater requiring disposal after treatment (Chapter 23).

The treatment of wastewater is largely dependent on biological processes, which are susceptible to toxic compounds. Table 14.5 gives typical inhibition threshold concentrations for municipal activated sludge systems. However, where such compounds are continuously discharged to a particular treatment plant then the micro-organisms gradually become acclimatized to significantly higher threshold concentrations. It is important that all wastewaters are screened before treatment or disposal to sewer to ensure that they are biodegradable and amenable to the treatment provided (Section 15.3).

While toxicity is the major factor associated with the treatment of industrial wastewaters, excessive temperatures or variable hydraulic discharge rates are also important. Excessive amounts of oil or fats, acidic or alkaline wastes, suspended solids, explosive or flammable materials, volatile and odorous compounds, and corrosive gases all need to be controlled on-site before further treatment is considered. Examples of pre-treatment options for organic-based industrial wastewaters prior to discharge are given in Table 14.6. The main physico-chemical unit processes are explored in Chapter 20 while the biological unit processes are discussed in Chapters 15–19.

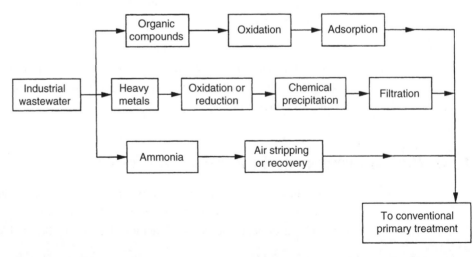

FIGURE 14.9
Main industrial
pre-treatment
options.

TABLE 14.5   Typical inhibition threshold concentrations of common ions and organic compounds for municipal activated sludge

Substance	Inhibition threshold for activated sludge $(mg\,l^{-1})$
Cadmium	1–10
Total chromium	1–100
Chromium III	10–50
Chromium VI	1
Copper	1
Lead	0.1–5.0, 10–100
Nickel	1.0–2.5
Zinc	0.3–5.0
Arsenic	0.1
Mercury	0.1–1, 2.5 as Hg(II)
Silver	0.25–5.0
Cyanide	0.1–5
Ammonia	480
Iodine	10
Sulphide	10
Anthracene	500
Benzene	100–500, 125–500
2-chlorophenol	5
Ethylbenzene	200
Pentochlorophenol	0.95, 50
Phenol	50–200, 200
Toluene	200
Surfactants	100–500

Reproduced from Water Environment Federation (1994) with permission of the Water Environment Federation, Alexandria, Virginia.

TABLE 14.6   Pre-treatment steps for typical industrial wastewaters

Industry	Typical pre-treatment technologies
Food-processing dairies	Equalization, biological treatment, removal of whey
Meat, poultry, and fish	Screening, gravity separation, flotation, coagulation/precipitation, biological treatment
Fruit and vegetable	Screening, equalization, gravity separation, neutralization, biological treatment, coagulation/precipitation
Breweries and distilleries	Screening, centrifugation, biological treatment
Pharmaceuticals	Equalization, neutralization, coagulation, solvent extraction, gravity separation, biological treatment
Organic chemicals	Gravity separation, flotation, equalization, neutralization, coagulation, oxidation, biological treatment, adsorption
Plastics and resins	Gravity separation, flotation, equalization, chemical oxidation, solvent extraction, adsorption, biological treatment
Leather tanning and finishing	Screening, gravity separation, flotation, coagulation, neutralization, biological treatment

Reproduced from Water Environment Federation (1994) with permission of the Water Environment Federation, Alexandria, Virginia.

14.4 Design of Wastewater Treatment Plants

14.4.1 SELECTION OF PROCESSES

The selection of specific unit processes depends not only on the nature of the waste-water, including degradability and treatability by selected processes, but also discharge requirements. Other important factors are environmental impact, land availability, projected life of plant design and cost. All the relevant factors in process selection should be considered, although their relative importance will vary due to social, environmental, political differences as well as technological availability and expertise. Considerations and selection procedure will vary according to whether the treatment plant serves an urban or rural catchment, or in a developed or non-developed country (Metcalf and Eddy, 1991). Suggested guidelines in the form of algorithms to aid process selection for physico-chemical, biological and solid separation processes are given in Figs 14.10–14.12. These are based on simple yes/no responses and so should be used in conjunction with more detailed accounts of the application and suitability of unit processes, such as Casey (1997) or Metcalf and Eddy (1991). There may also be special factors that fall outside these algorithms that will require tertiary unit processes to be installed in the design, which is why it is so important that all factors are considered at the start of the selection process. For example, discharges at coastal sites close to bathing beaches may require disinfection in order to comply with the EU Bathing Water Directive (Section 8.2).

Physico-chemical processes are used where the wastewater is not biodegradable and is usually employed by industries on-site prior to discharge to sewer (Fig. 14.10). The problem of toxic compounds in organic wastewaters is complex to solve. In this case biodegradability should be tested by carrying out treatability tests for both aerobic and anaerobic treatment. Toxic materials should be traced back to their source and removed prior to contamination by organic matter or other process streams. Figure 14.10 allows a very high degree of treatment to be achieved, therefore process selection is controlled by effluent quality objectives as the more complex and extensive the treatment the more expensive it becomes.

Solid separation is the most widely used approach to wastewater treatment and all treatment plants incorporate some solid separation processes. Process selection depends on particle size and density, suspended solid concentration, presence of oils or fats, and finally the organic nature of the wastewater and colloidal particles (Fig. 14.11).

Selection of biological treatment processes depends on the organic strength of the wastewater, its relative biodegradability and the location of the treatment plant (Fig. 14.12). Highly degradable wastewaters, such as those from sugar beet processing, confectionery manufacture, distilling and brewing are not successfully treated by conventional trickling filter or completely mixed activated sludge systems. Nutrient removal is required at plants discharging to sensitive receiving water, while compact systems may be required if land is at a premium (e.g. at coastal locations or high population density areas, such as Japan).

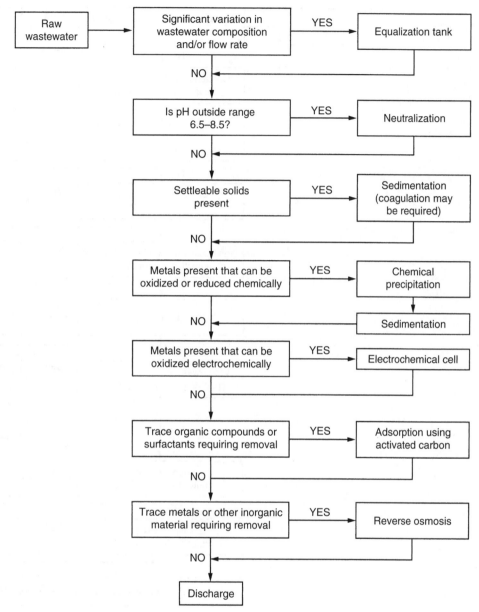

FIGURE 14.10
Process selection
algorithm for
physico-chemical
treatment.

Selection of treatment systems for small communities and individual houses is discussed in Chapter 22. Sludge treatment and disposal options are examined in Chapter 21.

14.4.2 EXAMPLE OF TREATMENT PLANT

The majority of wastewater treatment plants are purely domestic serving a population of <20 000 pe. They tend to follow the basic design criteria outlined in Fig. 14.13. Major cities are served by large municipal treatment plants often treating the wastes from

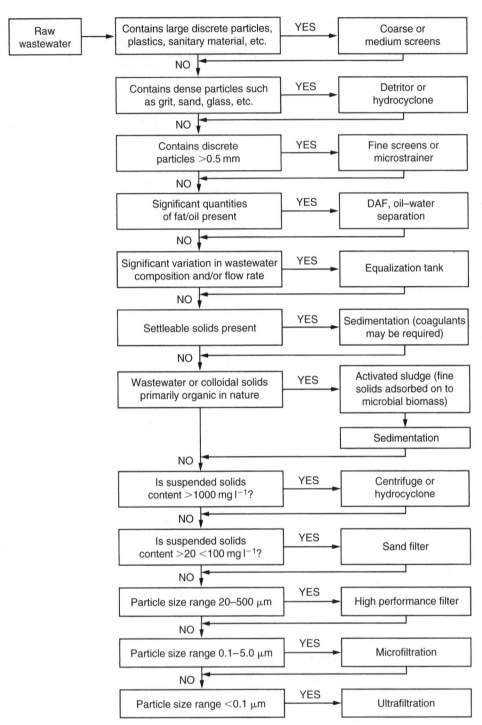

FIGURE 14.11
Process selection
algorithm for solids
removal in waste-
water treatment.

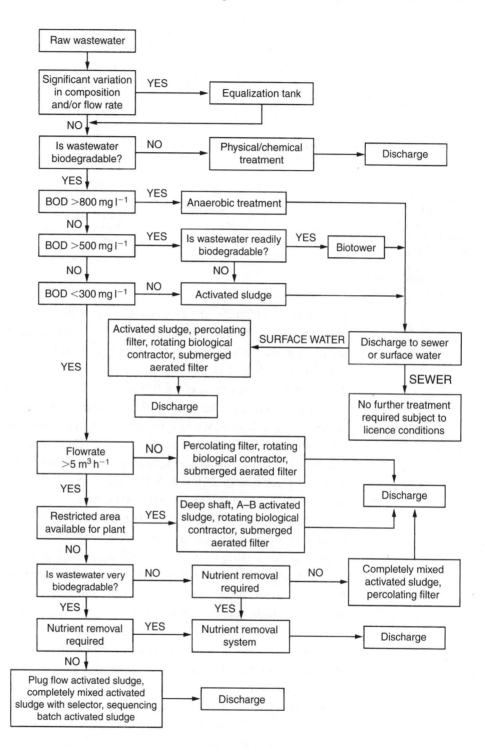

FIGURE 14.12
Process selection
algorithm for
biological waste-
water treatment.

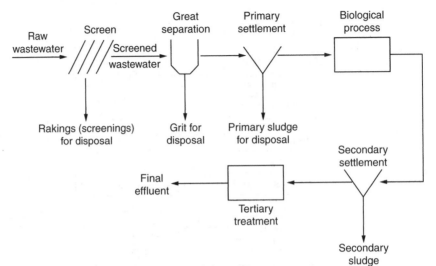

Figure 14.13
Typical layout of small sewage treatment plant serving a population of 20 000 or less.

populations $>1\,000\,000$. The Po-Sangone wastewater treatment plant (Torino, Italy) treats the sewage from a population of $1\,500\,000$ and the wastewater from over 1800 industries in the Piedmont region representing a total population equivalent of $3\,000\,000$ (Fig. 14.14a). The site covers 150 ha and serves a catchment area of 315 km^2 with an average daily flow of sewage of $550\,000\,m^3\,day^{-1}$. The wastewater passes through 20 mm spaced screens that are held in a specially constructed building so that the air can be deodorized by acid basic-washing to prevent odour problems at site (Fig. 14.14b), with a maximum of $25\,000\,m^3$ of air per hour treated. Combined grit removal and flotation follows prior to primary sedimentation by 12 radial-flow tanks each 52 m in diameter with a retention time of 2.5 h at 3DWF. Biological treatment is by 18 plug-flow activated sludge tanks using diffused air with 2000 diffusers per tank. Secondary settlement requires 18 radial-flow tanks with rapid sludge withdrawal using extraction bridges. Iron coagulants are added to the recirculated sludge for the removal of phosphorus. The effluent passes into the River Po after effluent polishing using multilayer filtration (i.e. gravel, sand and carbon). The plant is also equipped with effluent disinfection by sodium hypochlorite that can be used in emergencies. Sludge treatment includes thickening by six covered thickening (picket fence) tanks, which increases sludge solids from 2 to 5% (Fig. 14.14c). Sludge is then stabilized by anaerobic digestion using three heated mesophilic digesters (35°C) each with a volume of $12\,000\,m^3$, each followed by an unheated secondary digestion tank. After digestion the sludge is thickened in six covered thickening (picket fence) tanks before conditioning using lime and ferric chloride prior to dewatering using eight filter presses each with 140 plates. The sludge cake has a dry solids concentration 37–42% and is disposed to landfill. The air from the pre- and post-thickening tanks, as well as the filter press house, is deodorized using acid–base washing. The biogas from the digesters is used to generate electricity to operate the plant while the heat recovered from the cooling water and exhaust gases of the engines are used to heat the digesters. The performance characteristics of the plant are summarized in Table 14.7.

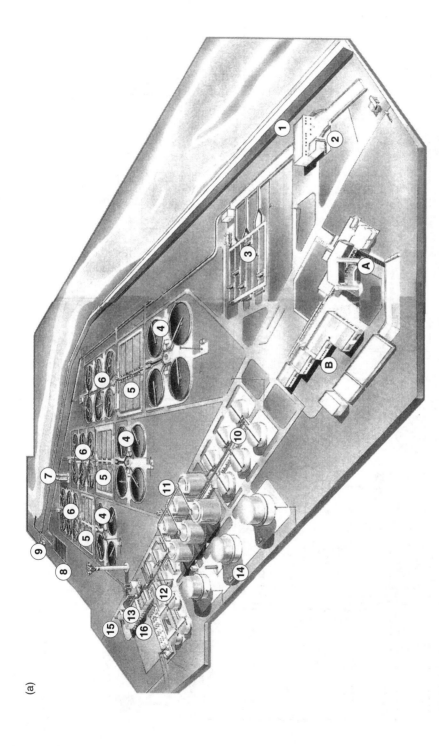

(a)

FIGURE **14.14** The Po-Sangone wastewater treatment plant in Torino, Italy. (a) Plan layout of plant, (b) water process flow diagrams and (c) sludge process flow diagram. *Wastewater Treatment*: 1. storm water overflow; 2. screening; 3. grit removal; 4. primary sedimentation; 5. aeration tanks; 6. secondary sedimentation; 7. emergency chlorination; 8. filtration; 9. effluent outfall. *Sludge treatment*: 10. raw sludge thickeners; 11. digestion tanks; 12. digested sludge thickeners; 13. power house; 14. biogas storage; 15. filter press house; 16. transformer station. A and B are administrative areas. (Reproduced by permission of Ing. Paolo Romano, General Manager, Azienda Po Sangone.)

Wastewater treatment

(b)

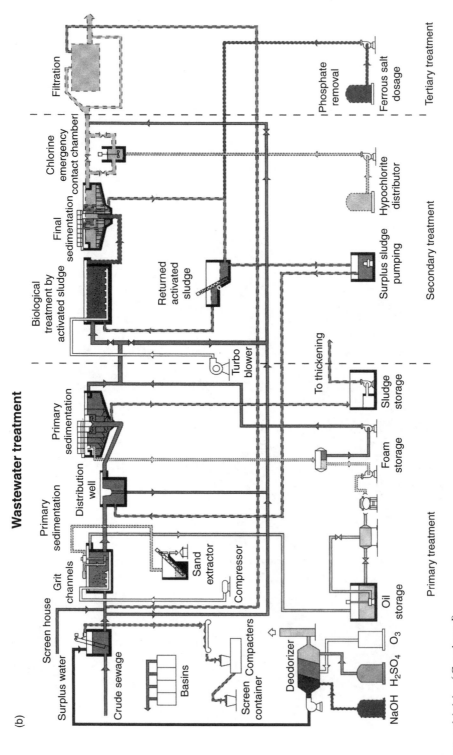

FIGURE **14.14** (*Continued*)

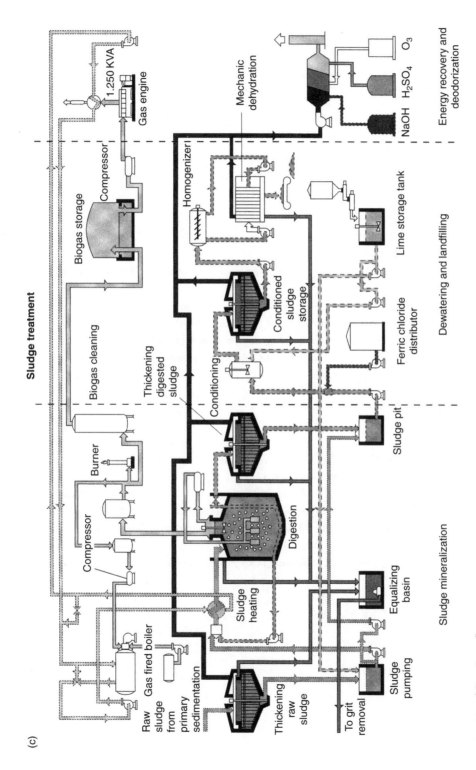

Sludge treatment

(c)

FIGURE **14.14** (*Continued*)

TABLE 14.7 The influent and treated (tertiary) effluent quality at the large municipal Po-Sangone wastewater treatment plant, Italy

Parameter (mg l⁻¹)	Influent		Final (tertiary) effluent	
	Average	Max.	Average	Max.
Suspended solids	200	980	9	30
BOD	200	500	8	24
Chemical oxygen demand	340	1300	26	81
Ammonia	26	41	3	9
Total phosphorus	3.9	7.8	0.6	1.0

REFERENCES

Casey, T. J., 1997 *Unit Treatment Processes in Water and Waste Water Engineering*, John Wiley and Sons, Chichester.

Metcalf and Eddy, 1991 *Waste Water Engineering: Treatment, Disposal and Reuse*, McGraw-Hill, New York.

Water Environment Federation, 1994 *Pre-treatment of Industrial Wastes*, Manual of Practice No. FD-3, Water Environment Federation, Alexandria, VA.

White, J. B., 1986 *Wastewater Engineering,* Arnold, London.

White, G. C., 1992 *Handbook on Chlorination and Alternative Disinfectants*, Van Nostrand Reinhold, New York.

FURTHER READING

IAWQ 1997 *Secondary Settling Tanks: Theory, Modelling, Design and Operation*, International Association on Water Quality, London.

Johnston, P. A., MacGarvin, M. and Stringer, R. C., 1991 Regulation of effluents and implications for environmental policy, *Water Science and Technology*, **24** (10), 19–27.

Nicoll, E. R., 1992 *Small Water Pollution Control Works: Design and Practice*, Ellis Horwood, Chichester.

Rendell, F., 1999 *Water and Wastewater Project Development.* Thomas Telford, London.

Chapter 15 Biological Aspects of Secondary Sewage Treatment

15.1 PROCESSES

After preliminary and primary treatment wastewater still contains significant amounts of colloidal and dissolved material that needs to be removed before discharge. The problem for the engineer is how to convert the dissolved material, or particles that are too small to settle unaided, into larger particles so that a separation process can remove them. This is achieved by secondary treatment. Chemical treatment using coagulants will deal with a portion of the colloidal solids, but a large portion of the polluting material will be unaffected. Also, the cost of continuous chemical addition and the problem of disposing of large quantities of chemical sludge makes this option normally unattractive (Section 20.2). Alternatively biological treatment can be used. This utilizes naturally occurring micro-organisms to convert the soluble and colloidal material into a dense microbial biomass that can be readily separated from the purified liquid using conventional sedimentation processes. As the micro-organisms are literally using the dissolved and colloidal organic matter as food (substrate), the total volume of sludge will be far less than for chemical coagulation. In practice, therefore, secondary treatment tends to be a biological process with chemical treatment used for toxic wastewater treatment.

A precise definition of secondary treatment is given by the Chartered Institution of Water and Environmental Management (IWEM) as:

> the treatment of sewage, usually after the removal of suspended solids, by bacteria under aerobic conditions during which organic matter in solution is oxidized or incorporated into cells which may be removed by settlement. This may be achieved by biological filtration or by the activated sludge process. Sometimes termed aerobic biological treatment.

(IWEM, 1992)

The Municipal Wastewater Treatment Directive (Section 8.3) defines secondary treatment more broadly as 'treatment of municipal wastewater by a process involving biological treatment with a secondary settlement'. So secondary treatment is a biological process where settled sewage enters a specially designed reactor where under aerobic conditions any remaining organic matter is utilized by micro-organisms. The reactor provides a suitable environment for the microbial population to develop and as long as oxygen and food, in the form of settled wastewater, are supplied then the biological oxidation process will continue. Biological treatment is primarily due to bacteria, which form the basic trophic level in the reactor food chain. The biological conversion of soluble material into dense microbial biomass has essentially purified the wastewater and all that is subsequently required is to separate the micro-organisms from the treated effluent by settlement. Secondary sedimentation is essentially the same as primary sedimentation except that the sludge is comprised of biological cells rather than gross faecal solids (Section 14.2.7).

Methods of purification in secondary treatment units are similar to the 'self-purification process' that occurs naturally in rivers and streams, and involves many of the same organisms (Section 5.3). Removal of organic matter from settled wastewaters is carried out by heterotrophic micro-organisms, predominately bacteria but also occasionally fungi. The micro-organisms breakdown the organic matter by two distinct processes, biological oxidation and biosynthesis, both of which result in the removal of organic matter from solution. Oxidation or respiration results in formation of mineralized end products which remain in solution and are discharged in the final effluent (Equation (15.1)), while biosynthesis converts the colloidal and soluble organic matter into particulate biomass (new cells) which can then be subsequently removed by settlement (Equation (15.2)). If the food supply, in the form of organic matter, becomes limiting then the microbial cell tissue will be endogenously respired (auto-oxidation) by the micro-organisms to obtain energy for cell maintenance (Equation (15.3)). All three processes occur simultaneously in the reactor and can be expressed stoichiometrically as:

(a) Oxidation:

$$COHNS + O_2 + Bacteria \rightarrow CO_2 + NH_3 + Other\ end\ products + Energy \tag{15.1}$$

(Organic matter)

(b) Biosynthesis:

$$COHNS + O_2 + Bacteria \rightarrow C_5H_7NO_2 \tag{15.2}$$

(Organic matter)　　　(New cells)

(c) Auto-oxidation:

$$C_5H_7NO_2 + 5O_2 \rightarrow 5CO_2 + NH_3 + 2H_2O + Energy \tag{15.3}$$

(Bacteria)

In natural waters soluble organic matter is principally removed by oxidation and biosynthesis, but in the intensified microbial ecosystem of the biological treatment plant adsorption is perhaps the major removal mechanism, with material adsorbed and agglomerated onto the dense microbial mass. The adsorptive property of the microbial biomass is particularly useful as it is also able to remove from solution non-biodegradable pollutants present in the wastewater, such as synthetic organics, metallic salts and even radioactive substances. The degree to which each removal mechanism contributes to overall purification depends on the treatment system used, its mode of operation and the composition of the wastewater.

In nature, heterotrophic micro-organisms occur either as thin films (periphyton) growing over rocks and plants, or in fact over any stable surface, or as individual or groups of organisms suspended in the water (Fig. 15.1). These natural habitats of aquatic heterotrophs have been utilized in wastewater treatment to produce two very different types of biological units, one using attached growths and the other suspended microbial growths. The design criteria for secondary treatment units are selected to create ideal habitats to support the appropriate community of organisms responsible for the purification of wastewater, so attached and suspended microbial growth systems will require fundamentally different types of reactors. Both treatment systems depend on a mixed culture of micro-organisms, but grazing organisms are also involved so that a complete ecosystem is formed within the

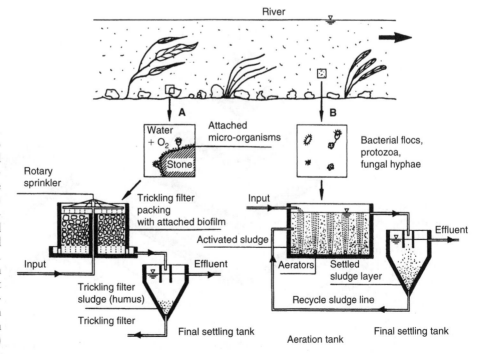

FIGURE 15.1 Relationship between the natural microbial communities in rivers and the development of (A) percolating (trickling) filters and (B) activated sludge processes. (Reproduced from Mudrack and Kunst (1987) with permission of Addison Wesley Longman Ltd, Harlow.)

reactor, each with distinct trophic levels. In its simplest form the reactor food chain comprises:

Heterotrophic bacteria and fungi

↓

Holozoic protozoa

↓

Rotifers and nematodes

↓

Insects and worms

Owing to the nature of the reactor, suspended growth processes have fewer trophic levels than attached growth systems (Fig. 15.2). These man-made ecosystems are completely controlled by operational practice, and are limited by food (organic loading) and oxygen (ventilation/aeration) availability.

Chemical and environmental engineers have been able to manipulate the natural process of self-purification, and, by supplying ideal conditions and unlimited opportunities for metabolism and growth, have intensified and accelerated this biological process to provide a range of secondary wastewater treatment systems. However, a number of basic criteria must be satisfied by the design. In order to achieve a rate of oxidation well above that found in nature, a much denser biomass in terms of cells per unit volume must be maintained within the reactor. This will result in increased oxygen demands, which must be met in full so as not to limit the rate of microbial oxidation. Essentially this is done by increasing the air–water interface. The wastewater

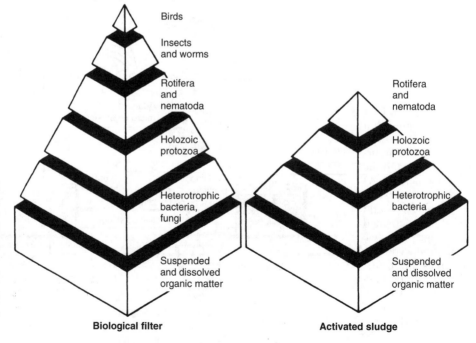

FIGURE 15.2 Comparison of the food pyramids for fixed film and completely mixed systems. (Reproduced from Wheatley (1985) with permission of Blackwell Science Ltd, Oxford.)

containing the polluting matter must be brought into contact for a sufficient period of time with a dense population of suitable micro-organisms in the presence of excess oxygen, to allow oxidation and removal of unwanted material to the desired degree. Finally, inhibitory and toxic substances must not be allowed to reach harmful concentrations within the reactor.

The main methods of biological treatment rely on aerobic oxidation, although anoxic and anaerobic treatment is also widely employed (Chapter 19). To ensure that oxidation proceeds quickly it is important that as much oxygen as possible comes into contact with the wastewater so that the aerobic micro-organisms can breakdown the organic matter at maximum efficiency. Secondary treatment units of wastewater treatment plants are designed to bring this about.

Oxidation is achieved by three main methods:

1. By spreading the sewage into a thin film of liquid with a large surface area so that all the required oxygen can be supplied by gaseous diffusion (e.g. percolating filter).
2. By aerating the sewage by pumping in bubbles of air or stirring vigorously (e.g. activated sludge).
3. By relying on algae present to produce oxygen by photosynthesis (e.g. stabilization pond).

In systems where the micro-organisms are attached, a stable surface must be available. Suitable surfaces are provided by a range of media such as graded aggregate or moulded plastic and even wooden slats, retained in a special reactor, on which a dense microbial biomass layer or film develops. These reactors are generally categorized as fixed-film reactors, the most widely used type being the percolating or trickling filter (Fig. 16.1). Organic matter is removed by the wastewater as it flows in a thin layer over the biological film covering the static medium. Oxygen is provided by natural ventilation, which moves through the bed of medium via the interstices supplying oxygen to all parts of the bed. The oxygen diffuses into the thin layer of wastewater to the aerobic micro-organisms below (Fig. 16.2). The final effluent not only contains the waste products of this biological activity, mainly mineralized compounds such as nitrates and phosphates, but also particles of displaced film and grazing organisms flushed from the medium. These are separated from the clarified effluent by settlement, and the separated biomass disposed as secondary sludge(Fig. 15.3). The design and operation of fixed-film reactors are dealt with in detail in Chapter 16.

In suspended growth or completely mixed system processes the micro-organisms are either free-living or flocculated to form small active particles or flocs which contain a variety of micro-organisms including bacteria, fungi and protozoa. These flocs are mixed with wastewater in a simple tank reactor, called an aeration basin or tank, by aerators that not only supply oxygen but also maintain the microbial biomass in suspension to ensure maximum contact between the micro-organisms and the nutrients

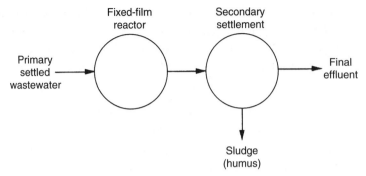

FIGURE 15.3
Schematic layout of a fixed-film reactor.

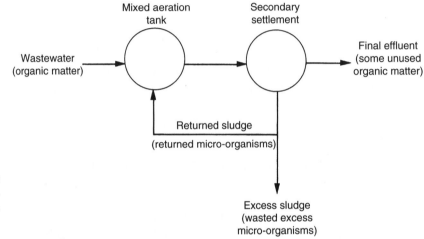

FIGURE 15.4
Schematic for a completely mixed (activated sludge) reactor.

in the wastewater. The organic matter in the wastewater is taken out of solution by contact with the active suspended biomass. The purified wastewater contains a large quantity of micro-organisms and flocs. The active biomass is separated from the clarified effluent by secondary settlement, but if the biomass was disposed of as sludge the concentration of active biomass in the aeration basin would rapidly fall to such a low density that little purification would occur. Therefore, the biomass, called activated sludge, is returned to the reactor to maintain a high density of active biological solids ensuring a maximum rate of biological oxidation (Fig. 15.4). Excess biomass, not required to maintain the optimum microbial density in the reactor, is disposed of as surplus sludge (Table 21.1). The design and operation of completely mixed systems is examined in detail in Chapter 17.

Apart from the oxidation of organic (carbonaceous) material by heterotrophs, ammonia can also be oxidized during wastewater treatment by autotrophic bacteria, a process known as nitrification. Nitrification can take place in the same reactor as carbonaceous oxidation, either at the same time or in different areas or zones of the reactor, or it can take place separately in a specially constructed reactor. The process is a two-step bacteriological reaction that requires considerable oxygen. In the first step ammonia is oxidized to nitrate by bacteria of the genus *Nitrosomonas*, which is further, oxidized to nitrate by bacteria of the genus *Nitrobacter*. This is considered in detail in Section 16.3.2.

$$2NH_4 + 3O_2 \rightarrow 2NO_2 + 4H^+ + 2H_2O + Energy \qquad (15.4)$$

$$2NO_3 + O_2 \rightarrow 2NO_3 + Energy \qquad (15.5)$$

The microbial biomass with the reactor transforms the organic matter in the settled sewage by biophysical processes (flocculation and adsorption) and biochemical processes (biological oxidation and synthesis) to produce a variety of end products. Some of these are soluble or gaseous end products such as nutrient salts or carbon dioxide from oxidation processes or soluble products from the lysis of the micro-organisms comprising the film. The remainder is present as solids that require separation from the final effluent. The mechanisms for adsorption and subsequent absorption are the same for all bacterial systems and are discussed in Section 17.1.

Biological wastewater treatment differs from more traditional fermentation processes, such as the production of bakers' yeast, in a number of important ways. Principally wastewater treatment is aimed at removing unwanted material while the commercial fermentation processes are all production systems. These production fermentations use highly developed, specialized strains of particular micro-organisms to synthesize the required end product, while in wastewater treatment a mixture of micro-organisms are used. These are largely self-selecting with nearly all the organisms that can contribute to substrate removal welcome. Unlike commercial fermenters, wastewater reactors are not aseptic and because production fermenters require highly controlled conditions they are more complex and comparatively more expensive than those used in wastewater treatment.

Although the IWEM definition stresses that aerobic systems should be used, and this has traditionally been the case, the EU places no such restrictions in its definition. The secondary treatment phase may comprise of other biological systems, both aerobic and anaerobic, or incorporate a mixture of several systems. Among the more unusual biological systems used for municipal treatment that fall outside the fixed film or completely mixed categories are aerobic and anaerobic lagoons, reed beds, wetlands and land treatment (Chapter 18). In practical terms the water industry is conservative in the selection of wastewater technology, choosing those systems for which there is plenty of design and operational data available, so that they can be assured that the process will be both adequate and reliable. For that reason treatment process selection is often limited to a straight choice between fixed-film and completely mixed systems (Table 15.1), which are dealt with fully in Chapters 16 and 17, respectively. However, there are many other processes available that are often selected for unusual locations or more difficult wastes (Chapters 18–20).

TABLE 15.1 Comparison of the activated sludge and percolating filter processes

	Activated sludge	*Percolating filtration*
Capital cost	Low	High
Area of land	Low: advantageous where land availability is restricted or expensive	Large: 10 times more area required
Operating cost	High	Low
Influence of weather	Works well in wet weather, slightly worse in dry weather, less affected by low winter temperatures	Works well in summer but possible ponding in winter
Technical control	High: the microbial activity can be closely controlled; requires skilled and continuous operation	Little possible except process modifications. Does not require continuous or skilled operation
Nature of wastewater	Sensitive to toxic shocks, changes in loading and trade wastewaters; leads to bulking problems	Strong wastewaters satisfactory, able to withstand changes in loading and toxic discharges
Hydrostatic head	Small: low pumping requirement, suitable for site where available hydraulic head is limited	Large: site must provide natural hydraulic head otherwise pumping is required
Nuisance	Low odour and no fly problems. Noise may be a problem both in urban and rural areas	Moderate odour and severe fly problem in summer possible. Quiet in operation
Final effluent quality	Poor nitrification but low in suspended solids except when bulking	Highly nitrified, relatively high suspended solids
Secondary sludge	Large volume, high water content, difficult to dewater, less stabilized	Small volume, less water. Highly stabilized
Energy requirement	High: required for aeration, mixing and maintaining sludge flocs in suspension and for recycling sludge	Low: natural ventilation, gravitational flow
Synthetic detergents	Possible foaming, especially with diffusers	Little foam
Robustness	Not very robust, high degree of maintenance on motors, not possible to operate without power supply	Very sturdy, low degree of maintenance, possible to operate without power

15.2 KINETICS

15.2.1 BACTERIAL KINETICS

In terms of substrate removal, the rate of carbonaceous oxidation, nitrification and denitrification depends on the rate of microbial growth, and in particular the rate of bacterial growth. This is best illustrated by observing the development of a microbial population with time in batch culture. When a small inoculum of viable bacterial cells are placed in a closed vessel with excess food and ideal environmental conditions, unrestricted growth occurs. Monod (1949) plotted the resultant microbial growth curve from which six discrete phases of bacterial development can be defined

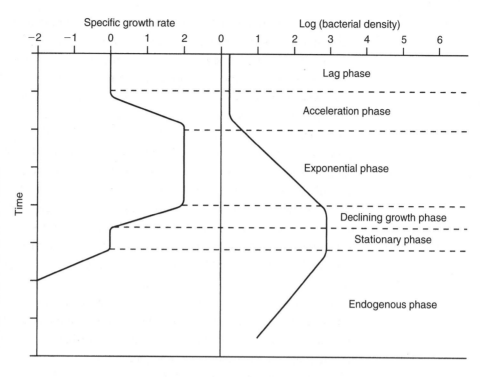

FIGURE 15.5 The microbial growth curve showing bacterial density and specific growth rate at various microbial growth phases. (Reproduced from Benefield and Randall (1980) with permission of Prentice-Hall Inc., New York.)

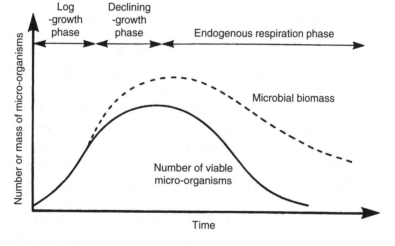

FIGURE 15.6 Microbial growth curves comparing total biomass and variable biomass. (Reproduced from Clark *et al.* (1977) with permission of Addison Wesley Longman Inc., Glenview, Illinois.)

(Fig. 15.5). The microbial concentration is usually expressed as either the number of cells per unit volume or mass of cells per unit volume of reactor. However, it is not possible to directly translate one set of units from the other since the size of the cells and the creation of storage products by cells dramatically affects the mass but not the number of cells (Fig. 15.6).

The lag phase represents the acclimatization of the organisms to the substrate; however, depending on the size and degree of adaptation of the inoculum to its new environment, the lag phase may be very short or even absent. Cells only begin to divide when a sufficient concentration of the appropriate enzymes have built up, but once division has started the population density of bacteria rapidly increases. In the acceleration phase the generation time decreases and there is a discernible increase in the growth rate leading to the exponential or log phase. In this phase the generation time is minimal, but constant, with a maximal and constant specific growth rate resulting in a rapid increase in the number and mass of micro-organisms. This is the period when the substrate conversion is at its maximum rate. The rate of metabolism and in particular the growth rate is limited only by the microbial generation and its ability to process substrate. The exponential phase continues until the substrate becomes limiting. This produces the declining growth phase where the rate of microbial growth rapidly declines as the generation time increases and the specific growth rate decreases as the substrate concentration is gradually diminished. It is at this stage that the total mass of the microbial protoplasm begins to exceed the mass of viable cells, since many of the micro-organisms have ceased reproducing due to substrate-limiting conditions (Fig. 15.6). In batch situations the accumulation of toxic metabolites or changes in the concentration of nutrients, or other environmental factors such as oxygen or pH can also be responsible for the onset of the declining growth phase. The microbial growth curve flattens out as the maximum microbial density is reached, with the rate of reproduction apparently balanced by the death rate. This is the stationary phase. Now the substrate and nutrients are exhausted and there is a high concentration of toxic metabolites. It has been suggested that the majority of cells remain viable during this phase but in a state of suspended animation, without the necessary substrate or environmental conditions to continue to reproduce. The final phase is the endogenous or the log-death phase. The substrate is now completely exhausted and the toxic metabolites have become unfavourable for cell survival. The microbial density decreases rapidly with a high micro-organism death rate resulting in the rate of metabolism and hence the rate of substrate removal also declines. The total mass of microbial protoplasm decreases as cells utilize their own protoplasm as an energy source (endogenous respiration), and as cells die and lyse they release nutrients back into solution that can be utilized by the remaining biomass; although there is a continued decrease in both the number and mass of micro-organisms.

The microbial growth curve is not a basic property of bacterial cells but is a response to their environmental conditions within a closed system. Because biological treatment processes are normally continuous flow systems it is not directly applicable, although it is possible to maintain such systems at a particular growth phase by controlling the ratio of substrate to microbial biomass, commonly referred to as the food to micro-organism ratio (f/m). The f/m ratio maintained in the aeration tank in an activated sludge system controls the rate of biological oxidation as well as the volume of microbial biomass produced by maintaining microbial growth either in the log, declining or endogenous growth phase. The type of activated sludge process can be defined by the f/m ratio as being high rate, conventional or extended. For example, at a high f/m ratio micro-organisms are in the log-growth phase, which is characterized by excess substrate and maximum rate of metabolism. At low f/m ratios the overall metabolic action in the aeration tank is endogenous, with the substrate limiting microbial growth so that cell lysis and resynthesis occurs. There is growing interest in the development of batch treatment systems that correspond more closely to the basic microbial growth curve model (Section 17.4).

15.2.2 RATES OF REACTION

The rate at which components of wastewater, such as organic matter, are removed and the rate at which biomass is produced within a reactor are important criteria in the design and calculation of the size of biological reactors. The most useful method of describing such chemical reactions within a biological reactor is by classifying the reaction on a kinetic basis by reaction order. Reaction orders can differ when there is variation in the micro-organisms, the substrate or environmental conditions, so they must be measured experimentally.

The relationship between rate of reaction, concentration of reactant and reaction order (n) is given by the expression:

$$\text{Rate} = (\text{Concentration})^n \tag{15.6}$$

or more commonly by taking the log of both sides of the equation:

$$\log \text{Rate} = n \log(\text{Concentration}) \tag{15.7}$$

Equation (15.7) is then used to establish the reaction order and rate of reaction. If the log of the instantaneous rate of change of the reactant concentration at any time is plotted as a function of the log of the reactant concentration at that instant, then a straight line will result for constant order reactions and the slope of the line will be the order of the reaction (Fig. 15.7). The rate of reaction is independent of the

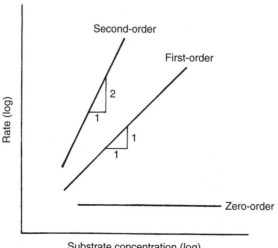

FIGURE 15.7
Log plots of reaction rates.

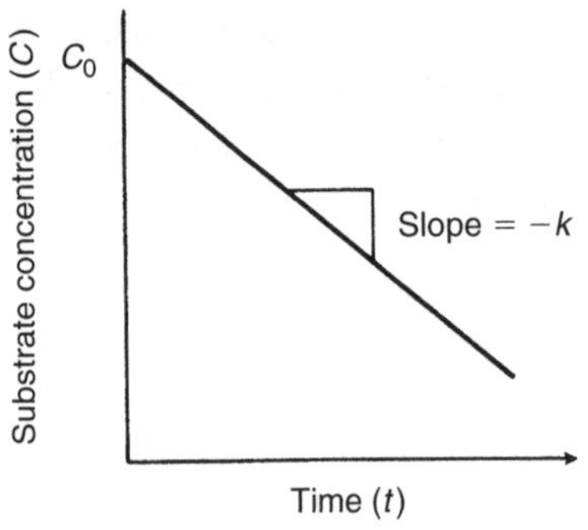

FIGURE 15.8
Arithmetical plot of
a zero-order
reaction.

reactant concentration in zero-order reactions, which results in a horizontal line when plotted. The rate of reaction in first-order reactions is directly proportional to the reactant concentration and with second-order reactions the rate is proportional to the concentration squared.

In practice, it is simplest to plot the concentration of reactant remaining against time in order to calculate the reaction rate. Zero-order reactions are linear when the plot is made on arithmetic paper (Fig. 15.8), following Equation (15.8):

$$C - C_0 = -kt \tag{15.8}$$

where C is the concentration of reactant ($mg\,l^{-1}$) at time t, C_0 is the constant of integration ($mg\,l^{-1}$) which is calculated as $C = C_0$ at $t = 0$ and k the reaction rate constant ($mg\,l^{-1}\,day^{-1}$). First-order reactions proceed at a rate directly proportional to the concentration of one reactant, so that the rate of reaction depends on the concentration remaining, which is decreasing with time. The plot of variation in concentration with time on arithmetic paper does not give a linear response but a curve. However, first-order reactions follow Equation (15.9):

$$\ln\left(\frac{C_0}{C}\right) = kt \tag{15.9}$$

or

$$\log\frac{[C_0]}{[C]} = \frac{kt}{2.3} \tag{15.10}$$

so that a plot of $\log C$ (the log of the concentration of the reactant remaining) against time will give a linear trace (Fig. 15.9). Second-order reactions proceed at a rate proportional to the second power of the concentration of a single reactant and obey the function:

$$\frac{1}{C} - \frac{1}{C_0} = kt \tag{15.11}$$

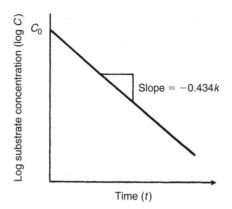

FIGURE 15.9
Semi-log plot of a
first-order reaction.

Slope $= -0.434k$

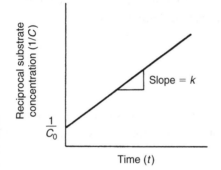

FIGURE 15.10
Arithmetical plot of
a second-order
reaction.

Slope $= k$

with arithmetic plot of the reciprocal of the reactant concentration remaining $(1/C)$ against time giving a linear trace, the slope of which gives the value of k (Fig. 15.10).

For any set of values C and t, such as the rate of removal of organic matter as measured by biochemical oxygen demand (BOD) at regular intervals, the reaction rate equations can be tested by making the appropriate concentration versus time plots and noting any deviation from linearity. It should be noted that while fractional reaction orders are possible it is normal for an integer value for reaction order to be assumed.

15.2.3 ENZYME REACTIONS

The organic matter in wastewater is utilized by micro-organisms in a series of enzymatic reactions. Enzymes are proteins or proteins combined with either an inorganic or low-molecular-weight organic molecule. They act as catalysts forming complexes with the organic substrate, which they convert to a specific product, releasing the original enzyme to repeat the same reaction. The enzymes have a high degree of substrate specificity, so a bacterial cell must produce different enzymes for each substrate utilized. Two types of enzymes are produced: extra-cellular enzymes which convert substrate extra-cellularly into a form that can be taken into the cell for further breakdown by the intra-cellular enzymes which are involved in synthesis and energy reactions within the cell. Normally the product of an enzyme-catalysed reaction immediately combines with another enzyme until the final end product required by the cell is reached after a sequence of enzyme–substrate reactions.

The overall rate of biological reaction within a reactor is dependent on the catalytic activity of the enzymes in the prominent reaction. If it is assumed that enzyme-catalysed reactions involve the reversible combination of an enzyme (E) and substrate (S) in the form of a complex (ES) with the irreversible decomposition of the complex to a product (P) and the free enzyme (E), then the overall reaction can be written:

$$E + S \xrightleftharpoons[k_2]{k_1} ES \xrightarrow{k_3} E + P \tag{15.12}$$

where k_1, k_2 and k_3 represent the rate of the reactions. Under steady-state conditions the various rate constants can be expressed as:

$$\frac{k_2 + k_3}{k_1} = K_m \tag{15.13}$$

where K_m is the saturation or Michaelis constant. The Michaelis–Menten equation allows the reaction rate of enzyme-catalysed reactions to be calculated:

$$r = \frac{R_{max}[S]}{K_m + [S]} \tag{15.14}$$

where r is the reaction rate, R_{max} is the maximum reaction rate at which the product is formed ($mg\,l^{-1}day^{-1}$) and S is the substrate concentration ($mg\,l^{-1}$). If this equation is plotted graphically (Fig. 15.11) it can be seen that the rate of an enzyme-catalysed reaction is proportional to the substrate concentration at low-substrate concentrations (first order) but as the substrate concentration increases the rate of

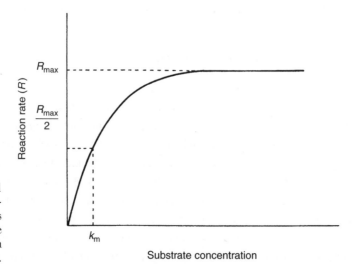

FIGURE 15.11
The rate of enzyme-catalysed reactions as presented by the Michaelis–Menten equation.

reaction declines, finally becoming constant and independent of the substrate concentration (zero order). In practical terms this is seen when a batch reactor is started and no further substrate is added. Initially the rate of reaction is only restricted by the ability of the enzymes to utilize the substrate, which is in excess, so the reaction kinetics are zero order. However, as the substrate is utilized the reaction begins to become substrate-limited resulting in fractional-order reactions until the substrate concentration is so low that the rate of reaction becomes totally limited by the substrate concentration and so first-order kinetics result. First-order kinetics are assumed when $[S] \leqslant K_m$. From Fig. 15.11, it can be seen that the saturation constant K_m is equal to the substrate concentration when the reaction rate is equal to $(R_{max}/2)$.

Most biological wastewater treatment systems are designed to operate with high microorganism concentrations, which results in substrate-limiting conditions. Therefore, the majority of wastewater treatment processes can be described by first-order kinetics.

15.2.4 ENVIRONMENTAL FACTORS AFFECTING GROWTH

A number of environmental factors affect the activity of wastewater microbial populations and the rate of biochemical reactions generally. Of particular importance are temperature, pH, dissolved oxygen, nutrient concentration and inhibition by toxic compounds. It is possible to control all these factors within a biological treatment system, except temperature, to ensure that microbial growth continues under optimum conditions. The majority of biological treatment systems operate in the mesophilic temperature range, growing best in the temperature range 20–40°C. Aeration tanks and percolating filters operate at the temperature of the wastewater, 12–25°C, although in percolating filters the air temperature and the rate of ventilation can have a profound effect on heat loss. The higher temperatures result in increased biological activity, which in turn increases the rate of substrate removal. The increased metabolism at the higher temperatures can also lead to problems of oxygen limitations. Generally activated sludge systems perform better than percolating filters below 5–10°C, although heterotrophic growth continues at these temperatures. However, the practice in colder climates of covering filters and controlling the rate of ventilation, thus reducing heat loss within the filter bed, has largely overcome this problem. Van't Hoff's rule states that the rate of biological activity doubles with every 10°C rise in temperature within the range 5–35°C. The variation in reaction rate with temperature is represented by the modified Arrhenius expression:

$$k_T = k_{20} \, \phi^{T-20} \tag{15.15}$$

where T is the temperature (°C), k is the reaction rate constant at temperature T (day^{-1}), k_{20} is the reaction rate constant at 20°C (day^{-1}) and ϕ is the temperature coefficient. There is a rapid decrease in growth rate as the temperature increases above 35°C which falls to zero as the temperature approaches 45°C. Anaerobic digestion tanks are normally heated as near to the optimum temperature of the mesophilic range as possible (35–37°C). The pH and dissolved oxygen concentration in a biological reactor can be controlled to any level by the operator. The optimum pH range for carbonaceous oxidation lies between 6.5 and 8.5. At pH > 9.0 microbial activity is inhibited while at pH < 6.5 fungi dominate over the bacteria in the competition for the substrate. Fluctuations in the influent pH are minimized by completely mixed aeration reactors that offer maximum buffering capacity. If the buffering capacity is not sufficient to maintain a pH within the acceptable range, then pH adjustment will be required. Anaerobic bacteria have a smaller pH tolerance ranging from pH 6.7–7.4 with optimum growth at pH 7.0–7.1. A dissolved oxygen concentration between 1 and $2\,mg\,l^{-1}$ is sufficient for active aerobic heterotrophic microbial activity, although optimum growth is dependent on sufficient essential nutrients and trace elements being present. Apart from organic carbon, nitrogen and phosphorus, true elements such as sulphur, iron, calcium, magnesium, potassium, manganese, copper, zinc and molybdenum must also be available for bacterial metabolism. Normally all these nutrients and elements are present in excess in sewage although they may have to be supplemented in industrial wastewaters where nitrogen and phosphorus are normally only present in low concentrations. It is desirable to maintain a BOD_5:N:P mass ratio of 100:5:1 to ensure maximum microbial growth (Section 13.1). Toxic compounds, such as phenol, cyanide, ammonia, sulphide, heavy metals and trace organics, can totally inhibit the microbial activity of a treatment plant if the concentration exceeds threshold limits for the micro-organisms. However, constant exposure to low concentrations of these substances result in the microbial community becoming acclimatized and increasing the concentration which they can tolerate. Completely mixed aeration tanks are able to dilute shock toxic loads reducing the influent concentration to the final effluent concentration, while in filters the contact time between the micro-organisms and the toxic material is relatively short.

15.2.5 KINETIC EQUATIONS OF BACTERIAL GROWTH

The various growth phases on the microbial growth curve (Fig. 15.5) can be represented quantitatively. The net microbial growth rate $(mg\,l^{-1}\,day^{-1})$ can be expressed as:

$$\frac{dX}{dt} = \mu X - k_d X \tag{15.16}$$

where X is the concentration of micro-organisms $(mg\,l^{-1})$, μ is the specific growth rate (day^{-1}), k_d is the endogenous decay coefficient (day^{-1}) and t is the time in days. The integrated form of this equation when plotted on semi-log arithmetic graph paper results in a straight line, hence the term logarithmic growth phase. The equation assumes that all the micro-organisms are viable and while this may be true for a test-tube culture, it cannot be the case for a wastewater treatment unit with long retention times. It is, however, assumed that a constant fraction of the organisms within the biological treatment unit will remain viable.

Exponential growth will continue so long as there is no change in the composition of the biomass and the environmental conditions remain constant. In a batch reactor a change in environmental conditions will inevitably occur, normally substrate limitation, which will cause a change from exponential growth into the declining growth phase. The most commonly used model, relating microbial growth to substrate utilization, is that of Monod. He observed that the growth rate dX/dt was not only a function of organism concentration but also of some limiting substrate or nutrient concentration. He described the relationship between the residual concentration of the growth-limiting substrate or nutrient and the specific growth rate of biomass (μ) by the classical function that takes his name:

$$\mu = \mu_m \frac{S}{K_s + S} - k_d \tag{15.17}$$

where μ_m is the maximum specific growth rate at saturation concentration of growth-limiting substrate (day^{-1}), S is the substrate concentration (mg l^{-1}) and K_s is the saturation constant (mg l^{-1}) which is the concentration of limiting substrate at which the specific growth rate equals half of the maximum specific growth rate ($\mu = \mu_m/2$).

From this relationship it can be seen that specific growth rate can have any value between zero and μ_m provided that the substrate can be held at a given constant value. This is the basis for all continuous flow treatment processes in biological wastewater treatment in which micro-organisms are continuously cultivated, but the overall rate of metabolism is controlled by the substrate concentration. When plotted, the Monod relationship between specific growth rate and the growth-limiting substrate concentration (Fig. 15.12) has the same form as the Michaelis–Menten equation which describes the rate of reaction of an enzyme with the substrate concentration. Figure 15.12 shows that the microbial growth rate increases as the availability of substrate increases until the maximum specific growth rate is achieved, at which point a factor other than substrate, such as generation rate or a specific nutrient, becomes growth limiting.

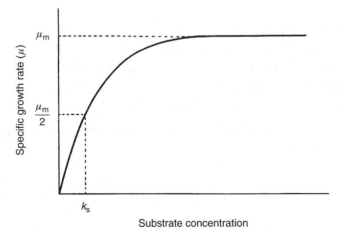

FIGURE 15.12 The relationship between specific growth rate and growth-limiting nutrient concentration.

The specific growth rate of the equation of microbial growth under exponential growth conditions $dX/dt = \mu$ can be replaced by the Monod function so that:

$$\frac{dX}{dt} = \mu_m \frac{SX}{K_s + S} \tag{15.18}$$

where the growth rate (dX/dt) is directly proportional to substrate concentration (first order). When the substrate concentration is much larger than K_s the expression for growth rate reduces to:

$$\frac{dX}{dt} = \mu_m X \tag{15.19}$$

where the growth rate is independent of substrate concentration (zero order).

The organic carbon as measured by BOD or chemical oxygen demand (COD) is usually considered to be the rate-limiting substrate in aerobic wastewater treatment systems, although the growth rate of the micro-organisms can also be controlled by other substances, such as ammonia, phosphate, sulphate, iron, oxygen, carbon dioxide and light. In low-substrate concentrations the rate of mass transfer into the cell may also control the rate of growth. The thickness of microbial flocs or film may be orders of magnitude greater than the size of an individual micro-organism, so the rate of mass transfer may limit growth under certain conditions. However, the Monod function can still be used if K_s is considered as a variable dependent upon the degree of mixing in the reactor.

In nearly all biological wastewater treatment systems the retention time of micro-organisms is such that the endogenous phase occurs. Some unit processes, such as oxidation ditches, are designed to operate specifically in this phase. The basic equation for microbial growth can be modified to incorporate endogenous decay:

$$\frac{dX}{dt} = (\mu - k_d)X \tag{15.20}$$

where k_d is the specific endogenous decay rate which includes endogenous respiration, death and subsequent lysis. The specific endogenous decay rate k_d is of little significance when the retention time is short, being an order of magnitude less than μ. However, when the system is operated in the endogenous growth phase, k_d is important in the calculation of the net amount of micro-organisms produced and the oxygen utilization rate.

The mass of organisms produced is related to the mass of substrate consumed, using the expression:

$$\frac{dX}{dt} = -Y \frac{[dS]}{[dt]} \tag{15.21}$$

where X is the concentration of micro-organisms, S is the concentration of substrate, Y is the yield coefficient and t is the time. The yield coefficient is a function of the species of micro-organisms present, the type of substrate and environmental conditions. However, it is usually assumed to be constant for a given biological treatment process treating a specific wastewater. The yield coefficient is determined experimentally, and so such factors as formation of storage products (i.e. glycogen and poly-β-hydroxybutrate), temperature, pH and the variation in the fraction of viable cells, all of which can significantly affect the coefficient, must be taken into account. This is done by ensuring that the experimental conditions under which the coefficient is measured are the same as those encountered in practice, or by taking into account the factors used in the model of yield. BOD or COD are usually used as a measure of substrate concentration and volatile suspended solids as an index of organism concentration. Yield coefficient is expressed as mass (or mole) of organism produced per mass (or mole) of substrate consumed. Similar relationships can be established for the utilization of other substances such as oxygen or light energy, or the formation of products such as methane or carbon dioxide.

The expression of yield can be combined with the constant growth rate equation $(dX/dt = \mu X)$ to give the rate of substrate utilization:

$$\frac{dS}{dt} = -\frac{\mu}{Y} X \qquad (15.22)$$

When the amount of substrate utilization for cell maintenance is expected to be significant, that is at high temperatures, long residence times and high cell concentrations, the above equation should be modified:

$$\frac{dS}{dt} = -\frac{\mu}{Y} X - K_m X \qquad (15.23)$$

where K_m is the specific maintenance coefficient.

Kinetic constants should be determined experimentally using bench or pilot scale methods. However, due to the biological and chemical nature of such experiments the wastewater engineer and designer often prefer to rely on data presented in the literature. The variation in the kinetic constants quoted in the literature, due to the variation in wastewater quality and environmental conditions, is large. For example, $Y = 0.35–0.45$ mg volatile suspended solids mg COD^{-1}, $k_d = 0.05–0.10$ day^{-1}, $k_s = 25–100$ mg l^{-1} and $\mu_m = 3.5–10.0$ day^{-1}, so clearly kinetic analysis can only be performed with any degree of confidence on experimentally collected data. The use of bacterial kinetics in the design of secondary wastewater treatment systems is excellently reviewed by Ramalho (1977).

15.3 TREATABILITY, TOXICITY AND BIODEGRADABILITY ASSESSMENT

It is vital to know that wastewaters discharged to the treatment plant are non-toxic to the biological phase of the process, and are able to be removed and broken down. New industrial wastewaters should always be fully investigated in terms of their treatability, toxicity and biodegradation before they are accepted for discharge to sewer.

Treatability of a substance is defined as the amenability of that substance to removal during biological wastewater treatment without adversely affecting the performance of the treatment plant.

Toxicity, or the extent to which a substance adversely affects micro-organisms, is most likely dependent on the concentration of micro-organisms present (i.e. f/m ratio). Therefore, tests must be conducted at various concentrations of inoculum and to quantify the toxic effects of a substance as an EC_{50} (i.e. the concentration of the substance giving 50% inhibition of the control) (Section 6.1).

Biodegradation can be defined as the breakdown of a substance by micro-organisms. Primary biodegradation is the alteration of the chemical structure of a substance resulting in loss of a specific property of the substance. Ultimate biodegradation is the complete breakdown of a substance to either fully oxidized or reduced simple molecules and the formation of new cells.

Screening or indicative methods for wastewaters can all be carried out relatively quickly using standard equipment in the wastewater laboratory. These methods are categorized as either biochemical or bacterial tests. They are alternative methods to continuous simulation tests using experimental pilot plants. In practice, experimental pilot plants are more time consuming and costly, but often provide information that can be more easily transferred to full-scale treatment plant operation (Kilroy and Gray, 1995).

Biochemical tests are based on the measurement of the activity of specific metabolic products of micro-organisms. While used for the determination of microbial biomass and microbial activity, they can also be used for toxicity screening. Enzymatic assays (e.g. dehydrogenase, beta-galactosidase and *in vitro* urease activity) are used specifically to measure cellular activity while adenosine triphosphate (ATP) assays give an indication of live biomass.

Bacterial tests to assess the effect of a toxicant rely on the inhibition of some easily measured microbial activity. These include:

1. Assays based on the measurement of growth inhibition and viability of bacterial cells. Since growth is the summation of cellular processes, it reflects toxic effects on numerous biological functions (e.g. *Sprillum volutans* bioassay, bacterial growth inhibition, bioluminescence).
2. Assays based on the inhibition of oxygen uptake by bacterial cells. Aerobic bacteria require oxygen for cell maintenance, growth and division. The rate of

oxygen uptake (i.e. the respiration rate) responds rapidly to the presence of inhibitors and measurements have the advantage of speed and simplicity. Loss of respiration signifies death of a cell and as such is a measure of acute toxicity. However, cells may still respire even after they have lost the ability to grow or divide (i.e. active but not viable). Therefore, short-term respiration assays do not identify effects on growth and cell division that may become obvious after several generations (e.g. manometry, electrolytic respirometry, direct measurement of dissolved oxygen, microbial oxygen electrode screening tests).

3. Assays based on substrate utilization (e.g. modified OECD screening test (MOST), modified semi-continuous activated sludge (SCAS) test, glucose uptake).
4. Assays based on carbon dioxide production (e.g. the Sturm test).

The most widely used toxicity and biodegradability tests in water technology are summarized below.

15.3.1 THE ACTIVATED SLUDGE RESPIRATION INHIBITION TEST

The activated sludge respiration inhibition test uses a mixed population of micro-organisms collected from the activated sludge sewage treatment process. Activated sludge respires rapidly in the presence of synthetic sewage, and the addition of a toxic concentration of a chemical results in a decrease in the respiration rate proportional to the toxicity of the toxicant. The activated sludge respiration inhibition test (OECD, 1984) involves incubating unacclimated samples of activated sludge mixed liquor from a municipal treatment plant. Activated sludge is continuously aerated with a feed substrate in the presence of varying concentrations of the test chemical for 30 min. The EC_{50} is the concentration of test chemical (as a percentage dilution) required to reduce the respiration rate (mg O_2 h^{-1}) by 50%, and is calculated by comparison with an uncontaminated control. The mixed liquor suspended solids (MLSS) concentration should be maintained at 3000 mg l^{-1} and the respiration rate expressed as the specific oxygen uptake rate in mg O_2 g^{-1} MLSS h^{-1}. Diluted synthetic sewage is added to aerated activated sludge and the oxygen utilization measured using an oxygen probe placed in the sample, which is continuously stirred in a standard BOD bottle. The oxygen concentration is recorded every 15 s or continuously using a chart recorder over a total period of 30 min.

The concentration of dissolved oxygen is plotted against time and a regression line fitted. The respiration rate of the test samples and controls are calculated from the slopes (concentration of oxygen over time). Percentage inhibition is plotted for each dilution using the equation:

$$\% \text{ inhibition} = \frac{R_c - (R_t - R_{pc})}{R_c} \times 100 \qquad (15.24)$$

where R_c is the respiration rate of the control, R_t is the respiration rate of the test sample and R_{pc} is the respiration rate of the physico-chemical uptake at the test dilution (i.e. no activated sludge added). The percentage inhibition values (x-axis) are then plotted against the various dilutions of the test chemical (y-axis) and the 50% inhibition point calculated from the slope of the regression line. Poor reproducibility due to the varying nature of the activated sludge inoculum used should be taken into consideration when using this test. To some extent this is a treatment-plant-specific assessment procedure. The 30-min Polytox® toxicity procedure overcomes this problem by using a standard bacterial preparation in place of the activated sludge, significantly improving reproducibility (Elnabarawy et al., 1988).

15.3.2 THE BOD INHIBITION TEST

One of the simplest ways of assessing the toxicity of a test chemical is to measure the effect on oxygen utilization by heterotrophic micro-organisms grown on a known, readily degraded substrate as measured by the standard 5-day BOD test. The most widely used standard substrate is a mixture of glutamic acid and glucose (UK Standing Committee of Analysts, 1982). The percentage inhibition is calculated for a range of dilutions of the test chemical by comparing the oxygen uptake to the test chemical-free controls. The BOD test is carried out using standard methods (UK Standing Committee of Analysts, 1988). The test substances and standard substances are dissolved in the BOD dilution water. The standard substrate is made up of $0.15\,g\,l^{-1}$ glucose and $0.15\,g\,l^{-1}$ glutamic acid, which has a calculated BOD of $220\,mg\,l^{-1}$. A standard bacterial seed (e.g. Bioseed®) should be used instead of settled sewage. Percentage inhibition is calculated using the following equation:

$$\% \text{ inhibition} = \frac{\text{BOD}_s - \text{BOD}_t}{\text{BOD}_s} \times 10 \tag{15.25}$$

where BOD_s is the BOD of the control solution and BOD_t is the BOD of the test solution. Percentage inhibition is plotted against the dilution of the test chemical with the EC_{50} calculated from the slope of the plot.

15.3.3 THE MICROTOX® BIOASSAY TEST

The Microtox® bioassay test, developed by Beckman Instruments Incorporated, is based on the measurement of the light emitted by a lyophilized luminescent bacterium *Photobacterium phosphoreum*. On exposure to toxicants the light output is reduced in proportion to the toxicity of the test substance. Intensity of the light output is also dependent on several external factors including temperature, pH and salinity. In order to minimize these factors the test is conducted to a high level of standardization. This is achieved by using a fresh freeze-dried bacterial inoculum for each set of tests, standard sterile diluent and chemicals, and a standard test apparatus that controls the test temperature at 15°C. The end point of the Microtox® test is the concentration of the test chemical, expressed as the quantity required to reduce light production by *P. phosphoreum* by 50% (EC_{50}) from the base level after incubation intervals of 5 and 15 min with reference to the light emission from the test chemical-free control. The whole assay is highly automated and produces results with low variability. Initial screening tests are carried out in order to determine the most suitable concentrations of the test chemical to be used (Kilroy and Gray, 1995). Microtox® is used at wastewater treatment plants to screen wastewater entering the plant for toxicity, to identify sources of potential contaminants and to check the treatability of new discharges.

15.3.4 DEHYDROGENASE ENZYMATIC ASSAY

Dehydrogenases are intra-cellular enzymes linked to cell respiration and so are good indicators of cellular viability. The tetrazolium salt 2,3,5-triphenyl tetrazolium chloride (TTC) is widely used to measure the inhibition of dehydrogenase activity by chemical toxicants. TTC is reduced through the action of dehydrogenases to triphenyl formazan, a red insoluble precipitate. Following incubation of the test sample in the presence of TTC, the resulting red formazan is dissolved in an organic solvent and the absorbance is read at 485 nm. Percentage inhibition is plotted against the concentration of the test chemical to measure the EC_{50} (Kilroy and Gray, 1995).

REFERENCES

Benefield, L. D. and Randall, C. W., 1980 *Biological Process Design for Wastewater Treatment*, Prentice-Hall, New York.

Clark, J. W., Viessman, W. and Hammer, M. J., 1977 *Water Supply and Pollution Control*, Harper and Row, New York.

Elnabarawy, M. T., Robideau, R. R. and Beach, S. A., 1988 Comparison of three rapid toxicity test procedures: Microtox®, Polytox® and activated sludge respiration inhibition. *Toxicity Assessment*, **3**, 361–73.

IWEM, 1992 *Glossary of Terms Used in Water Pollution Control*, Institution of Water and Environmental Management, London.

Kilroy, A. and Gray, N. F., 1995 Treatability, toxicity and biodegradability test methods, *Biological Reviews of the Cambridge Philosophical Society*, **70**, 243–75.

Monod, J., 1949 The growth of bacterial cultures, *Annual Reviews of Microbiology*, **3**, 371–94.

Mudrack, K. and Kunst, S., 1987 *Biology of Sewage Treatment and Water Pollution Control*, Ellis-Horwood, Chichester.

OECD, 1984 *Guidelines for the Testing of Chemicals: Activated Sludge Respiration Inhibition Test.* Method: 209, Organization for Economic Co-operation and Development, Paris.

Ramalho, R. S., 1977 *Introduction to Wastewater Treatment Processes*, Academic Press, New York.

UK Standing Committee of Analysts, 1982 *Methods for Assessing the Treatability of Chemicals and Industrial Waste Waters and Their Toxicity to Sewage Treatment Processes*, Methods for the Examination of Waters and Associated Materials, HMSO, London.

UK Standing Committee of Analysts, 1988 *Five-Day BOD (BOD5) Test and Dissolved Oxygen in Water*, 2nd edn, Methods for the Examination of Waters and Associated Materials, HMSO, London.

Wheatley, A. D., 1985 Wastewater treatment and by-product recovery. In: Sidwick, J. M., ed., *Topics in Wastewater Treatment, Critical Reports on Applied Chemistry*: *11*, Blackwell Scientific Publications, Oxford, pp 68–106.

FURTHER READING

Gray, N. F., 2004 *Biology of Wastewater Treatment*, 2nd edn, Imperial College Press, London.

Metcalf and Eddy, 1991 *Wastewater Engineering*, 3rd edn, McGraw-Hill, New York.

Chapter 16 Fixed-film Systems

16.1 THE BASIS OF THE PROCESS

In fixed-film reactors, the microbial biomass is present as a film, which grows over the surface of an inert and solid medium. Purification is achieved when the wastewater is brought into contact with this microbial film. Because the active biomass is largely retained within the reactor, there is no need to recirculate any displaced biomass to maintain a sufficient density of micro-organisms, as is the case with completely mixed systems (Section 15.1). The required contact between the film and wastewater is achieved in most fixed-film reactors by allowing the wastewater to pass over the stationary medium on which the film has developed. However, it is not essential for the medium to be stationary, and in more recently developed reactor systems (e.g. rotating biological contactors (RBCs), fluidized beds, etc.) the medium itself moves through the wastewater (Section 16.4.1). Fixed-film reactors can be designed as secondary treatment processes to partially treat (high-rate systems) or fully treat (low-rate systems) either screened or more generally settled wastewater (Bruce and Hawkes, 1983). While they are normally aerobic, for carbonaceous removal and nitrification, anoxic and anaerobic filters are used for denitrification (Section 18.4.1) and for treating moderately strong organic wastewaters, respectively (Section 19.3). While anoxic and anaerobic filters are submerged reactors, that is, the media is permanently submerged under the wastewater, submerged aerobic filters are less common as they require a diffused aeration system to maintain a sufficient oxygen concentration within the reactor. However, submerged aerobic filters are particularly useful where loadings are intermittent and where difficulty occurs in maintaining the minimum wetting loading in order to prevent the filter drying up and so the film dying. Fixed-film systems are now primarily used for the treatment of industrial wastewaters, they have a number of distinct advantages over the activated sludge process for the treatment of domestic wastewaters, especially for smaller populations (Table 15.1).

16.2 PERCOLATING FILTERS

The most widely used fixed-film reactor is the percolating filter, also known as the biological or trickling filter. In its simplest form it comprises of a bed of graded hard material, the medium, about 2.0 m in depth (Fig. 16.1). The settled wastewater is spread evenly over the surface of the bed by a distribution system, which can be used

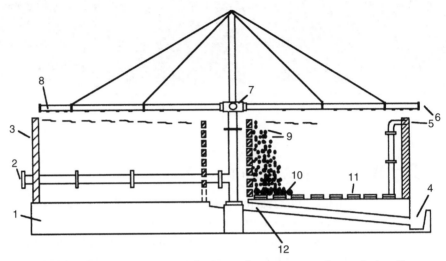

FIGURE 16.1   Basic constructional features of a conventional percolating filter:
1. foundation floor; 2. feed pipe; 3. retaining wall; 4. effluent channel; 5. ventilation
pipe; 6. distributor arm; 7. rotary seal; 8. jets; 9. main bed of medium; 10. base layer
of coarser medium; 11. drainage tiles; 12. central well for effluent collection.
(Reproduced from Bruce and Hawkes (1983) with permission of Academic Press
Ltd., London.)

to regulate the volume and frequency of application of the wastewater. The filter has
a ventilating system to ensure free access of air to the bed, which passes through the
interstices or voids of the medium ensuring all parts of the filter have sufficient oxygen.
The movement of air is normally by natural draft, and, depending on the relative temperature difference between the air and inside the filter, can be in an upward or downward direction. The treated effluent passes through a layer of drainage tiles, which
supports the medium and flows away to the secondary settlement tank (Gray, 2004).

16.2.1 FILM DEVELOPMENT

The microbial film takes 3–4 weeks to become established during the summer and up
to 2 months in the winter. Only then has the filter reached its maximum purification
capacity, although nitrification will take longer to become established. Unlike many
industrial wastewaters, it is not necessary to seed domestic wastewaters treated by percolating filters as all the necessary micro-organisms are present in the sewage itself,
with the macro-invertebrate grazers either flying onto the filter or entering the sewer
via storm water run-off and subsequently colonizing it. The micro-organisms quickly
form a film over the available medium. However, the film only develops on the surfaces that are receiving a constant supply of nutrients, so the effectiveness of media
to redistribute the wastewater within the filter, to prevent channelling and to promote maximum wetting of the medium is an important factor affecting performance.
The film is a complex community of bacteria, fungi, protozoa and other mesofauna,
plus a wide diversity of macro-invertebrates, such as enchytraeids and lumbricid worms,

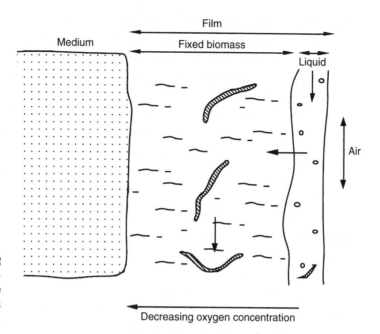

FIGURE 16.2
Diagrammatic rep-
resentation of the
structure of film in a
fixed-film reactor.

dipteran fly larvae and a host of other groups which all actively graze the film (Fig. 15.2) (Curds and Hawkes, 1975; Hawkes, 1983).

The organic matter in the wastewater is degraded aerobically by heterotrophic micro-organisms which dominate the film. The film has a spongy structure, which is made more porous by the feeding activities of the grazing fauna that are continually burrowing through the film. The wastewater passes over the surface of the film and to some extent through it, although this depends on film thickness and the hydraulic loading (Fig. 16.2). In low-rate filters a large proportion of the wastewater may be flowing through the film matrix at any one time, and it is the physical straining action of this matrix, which allows such systems to produce extremely clear effluents. Another advantage is that the greater the proportion of wastewater that flows through the film the greater the micro-organism–wastewater contact time, which is known as the hydraulic retention time (HRT). The higher the hydraulic loading the greater the proportion of the wastewater passing over the surface of the film, which results in a lower HRT and a slightly inferior final effluent. The first stage of purification is the adsorption of organic nutrients onto the film. Fine particles are flocculated by extra-cellular polymers secreted by the micro-organisms and adsorbed onto the surface of the film, where, along with organic nutrients which have been physically trapped, they are broken down by extra-cellular enzymes secreted by heterotrophic bacteria and fungi. The soluble nutrients in the wastewater and resulting from this extra-cellular enzymatic activity are directly absorbed by the film micro-organisms and synthesized.

Oxygen diffuses from the air in the interstices, first into the liquid and then into the film. Conversely, carbon dioxide and the end products of aerobic metabolism diffuse in the other direction. The thickness of the film is critical, as the oxygen can only diffuse for a certain distance through the film before being utilized, leaving the deeper areas of the film either anoxic or anaerobic (Fig. 16.2). The depth to which oxygen will penetrate depends on a number of factors, such as composition of the film, its density and the rate of respiration within the film itself, and has been estimated to be between 0.06 and 4.00 mm. The critical depth is approximately 0.2 mm for a predominately bacterial film increasing to between 3 and 4 mm for a fungal film. Only the surface layer of the film is efficient in terms of oxidation, so only a thin layer of film is required for efficient purification; in fact the optimum thickness in terms of performance efficiency is only 0.15 mm. This means that in design terms it is the total surface area of active film that is important and not the total biomass of the film (Hawkes, 1983).

The film accumulates and generally becomes thicker over time during filter operation. This is due to an increase in microbial biomass from synthesis of the waste and also due to accumulation of particulate material by flocculation and physical entrainment, where the accumulation rate exceeds the rate of solubilization and assimilation by the micro-organisms. Where the wastewater is largely soluble, such as food-processing wastewaters, most of the film increase will be due to microbial growth. Whereas with domestic wastewater that has been poorly settled the accumulation of solids may account for the major portion of the film accumulation. Once the film exceeds the critical thickness an anoxic and subsequently an anaerobic environment is established below the aerobic zone. As the thickness continues to increase most of the soluble nutrients will have been utilized before they can reach the lower micro-organisms, forcing them into an endogenous phase of growth. This has the effect of destabilizing the film as the lower micro-organisms break up, resulting in portions of the film within the filter becoming detached and washed away in the wastewater flow, a process known as sloughing. Although thick-film growths do not reduce the efficiency of the filter, excessive growths can reduce the volume of the interstices, reducing ventilation and even blocking them completely, preventing the movement of wastewater. Severe clogging of the interstices is known as ponding and is normally associated with the surface of the filter causing large areas of the surface to become flooded.

The accumulation of film within a filter bed follows a seasonal pattern, being low in summer due to high metabolic and grazing rates but high in the winter due to a reduced microbial metabolic rate and a reduction in the activity of the grazing fauna. As the temperature increases in the spring there is a discernible sloughing of the film, which has accumulated over the winter months (Fig. 16.3). Temperature has been shown to be an important factor in film accumulation and below 10°C the rate of film accumulation increases rapidly. At higher temperatures a greater proportion of the biochemical oxygen demand (BOD) removed by adsorption is oxidized so fewer solids accumulate. The rate of oxidation decreases as the temperature falls, although the rate of adsorption remains unaltered. Therefore, at the lower temperatures there is a gradual increase in solids accumulation that eventually results in the filters becoming clogged (Fig. 16.4). In the warmest months the high microbial activity may exceed the rate of adsorption, thus reducing the overall film biomass. The grazing fauna also play a significant

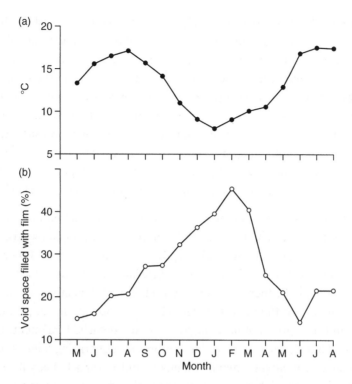

FIGURE 16.3   Seasonal variation in the temperature of sewage applied to a percolating filter and the proportion of void space in the filter filled with film and water. (a) Monthly average temperature of applied sewage, (b) monthly average proportion of voids filled with film measured using neutron scattering. (Reproduced from Bruce *et al.* (1967) with permission of the Chartered Institution of Water and Environmental Management, London.)

FIGURE 16.4
Rate of film accu-
mulation in perco-
lating filters showing
that as the rate of
adsorption is con-
stant, the rate of
oxidation is propor-
tional to the temper-
ature of the film.

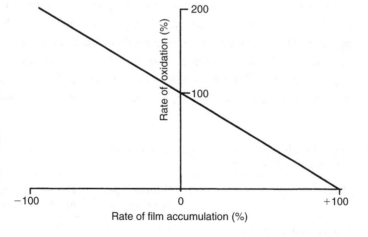

role in reducing the overall film biomass by directly feeding on the film, converting it into dense faeces that are flushed from the filter in the effluent. The spring sloughing of the accumulated film is pre-empted by the increased activity of the grazers, which loosen the thick film from the medium, and once the film has become detached it strips accumulated film from other regions of the bed as it is washed out. Although grazers suppress maximum film accumulation and maintain minimum film accumulation for a long period after sloughing, it is temperature that primarily controls film accumulation. While hydraulic loading in low-rate filters is of little significance compared to the action of macro-invertebrate grazers in controlling film, as the hydraulic loading increases, as is the case after modifications to the process, such as recirculation or double filtration, then physical scouring of the film by the wastewater becomes increasingly important. In high-rate filters, especially those employing modular plastic media, the high-hydraulic loading controls the film development by scouring the film from the smooth surfaced media as it reaches critical thickness (Hawkes, 1983).

The final effluent contains a small amount of solids that are removed by settlement prior to final discharge. The solids comprise of flocculated solids, detached fragments of the accumulated film and the grazing fauna, and finally fragments of their bodies and faeces. These solids are collectively known as humus and like all secondary solids they require further treatment after settlement. The mode of operation will influence the nature of the humus sludge. High-rate operation will produce a humus sludge mainly comprised of flocculated solids and detached fragments of film, while a low-rate filter sludge contains a large proportion of grazing fauna and animal fragments. Sludges containing animal fragments and grazing fauna will be more stable than sludges from high-rate systems where the grazing fauna is absent or reduced. The production of humus varies seasonally with mean production rates for low-rate filters between 0.20 and 0.25 kg humus per kg BOD removed, varying from $0.1 \, \mathrm{kg \, kg^{-1}}$ in the summer to a maximum of $0.5 \, \mathrm{kg \, kg^{-1}}$ during the spring sloughing period. In high-rate systems a greater volume of sludge is produced due to the shorter HRT resulting in less mineralization. The humus production does not vary seasonally to the same extent with a mean production rate of $0.35 \, \mathrm{kg \, kg^{-1}}$, although this is dependent on the nature of the influent wastewater. Sludge production is much less compared with activated sludge, being more stabilized and containing less water (Table 16.1), although as loading increases the mode of purification in percolating filters approaches that of the activated sludge process and the sludge alters accordingly, both in quality and quantity. Separation takes place in a secondary settlement tank

TABLE 16.1 Typical sludge volumes and characteristics from secondary treatment processes compared with primary sludge

Source	Volume (l per head per day)	Dry solids (kg per head per day)	Moisture content (%)
Primary sedimentation	1.1	0.05	95.5
Low-rate percolating filtration	0.23	0.014	93.9
High-rate percolating filtration	0.30	0.018	94.0
Activated sludge (wasted)	2.4	0.036	98.5

Reproduced from Open University (1975) with permission of the Open University Press, Milton Keynes.

traditionally called a humus tank. They are similar in principle to primary settlement tanks, although often of different design. For example, deep square tanks of small cross-sectional area with an inverted pyramid bottom with a maximum upflow rate of $<1\,m\,h^{-1}$ are common (Section 14.2). Percolating filters usually achieve high levels of nitrification, so a long sludge retention time within the humus tank may give rise to anoxic conditions and problems of denitrification resulting in the carry over of sludge in the final effluent. High-rate sludges are more susceptible to denitrification than low-rate sludges that are more stabilized and so exert less of an oxygen demand within the settlement tank. Temperature is also important with gas production much heavier during the summer. While the bulk of the solids settle easily, there is a fraction of fine solids, which do not and are carried out of the humus tank in the final effluent. These fine solids are responsible for a significant portion of the residual BOD in the final effluent, so if a high-quality effluent is required some form of tertiary treatment may be necessary. The residual BOD can be successfully removed by any tertiary treatment process, such as microstraining, sand filtration, upflow clarification or land treatment (Chapter 20) (Gray, 2004).

16.2.2 DESIGN AND MODIFICATIONS

The loading to percolating filters depends on the physical characteristics of the medium and the degree of film accumulation. Hydraulic loading is expressed in cubic metres of wastewater per cubic metre of filter medium per day, although surface loadings or irrigation rates in cubic metres of wastewater per square metre of filter surface are occasionally used instead of volumetric loadings. Organic loading is expressed as kg of BOD per cubic metre of filter medium per day. These loadings are used to differentiate between low-rate ($<3\,m^3\,m^{-3}\,day^{-1}$, $<0.6\,kg\,BOD\,m^{-3}\,day^{-1}$) and high-rate ($>3\,m^3\,m^{-3}\,day^{-1}$, $>0.6\,kg\,BOD\,m^{-3}\,day^{-1}$) filters. The original design was for complete treatment using one filter with the wastewater passing through it only once, producing a 20:30 effluent and full nitrification. This is achieved by using mineral medium of a nominal size of 37.5–50 mm loaded organically at $<0.1\,kg\,BOD\,m^{-3}\,day^{-1}$ and hydraulically at $<0.4\,m^3\,m^{-3}\,day^{-1}$ (Fig. 16.5a). Conventional single-pass filtration can also be used to partially treat wastewater by using a larger nominal sized mineral filter medium (>63 mm), random plastic medium, or more commonly modular plastic medium and loading at very high organic and hydraulic loadings. The high voidage of such medium prevents ponding and heavy accumulation of film. Almost any degree of partial treatment is possible (e.g. $1\,kg\,BOD\,m^{-3}\,day^{-1}$ will permit 80–90% BOD removal, while 3–$6\,kg\,BOD\,m^{-3}\,day^{-1}$ will permit 50% BOD removal), although there is no nitrification. Often referred to as roughing or high-rate treatment it is used to remove large weights of BOD prior to discharge where there is a large dilution available or to a low-rate secondary treatment process. Where it is followed by a low-rate filter, it is known as two-stage filtration (Fig. 16.5d). This system is used for the treatment of strong organic wastewaters where a high-quality final effluent is required. The first stage removes 70% of the BOD while

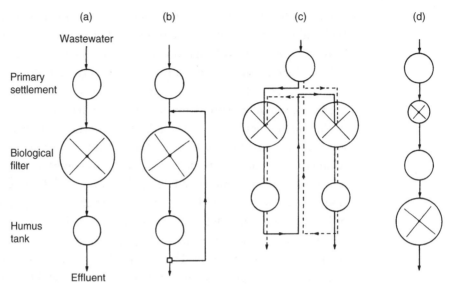

FIGURE 16.5
The main systems
of operation of
percolating filters.
(a) Single filtration,
(b) recirculation,
(c) ADF and
(d) two-stage filtra-
tion with high-rate
primary biotower.

the second stage polishes the effluent producing a 20:30 fully nitrified effluent. Loading rates for the first stage are between 1.6 and 2.3 kg BOD m^{-3} day^{-1} and 0.04–0.12 kg BOD m^{-3} day^{-1} for the second stage.

There are a number of modifications to the original design. Recirculation dilutes the incoming wastewater with returned final effluent. This is used to dilute strong industrial wastewaters, and the higher hydraulic loading has a flushing effect on the top section of the filter encouraging a thinner more active film and ensuring better utilization of the medium. This allows the loading to be increased to between 0.15 and 0.2 kg BOD m^{-3} day^{-1} with little effect on final effluent quality and only a slight reduced nitrification efficiency. When the organic material in the influent is soluble and readily oxidized, such as food-processing wastes, then the organic loadings can be as high as 0.5–0.6 kg BOD m^{-3} day^{-1}. Recirculation ratios of effluent to influent for domestic sewage is 1:1 or 2:1, but for strong industrial wastewaters it can be as high as 25:1 (Fig. 16.5b). The actual recirculation ratio can be calculated by estimating the BOD of the diluted influent fed to the filter (S_o):

$$S_o = \frac{(Q_i S_i + Q_r S_r)}{(Q_i + Q_r)}$$ (16.1)

where the flow rate (Q) and the BOD (S) are known for the influent (i) and the treated effluent (r). The recycle ratio R_r is Q_r:Q_i, and the equation can be rearranged as:

$$S_o = \frac{(S_i + R_r S_r)}{(1 + R_r)}$$ (16.2)

As the effluent BOD in low-rate systems is usually very low, S_o can be estimated as:

$$S_o = \frac{S_i}{(1 + R_r)} \tag{16.3}$$

Recirculation of unsettled effluent from the filter is common otherwise excessive humus tank capacity would be required. For example, a 1:1 recirculation ratio would double the flow to the humus tank. The low substrate conditions at the top of the filter encourage bacterial growth rather than fungal development (Pike, 1978).

There are a number of simple empirical models that can be employed in the design of percolating filters (CIWEM, 2000). The most widely used model is that prepared by the US National Research Council (National Research Council, 1946):

$$E = \frac{100}{1 + 0.448\sqrt{W/VF}} \tag{16.4}$$

where E is the efficiency of BOD removal inclusive of recirculation and sedimentation (%), W is the BOD loading rate ($kg\,day^{-1}$), V the filter volume (m^3) and F the recirculation factor (i.e. the average number of passes through the filter) (Equation (16.5)):

$$F = \frac{1 + R}{(1 + R/10)^2} \tag{16.5}$$

where:

$$R = \frac{Q_r}{Q} \tag{16.6}$$

and Q_r is the recirculation flow ($l\,min^{-1}$) and Q is the wastewater flow ($l\,min^{-1}$).

Equation (16.4) can be used to calculate the volume of mineral filter media required by:

(a) calculating the required BOD removal efficiency (E):

$$E = \frac{\text{influent BOD} - \text{required effluent BOD}}{\text{influent BOD}} \tag{16.7}$$

(b) calculating the recirculation factor (F) using Equation (16.5);
(c) calculating the influent BOD loading rate (W):

$$W = \text{BOD}_{inf} \times P \times \text{population served} \times 10^{-6} \; (kg\,day^{-1}) \tag{16.8}$$

where BOD_{inf} is the influent BOD ($mg\,l^{-1}$) and P the per capita flow rate ($l\,ca^{-1}day^{-1}$).

Using Equation (16.4) the volume of filter media (V) required for the desired BOD removal efficiency using a certain recirculation factor can be calculated. Where a second filtration stage is used the removal efficiency of this stage (E_2) is calculated using Equation (16.9), where W' is the second-stage BOD loading rate (kg day^{-1}) and E_1 is the fractional efficiency of BOD removal of the first filtration stage inclusive of recirculation and sedimentation.

$$E_2 = \frac{100}{1 + (0.448/1 - E_1)\sqrt{W'/VF}} \tag{16.9}$$

Alternating double filtration (ADF) employs two identical filters loaded in series (Fig. 16.5c). Using a coarser medium (63–75 mm) to allow for higher hydraulic and organic loadings there is a rapid and heavy development of film in the first filter. After 1–2 weeks the sequence of loading is reversed to prevent clogging in the first filter, putting the micro-organisms under food-limiting conditions forcing them into endogenous respiration reducing overall film biomass but at the same time retaining a healthy and active film. In this way both filters have periods of high and low organic loading allowing higher rates of treatment, reducing sludge production and preventing ponding. For domestic sewage a 20:30 effluent can be achieved at loadings of $<1.5\,\mathrm{m^3\,m^{-3}\,day^{-1}}$ and $<0.24\,\mathrm{kg\,BOD\,m^{-3}\,day^{-1}}$, although much higher loadings are normal. No nitrification is possible.

16.3 OPERATIONAL PROBLEMS OF PERCOLATING FILTERS

All systems have problems, although percolating filters are surprisingly reliable and problem free. However, the most reported problems are media-related, such as ponding and poor performances during winter, also inhibition of nitrification and fly nuisance (Hawkes, 1983). Nitrification is also discussed in Section 18.4.

16.3.1 PONDING

Performance is generally related to the specific surface area of the medium provided that film accumulation does not result in blocking of the interstices. The ideal medium should have a high specific surface area for the maximum development of film, high voidage to permit the movement of air and large interstices to prevent ponding (Table 16.2). However, selecting a medium is a compromise between the conflicting requirements of a high surface area, which requires a relatively fine grade of filter media, and large voids which are obtained with coarser grades of media. Smaller grades of medium produce better effluents but the interstices tend to fill with film in the winter and block the flow of wastewater through the filter (ponding), which results in a loss of performance. Surface texture is also important with rough mineral

TABLE 16.2   Main requirements for a percolating filter medium

Strong enough to take weight of medium and accumulated
  film without deforming or being crushed
Chemically non-reactive
Durability (minimum life of 20 years)
High surface area
High voidage
Large interstices
Varied ecological niches

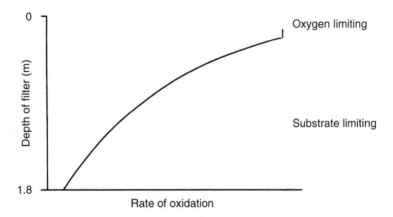

FIGURE 16.6   Rate of oxidation in relation to the depth of a percolating filter indicating where oxygen and substrate concentration are major limiting factors. (Reproduced from Tebbutt (1979) with permission of Butterworth-Heinemann Ltd., Oxford.)

medium, such as clinker and blast furnace slag having a specific surface area up to 17% greater than smooth mineral medium, such as granite or gravel. In performance terms smoother medium is marginally less effective than rough surfaced medium, although the latter is more susceptible to excessive accumulation of solids which shears from smooth medium more readily (Gray, 2004). In practice, as the film develops many of the smaller holes in rough textured medium become filled with film and organic debris, so that the surface area of such medium tends to approach that for smooth medium types. Rougher surfaced mineral medium does provide a greater range of niches for the flora and fauna of filters, resulting in a greater diversity of species, as well as allowing greater hydraulic loadings to be applied before the film is sheared off. So performance will vary seasonally according to the medium used, with larger grades producing a better effluent in the winter, whereas the smaller grades perform best in the summer. Because the maximum availability of food occurs at the surface of a filter, it is here that maximum film accumulation and oxidation occurs. In fact the bulk of the active biomass of the filter is located in the top third of the reactor. Consequently, the rate of oxidation is limited at the top of the filter by the availability of oxygen, with the food in excess, while lower down the food becomes limiting (Fig. 16.6). Using terms more associated with the activated sludge process, the food/micro-organism (f/m) ratio decreases with depth, although the rate of decrease rapidly slows as the density of active micro-organisms also falls. In addition, the micro-organisms alter their growth rate with depth from rapid logarithmic growth at the top to endogenous respiration at the base. Filters are plug-flow reactors and so bands of different micro-organisms occur through the bed, with maximum heterotrophic activity at the top and nitrification at the bottom.

Random plastic medium overcomes many of the problems of conventional mineral medium by having a high specific surface area (up to $330 \, m^2 m^{-3}$ for Flocor RC), large interstices and a high voidage (>90%), almost double that for similar grades of blast furnace slag. This allows much higher organic loadings to be applied to filters without the risk of ponding. Such media has also been successfully used to up-rate existing filters by replacing the top 500–900 mm layer of mineral with plastic medium, although ponding can occur at the interface (Gray and Learner, 1984). Several problems are associated with plastic medium: first, the lack of bulk density in relation to mineral medium, resulting in plastic filters unable to retain heat (rather like a storage heater). In a random plastic filter the temperature inside the bed is very close to the air temperature, while mineral medium can be over 10°C higher than the air temperature during winter. This rapid temperature fluctuation in random plastic filters reduces the rate of biological activity and suppresses nitrification. A fungus *Subbaromyces splendens* (*S. splendens*), not yet isolated from the natural environment, colonizes plastic medium. Fungi are a common component of the film in the upper regions of percolating filters and are associated with strong sewages and industrial wastewaters. However, *S. splendens* grows extensively at all depths on random plastic medium, especially those treating weak domestic sewage and some industrial wastewaters. It forms rather large dense growths, that can block the large interstices in random and modular plastic medium at the surface, but more commonly cause ponding at the centre or towards the base of the filter. As random and modular plastic medium are often housed in prefabricated structures above ground, and which are often much taller than conventional mineral filters, when ponding occurs at depth and the filter fills with liquid, the increase in total bulk density causes such units to structurally fail (Gray, 1983). This has been a particular problem in food and pharmaceutical wastewater treatment installations. Film accumulation and ponding can also be controlled by changing the mode of operation by using recirculation or ADF. Controlling the frequency and duration of dosing can also control ponding effectively (Hawkes, 1983).

16.3.2 NITRIFICATION

In low-rate single-pass filters containing 50 mm stone medium, virtually full nitrification can be obtained throughout the year with a specific ammonia removal rate of between 120 and $180 \, mg \, m^{-2} day^{-1}$ when loaded at $0.1 \, kg \, BOD \, m^{-3} day^{-1}$, resulting in a final effluent low in ammonia but rich in nitrate. In a percolating filter nitrifying bacteria tend to become established later than heterotrophs, with *Nitrosomonas* spp. established before *Nitrobacter* spp., as ammonia is abundant (Section 15.1). The first sign of nitrification in a filter is the production of nitrite rather than nitrate. The number of nitrifying bacteria and the level of nitrifying activity both increase with depth, resulting in the upper level of single-pass filters being dominated by heterotrophs and the lower section containing a proportionately higher number of nitrifying bacteria. The reason for this apparent stratification is due to a number of factors. The autotrophic bacteria responsible for nitrification are slow growing compared to heterotrophs and have an even more reduced growth rate in competitive situations. So in the upper layers of the filter, where there is abundant organic matter, the heterotrophs will dominate. Nitrifiers are extremely sensitive to toxic compounds in the wastewater, especially heavy metals, and so the presence of such compounds will limit the growth of the bacteria until the compounds have been removed from the wastewater by adsorption by the heterotrophic film as it passes through the filter bed. Research has shown that the process is also inhibited when the oxygen concentration in the influent wastewater is limited. Nitrifying bacteria are strict aerobes and so will be inhibited by reduced aerobic conditions caused by high heterotrophic activity and, as nitrification is a high-oxygen-consuming process, adequate supplies of air are required. While organic matter does not directly inhibit nitrification, the nitrifying bacteria need to be attached to a stable surface, which suggests inhibition may be due to competition for space. When loadings are increased, extending the depth of heterotrophic activity, nitrifying bacteria are overgrown and eliminated by the quicker growing heterotrophic bacteria. Nitrification is virtually eliminated by hydraulic loadings of domestic wastewater of $>2.5 \, \mathrm{m^3 m^{-3} day^{-1}}$ due to enhanced heterotrophic growth extending throughout the depth of the bed and effectively pushing the nitrifying organisms out of the filter. Temperature also has a marked influence on nitrification and the large fluctuations in temperature seen in random plastic filters account for the low degree of nitrification which occurs with such media. Although the threshold temperature for the process is 10°C, with domestic wastewater rarely falling to below 12°C (normal annual range 12–18°C), a few degrees reduction in the temperature below 10°C is likely to have a disproportionate reduction on nitrification. This is clearly seen in comparative studies between mineral and plastic medium filters where the high voidage of the plastic medium allows a greater degree of ventilation and large diurnal changes in temperature similar to the air temperature (Barnes and Bliss, 1983).

Ammonia is not only removed by nitrification, but also by volatilization of free ammonia and by metabolism into new cellular material. At times of low film accumulation, when the grazing fauna population is still large, the ammonia concentration in the final effluent will be high. This increase in the ammonia concentration is due to the excretion products of the grazing fauna. The oxygen profile formed by diffusion through the film results in nitrate ions being lost by denitrification as gaseous nitrogen from the anoxic zone near the biomass–medium interface.

16.3.3 FLY NUISANCE

Many of the dipteran flies form dense swarms as they emerge from the filter bed, and for most chironomids swarming is a prelude to mating. The numbers of flies emerging can be so large that from a distance it has the appearance of smoke coming from the filters. Emergence is affected by temperature, light intensity and wind velocity. Insects will not fly unless the temperature is above a threshold value, for example, 10°C for *Psychoda alternata*, 7–8°C for *Hydrobaenus minimus*, 4°C for both *Metriocnemus* spp. and *Sylvicola fenestralis*. For every 1.2°C rise in temperature above the threshold value up to 24°C the number of *Sylvicola* flies in flight doubles. Diel periodicity is observed for all the species with the peak of emergence for *P. alternata* and *Psychoda severini* in the early afternoon, *S. fenestralis* at dusk and a smaller peak at dawn. Wind velocity is an important factor with swarms quickly broken up and individuals unable to fly if the wind is too strong. In essence the stronger the fly the greater the wind velocity it can withstand, with *S. fenestralis* able to withstand velocities up to $6.7 \, \mathrm{m \, s^{-1}}$. At the treatment plant the density of flies can be problematic making working conditions unpleasant as flies are drawn into the mouth and nostrils, and are caught in the eyes. None of the commonly occurring flies, including *S. fenestralis*, bite. However, although harmless the large size and intimidating appearance of *S. fenestralis* does alarm people resulting in frequent complaints from nearby residents. Psychodid flies are found up to 1.6 km away from the treatment plant, although this does depend on the direction of the prevailing wind. *Sylvicola* are found in quite large numbers up to 1.2 km, while few reach farther than 2.4 km from the plant. However, flies are well known to cause both aesthetic and public health problems to those living or working close to treatment plants with *P. alternata, P. severini, S. fenestralis, H. minimus* and *Metriocnemus hygropetricus* the main nuisance species. In general *P. alternata* comprises on average >80% of the total annual emergence of flies from percolating filters (Hawkes, 1983).

Three control options are available for remedying fly nuisances. Physical methods have been least successful and are not recommended. These include flooding filters to eliminate fly species; covering the surface of the medium with a layer of finer media (13–19 mm) to a depth of 250 mm which reduces the numbers of adult *Psychoda* and *Sylvicola* emerging, but results in severe ponding during the winter; and finally enclosing filters which is rather expensive and may not always be successful due to the need for air vents that allow flies access and escape. The most widely adopted control method is the use of chemical insecticides. Currently the two insecticides most widely used for controlling filter flies are Actellic (pirimiphos methyl) and Dimilin (difluben-zuran) which are 0-2-diethylamine-6-methylpyrimidin-4-yl, 0-0-dimethyl phospho-rothioate and 1-(4-chlorophenyl)-3-(2,6-difluorobenzoyl) urea, respectively. However, neither appears particularly effective against *S. fenestralis*. Limited control can be obtained by changing the operational practice of the plant; for example, by limiting the amount of film accumulation, especially at the surface, by reducing the f/m ratio or by using one of the modifications, such as recirculation, double filtration or ADF.

Increasing the hydraulic loading may also make it more difficult for dipterans to complete their life cycles in the filter. Continuous dosing using nozzles prevents emergence and severely reduces the available surface area for the flies to lay eggs. Biological control using the entomopathogenic bacterium *Bacillus thuringiensis* var. *israelensis* is successfully used to control filter flies. The bacterium, which is commercially available as Teknor®, is effective against fly larvae only and does not affect other filter fauna or performance, although it is comparatively expensive.

16.4 OTHER FIXED-FILM REACTORS

16.4.1 ROTATING BIOLOGICAL CONTACTORS

The basic design of rotating biological contactors (RBCs) consists of a series of flat or corrugated discs, normally made out of plastic, 1–3 m in diameter and mounted on a horizontal shaft. This is driven mechanically so that the discs rotate at right angles to the flow of settled sewage. The discs, which are spaced 20–30 mm apart, are placed in a contoured tank that fits fairly closely to the rotating medium so that up to 40% of their area is immersed and are slowly, but continuously, rotated. Some RBC designs incorporate primary settlement, but all allow for secondary settlement (Fig. 16.7).

The flow of wastewater through the tank, and the action of the rotating medium, produces a high hydraulic shear on the film ensuring efficient mass transfer from the liquid into the film as well as preventing excessive film accumulation. Discs are arranged in groups separated by baffles to minimize short circuiting, to reduce the effect of surges in flow and to simulate plug flow. The spacing of discs is tapered along the shaft to ensure that the higher film development in the initial stages of the processes do not cause the gaps between discs to become blocked. The speed of rotation varies

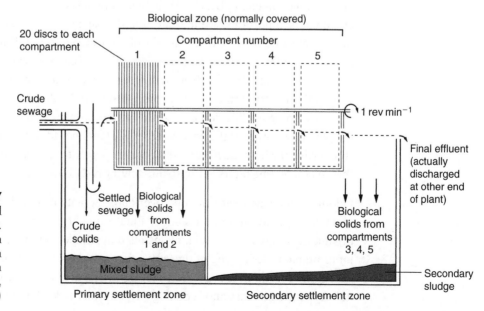

FIGURE 16.7
Rotating biological contactor. (Reproduced from Mann (1979) with permission of WRc plc, Medmenham.)

from 0.75 to 1.0 rev min^{-1}, although the velocity of the medium through the wastewater should not exceed 0.35 m s^{-1}. Fast rotation speeds may cause excessive sloughing of the film, while too slow rotation may lead to insufficient aeration and allow settlement of film in the aeration section.

There is usually a small head loss of between 10 and 20 mm between each compartment. As the discs rotate they alternatively adsorb organic nutrients from the wastewater and then oxygen from the atmosphere for oxidation. Any solids in the treated wastewater are separated out in the settlement zone before discharge.

The discs being well balanced along the drive shaft, and being of light-weight plastic, require little energy to rotate them, so that only small motors are required. For example, a 0.3 kW motor is sufficient to operate a 300 population equivalent unit. An electric motor is not always necessary to drive the rotor. Air can be sprayed into the rotating medium from below and to one side of the horizontal shaft to induce movement. The liquid movement and asymmetric buoyancy from the air collecting on one side of the medium induces it to rotate. This system also prevents the risk of anaerobiosis in the first compartment or RBC unit, where they are used in series for large populations.

RBC units are covered with glass-reinforced plastic (GRP) or plastic shrouds to protect the medium and film from the weather, as well as to reduce the load on the motor turning the rotor due to wind. Covering insulates the system reducing heat loss and increasing the rate of oxidation, allowing their use even in the coldest climates. The cover also reduces noise, eliminates fly nuisance and controls odour to some extent. RBCs can serve single families and huge cities. Larger units are generally not covered but housed in vented buildings and are operated in series.

Loading varies according to manufacturers' specifications and is expressed as a surface organic loading (g BOD m^{-2} day^{-1}). Both sides of the discs, including the free and immersed portions, are measured in the calculation. To ensure a 20:30 effluent the loading rate of settled sewage should not exceed 6 g BOD m^{-2} day^{-1}. For good roughing treatment this loading rate can be increased by a factor of 10. Low-rate percolating filters require significantly lower surface loading rates of between 1.0 and 1.2 g BOD m^{-2} day^{-1} in order to achieve a similar 20:30 effluent. Depending on the organic loading nitrification can occur, but towards the end of the system. Owing to the plug-flow nature of the process there is a tendency for nitrite to accumulate in the compartment where nitrification first occurs. The nitrogen oxidation rate is similar to that in percolating filters at approximately 1 g NH$_3$-N m^{-2} day^{-1} (Pike, 1978).

Odour is associated with organic overload, septicity caused by thick-film development or septicity of settled sludge. Final effluent can be recirculated to reduce organic overload or a second tank installed in parallel. Septicity can be controlled by more frequent desludging of the settlement zone. A common problem, especially at hotels and restaurants, is grease covering the discs and film reducing the biological activity of the biomass and reducing oxygen transfer leading to septicity and odour.

Efficient grease traps only partially control this problem, which requires more careful usage and disposal of fats at source.

Like septic tanks, RBCs have a small head loss (<100 mm) and so are low form structures. This combined with green covers makes them inconspicuous and ideal for hotels, golf courses and other amenity areas where screening may be a problem. The main advantages of RBCs over traditional fixed-film reactors are no ponding, complete wetting of the medium, unlimited oxygen, low land area requirement (<10% than for low-rate filtration), low sludge production, excellent process control, ease of operation, high BOD removal, good nitrification, no distribution problems, no recirculation required, good settleability of sludge, low fly nuisance, inconspicuous, low noise and low odour. Disadvantages include cost, loss of treatment during power loss, frequent motor and bearing maintenance, and problem of excessive film build up on discs after power failure resulting in damage and possible failure of the motor when restarted.

16.4.2 SUBMERGED AERATED FILTERS

More widely known as biological aerated filters (BAFs) the active film grows over a support medium that is completely submerged in wastewater with oxygen supplied by diffusers at the base of the reactor. Owing to the high specific surface area of the media, large weights of film can develop making these filters very efficient. The low shear exerted on the film results in little natural sloughing so the film builds up rapidly requiring regular backwashing to prevent the medium becoming blocked. Flow in BAFs is not dependent on gravity and so can be either in an upward or downward direction.

Like percolating filters media selection is based on specific surface area, surface roughness, durability and cost. A wide range of media is available although most is plastic. Modular plastic media similar to that used in high-rate percolating filters (biotowers) can be used in both upflow and downflow systems, as can random granular media which forms a layer at the base of the reactor (Fig. 16.8). This media ranges from 2 to 10 mm in diameter and includes pumice stone, expanded shale and plastics. Larger media is used for carbonaceous oxidation as larger interstices are needed for film development. Smaller media can be used for nitrification where the film development is significantly less. A rough surface ensures adequate biomass is retained during backwashing for rapid recovery of full biological activity. Floating media can only be used in upflow reactors and is made out of plastics with a specific gravity of <1.

Aeration is provided by lateral aeration pipes with sparge holes up to 2 mm in diameter. Additional aeration capacity is required for air scouring the medium during backwashing. Owing to the pressure, the air supplied is usually warm which increases the rate of oxidation, especially during the winter.

As the biomass accumulates on the random media the head loss across the filter increases and may eventually become completely blocked. Regular backwashing is

Sunken granular media, downward flow

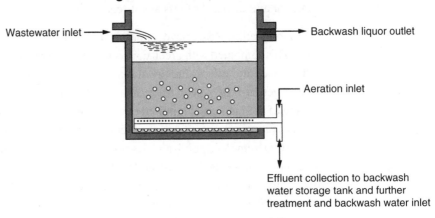

Sunken granular media, upward flow

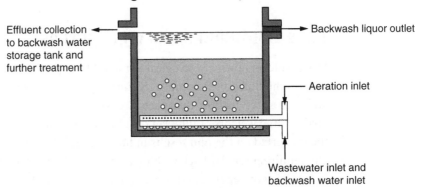

Floating granular media, upward flow

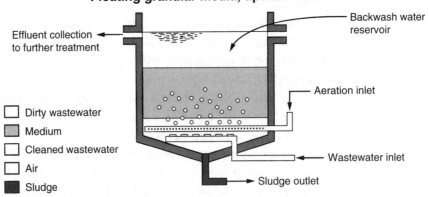

FIGURE 16.8
Commonest designs
of submerged BAFs.

required to remove the excessive film (daily for carbonaceous and monthly for nitrifying BAF systems). The influent is cut off and the media expanded to release excessive film and interstitial solids by increasing the air flow by two to three times the normal rate. Final effluent is introduced with the air to scour and wash away the loose biomass and solids. The air is then switched off and the water continues to backwash the filter medium. Finally the water is also switched off and the medium is allowed to settle before the reactor is restarted. There is a short recovery period during which the effluent is normally turbid due to solids loss. Modular media does not require backwashing although settled solids are occasionally drained through a sump trap in the base of the reactor. This is also necessary with floating media systems.

The high concentration of biomass and high oxidation rates can produce high-quality effluents with low suspended solids concentrations, as well as removing large weights of BOD per unit volume. Normal organic loading rates fall between 0.25 and 2.0 kg BOD m^{-3} day^{-1}. Nitrification is possible at low organic loadings, although, due to the regular backwashing required for carbonaceous systems, it is normal to have separate reactors for nitrification. Solids removal from effluents is usually achieved by the use of physical filters of similar design using a smaller smooth medium, which also needs regular backwashing (e.g. twice a day).

These fully enclosed systems allow a high degree of operational control, and there is little odour or fly nuisance associated with them. They are normally used for the treatment of industrial wastewaters but, due to their compactness and high efficiency, they are becoming increasingly popular for domestic sewage treatment.

16.4.3 FLUIDIZED BEDS

These processes use even smaller media, such as sand (0.2–2.0 mm diameter), glass or anthracite, which provides a very high specific surface area compared to other media ($>3000 m^2 m^{-3}$), allowing considerable biomass to develop equivalent to a mixed liquor suspended solids (MLSS) concentration of up to 40 000 mg l^{-1} (Section 17.2.1). Porous media is also used to allow the biomass to grow inside as well on the surface of the media (e.g. reticulated polyester foam pads). A high density of biomass is maintained without the need for recirculation, however this results in a very high oxygen demand that can only be satisfied by the use of pure oxygen.

Oxygen is injected into the influent stream as it enters the base of the reactor giving an oxygen concentration approaching 100 mg l^{-1}. The expansion of the media is controlled by the rate of wastewater input, ensuring that treated effluent can be removed from the reactor above the expanded layer without loss of media or solids. A secondary settlement tank is normally employed but is not absolutely necessary as all the solids can be retained within the reactor by careful control of the upflow system (Fig. 16.9). The film-coated medium is regularly removed and cleaned by passing it through a hydroclone where a high shear strips away any attached microbial growth. The clean medium is then returned but the sludge must be treated further before disposal. Designed for strong wastewaters, especially where the loading may be variable, the system has a short HRT (15–20 min) and a high BOD removal rate ($>95\%$). When used for nitrification only, then a wastewater with a 20 mg l^{-1} ammonia concentration can be reduced by 99% in 11 min at 24°C.

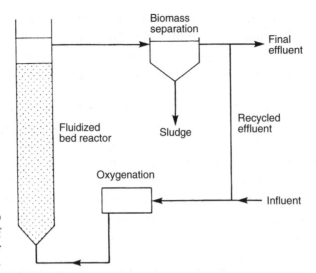

FIGURE 16.9
Typical design of
fluidized bed reactor
system.

Fluidized beds are highly compact systems, but expensive to run due to the cost of pure oxygen and the greater use of pumps. They produce no odour or fly nuisance and produce a concentrated sludge (10% dry solids). Apart from aerobic carbonaceous oxidation and nitrification, such beds can also be operated in an anoxic or anaerobic mode. Anoxic beds are used for denitrification while anaerobic beds, which expand rather than fluidize, are used to convert carbonaceous wastes into gaseous end products that require a gas–liquid separation stage.

16.4.4 NITRIFYING FILTERS

All the above designs of fixed-film reactors can be used solely for the oxidation of ammonia (nitrification). They are similar to normal filter design except smaller grades of media can be used as minimum development of heterotrophic film will occur as the wastewater has already passed through a carbonaceous oxidation stage. The nitrifying bacteria are very slow growing and produce minimal microbial biomass so that ponding or excessive film accumulation does not occur (Section 16.3.2). Where a standard percolating filter design is used then up to 120 g of ammonia–nitrogen can be oxidized per m^3 of medium per day depending on the specific surface area of the medium (Barnes and Bliss, 1983).

REFERENCES

Barnes, D. and Bliss, P. J., 1983 *Biological Control of Nitrogen in Wastewater Treatment*, E. and F. N. Spon, London.

Bruce, A. M. and Hawkes, H. A., 1983 Biological filters. In: Curds, C. R. and Hawkes, H. A., eds, *Ecological Aspects of Used-water Treatment*: 3. *The Processes and Their Ecology*, Academic Press, London, pp 1–111.

Bruce, A. M., Truesdale, G. A. and Mann, H. T., 1967 The comparative behaviour of replicate pilot-scale percolating filters, *Journal of the Institution of Public Health Engineers*, **6**, 151–75.

CIWEM, 2000 *Biological Filtration and Fixed Film Processes*. Chartered Institution of Water and Environmental Management, London.

Curds, C. R. and Hawkes, H. A. (eds), 1975 *Ecological Aspects of Used-water Treatment*: 1. *The Organisms and their Ecology,* Academic Press, London.

Gray, N. F., 1983 Ponding of a random plastic filter medium due to the fungus *Subbaromyces splendens* Hesseltine in the treatment of sewage, *Water Research*, **17**, 1295–302.

Gray, N. F., 2004 *Biology of Wastewater Treatment*, 2nd edn, Imperial College Press, London.

Gray, N. F. and Learner, M. A., 1984 Comparative pilot-scale investigation into uprating the performance of percolating filters by partial medium replacement, *Water Research,* **18**, 409–22.

Hawkes, H. A., 1983 The applied significance of ecological studies of aerobic processes. In: Curds, C. R. and Hawkes, H. A., eds, *Ecological Aspects of Used-water Treatment*: 3. *The Processes and their Ecology*, Academic Press, London, pp 173–334.

Mann, H. T., 1979 *Septic Tanks and Small Sewage Treatment Plants*, Technical Report 107, Water Research Centre, Stevenage.

National Research Council, 1946 Trickling filters (in sewage treatment at military installations), *Sewage Works Journal*, **18** (5).

Open University, 1975 *Environmental Control and Public Health*, Unit 2, PT272, Open University Press, Milton Keynes.

Pike, E. B., 1978 *The Design of Percolating Filters and Rotary Biological Contactors, Including Details of International Practice*, Technical Report TR93, Water Research Centre, Stevenage.

Tebbutt, T. H. Y., 1979 *Principles of Water Quality Control*, Pergamon Press, London.

FURTHER READING

Gray, N. F., 2004 *Biology of Wastewater Treatment*, 2nd edn, Imperial College Press, London.

Qasim, S. R., 1994 *Wastewater Treatment Plants: Planning, Design and Operation*, Technomic Publishing, Lancaster, PA.

Chapter 17 Activated Sludge

17.1 THE PROCESS

The activated sludge process relies on a dense microbial population being in mixed suspension with the wastewater under aerobic conditions. With unlimited food and oxygen, extremely high rates of microbial growth and respiration can be achieved resulting in the utilization of the organic matter present to either oxidized end products (i.e. CO_2, NO_3, SO_4 and PO_4) or the biosynthesis of new micro-organisms (Section 15.1). The activated sludge process relies on five inter-related components: the reactor, the activated sludge itself, the aeration/mixing system, the sedimentation tank and the returned sludge (Table 17.1 and Fig. 17.1). However, there is an increasing interest in replacing the sedimentation tank unit with either an internal or

TABLE 17.1   Main components of all activated sludge systems

1. *The reactor*: This can be a tank, lagoon or ditch. The main criteria of a reactor are that the contents can be adequately mixed and aerated. The reactor is also known as the aeration tank or basin
2. *Activated sludge*: This is the microbial biomass within the reactor which is comprised mainly of bacteria and other microfauna and flora. The sludge is a flocculant suspension of these organisms and is often referred to as the mixed liquor. The normal concentration of mixed liquor expressed as suspended solids (MLSS) is between 2000 and 5000 mg l^{-1}
3. *Aeration/mixing system*: Aeration and mixing of the activated sludge and incoming wastewater are essential. While these tasks can be performed independently they are normally carried out using a single system, either surface aeration or diffused air is used
4. *Sedimentation tank*: Final settlement (or clarification) of the activated sludge displaced from the aeration tank by the incoming wastewater is required. This separates the microbial biomass from the treated effluent
5. *Returned sludge*: The settled activated sludge in the sedimentation tank is recycled back to the reactor to maintain the microbial population at a required concentration in order to ensure continuation of treatment

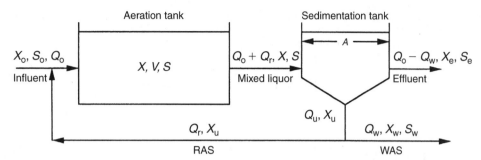

FIGURE 17.1   Schematic diagram of the activated sludge process where A is the surface area of the sedimentation tank, V is the aeration tank volume, S is the substrate (food) concentration, Q_o is the influent flow rate, Q_r is the returned activated sludge (RAS) flow rate, Q_w is the waste activated sludge (WAS) flow rate, X is the aeration tank MLSS concentration, X_u is the RAS suspended solids concentration, X_e is the effluent suspended solids concentration and X_w is the WAS suspended solids concentration, which is normally equal to X_u.

external membrane filtration unit to separate the solids from the final effluent (Section 20.8.5).

Removal of organic matter (substrate) in the activated sludge process comprises of three mechanisms:

(a) adsorption and also agglomeration onto microbial flocs,
(b) assimilation which is the conversion to new microbial cell material,
(c) mineralization which is complete oxidation.

The predominant removal mechanism can be chosen by specific operating conditions. For example, conditions favouring assimilation removes substrate by precipitating it in the form of biomass, which results in a higher proportion of the cost required for sludge separation and disposal (high-rate activated sludge). Under conditions favouring mineralization the volume of biomass is reduced under endogenous respiratory conditions. This results in lower sludge handling costs but higher aeration costs. Currently the higher cost of sludge treatment and disposal favours plants operating with low sludge production. The relationship between substrate (food) concentration and sludge biomass (micro-organisms) concentration is a fundamental one in activated sludge operation (Fig. 17.2). In the activated sludge plant the mass of micro-organisms multiply rapidly in presence of oxygen, food and nutrients. After maturation, the micro-organisms are developed to assimilate specific waste (log growth phase), which is the period of maximum removal. Then under substrate-limiting conditions, the micro-organisms enter a declining growth phase leading eventually to auto-oxidation (or endogenous respiration). In practice activated sludge processes operate towards the end of the log phase and in the declining/stationary growth phases (Section 15.2.1).

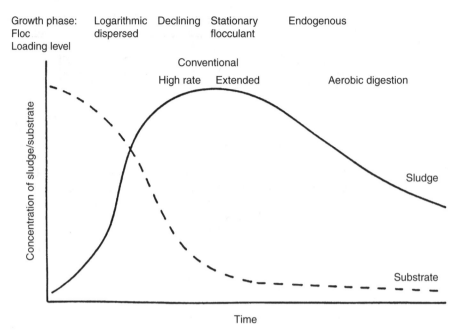

FIGURE 17.2
The microbial growth curve showing the operational growth phase used in different modes of activated sludge. (Reproduced from Winkler (1981) with permission of Ellis Horwood Ltd, Chichester.)

17.2 PROCESS CONTROL

A number of parameters are used to operate activated sludge plants. Those most important in process control can be categorized as:

(a) biomass control,
(b) plant loading,
(c) sludge settleability,
(d) sludge activity.

These are considered in detail below.

17.2.1 BIOMASS CONTROL

Mixed liquor suspended solids

The concentration of suspended solids in the aeration tank, commonly known as the mixed liquor suspended solids (MLSS) concentration, is a crude measure of the biomass available for substrate removal. It is the most basic operational parameter and is used to calculate other important operating parameters. Expressed either in $mg\,l^{-1}$ or $g\,m^{-3}$, some of the MLSS may be inorganic, so by burning the dried sludge at 500°C in a muffle furnace the MLSS can be expressed as the mixed liquor volatile suspended solids (MLVSS) which is a more accurate assessment of the organic fraction and hence of the microbial biomass. However, neither the MLSS nor the MLVSS can distinguish between the active and non-active microbial fraction, or the level of sludge activity.

TABLE 17.2   Comparison of loading and operational parameters for different activated sludge treatment rates

Treatment rate	Retention period (h)	BOD loading per capacity (kg BOD m⁻³d⁻¹)	Sludge loading (f/m) (kg BOD kg⁻¹d⁻¹)	Sludge age (d)	Sludge production kg dry sludge per kg BOD removed	Application
Conventional	5–14	0.4–1.2	0.2–0.5	3–4	0.5–.08	Conventional treatment for medium and large works to produce 20–30 effluent with or without nitrification depending on loading within range
High	1–2	>2.5	>1	0.2–0.5	0.8–1.0	For pre-treatment or partial treatment
Low	24–72	<0.3	< 0.1	>5–6	0.4	Extended aeration for full treatment on small works. Effluent highly stabilized but may contain fine solids

Reproduced from Hawkes (1983) with permission of Academic Press Ltd, London.

Other methods must be used if an accurate estimation of microbial activity is required (e.g. biochemical assessment), although for routine operational control the MLSS is sufficient. The normal MLSS range is 1500–3500 mg l⁻¹ for conventional activated sludge units, rising to 8000 mg l⁻¹ for high-rate systems (Table 17.2). The MLSS concentration is controlled by altering the sludge wastage rate. In theory the higher the MLSS concentration in the aeration tank, then the greater the efficiency of the process as there is a greater biomass to utilize the available substrate. However, in practice, high operating values of MLSS are limited by the availability of oxygen in the aeration tank and the ability of the sedimentation unit to separate and recycle activated sludge.

Sludge residence time or sludge age

The sludge residence time (SRT) is the time in days that the microbial sludge spends in the activated sludge process (Equation (17.1), both within the aeration tank and while being separated or returned. The SRT (t_s) is calculated by using Equation (17.2):

$$\text{SRT} = \frac{\text{Total amount of sludge solids in the system}}{\text{Rate of loss of sludge solids from system}} \qquad (17.1)$$

$$t_s = \frac{V X}{(Q_w X_w) + (Q_e X_e)} \qquad (17.2)$$

where V is the volume of liquid in the aeration tank (m³), X_w is the MLSS in waste stream (mg l⁻¹), X is the MLSS (mg l⁻¹), Q_e is the effluent discharge rate (m³ day⁻¹),

Q_w is the sludge wastage rate (m^3 day^{-1}) and X_e is the suspended solids concentration in the effluent ($mg\,l^{-1}$). If the proportion of microbial cells in the MLSS is assumed constant, then the SRT can be referred to as either sludge age or the mean cell residence time (MCRT). In practice, if the amount of sludge in the aeration tank is constant (VX) then the sludge wastage rate ($Q_w X_w$) is the net sludge production of the system in kg DS day^{-1}.

SRT is the operational factor giving control over sludge activity and is controlled by altering the sludge wastage rate. A low SRT (<0.5 days) produces a sludge with a high growth rate as used in high-rate units for pre-treatment or partial treatment, which produces a high volume of waste sludge. A high SRT (>0.5 days) produces a low growth rate sludge, as in extended aeration systems, which produces a low volume of sludge which is more stabilized. Conventional activated sludge has an SRT of 3–4 days, which gives good settling properties. Long sludge ages in excess of 6 days result in poor settling properties.

17.2.2 PLANT LOADING

Four loadings are used in the design and operation of activated sludge.

Volumetric loading is the flow of wastewater in relation to the aeration tank capacity. This is also known as the hydraulic retention time (HRT):

$$HRT = \frac{V \times 24}{Q} \tag{17.3}$$

where V is the total liquid capacity of the aeration tank (m^3) and Q is the rate of flow of influent wastewater to the tank ($m^3 day^{-1}$). Expressed in hours, HRT does not take into account the flow of recycled activated sludge to the aeration tank, which can be 25–50% of the overall flow. Therefore, the actual HRT is much less than calculated by Equation (17.3). For that reason it is known as the nominal retention time.

The HRT must be long enough to allow the required degree of adsorption, flocculation and mineralization to occur. In conventional plants the HRT is 5 h at dry weather flow (DWF). During a storm when the loading increases to 3 DWF with maximum recycle of sludge (1.5 DWF), the actual HRT may be as short as 1 h.

The recycle ratio (r) is the returned activated sludge flow rate (Q_r) divided by the influent flow rate (Q):

$$r = \frac{Q_r}{Q} \tag{17.4}$$

Typical ratios are 0.25–0.5 for conventional systems, and between 0.75 and 1.5 for extended aeration.

Organic loading is the biochemical oxygen demand (BOD) load in relation to the tank capacity. As the organic content of wastewaters vary then loading (OL) should be expressed as kg BOD per unit aeration tank volume:

$$\text{OL} = \frac{Q \times \text{BOD}}{V \times 1000} \text{ kg BOD m}^{-3} \text{ day}^{-1} \quad (17.5)$$

For conventional units this is 0.4–1.2 kg BOD m^{-3}day^{-1}, while for high rate >2.5, and for extended aeration <0.3 kg BOD m^{-3}day^{-1}.

As the biomass is actively removing the organic substrate in the wastewater, it follows that the BOD loading should be related to the volume of biomass in the aeration tank (i.e. sludge loading). The sludge loading is normally referred to as the food (f) to micro-organism (m) ratio:

$$\frac{f}{m} = \frac{\text{Organic loading rate}}{\text{Volume of biomass}} = \frac{Q \times \text{BOD}}{VX} \text{ kg (BOD) kg (MLSS) day}^{-1} \quad (17.6)$$

When the f/m ratio is high the micro-organisms are in the log growth phase with excess food, maximum rate of metabolism and large removal of BOD. However, under these conditions, the micro-organisms do not form flocs but are dispersed making it difficult to settle and recycle the biomass (sludge). Because food is in excess, not all the substrate is used and so will be lost in the final effluent resulting in a high BOD (e.g. high-rate systems). In contrast, with low f/m ratios the micro-organisms are in a food-limited environment, even though the rate of metabolism may be high when the recycled micro-organisms are first mixed with the incoming wastewater. Once food is limiting, the rate of metabolism rapidly declines until the micro-organisms are in the endogenous respiration phase with cell lysis and resynthesis taking place. Therefore, almost complete oxidation of substrate occurs producing a high-quality effluent, low BOD, good flocculation and sludge settlement.

Overall, the lower the f/m ratio, the lower the rate of metabolism and the greater the BOD removal and sludge settleability (Fig. 17.3). However, as removal efficiency increases so does the overall oxygen demand of the system and so the overall cost of BOD removal.

Floc load is used to measure the organic loading over a specified period of time at the point where the influent wastewater to the aeration tank mixes with the returned bio-mass. It can be expressed in terms of BOD or chemical oxygen demand (COD) (mg l^{-1}) in relation to the MLSS (g l^{-1}) concentration:

$$\text{Floc load} = \frac{\text{COD per unit of time at point of mixing}}{\text{MLSS per unit of time at point of mixing}} \text{ mg COD g}^{-1} \text{ MLSS}$$

$$(17.7)$$

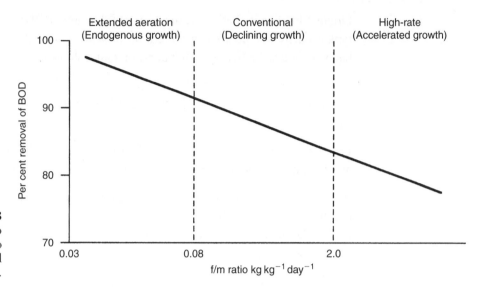

FIGURE 17.3
The relationship
between f/m ratio
and microbial
growth.

17.2.3 SLUDGE SETTLEABILITY

Most problems associated with the activated sludge process involve poor settleability. Therefore, a rapid assessment method is vital to ensure good separation in secondary settlement so that the final effluent has a low suspended solids concentration, and that sufficient biomass is returned to the aeration tank. Two indices are widely used, the sludge volume index (SVI) and the stirred specific volume index (SSVI). The SVI is measured by filling a 1-1 graduated cylinder with mixed liquor and allowing it to settle for 30 min. The volume of settled sludge (V) is then measured in ml and MLSS in $mg\,l^{-1}$:

$$SVI = \frac{V \times 1000}{MLSS}\,ml\,g^{-1} \tag{17.8}$$

A good sludge should have an SVI $<80\,ml\,g^{-1}$ and a very good one around $50\,ml\,g^{-1}$. An SVI $> 120\,ml\,g^{-1}$ indicates poor settling properties. The SVI test is severely affected by high MLSS concentrations with virtually no settlement occurring at concentrations $>4000\,mg\,l^{-1}$. Also, due to the quiescent conditions in the SVI test, sedimentation proceeds past the hindered settlement stage in which all the sludge flocs are evenly distributed (type III settlement); transitional or even compression settlement phases may have started (type IV) making the test less representative of actual sedimentation tank conditions (Gray, 2004). These problems are overcome by using the SSVI, which is now widely used, especially where a more accurate assessment of sludge settlement is required. It is measured using a special settling column 0.5 m deep and 0.1 m in diameter, with settlement impeded by a wire stirrer rotating at $1\,rev\,min^{-1}$ (Fig. 17.4). This test reproduces the non-ideal situation found in sedimentation tanks where the SVI is measured under complete quiescence (White, 1975). The SSVI is calculated by pouring approximately 3.5 l of homogeneous mixed liquor

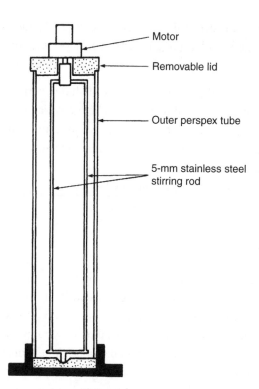

into the cylinder to the 50 cm level. The stirrer is connected and the height of the sludge interface in the column measured (h_0). After 30 min the height of the sludge interface is measured again (h_i) and the SSVI calculated as:

$$\text{SSVI} = \frac{100\,h_i}{C_0 h_0}\,\text{ml g}^{-1} \qquad (17.9)$$

where C_0 is the initial concentration of suspended solids (per cent w/w) (i.e ($\text{MLSS (g l}^{-1})/1000) \times 100$) (DOE, 1985). In terms of SSVI a good sludge has a value <120 ml g^{-1}, while a sludge with poor settleability has a value >200 ml g^{-1}.

17.2.4 SLUDGE ACTIVITY

It has not proved possible to model activated sludge systems accurately due to being unable to relate bacterial numbers directly to performance. The problem has been in differentiating viable and non-viable cells, and estimating levels of activity due to the age of the cell. Biochemical analyses are used for research purposes to assess the effect of overloading and toxic substances. Analyses used include adenosine triphosphate (ATP), which measures number of viable cells, and enzymatic activity, which measures the level of activity (Wagner and Amann, 1997). Dehydrogenase activity using triphenyl tetrazolium chloride (TTC) that is reduced to a red dye triphenyl formazin (TF) is also widely used, as dehydrogenase enzymes are readily extracted and measured using a spectrophotometer (Section 15.3.4). The response of dehydrogenase to changes in the activated sludge is both rapid and sensitive. However, none of these methods are simple to carry out and are generally expensive (Coackley and O'Neill, 1975).

The simplest and most direct measurement of sludge activity is the specific oxygen consumption or uptake rate (SOUR). It can be measured in either in the field or the laboratory by filling a BOD bottle with mixed liquor and using an oxygen electrode fitted in the bottle to measure the fall in oxygen concentration over time:

$$SOUR = \frac{\text{Oxygen depletion (mg l}^{-1}\text{ min}^{-1})}{\text{MLVSS (g)}} \times 60 \text{ mg O}_2 \text{ g}^{-1} \text{ h}^{-1} \quad (17.10)$$

17.3 MODES OF OPERATION

By using different combinations of the main operating parameters, various different rates and degree of treatment are possible. This flexibility in design, allowing operation over a wide range of loadings to suit specific treatment objectives, is the major advantage of the activated sludge process over other treatment systems. While primarily designed to remove carbonaceous BOD, with suitable operational control and plant modifications it can also achieve nitrification, denitrification and phosphorus control (Section 18.4). Depending on sludge loading then activated sludge plants can be categorized as high-rate, conventional or extended aeration, although the delineation between these categories is by no means precise and these terms are used in their broadest sense.

17.3.1 AERATION

This is a major operational variable. The main functions of aeration are:

(a) to ensure an adequate and continuous supply of dissolved oxygen (DO) for biomass;
(b) to keep the biomass in suspension;
(c) to mix the incoming wastewater with the biomass, and to remove from solution excess carbon dioxide resulting from oxidation of organic matter.

Two aeration systems are generally employed, although there are other methods available (e.g. sparges). Surface aeration is where aeration and mixing is achieved by the use of blades or vanes that are rotated at speed. The aerator rotates either about a vertical or horizontal shaft (Fig. 17.5) and is positioned at or near the surface of the liquid. The action of the blades causes considerable turbulence resulting in enhanced oxygen transfer. Diffused aerators supply oxygen, which is supplied directly to the aeration tank via a series of diffuser domes (Fig. 17.6). The air is pumped under pressure that causes considerable turbulence and hence mixing. The size of the pores in the diffuser controls both the size and the number of the air bubbles and so the rate of oxygen transfer (Gray, 2004).

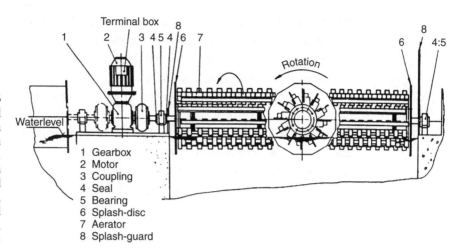

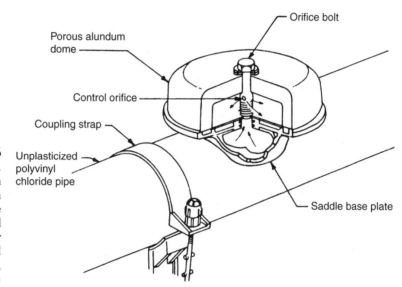

FIGURE 17.5
General
arrangement of a
Mammoth
horizontal rotor.
(Reproduced with
permission of Gerald
O'Leary,
Environmental
Protection Agency,
Wexford.)

1 Gearbox
2 Motor
3 Coupling
4 Seal
5 Bearing
6 Splash-disc
7 Aerator
8 Splash-guard

FIGURE 17.6
Dome diffuser.
(Reproduced from
IWPC (1987) with
permission of the
Chartered
Institution of Water
and Environmental
Management,
London.)

17.3.2 AERATION MANAGEMENT

There is an optimum immersion depth for maximum aeration efficiency. The power consumption increases linearly with immersion of the aerator, but the oxygenation does not. After 65% immersion at most plants the aerator is not able to give any appreciable increase in oxygenation capacity (OC), although if fully immersed the power consumption will increase by a further 30% (Fig. 17.7). With the influx of modern aerators there is less technical data available. Therefore, although expensive, aeration equipment should be tested, using the unsteady-state sulphite test, when commissioned to ensure that it complies with the specification and to determine the optimum immersion depth. This means that the stop limit on the overflow weir can be adjusted to coincide with the apex of the curve.

The oxygen demand of the mixed liquor varies during the day, so it is vital that oxygen (or redox) control is used within the aeration tank. In this way the oxygen supplied can be matched to the demand preventing overaeration. Where the plant is operated at design loading, the inclusion of oxygen control can reduce the overall power consumption by as much as 40% (Fig. 17.8). An undesirable design feature is the use of a single vertical aerator in a deep tank, resulting in the aerator never being able to be turned off due to the problem of re-suspending the mixed liquor once it has settled. The separation of the mixing and aeration functions by using a paddle system and diffusers, respectively, is the preferred design for most plants allowing the costly aeration to be switched off while still maintaining the mixed liquor in suspension, unless several mechanical aerators can be included within the aeration tank.

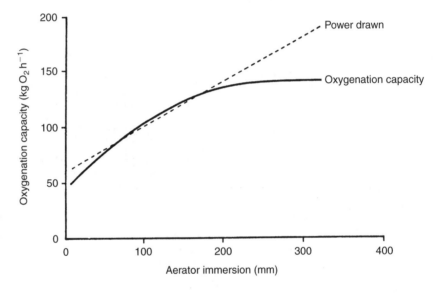

FIGURE 17.7
The OC and power used by a cone aerator at various depths of immersion. (Reproduced from Rachwal and Waller (1982) with permission of C.E.P. Consultants, Edinburgh.)

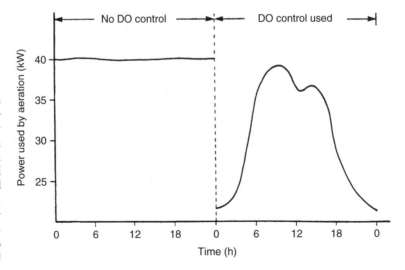

FIGURE 17.8
The effect on power consumption by aerators in an oxidation ditch when DO control is used. (Reproduced from Cox *et al.* (1982) with permission of C.E.P. Consultants, Edinburgh.)

17.3.3 OXYGEN TRANSFER

There are two methods employed to test the oxygen transfer rate and overall efficiency of aeration systems. Steady-state techniques measure the amount of oxygen required to achieve a mass balance between supply and utilization by the mixed liquor during normal operation, although constant operating and loading conditions are required. Steady-state methods are only used for completely mixed systems. Unsteady-state techniques are more commonly used and measure the rate of change of DO concentration during the reaeration of deoxygenated clean water. The water is deoxygenated by adding sodium sulphate in solution with cobalt chloride to catalyse the deoxgenation reaction. Once all the unreacted sulphite has been utilized the test can be carried out. Using a number of DO probes positioned throughout the tank, the aerator is switched on and the DO concentration monitored over time until the water is saturated with oxygen or steady-state DO conditions are achieved (Fig. 17.9).

Using the DO readings, a graph of $\log_e$ of the DO deficit $(C_s - C)$ is plotted against time for each probe, only using DO values between 20% and 80% saturation. The slopes of these graphs are then used to calculate separate values for the oxygen transfer rate $(K_L a)$ at temperature T:

$$K_L a(T) = \frac{60}{t_2 - t_1} \log_e \frac{(C_s - C_1)}{(C_s - C_2)} \ \text{h}^{-1} \tag{17.11}$$

where C_s is the saturated DO concentration $(\text{mg} \, \text{l}^{-1})$, C_1 and C_2 are the DO concentrations $(\text{mg} \, \text{l}^{-1})$ at times t_1 and t_2 (min), respectively. The mean $(K_L a)$ value is taken as the $(K_L a)$ for the whole system. The OC of the aeration system is calculated using the following equation:

$$\text{OC} = K_L a(T) \times V \times C_s(T) \times 10^{-3} \ \text{kg} \, \text{h}^{-1} \tag{17.12}$$

FIGURE 17.9
The relationship between the saturated DO concentration (C_s), the DO concentration at time zero (C_0), the concentration at time t (C_L) and the DO deficit $(C_s - C_L)$.

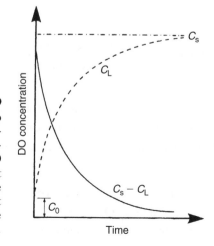

where V is the volume of water used in the test, T is the temperature of the water and $C_s(T)$ is the oxygen saturation concentration of clean water at test temperature T. The transfer rate coefficient (K_La) is dependent on temperature and is expressed at a standard temperature of 20°C (i.e. $(K_La)_{20}$). Therefore, it must be converted using:

$$(K_La)_T = (K_La)_{20}(1.024)^{T-20} \qquad (17.13)$$

where T is the temperature at which (K_La) is measured (Section 6.3).

The values for (K_La) obtained using clean water by this method will be higher than those obtained using activated sludge. The impurities in wastewater have significant effects on K_La. For example, both fatty and surface active materials such as detergents reduce the rate of oxygen transfer. However, detergents and fatty acids when present as soaps (pH > 6) are able to increase K_La by preventing bubble coalescence, thus maintaining the mean bubble size at a lower level than if such chemicals were absent. This increases the total interfacial area thus increasing the overall mass transfer of oxygen. These two opposing effects rarely cancel each other out and so must be considered in the calculation of K_La. Other impurities in water can also alter K_La; therefore the effects of all impurities must be calculated together by measuring the α factor (IWPC, 1987):

$$\alpha \text{ factor} = \frac{(K_La) \text{ wastewater}}{(K_La) \text{ clean water}} \qquad (17.14)$$

The value of K_La varies for each type of wastewater and also the duration of aeration. For example, α varies from 0.3 at the beginning of the aeration period for domestic sewage to 0.8 after 4 h aeration. Typical α values for mixed liquor vary from 0.46 to 0.62. The impurities in wastewaters will also affect the oxygen saturation concentration compared with clean water at the same temperature. This is adjusted by the β factor where:

$$\beta \text{ factor} = \frac{C_s \text{ in wastewater}}{C_s \text{ in clean water}} \qquad (17.15)$$

The β value normally approximates to 0.9. Once the α and β factors are known for a particular wastewater, then the clean water test results can be used to predict the expected field results. Details of testing aerators, including a worked example, are given in IWPC (1987).

Aeration efficiency in terms of mass of oxygen transferred to the mixed liquor per unit of energy expended is expressed as kg O_2 kW h^{-1}. An estimate of aeration efficiency can be made by measuring the oxygen demand (OD) exerted by carbonaceous oxidation and nitrification, taking into account the flow rate, influent and effluent BOD concentrations using the following equation:

$$OD = 0.0864q_s[0.75(BOD_i - BOD_e)]$$
$$+ \frac{5.25 \times 10^{-4} \times C_{MLSS} \times V}{q_s} + 4.3(N_i - N_e)\, kg\, day^{-1} \quad (17.16)$$

where q_s is the mean flow rate of settled influent ($l\,s^{-1}$), BOD is the mean BOD of the influent (i) and effluent (e) ($mg\,l^{-1}$), C_{MLSS} is the mean MLSS concentration ($mg\,l^{-1}$), V is the aeration tank volume (m^3), and N is the mean ammonia concentration in the influent (i) and effluent (e) ($mg\,l^{-1}$). The calculated oxygen demand can be converted to aeration efficiency by dividing by the daily power consumption ($kW\,h\,day^{-1}$). While this assessment is not as accurate as standard performance tests, it does allow an excellent comparison between plants without the prohibitive expense of *in situ* tests (Chambers and Jones, 1988).

17.3.4 THE USE OF PURE OXYGEN

The use of pure oxygen instead of air results in oxygen transfer occurring much more quickly increasing the potential rate of BOD removal. The amount of oxygen in mixed liquor can be increased by a factor of 5 by using pure oxygen. Two systems are available, enclosed systems using surface aerators (e.g. Unox) or open systems using diffusers. The aeration tank capacity is reduced and the HRT reduced from 6 to 3 h. The best system in the non-industrial context is Vitox developed by BOC Ltd. Oxygen is pumped directly into the aeration tank via an expansion nozzle or is added to the influent line using a venturi. Advantages are that it can be used with existing aeration tanks without any extra tank construction or modification. It can be used to replace an existing aeration system or be used to supplement it by providing extra oxygen at peak flows or when seasonal overloading occurs. So it is ideal for uprating overloaded plants on a temporary or permanent basis. In comparison to expanding a plant by the construction of new aeration tanks, uprating the existing tanks using Vitox can be as little as 15% of the cost. The saving in capital cost is likely to be offset by significantly higher operating costs, that is why these systems are generally employed where land availability is restricted or land costs are prohibitively high.

17.4 AERATION TANK DESIGN

17.4.1 TANK CONFIGURATION

Aeration tank design tends towards either plug flow or completely mixed (Fig. 17.10). In a plug-flow reactor the influent and returned sludge are added at one end of an

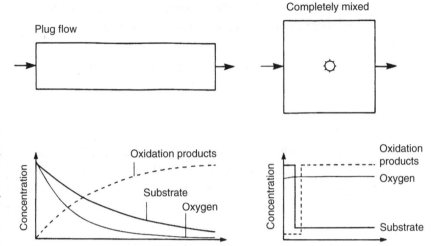

FIGURE 17.10 Comparison of the utilization of substrate and oxygen, and the production of oxidation products through a plug flow and completely mixed reactor.

elongated rectangular aeration tank. Typical dimensions are 6–10 m wide, 30–100 m long, 4–5 m deep. As biomass proceeds down the length of the tank:

(a) treatment occurs, the degree depending on retention time;
(b) the microbial growth curve is discernible as in batch systems with sludge activity high at the inlet, but low at the outlet;
(c) there is rapid removal initially but becoming progressively slower;
(d) a discernible BOD gradient is formed.

Disadvantages of plug-flow systems include:
(a) DO deficiency occurring at the inlet where oxygen demand is greatest;
(b) oxygen is usually in excess at the outlet where oxygen demand is least;
(c) as toxic and shock loads are not diluted or buffered, they pass through the tank as a discrete plug, often resulting in serious effects on performance.

Advantages include:
(a) no short circuiting within the aeration tank, although baffles can increase the HRT even more;
(b) plug-flow systems produce sludges with good settleability.

In completely mixed systems, the influent wastewater and returned sludge are immediately mixed and instantly diluted giving a uniform loading throughout the aeration tank.

Advantages include:
(a) minimizing the effects of toxic and shock loads;
(b) plug-flow MLSS concentrations are between 2000 and 3000 mg l^{-1}, while completely mixed MLSS concentrations are much higher at 3000–6000 mg l^{-1} permitting higher BOD loadings.

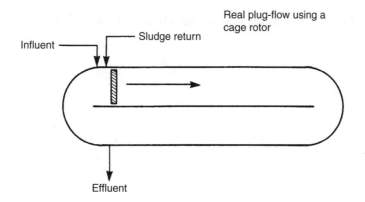

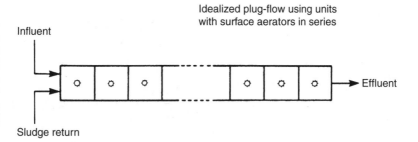

FIGURE 17.11
The oxidation ditch with a single rotor is close to the idealized design of a plug-flow reactor generally used in the treatment of wastewater.

Disadvantages include:
(a) possible short circuiting in the aeration tank;
(b) a low density sludge is produced;
(c) poor settleability;
(d) no nitrification possible.

These disadvantages can be overcome by using aeration tanks in series, which is approaching plug flow in design (Fig. 17.11).

17.4.2 COMPLETELY MIXED REACTORS

Using Fig. 17.1, mass balance equations for biomass production (Equation (17.17)) and substrate (food) utilization (Equation (17.20)) in activated sludge can be developed (Benefield and Randall, 1980):

$$Q_0 X_0 \;+\; V\frac{\mathrm{d}X}{\mathrm{d}t} \;=\; (Q_0 - Q_\mathrm{W})X_\mathrm{e} \;+\; Q_\mathrm{W} X_\mathrm{w} \qquad (17.17)$$

$\underset{\text{Influent biomass}}{} \quad \underset{\text{Biomass production}}{} \quad \underset{\text{Effluent biomass}}{} \quad \underset{\text{Wasted biomass}}{}$

Using the Monod function (Equation (15.17)) for the specific growth rate of biomass, Equation (17.17) can be rewritten as:

$$Q_0 X_0 + V\left(\frac{\mu_\mathrm{m} S X}{K_\mathrm{s} + S} - k_\mathrm{d} X\right) = (Q_0 - Q_\mathrm{W})X_\mathrm{e} + Q_\mathrm{W} X_\mathrm{w} \qquad (17.18)$$

If the biomass in the influent and effluent are negligible (i.e. $X_0 = X_e = 0$), then:

$$\frac{\mu_m S}{K_s + S} = \frac{Q_w X_w}{VX} + k_d \qquad (17.19)$$

$$\underset{\substack{\text{Influent} \\ \text{substrate}}}{Q_0 S_0} + \underset{\substack{\text{Substrate} \\ \text{consumed}}}{V \frac{dS}{dt}} = \underset{\substack{\text{Effluent} \\ \text{substrate}}}{(Q_0 - Q_w) S_e} + \underset{\substack{\text{Wasted} \\ \text{substrate}}}{Q_w X_w} \qquad (17.20)$$

Using the equation for substrate utilization from Section 15.2:

$$-\frac{dS}{dt} = \frac{1}{Y} \frac{\mu_m SX}{K_s + S} \qquad (17.21)$$

$$Q_0 S_0 - V \left(\frac{1}{Y} \frac{\mu_m SX}{K_s + S} \right) = (Q_0 - Q_w) S_e + Q_w S_w \qquad (17.22)$$

The substrate in the aeration basin is equal to that in the secondary settlement tank and in the effluent (i.e. $S = S_w = S_e$), so Equation (17.22) can be rearranged as:

$$\frac{\mu_m S}{K_s + S} = \frac{Q_0 Y}{VX} (S_0 - S) \qquad (17.23)$$

By combining Equations (17.19) and (17.23) then:

$$\frac{Q_w X_w}{VX} + k_d = \frac{Q_0 Y}{VX} (S_0 - S) \qquad (17.24)$$

The HRT of the influent in the aeration basin (ϕ) is:

$$\phi = \frac{V}{Q_0} \qquad (17.25)$$

The MCRT in the aeration basin (ϕ_c) is derived from Equation (17.2) as:

$$(\phi_c) = \frac{VX}{Q_w X_w} \qquad (17.26)$$

As a fraction of the biomass from the sedimentation tank is returned to the aeration basin $\phi_c > \phi$, Equation (17.24) becomes:

$$\frac{1}{\phi_c} + k_d = \frac{1}{\phi}\frac{Y}{X}(S_0 - S) \tag{17.27}$$

So the MLSS in the aeration basin (X) can be calculated as:

$$X = \frac{\phi_c}{\phi}Y\left(\frac{S_0 - S}{1 + k_d\phi_c}\right) \tag{17.28}$$

The f/m ratio is determined from Equation (17.6) using the notation in Fig. 17.1 as:

$$\frac{f}{m} = \frac{S_0}{\phi X} \tag{17.29}$$

$$\frac{f}{m} = \frac{S_0}{(V/Q_0)X} = \frac{Q_0 S_0}{VX} \tag{17.30}$$

EXAMPLE 17.1   Calculate the aeration basin volume (V), the HRT (ϕ), the volume of sludge wasted each day (Q_w), the mass of sludge wasted each day ($Q_w X_w$), the fraction of sludge recycled Q_r/Q_0 and the f/m ratio for a completely mixed activated sludge system.

Population equivalent served is 60 000 with a per capita water use of 225 l day^{-1}, influent BOD (S_0) is 280 mg l^{-1}, required effluent BOD is 20 mg l^{-1}, yield coefficient (Y) is 0.6, decay rate (k_d) is 0.06 day^{-1}. Assumed optimum operational factors are aeration tank MLSS (X) 3500 mg l^{-1}, WAS MLSS (X_w) 14 000 mg l^{-1} and a MCRT (ϕ) of 10 day^{-1}. Express values in m^3, kg and day^{-1}.

(a) Calculate the aeration tank volume (V):
   By combining Equations (17.25) and (17.28):

$$X = \frac{\phi_c}{V}Q_0 Y\left(\frac{S_0 - S}{1 + k_d\phi_c}\right) \tag{17.31}$$

$$V = \frac{\phi_c Q_0 Y}{X}\left(\frac{S_0 - S}{1 + k_d\phi_c}\right) \tag{17.32}$$

$$V = \frac{10 \times 13\,500 \times 0.6}{3.5}\left(\frac{0.28 - 0.02}{1 + 0.06 \times 10}\right)$$

$$V = 3761 \text{ m}^3$$

(b) The HRT is calculated using Equation (17.25):

$$\phi = \frac{3761}{13\ 500} = 0.279\ \text{day}^{-1}$$

(c) Volume of wasted sludge per day (Q_w):
Using Equation (17.26) then:

$$Q_w = \frac{VX}{\phi_c X_w} \qquad (17.33)$$

$$= \frac{3761 \times 3.5}{10 \times 14} = \frac{13\ 164}{140} = 94.0\ \text{m}^3\ \text{day}^{-1}$$

(d) Mass of sludge wasted each day ($Q_w X_w$):

$$Q_w X_w = 94.0 \times 14 = 1316\ \text{kg\ day}^{-1}$$

(e) Fraction of sludge recycled (Q_r/Q_0) (Equation (17.4)) is calculated from bio-mass mass balance of secondary settlement tank:

$$(Q_0 + Q_r)X = (Q_0 - Q_w)X_e + (Q_r + Q_w)X_w \qquad (17.34)$$

$$Q_r = \frac{Q_0 X - Q_w X_w}{X_w - X}$$

$$Q_r = \frac{(13\ 500 \times 3.5) - 1316}{14 - 3.5} = \frac{45\ 934}{10.5} = 4374.7\ \text{m}^3\ \text{day}^{-1} \qquad (17.35)$$

$$\frac{Q_r}{Q_0} = \frac{4374.7}{13\ 500} = 0.324\ (32.4\%)$$

(f) Calculate the f/m ratio using Equation (17.30):

$$\frac{f}{m} = \frac{Q_0 S_0}{VX} = \frac{13\ 500 \times 0.280}{3761 \times 3.5} = \frac{3780}{13\ 163.5} = 0.29$$

Different design and operating parameters are used for plug-flow systems such as oxidation ditches, and these are reviewed by Benefield and Randall (1980).

Completely mixed systems are widely used as they are comparatively cheap to construct. They tend to employ vertical surface aerators. Tanks are normally square in cross section with a flat base, 3–6 m in depth with a free board of 1.0–1.5 m to contain spray. Plug-flow systems are generally of the oxidation ditch type (Fig. 17.11).

17.5 COMMON MODIFICATIONS

17.5.1 TAPERED AERATION (PLUG FLOW)

When oxygen is supplied evenly along the length of the tank, DO concentration will be low at the inlet area but high at the exit. This problem is overcome by tapering the supply of oxygen according to respiratory requirement of the biomass (Fig. 17.12). In plug-flow aeration tanks 45–50% of the oxygen is required in the first third of its length, 30% in the second third of its length and 20–25% is required in the last third. Aeration in the last section must always be sufficient to maintain mixed liquor in suspension.

17.5.2 CONTACT STABILIZATION (BOTH PLUG AND COMPLETELY MIXED)

In this process, which is often referred to as biosorption, the removal processes of adsorption and oxidation occur in separate tanks. The influent enters a contact tank and is mixed with the returned biomass at a MLSS of 2000–3000 mg l^{-1} for 0.5–1.0 h during which time a high degree of adsorption on to the flocs occurs. Settlement follows and all the separated solids go to the aeration tank where they are aerated for 5–6 h at 4000–10 000 mg l^{-1} MLSS for oxidation (Fig. 17.13). By having such a short HRT in the contact tank, it is not necessary to provide aeration tank capacity for the entire

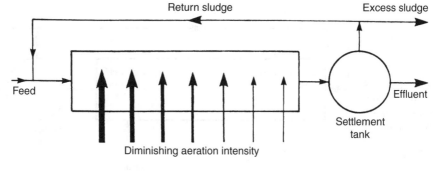

FIGURE 17.12 Schematic flow diagram of tapered aeration in a plug-flow activated sludge system.

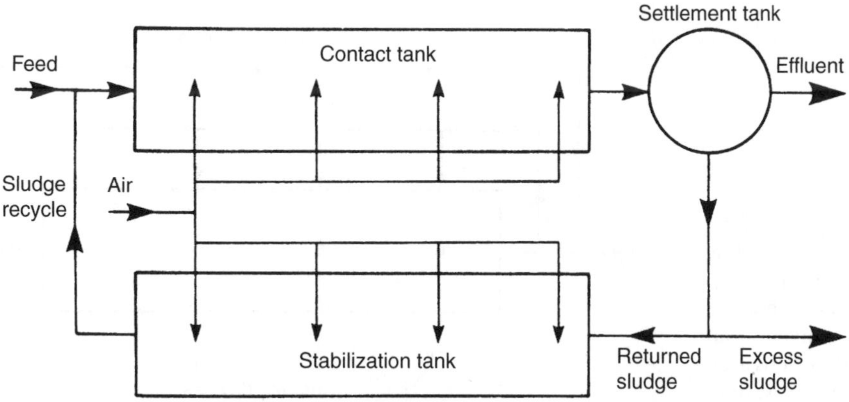

FIGURE 17.13 Schematic flow diagram of contact stabilization activated sludge system.

flow as is the case with other systems and this results in up to 50% less aeration capacity required. This leads to a significant saving in both capital and aeration costs. There is less sludge produced by the process due to endogenous respiration and the sludge is of better quality. Once oxidation is complete the biomass is returned to the contact tank (Section 17.7). A separate tank is required for completely mixed systems or a zone for plug-flow systems (Fig. 17.14).

17.5.3 INCREMENTAL FEEDING OR STEP AERATION (PLUG FLOW)

Oxygen demand is evened out along the length of the aeration tank by adding food and oxygen at various points along its length, thus evening out the BOD loading, the oxygen demand and the oxygen utilization rate (Fig. 17.15). The operator can alter the proportion of influent entering each stage to take account of changes in either the organic or hydraulic loading, providing a high degree of operational flexibility. The process is intermediate between plug and completely mixed flow regimes.

17.5.4 SEQUENCING BATCH REACTORS (SBR)

The original design of activated sludge was a single tank batch reactor where aeration and settlement took place in the same tank (Alleman and Prakasam, 1983). This design was developed into the conventional systems used today where aeration and settlement

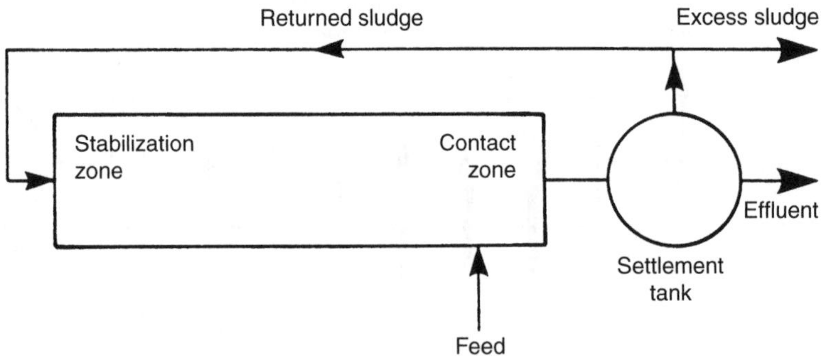

FIGURE 17.14 Schematic flow diagram of a single basin contact stabilization activated sludge system.

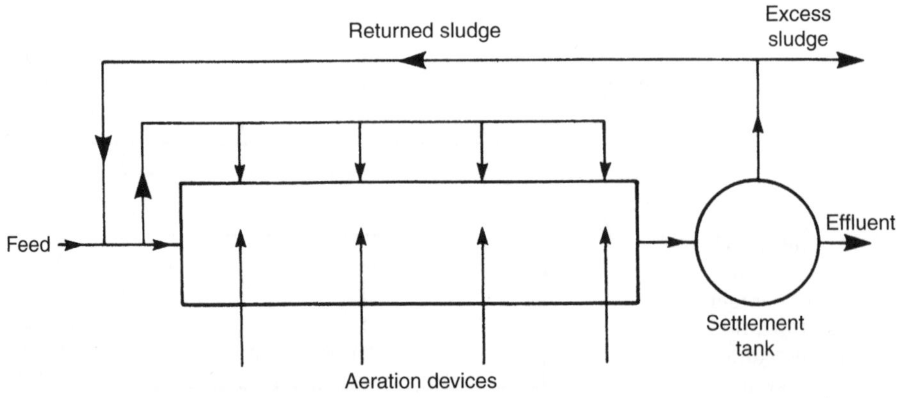

FIGURE 17.15 Schematic flow diagram of step aeration in a conventional plug-flow activated sludge system.

occur in separate tanks. However, renewed interest in batch operation has led to the introduction of sequencing batch reactors (SBR) systems. Using a single tank the sequence of events outlined in Fig. 17.16 is carried out repeatedly, with each complete cycle taking between 4 and 48 h with the SRT varying from 15 to 70 days. The f/m ratio will vary according to the length of the cycle but typically ranges from 0.03 to 0.18. The tanks permit better operation and allow precise control over periods when the mixed liquor can be allowed to become anoxic to permit denitrification (Section 18.4).

17.5.5 LOW-AREA HIGH-RATE SYSTEMS

The EU Urban Wastewater Treatment Directive requires that all sewages receive at least secondary treatment. Logistically this proves very difficult for larger towns where the construction of new pumping stations, primary treatment tanks and out-falls has left very little space for the development of conventional secondary treatment

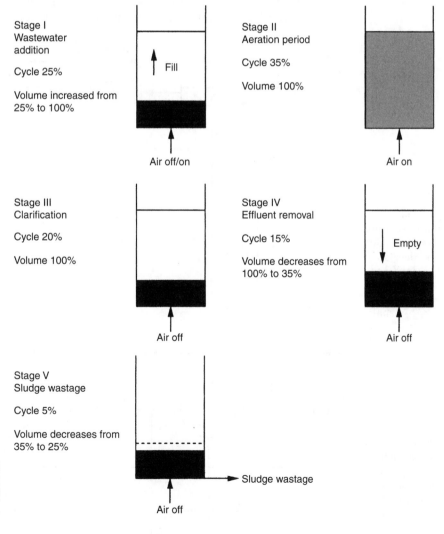

FIGURE 17.16
Typical cycle and
configuration of
a SBR.

systems. The ICI deep shaft process requires only 20% of the land space that a conventional activated sludge system requires (Fig. 17.17). For example, at Tilbury in London, a deep shaft system treats a population equivalent to 350 000 on an area of 100×200 m. The process consists of a deep well of between 30 and 220 m and 0.8–6 m in diameter. The advantage is the high hydrostatic pressure at the base of the shaft gives a very high oxygen solubility with little power requirement. At a sludge loading of 0.8–1.0 kg kg^{-1} day^{-1}, the process achieves a BOD removal of 95% although solids are high, a factor that is less important in coastal disposal. No primary sedimentation is required, although the influent must be carefully screened before entering the system due to the potential accumulation of dense solids at the base of the shaft. The HRT is only 1.1–1.8 h, making this process capable of treating high hydraulic load.

The A–B process is widely used in the Netherlands and Germany. It is two separate activated sludge plants operated in series. A highly loaded first (A) stage at a sludge loading of 3–6 kg kg^{-1} day^{-1} to give 70% BOD removal. This is followed by a low-loaded second (B) stage at a loading of 0.15–0.3 kg kg^{-1}day^{-1} to give a high purity fully nitrified effluent (Fig. 17.18). It uses only 60% of the conventional aeration tank

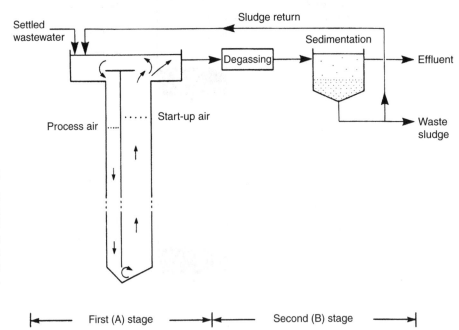

FIGURE 17.17
Schematic diagram of the deep shaft process developed by ICI plc employing a U-shaped reaction shaft.

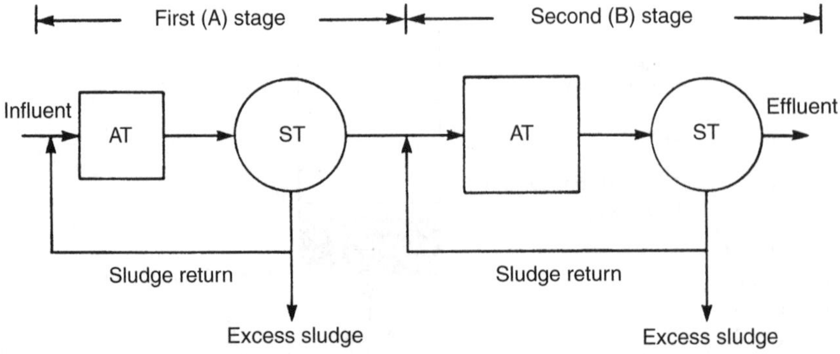

FIGURE 17.18
The A–B activated sludge process. AT: aeration tank; ST: sedimentation tank.

capacity resulting in a capital saving of 30%, and 20% lower aeration costs. Owing to their size both plants can be covered, and odours and pathogens removed by scrubbing, making them ideal for residential areas.

17.6 ECOLOGY

Like all biological treatment processes the activated sludge system relies on a mixed culture of bacteria to carry out the basic oxidation of the substrate with higher grazing organisms such as protozoa and rotifers also present, forming a complete ecosystem with several trophic levels (Fig. 17.19). The activated sludge aeration tank is a truly aquatic environment, although the high substrate concentration and high level of bacterial activity make it unlike any natural environment. The constant aeration and re-circulation of sludge makes it inhospitable for most aquatic species, especially those larger than the smaller mesofauna such as rotifers and nematodes, or animals with long life cycles. In comparison to a percolating filter, where the substrate flows over a thin film attached to an inert medium, and where a diverse macro-invertebrate grazing fauna is found, the activated sludge system is much simpler, with fewer trophic levels forming simple food webs.

As the floc ages it becomes colonized by bacterial feeding organisms such as ciliate protozoa, nematodes and rotifers. Active bacteria are restricted to the outer surface of the flocs with distinct oxygen and substrate gradients towards the centre of each floc. With long sludge ages the slower growing bacteria such as the nitrifiers will also colonize the

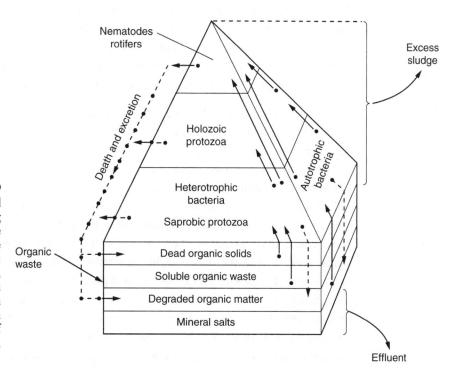

FIGURE 17.19
Food pyramid representing feeding relationships in the activated sludge processes: →synthesis, ---→degradation (reproduced from Hawkes (1983) with permission of Academic Press Ltd, London).

flocs. As the flocs age the rate of oxidation gradually declines and in older flocs most of the bacteria are non-viable, although they retain active enzyme systems and continue to excrete extra-cellular enzymes. Material continues to be adsorbed onto older flocs. An important selective factor is that only the species of bacteria capable of forming compact flocs will settle out in the sedimentation tank, and so will be returned as sludge to the aeration tank. Those bacteria unable to flocculate are lost with the final effluent.

17.7 OPERATIONAL PROBLEMS

17.7.1 FLOCCULATION

Under the microscope activated sludge comprises of discrete clumps of micro-organisms known as flocs. The floc is the basic operational unit of the process. Good flocculant growth is important for the successful operation of the process, so that suspended, colloidal and ionic matter can be removed by adsorption, absorption and agglomeration in the aeration tank. Also in the sedimentation tank, good floc structure is vital for rapid and efficient separation of sludge from the treated effluent. Flocs are clusters of several million heterotrophic bacteria bound together with some inert organic and inorganic material. The binding polymer (extra-cellular polymers or ECPs) not only bind bacteria together to form the floc, but it also allows adsorption to occur. The binding polymer can be up to 10% of the dry weight of the floc. It is composed of a number of compounds such as polysaccharides, amino-polysaccharides and protein. The exact nature of the polymer is dependent on the bacteria producing it. Flocs range in size from 0.5 μm (individual bacteria) to 1000 μm (1 mm) in diameter. The surface area of flocs is extremely high as the flocs are porous, rather like sponges, allowing high rates of adsorption and mineralization. Rapid agglomeration of suspended solids and colloidal matter onto flocs occurs as soon as the sludge is mixed with the wastewater (enmeshment). This results in a sharp fall in the residual BOD of the wastewater, which is the basis of the contact stabilization process.

The adsorption capacity of flocs depends on the availability of suitable cell surfaces, so once all the adsorption sites are occupied then the floc has a very reduced capacity to adsorb further material until it has metabolized that already adsorbed. Breakdown of adsorbed material is by extra-cellular enzymes, while subsequent assimilation of adsorbed material proceeds more slowly (stabilization). If the HRT is too short there will be a progressive reduction in BOD removal as there will have been insufficient time for adsorbed material to be stabilized. Only well-formed flocs settle out of suspension in the sedimentation tank, with dispersed micro-organisms and suspended solids carried out of the tank with the final effluent. Removal efficiency in the sedimentation tank is significantly lower for small particles, or those with a low density. The process depends on continuous re-circulation of settled sludge (biomass) to maintain an optimum MLSS concentration in the aeration tank. The system tends to return those micro-organisms which have formed flocs that rapidly settle in the sedimentation tank, so the process is microbially self-regulating with the best flocs re-circulated.

17.7.2 FLOC MORPHOLOGY

While all the operational parameters affect floc morphology it is sludge age which has the most profound effect. Young flocs contain actively growing and dividing heterotrophic bacteria, and have a high rate of metabolism. Older flocs have a lower proportion of viable cells, being composed mainly of non-viable cells that have retained active enzyme systems, surrounded by a viable bacterial layer. They have a reduced rate of metabolism but settle far more readily in contrast to young flocs which have poor settleability. As flocs age:

(a) slower growing autotrophs become established (e.g. nitrification), so the concept of sludge age is important in terms of overall efficiency;
(b) they undergo secondary colonization by other micro-organisms such as protozoa nematodes and rotifers – the ciliate protozoa are particularly important as they feed on dispersed bacteria reducing the turbidity of the final effluent, while higher trophic levels graze the floc itself reducing the overall biomass;
(c) non-degradable material accumulates in the flocs.

So flocs from lightly loaded plants (e.g. extended aeration units with a long sludge age) are compact with dark central cores (inorganic material such as iron hydroxide, calcium phosphate, aluminium hydroxide and recalcitrant organic debris). They have an active outer layer and an inert inner core. There is good secondary colonization indicating a long sludge age allowing these higher organisms to complete life cycles. In contrast, flocs from highly loaded plants (high-rate or overloaded conventional plant) produce open flocs with a loose structure, with dispersed bacteria also present and most bacteria zoogloeal in nature. As sludge age is so short flocs are comprised solely of bacteria, with poor secondary colonization.

17.7.3 SLUDGE PROBLEMS

The most common operational difficulty associated with activated sludge operation is concerned with the separation of sludge from the clarified wastewater in the sedimentation tank. The ability of activated sludge to separate is normally measured by an index of settleability such as the SVI or the stirred specific volume index (SSVI) (Section 17.2.3). The main problems are bulking, deflocculation, pinpoint floc, foaming and denitrification (Table 17.3). With the exception of denitrification, all settleability problems can be related to floc morphology. Floc structure is divided into two distinct categories micro- and macrostructure (Fig. 17.20). Microstructure is where the flocs are small ($<75\,\mu$m diameter), spherical, compact and weak. Flocs are composed solely of floc-forming bacteria by aggregation and bioflocculation with individual bacteria adhering to each other. The structure of these flocs is weak because turbulence can easily shear small pieces of the main floc. The maximum size of a floc depends on its cohesive strength and the degree of shear exerted by the turbulence within the

TABLE 17.3   Causes and effects of activated sludge separation problems

Name of problem	Cause of problem	Effect of problem
Dispersed growth	Micro-organisms do not form flocs but are dispersed forming only small clumps or single cells	Turbid effluent. No zone settling of sludge
Slime (jelly). Viscous bulking (also referred to as non-filamentous bulking)	Micro-organisms are present in large amounts of exo-cellular slime. In severe cases the slime imparts a jelly-like consistency to the activated sludge	Reduced settling and compaction rates. Virtually no solids separation in severe cases resulting in overflow of sludge blanket from secondary clarifier. In less severe cases a viscous foam is often present
Pin floc or pinpoint floc	Small, compact, weak, roughly spherical flocs are formed the larger of which settle rapidly. Smaller aggregates settle slowly	Low SVI and a cloudy, turbid effluent
Bulking	Filamentous organisms extend from flocs into the bulk solution and interfere with compaction and settling of activated sludge	High SVI – very clear supernatant. Low RAS and WAS solids concentration. In severe cases overflow of sludge blanket occurs. Solids handling processes become hydraulically overloaded
Blanket rising	Denitrification in secondary clarifier releases poorly soluble nitrogen gas which attaches to activated sludge flocs and floats them to the secondary clarifier surface	A scum of activated sludge forms on surface of secondary clarifier
Foaming scum formation	Caused by (i) non-degradable surfactants and (ii) by the presence of *Nocardia* sp. and other foam-associated species	Foams float large amounts of activated sludge solids to surface of treatment units. *Nocardia* and *Microthrix* foams are persistent, and difficult to break mechanically. Foams accumulate and can purify. Solids can overflow into secondary effluent or even overflow out of the tank free board onto walkways

Reproduced from Jenkins *et al.* (1984) with permission of Water Research Commission, Pretoria, South Africa.

aeration tank. Flocs generally settle quickly while the smaller sheared pieces are lost with the final effluent. Macrostructure is where floc bacteria aggregate around filaments that provide a type of backbone or support. This makes flocs larger, stronger and irregular in shape.

TABLE 17.4   Distinguishing sludge problems from other solid–liquid separation problems

	Bulking	*Pin floc*	*Deflocculation*	*Denitrification*
Flocs formed	Yes	Yes	No	Yes
Filaments present	Yes	No	No	No
Supernatant clear	Yes	No	No	Possibly
High SVI	Yes	Possibly	Possibly	Possibly
Rising sludge that covers surface	No	No	No	Yes
Turbid effluent due to small particles	No	Yes	Yes	No

The main sludge problem involving microstructure is deflocculation where bacteria fail to form flocs, or weak flocs are broken up by overaerating causing excessive shear. If not controlled then it becomes progressively difficult to maintain the desired MLSS concentration in the aeration tank, with the f/m ratio increasing as a consequence. When sludge is essentially microstructural in nature then deflocculation is caused by low DO, low pH, shock loadings or sludge loadings $>0.4 \, \text{kg} \, \text{kg}^{-1} \text{day}^{-1}$. If a sludge has a microstructure then a high sludge loading will eventually cause deflocculation to occur. Deflocculation is characteristic of toxic or inhibitory wastewaters, especially those in the pharmaceutical and allied chemical industries.

The common macrostructural problem is bulking which occurs when filamentous micro-organisms extend from the floc into the bulk solution (Table 17.4). This interferes with the settlement of flocs and subsequent compaction of the sludge so that a much thinner sludge is returned to the aeration tank with a low MLSS. It becomes progressively more difficult to maintain the desired MLSS concentration in the aeration tank, which will gradually fall leading to a decline in effluent quality as the f/m ratio increases.

The ideal floc has an SVI of between 50 and $120 \, \text{ml g}^{-1}$ and the final effluent is free from suspended solids and turbidity. The filaments and floc-forming species are balanced with filaments largely contained within the floc giving it strength and a definite structure. While there may be discernible filaments protruding from flocs these will be sufficiently scarce and of reduced length not to interfere with settlement. In contrast, bulking flocs, where the SVI is $>150 \, \text{ml g}^{-1}$ and the final effluent is gradually increasing in turbidity and suspended solids, large numbers of filaments will protrude from flocs. These will reduce settling velocity in the sedimentation tank, raise the level of the sludge blanket so that the sludge is uncompacted and thin, and eventually result in flocs being washed out of the sedimentation tank with the final effluent.

Two types of bulking flocs are discernible: fairly compact flocs with long filaments growing out of the flocs and linking individual flocs together (bridging), forming a mesh-work of filaments and flocs and flocs with a more open (diffuse) structure which is formed by bacteria agglomerating along the length of the filaments forming thin, spindly flocs of a large size (Fig. 17.20).

The type of floc formed, the type of compaction and settling interference caused depends on the type of filamentous micro-organisms present. Bulking is predominately

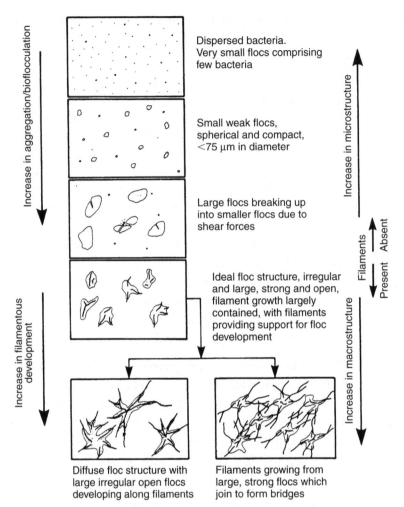

FIGURE 17.20 Diagrammatic representation of floc structure in terms of micro- and macrostructure.

Dispersed bacteria. Very small flocs comprising few bacteria

Small weak flocs, spherical and compact, <75 µm in diameter

Large flocs breaking up into smaller flocs due to shear forces

Ideal floc structure, irregular and large, strong and open, filament growth largely contained, with filaments providing support for floc development

Increase in aggregation/bioflocculation

Increase in microstructure

Increase in filamentous development

Increase in macrostructure

Filaments Present / Absent

Diffuse floc structure with large irregular open flocs developing along filaments

Filaments growing from large, strong flocs which join to form bridges

caused by bacterial filaments with bridging caused by types 02IN, 0961, 0805, 0041, *Sphaerotilus natans*, *Thiothrix* sp. and *Haliscomenobacter hydrossis*. While open floc structure is associated with type 1701, 0041, 0675, *Nostocoida limicola* and *Microthrix parvicella*. Twenty-five different bacteria are commonly recorded, also fungi and algae are also known to occasionally cause bulking. Types 0041, 021N, *Nocardia*, *M. parvicella* and *H. hydrossis* are particularly common bacterial filaments with a world-wide distribution (Table 17.5).

17.7.4 IDENTIFYING PROBLEMS

It is desirable to identify the extent and characterize existing problems, and to detect the onset of problems so that remedial action can be taken to prevent serious operational problems developing. This is done by various chemical, physical and microscopic analyses. While regular chemical analysis will indicate a reduction in treatment efficiency, loss of efficiency is generally the final symptom of a problem. Earlier signs

TABLE 17.5 The most frequently occurring filamentous micro-organisms recorded in activated sludge plants in various countries

Rank	USA	The Netherlands	Germany	South Africa	Ireland
1	*Nocardia*	*M. parvicella*	021N	0092	0041
2	1701	021N	*M. parvicella*	0041	021N
3	021N	*H. hydrossis*	0041	0675	*M. parvicella*
4	0041	0092	*S. natans*	*Nocardia*	0675
5	*Thiothrix*	1701	*Nocardia*	*M. parvicella*	*N. limicola*
6	*S. natans*	0041	*H. hydrossis*	1851	1851
7	*M. parvicella*	*S. natans*	*N. limicola*	0914	*Fungi*
8	0092	0581	1701	0803	1701
9	*H. hydrossis*	0803	0961	*N. limicola*	*H. hydrossis*
10	0675	0961	0803	021N	0803

of an impending problem can be obtained by the presence of filaments in the sludge, change in sludge morphology and ecology, a gradual increase in the SVI or an increased turbidity of the final effluent.

Sludge settleability

If a steady increase in the measured SVI (SSVI) occurs over time moving towards and exceeding $120\,\mathrm{ml\,g^{-1}}$ ($180\,\mathrm{ml\,g^{-1}}$) then bulking is imminent, and remedial action should be taken.

Cone test

During the measurement of SVI or SSVI the supernatant should be examined. The sludge should form a discrete blanket with a clear supernatant. If the sludge volume is $>650\,\mathrm{ml}$ (SVI only) with a clear supernatant then bulking should be suspected, but if the supernatant is turbid then another sludge problem should be considered (e.g. deflocculation, pinpoint floc) (Table 17.4).

Floc morphology

The floc morphology can tell us a lot about the state of the sludge (e.g. micro- or macrostructure). The protozoa may also help to rapidly identify such operational problems as organic overloading, underaeration or insufficient SRT (Table 17.6). It is very difficult to identify the bacteria in activated sludge to species level, for this reason typing is used. The filaments causing bulking may have more than one growth form and so are considered different types, while some types may contain two or more distinct species. The filamentous bacteria associated with activated sludge and with bulking and foaming can be readily identified using the detailed key by Eikelboom and van Buijsen (1981) which uses just three staining techniques, the Gram, Neisser and sulphur storage tests,

TABLE 17.6 Biological summary of mixed liquor

Heavily loaded (temporary overloaded/starting)	Normally loaded	Lightly loaded
Flagellates dominate Amoebae (naked)	Irregular flocs Ciliates present in large numbers Typical species *Vorticella convallaria* *Opercularia coarctata* *Aspidisca costata* *Euplotes affinis* Flagellates present in small number Amoebae present in small numbers	Small compact flocs Ciliates present in small numbers Typical species *Vorticella communis* *Epistylis rotans* *Stentor roeseli* Amoebae (testate) Rotifers and nematodes present Suctorians abundant

along with morphological features (Fig. 17.21). Jenkins *et al.* (1984) have extended this key making use of three more simple stains, the Indian ink reverse, poly-β-hydroxybutrate and crystal violet sheath stains to aid identification. Identification and assessment of micro-organisms can be aided by the use of a simple record sheet. A standard record sheet for observations, not only of the bacterial types present but also of the other flora and fauna, as well as floc morphology, is also recommended (Jenkins *et al.*, 1984). By examining a large number of bulking sludges, dominant bulking organisms have been related to various operational parameters. Clear associations have become evident so that the presence of various species can be used with some degree of accuracy to indicate causative operational problems of bulking. For example, fungi is associated with wastewater which has a strong acidic discharge that reduces the pH in the aeration tank; type 1701 and *S. natans* indicates low DO due to a high sludge loading; *M. parvicella* thrives during periods or in regions of low DO, but not anoxic or anaerobic, under long sludge age conditions where the f/m ratio is low (Table 17.7).

Filament counting

For rapid assessment a subjective index should be used. The degree of filamentous development is ranked against a series of drawings/photographs at 100× magnification (Fig. 17.22). Counting numbers or lengths of filaments are more quantitative but no more accurate in terms of prediction, but very much more time consuming. The total extended filament length (TEFL) is measured by taking 2 ml of mixed liquor, which is diluted and mixed with 1 l of distilled water, a dilution factor of 500. Sub-samples of 1 ml are placed on a slide and all filaments in that sample are measured and counted. The length of filaments can be estimated rapidly by placing them into various size categories, for example 0–10, 10–25, 25–50, 50–100, 100–200, 200–400 and 400–800 μm, while all filaments >800 μm in length should be measured individually. The TEFL is then calculated as either:

Total length of filaments × dilution factor μm ml^{-1} (TEFL per ml MLSS)

(17.36)

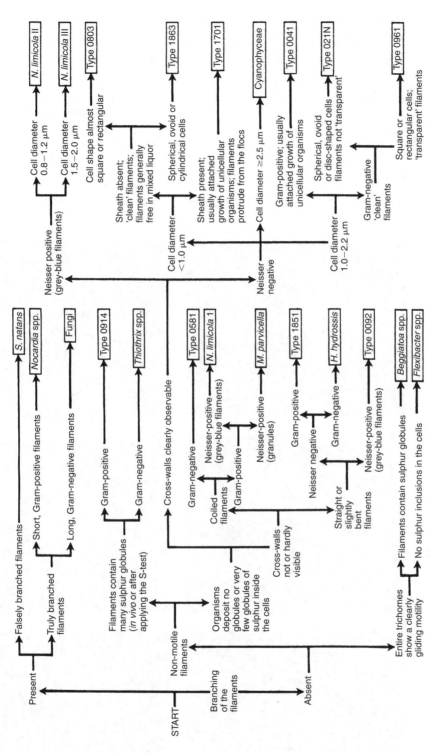

FIGURE 17.21 Identification key to filamentous micro-organisms in activated sludge. (Reproduced from Eikelboom and Buijsen (1981) with permission of TNO, Delft, The Netherlands.)

TABLE 17.7   The use of dominant filament type as indicators of conditions causing activated sludge bulking

Suggested causative conditions	Indicative filament types
Low DO	Type 1701, *S. natans, H. hydrossis, M. parvicella*
Low f/m	*M. parvicella, H. hydrossis, Nocardia* sp. types 021N, 0041, 0675, 0092, 0581, 0961, 0803
Septic wastewater/sulphide	*Thiothrix* sp., *Beggiatoa* sp. and type 021N
Nutrient deficiency	*Thiothrix* sp., *S. natans* type 021N and possibly *H. hydrossis and types* 0041 and 0675
Low pH	Fungi

Reproduced from Strom and Jenkins (1984) with permission of the Water Pollution Control Federation, Washington, DC.

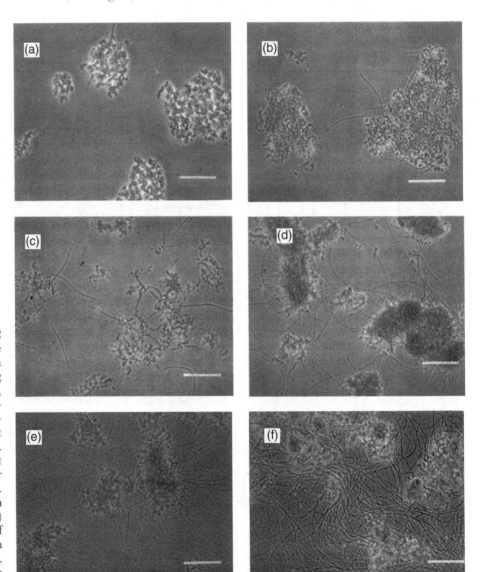

FIGURE 17.22 Filament abundance categories using a subjective scoring index where (a) few, (b) some, (c) common, (d) very common, (e) abundant and (f) excessive. Photographs are at 100× with the bar indicating 100 μm. (Reproduced from Jenkins *et al.* (1984) with permission of the Water Research Commission, Pretoria, South Africa).

or

$$\frac{\text{Total length of filaments} \times \text{dilution factor}}{\text{MLSS (g l}^{-1})} \ \mu\text{m g}^{-1} \ \text{l}^{-1} \ (\text{TEFL per g MLSS})$$

$$(17.37)$$

Good correlations have been obtained between SVI and TEFL.

A simpler and quicker method uses a 40 μl sub-sample of mixed liquor that is placed on a standard glass slide and covered with a 22×22 mm cover slip. The number of separate fields required to examine the entire area under the cover slip once at $100\times$ is calculated, which remains constant for any specific microscope (e.g. 144). Then using an eyepiece containing a single hairline graticule, the number of times any filament intersects with the hairline is counted (Equation (17.38)). This is repeated at least 20 times although the exact number depends on time and the desired level of precision:

$$\text{Total filament length} = \frac{C \times 144}{40 \times 20} \ \mu\text{m} \qquad (17.38)$$

where C is the sum of all the intersection, 40 is the sub-sample size in μl, 144 the number of fields of vision per slide and 20 is the number of times the count is repeated.

17.8 REMEDIAL MEASURES

In essence some alteration in operational practice is required to control sludge bulking. Specific control measures include operational control, chemical addition or process modification (in increasing order of investment and cost).

17.8.1 OPERATIONAL CONTROL

Operational control is cheapest, but requires a period of trial and error while adjustments are made to certain operating factors. Main problems are as discussed below.

Dissolved oxygen

Underaeration will result in bulking so the aeration tank requires automatic DO control to ensure at least 10% oxygen saturation for carbonaceous or 20% saturation for nitrification. The greatest demand will be at the inlet. Overaeration results in deflocculation due to shear.

Septicity of sewage

This is due to anaerobosis during passage in the sewer and can be overcome by oxygen injection or the revision of sewer design. Anaerobosis in primary sedimentation tanks is controlled by increasing the desludging rate.

Nutrient balance

A BOD:N:P ratio of 100:5:1 is needed to prevent bulking. Raw sewage is rich in N and P, while pharmaceutical, chemical and some food-processing industries are low in nutrients. Agricultural fertilizers are generally used to supplement N and P, while urea is used for N only.

Nature of substrate

If toxic substances are present then pre-treatment is required (Section 14.3). Low-molecular-weight carbohydrates encourage bulking, so roughing treatment using biotowers or biological aerated filters (Section 14.4) before the aeration tank is required.

Controlling sludge loading ratio

To avoid bulking the f/m ratio should be maintained between 0.2 and 0.45 kg kg day^{-1}. The f/m can be altered by adjusting:

(a) the BOD strength and flow rate;
(b) aeration tank volume;
(c) the MLSS concentration.

In practice only the MLSS concentration is easily altered. So reducing the MLSS by wasting sludge increases the f/m, while increasing the MLSS in the aeration tank by not wasting sludge will reduce the f/m ratio. A minimum MLSS concentration of 2000 mg l^{-1} is required, the upper limit depending on the aeration capacity and ability of the sedimentation tank to handle the sludge.

17.8.2 CHEMICAL ADDITION

Chemical addition is an expensive option and long term. It does not generally solve problems which will reappear once chemical addition is stopped. Two approaches are generally used.

Biocides

Biocides which are toxic are used to kill the filaments but not the floc formers. Biocides are not selective as such, but as the filaments grow into the surrounding liquid then the biocide will have maximum effect on filaments and minimum effect on the aggregated floc-forming bacteria. The dosing rate is critical, so it is best to start at a low concentration and slowly increase the dose. The most widely used biocides are chlorine or its derivatives (e.g. sodium hypochlorite) that are toxic at low concentrations, while hydrogen peroxide and ozone are very reactive and add oxygen to the aeration tank as they decompose.

Flocculants

Flocculants are metal salts or polyelectrolytes used to improve flocculation and set-tleability, as well as floc strength. If a metal salt is used (e.g. aluminium sulphate (alum) or ferric chloride) this can be added directly to tank. Some phosphorus removal also occurs, but there is a significant increase in metal salts. Polyelectrolytes have such a rapid response they must be added to the overflow stream as it enters the sedimentation tank (Section 20.5).

17.8.3 PROCESS MODIFICATION

Process modification requires capital expenditure and so should be considered only as a last resort. Three approaches are available.

Mixing pattern

The more plug flow the configuration of the system, then the better the sludge settleabil-ity. Plug flow reduces the competitive advantage that filaments have over floc-forming bacteria due to their high surface area to volume ratio, which makes them more efficient when substrate, nutrients or DO concentrations are low (i.e. accumulation–regeneration theory). A concentration gradient in a plug-flow reactor ensures that initially soluble substrate is rapidly absorbed and stored within cells in the high-concentration area of the reactor (accumulation). Bacteria can then subsequently proceed to digest the mater-ial and produce new cells (regeneration). Linked to a high specific growth rate and selection in the sedimentation tank, this ensures that floc formers dominate. In com-pletely mixed reactors where there is no concentration gradient and the substrate concentration is uniformly low, filaments have the competitive advantage over floc formers. The optimum size for a longitudinal tank is a length to width ratio of 20:1.

Selector

A selector also uses the accumulation–regeneration system. A special tank or zone is provided that is rich in nutrients and oxygen where the most efficient micro-organisms absorb the bulk of the substrate under high substrate concentration. This favours the floc formers. The main aeration tank has a much lower substrate concentration, ensuring a concentration gradient (Fig. 17.23) (Gabb et al., 1991).

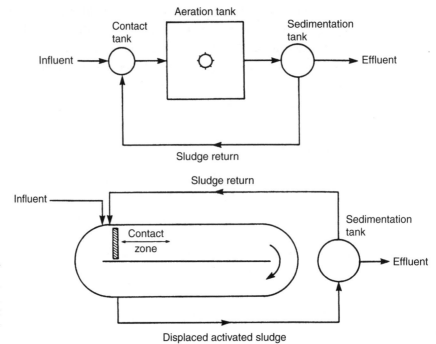

FIGURE 17.23
The incorporation of a contact tank in a completely mixed system or a contact zone in a plug-flow system.

Anoxic zones

Anoxic zones are mainly used for denitrification but have been found to improve sludge settleability (Section 18.4.1).

REFERENCES

Alleman, J. E. and Prakasam, T. B. S., 1983 Reflections on seven decades of activated sludge history, *Journal of the Water Pollution Control Federation*, **55**, 436–43.

Benefield, L. D. and Randall, C. W., 1980 *Biological Process Design for Wastewater Treatment*, Prentice-Hall, New York.

Chambers, B. and Jones, G. L., 1988 Optimization and uprating of activated sludge plants by efficient process design, *Water Science and Technology*, **20** (4/5), 121–32.

Cloete, T. E. and Muyima, N. Y. O. (eds), 1997 *Microbial Community Analysis: The Key to the Design of Biological Wastewater Treatment Systems*, Scientific and Technical Report No. 5, International Association on Water Quality, London.

Coackley, P. and O'Neill, J., 1975 Sludge activity and full-scale plant control, *Water Pollution Control*, **74**, 404–12.

Cox, G. C., Collins, O. C. and Everett, D. A. H., 1982 Works construction and operation in the UK. In: *Oxidation Ditch Technology*. International Conference, Amsterdam, C.E.P. Consultants, Edinburgh, pp 98–110.

DOE, 1985 *The Conditionability, Filterability, Settleability and Solids Content of Sludges 1984. (A Compendium of Methods and Tests)*, Methods for the Examination of Waters and Associated Materials, Department of the Environment, HMSO, London.

Eikelboom, D. H. and van Buijsen, H. J. J., 1981 *Microscopic Sludge Investigation Manual*, TNO Research Institute for Environmental Hygiene, Delft.

Gabb, G. M. D., Still, D. A., Ekama, G. A., Jenkins, D. and Marais, G. V. R., 1991 The selector effect on filamentous bulking in long sludge age activated sludge systems, *Water Science and Technology*, **23** (4/5), 867–77.

Gray, N. F., 2004 *Biology of Wastewater Treatment*, 2nd edn, Imperial College Press, London.

Hawkes, H. A., 1983 Activated sludge. In: Curds, C. R. and Hawkes, H. A., eds, *Ecological Aspects of Used Water Treatment: 2. Biological Activities and Treatment Processes*, Academic Press, London, pp 77–162.

IAWQ, 1995 *Activated Sludge Model No. 2*, Scientific and Technical Report No. 3, International Association on Water Quality, London.

IWPC, 1987 *Activated Sludge*, Institute of Water Pollution Control, Maidstone.

Jenkins, D., Richard, M. G. and Daigger, G. T., 1984 *Manual on the Causes and Control of Activated Sludge Bulking and Foaming*, Water Research Commission, Pretoria, South Africa.

Rachwal, A. J. and Waller, C. B., 1982 Towards greater efficiency. In: *Oxidation Ditch Technology*. International Conference, Amsterdam, October 1982, C.E.P. Consultants, Edinburgh, pp 151–60.

Strom, P. F. and Jenkins, D., 1984 Identification and significance of filamentous micro-organisms in activated sludge, *Journal of the Water Pollution Control Federation*, **51**, 2829–40.

Wagner, M. and Amann, R., 1997 Molecular techniques for determining microbial community structures in activated sludge. In: Cloete, T. E. and Muyima, N. Y. O., eds, *Microbial Community Analysis: The Key to the Design of Biological Wastewater Treatment Systems*, Scientific and Technical Report No. 5, International Association on Water Quality, London, pp 61–72.

White, M. J. D., 1975 *Settling of Activated Sludge*, Technical Report No. 11, Water Research Centre, Stevenage.

Winkler, M. A., 1981 *Biological Treatment of Waste Water*, Ellis Horwood, Chichester.

FURTHER READING

Albertson, O. E., 1991 Bulking sludge control – progress, practice and problems, *Water Science and Technology*, **23** (4/5), 835–46.

Eckenfelder, W. W. and Grau, P., eds, *1992 Activated Sludge Process Design: Theory and Practice*, Technomic Publishing, Lancaster, PA.

Gray, N. F., 2004 *Biology of Wastewater Treatment*, 2nd edn, Imperial College Press, London.

Jenkins, D., Richard, M. G. and Daigger, G. T., 1993 *Manual on the Causes and Control of Activated Sludge Bulking and Foaming*, Lewis Publishers, Boca Raton, FL.

Madoni, P. (ed.), 1991 *Biological Approach to Sewage Treatment Processes: Current Status and Perspectives*, Lugi Bazzucchi Center, Perugia, Italy.

Mara, D. and Horan, N. (eds), 2003 *The Handbook of Water and Wastewater Microbiology*. Academic Press, London.

Mines, R. O. and Sherrard, J. H., 1999 Temperature interactions in the activated sludge process, *Journal of Environmental Science and Health*, **A34**, 329–40.

INTRODUCTION

There are many innovative biological technologies for the treatment of wastewaters and sludges. Four such technologies that have now become widely accepted worldwide are considered below: these are stabilization ponds, the use of plants and wetlands, biological nutrient removal (BNR) systems and composting. However, it is in the area of biotechnology that many of the new technologies are evolving. Biotechnology is the exploitation of living organisms, generally micro-organisms, or biological processes in an industrial or commercial situation to provide desired goods or services. Biotechnology is not a distinct discipline but rather a result of four traditionally separate disciplines: biochemistry, microbiology, engineering and chemistry, interacting with each other. Although the term biotechnology has only become widely studied as a subject in its own right since the mid-1970s, humans have used many biotechnological processes for centuries, such as the exploitation of yeasts in baking and brewing, or bacteria to ripen cheese. New strains of micro-organisms can be developed which produce a desired substance in far greater quantities or degrade a particular substrate far more efficiently than the original parent strain. While this can be achieved by careful culturing and selection, the most important advance is the development of genetic manipulation, known ubiquitously as genetic engineering. This involves the transfer of genetic material from one cell to another to modify that cell's behavior to benefit some industrial process. Turning the bacterial cell into a factory capable of producing enzymes, vaccines, amino acids, steroids or other cellular products such as insulin and interferon has revolutionized the biomedical and food industries. The application of genetic manipulation to wastewater treatment is varied, ranging from the inoculation of specially cultured bacteria to enhance the performance of an existing conventional effluent treatment plant to the use of enzymes for the biodegradation of recalcitrant compounds (Gray, 2004).

BIOAUGMENTATION

Bioaugmentation, or the use of commercially grown micro-organisms to supplement or replace naturally occurring microbes, is now widely practiced in water and wastewater treatment (Huban and Plowman, 1997). The micro-organisms used are usually naturally occurring species collected from special sites where natural selection has already favoured microbes adapted to unusual conditions. For example, up to half the micro-organisms in coastal waters can degrade hydrocarbons, while meat-processing plants and even septic tanks can yield microbes capable of degrading fats and proteins. These organisms are collected and grown in the laboratory, being cultured on a medium rich in the pollutant they are required to degrade. From this the best strains can be isolated and grown. This enrichment technique can be developed by using mutagenesis to find the most efficient degraders or by producing new custom-built microbes by gene manipulation (Powledge, 1983). In practice, only organisms selected from the environment are used, with mutation or gene manipulation only used for enhanced industrial processes (e.g. the production of enzymes), so that there are no problems about releasing new genetically engineered species into the environment. The microbial culture that is eventually produced is grown in special reactors in order to produce the large quantities needed, then freeze-dried and packaged ready for sale. Most companies market their microbes as a powder that also contains important additives, such as wetting agents, emulsifiers to aid dispersion, and nutrients to aid reactivation and growth. To activate the microbes all that is required is to add warm water and stir. Bioaugmentation is used for the following key applications:

(a) improve removal efficiencies in terms of biochemical oxygen demand (BOD) or chemical oxygen demand (COD) removal;
(b) to improve the degradation of target substances that are often recalcitrant or cause operational problems;
(c) to reduce process instability often caused by fluctuation in the organic loading;
(d) to restart or commission treatment plants;
(e) to reduce the inhibitory effects of toxic compounds in wastewaters;
(f) to improve mixed liquor flocculation and separation;
(g) to induce and enhance nitrification;
(h) to reduce the accumulation of sludge and scum from aerobic and anaerobic digesters and lagoons by ensuring complete stabilization.

For most of the above applications the commercial product, either as micro-organisms or enzymes, is added directly to the biological reactor. The amount of product used is based on the activity level of the product, the concentration of the target substrate and the desired effluent quality. In some circumstances (e.g. applications (b) and (d)), the most efficient micro-organisms must be isolated and selected from the existing treatment plant on site, then grown on site in a separate reactor and added to the main influent waste stream either continuously or when required (Glancer and Soljan, 1998).

While most of these products are supplements for wastewater treatment a wide range of other pollution control products, including a dried microbial seed for the BOD test, have also been produced. There has been considerable success in using specialized cultures to clean up chemical spillages and also decontaminating soil. Commercial micro-organisms for cleaning up spillages of crude oil or hydrocarbons are produced by most companies. However, there is a wide range of microbial cultures now available from companies worldwide that are able to deal with most of the major chemicals which are transported in bulk, including pesticides and even Arochlor 1260 which is one of the most highly chlorinated of the polychlorinated biphenyls (PCBs). An interesting example occurred in northern California where a spillage of 21 000 gallons of 50% formaldehyde solution was successfully treated by containing the liquid in a sealed drainage ditch and pumping it through a temporary bioreactor where it was aerated and mixed with a hydrocarbon degrader made up of several mutant bacteria produced by the Polybac Corporation. The formaldehyde concentration was reduced from $1400\,\mathrm{mg\,l^{-1}}$ to $<1\,\mathrm{mg\,l^{-1}}$ after just 14 days, with the whole operation costing about 10% of physically removing the spillage. As well as dealing with spillages to surface water, opportunities also exist for cleaning up contaminated wells and aquifers. Fat- and grease-degrading bacteria can be added to drains and sewers to prevent blockages that would normally have to be removed manually.

There can be no doubt that commercially produced micro-organisms can achieve success in treating many difficult and recalcitrant wastes as well as uprating and seeding wastewater treatment plants, providing the conditions are suitable and the correct culture of microbes is utilized. However, these micro-organisms and their enzymes have to survive and function in the sub-optimal conditions of a complex ecosystem. The strains which are marketed are generally selected on their ability to metabolize certain pollutants, rather on their ability to survive environmental stress. While they are ill adapted to cope with predation by other organisms, like all natural populations they will adjust to the predation level. However, other factors, such as abiotic stresses, including sunlight and temperature, may have a significant effect on numbers and hence efficiency. The use of these products is very expensive and their use should be limited to those situations where normal treatment operation is at risk. They should not be used as a substitute for good operational management of existing biological reactors.

18.1 STABILIZATION PONDS

18.1.1 INTRODUCTION

A stabilization pond is any natural or man-made lentic (enclosed) body of water where organic waste is oxidized by natural activity. In practice, most constructed ponds are simply large shallow lagoons enclosed by earthen embankments. Used on their own or in series they can be operated as either anaerobic- or aerobic-based systems that can be loosely classified into three groups: anaerobic ponds and lagoons, oxidation ponds and aeration lagoons (Hawkes, 1983; Metcalf and Eddy, 1991). Stabilization ponds have been used since ancient times including as fish ponds in China. Today they are used to treat a wide variety of wastewaters including sewage. Such systems are common in areas where land is readily available and cheap, and there is plenty of sunshine (e.g. Australia, India, central and South Africa, USA).

In developing countries stabilization ponds are the most widely used system of sewage treatment being cheap to construct and maintain, being simple to operate and highly effective in removing pathogens (Mara, 1997). They are increasingly common in Europe, especially in France and Germany, where they are used for the initial treatment of strong industrial wastewaters or for the tertiary treatment of domestic wastewaters.

18.1.2 ANAEROBIC LAGOONS

Anaerobic ponds and lagoons are used for the preliminary treatment of strong organic wastes with digestion usually associated with the accumulated sludge in the bottom of the lagoon. Shallower lagoons are also used in some vegetable-processing industries combining the settlement of soil and vegetable waste matter with partial organic treatment. Such lagoons are commonly used in the sugar beet industry and while a 80–90% removal of suspended solids and a 60–70% reduction in BOD is possible, the final effluent produced will still be much stronger than a domestic wastewater ($>500\,mg\,l^{-1}$ BOD, $>200\,mg\,l^{-1}$ suspended solids) requiring further treatment. A number of problems are associated with these lagoons. Vegetable processing is generally seasonal and so the operational period of the lagoons will be short, <100 days per annum, leaving little time for an active anaerobic biomass to build up. Retention time in the lagoons is too short to ensure complete anaerobic breakdown and in plug-flow systems it is not possible for a sufficient mean cell residence time (MCRT) to develop. Complete degradation to methane is often inhibited in shallow lagoons due to partial reaeration of the water, and low temperatures. With vegetable wastes inhibition is also often due to a high C:N ratio.

Anaerobic lagoons are particularly effective for stronger and thicker wastes, and are widely used on farms for the storage and treatment of animal slurry. Distinct layers form in such lagoons with a bottom layer of settled solids forming a sludge, a liquid layer and a floating crust. The crust is up to 40–60 mm thick and insulates the lagoon preventing heat loss, suppressing odours and maintaining anaerobic conditions by preventing oxygen transfer. Influent wastewater enters near the bottom through the sludge layer and the outlet positioned at other end just below scum layer (Fig. 18.1). This induces an upward flow through the lagoon thus ensuring that most solids settle

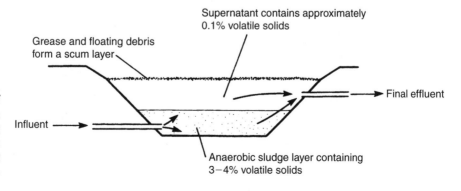

FIGURE 18.1
The basic design of an anaerobic lagoon used for treating strong wastes, such as meat-processing wastewater.

before discharge. A high grease or fibre component in the wastewater is necessary to produce a scum. Where no scum formation occurs, then the lagoon will behave as a facultative pond (Section 18.2.3). Cattle slurry rapidly forms a crust due to the high fibre content of their faeces, excluding air and producing ideal anaerobic conditions. Pig slurry in contrast does not readily crust over allowing oxygen to diffuse into the lagoon so that the top 100–150 mm of the liquid is aerobic. Crust formation is quicker in summer, and in a well-constructed anaerobic lagoon the BOD and suspended solids of the slurry will be reduced by 80% and 90%, respectively. During the summer methane production can be high and large pockets of gas will be seen breaking through the crust. Desludging of farm lagoons should be done every 3–5 years depending on depth, total capacity and degree of anaerobic degradation achieved, with the sludge spread directly onto the land without any of the associated pathogen problems. Most of these lagoons are between 2 and 4 m deep, in order to reduce the surface area to volume ratio, making them potentially dangerous. Also the crust can become covered with grass, weeds and even small bushes giving the surface a false appearance of stability. Loading is expressed as BOD per unit volume ($g\,BOD\,m^{-3}\,day^{-1}$) or as a hydraulic retention time (HRT). A reasonably high temperature is required to ensure complete anaerobic breakdown at a reasonable rate. The maximum permissible organic loading rate for anaerobic lagoons varies with temperature (Table 18.1). Using this information Equation (18.1) can be used to calculate the volume of the lagoon:

$$\lambda_v = \frac{L_i Q}{V_a} \qquad (18.1)$$

TABLE 18.1 Design values for the volumetric BOD loading rate on, and percentage BOD removal in, anaerobic ponds at various temperatures

Temperature (°C)	Volumetric loading rate ($g\,BOD\,m^{-3}\,day^{-1}$)	% BOD removal
≤10	100	40
11	120	42
12	140	44
13	160	46
14	180	48
15	200	50
16	220	52
17	240	54
18	260	56
19	280	58
20	300	60
21	300	62
22	300	64
23	300	66
24	300	68
≥25	300	70

Reproduced from Mara (1996) with permission of John Wiley and Sons Ltd, Chichester.

where λ_v is the organic loading rate $(\text{g BOD m}^{-3}\text{day}^{-1})$, L_i the BOD of the influent wastewater (g m^{-1}), Q the flow rate $(\text{m}^3\text{day}^{-1})$ and V_a the volume of the lagoon (m^3).

In temperate zones anaerobic lagoons only act as settlement ponds in winter with the accumulated sludge being degraded in the summer. The construction of a lagoon of this depth is not straightforward and advice will be required as it may need to be lined with welded butyl rubber or polyvinyl chloride (PVC) sheeting to prevent seepage causing groundwater contamination. Emptying such a deep lagoon requires extremely stable banks from which heavy plant can operate, so earthen banks may not always be suitable.

EXAMPLE 18.1   Design an anaerobic lagoon for influent wastewater with a BOD of $800\,\text{mg}\,\text{l}^{-1}$ and a flow rate of $20\,\text{m}^3\text{day}^{-1}$. The average temperature is 12°C.

(a) The organic loading at 12°C is $140\,\text{g BOD m}^{-3}\text{day}^{-1}$ (Table 18.1) and gives an average percentage BOD removal of 44%.
(b) Using Equation (18.1) to calculate the required volume of the lagoon:

$$V_a = \frac{L_i Q}{\lambda_v} \tag{18.2}$$

$$V_a = \frac{800 \times 20}{140} = 114.3 \text{ m}^3.$$

(c) Using a depth of 2 m, the area of the lagoon required is $57\,\text{m}^2$.
(d) Calculate the HRT (ϕ_a):

$$\phi_a = \frac{V_a}{Q} = \frac{114.3}{20} = 5.7 \text{ day}^{-1} \tag{18.3}$$

In developing countries anaerobic lagoons are normally used in series with a facultative pond followed by at least one maturation pond. In calculating the treatment of the effluent from an anaerobic lagoon by a facultative pond, the influent BOD to the facultative pond is calculated by using the percent removal value in Table 18.1 for the anaerobic lagoon. In the above example the BOD concentration of the influent would be $448\,\text{mg}\,\text{l}^{-1}$.

18.1.3 OXIDATION PONDS

Oxidation ponds are aerobic systems where the oxygen required by the heterotrophic bacteria is provided not only by transfer from the atmosphere but also by photosynthetic algae. There are four different types of stabilization ponds in this category: facultative ponds, maturation ponds, river purification lakes and high-rate aerobic stabilization ponds.

In facultative ponds the algae use the inorganic compounds (N, P, CO_2) released by aerobic and facultative bacteria for growth using sunlight for energy. They release oxygen into solution that in turn is utilized by the bacteria completing the symbiotic cycle (Fig. 18.2). Oxygen also occurs naturally by oxygen transfer, which is increased by turbulence. There are two distinct zones in facultative ponds: the upper aerobic zone where bacterial (facultative) activity occurs and a lower anaerobic zone where solids settle out of suspension to form a sludge that is degraded anaerobically. The algae are restricted to the euphotic zone, which is often only a few centimetres deep depending on the organic loading and whether it is day or night (Fig. 18.3). Such

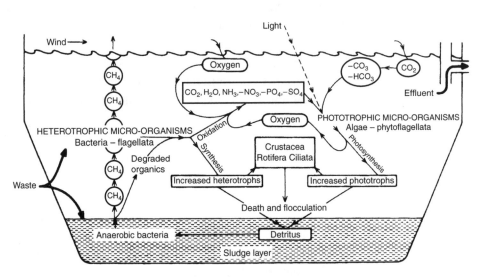

FIGURE 18.2 The key heterotrophic and phototrophic activities in a facultative pond that result in the complete stabilization of organic waste. (Reproduced from Hawkes (1983) with permission of Academic Press Ltd, London.)

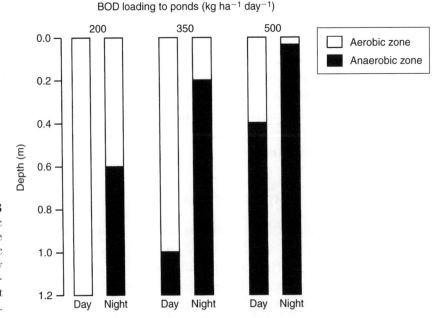

FIGURE 18.3 The effect of organic loading and the depth of the aerobic zone during the day and night in facultative ponds in hot climates.

ponds have a complex ecology with many predator–prey associations, with phyto- and zooplantonic forms predominating. Ponds can be used to produce fish, which feed either directly on the algal biomass or on the intermediate grazers.

Ponds are constructed to a depth of between 1.2 and 1.8 m to ensure maximum penetration of sunlight, and appear dark green in colour due to dense algal development. The concentration of algal is measured as chlorophyll a (i.e. the most important pigment in photosynthesis) and varies between 500 and 1500 $\mu g\,l^{-1}$. Owing to the importance of sunlight to provide oxygen from photosynthesis, organic loading is expressed as a surface loading rate either as $kg\,BOD\,ha^{-1}day^{-1}$ or $g\,BOD\,m^{-2}day^{-1}$ that varies significantly with temperature (Table 18.2). For example, the maximum permissible loading of facultative ponds in northern states of the USA is $2.2\,g\,BOD\,m^{-2}day^{-1}$ compared to $5.6\,g\,BOD\,m^{-2}day^{-1}$ in southern states. The maximum loading rates in facultative ponds are very low compared to other biological systems. The final effluent is rich in algae that have a high BOD (i.e. up to 70–90% of the total BOD). Overall facultative ponds achieve a BOD removal of about 50%, however when the algae is filtered out in the laboratory this rises to >90%. The BOD contribution of the algae in the effluent can be approximated as 1 mg of BOD for every 35 μg of chlorophyll a present.

TABLE 18.2   Design values for the surface BOD loading rate on facultative ponds at various temperatures

Temperatures (°C)	Surface loading rate ($kg\,BOD\,ha^{-1}day^{-1}$)
≤10	100
11	112
12	124
13	137
14	152
15	167
16	183
17	199
18	217
19	235
20	253
21	272
22	292
23	311
24	331
25	350
26	369
27	389
28	406
29	424

The BOD surface loading is calculated using Equation (18.4):

$$\lambda_s = 10 \frac{L_i Q}{A_f} \tag{18.4}$$

where λ_s is the surface loading rate ($kg\,ha^{-1}day^{-1}$), L_i the BOD of the influent wastewater ($g\,m^{-3}$), Q the flow rate ($m^3\,day^{-1}$) and A_f the area of the pond (m^2).

EXAMPLE 18.2 Design a facultative pond for an influent wastewater BOD of $550\,mg\,l^{-1}$ and a flow rate of $120\,m^3\,day^{-1}$ where the average temperature is 20°C.

(a) The BOD surface loading rate at 20°C is given in Table 18.2 as $253\,kg\,BOD\,ha^{-1}day^{-1}$.
(b) The surface area of the pond required is calculated using Equation (18.4):

$$A_f = 10 \frac{L_i Q}{\lambda_s} = \frac{10 \times 550 \times 120}{253} = 2608.7\ m^2 \tag{18.5}$$

(c) The HRT of the facultative pond (ϕ_f) assuming a depth (D_f) of 1.5 m is:

$$\phi_f = \frac{A_t D_f}{Q} = \frac{2609 \times 1.5}{120} = 32.6\ day^{-1} \tag{18.6}$$

Facultative ponds are generally used in series after an anaerobic lagoon and both of these systems are used primarily for the removal of organic matter. Although some pathogen removal occurs in these tanks, the effluent from an anaerobic–facultative system does not produce water that can be safely used for irrigation, especially in warmer climates, due to a high level of pathogen contamination. Removal of parasites (including helminths) and other pathogens (bacterial, viral and protozoan), is significantly enhanced by using maturation ponds in series after the facultative pond (Mara, 1997).

Maturation (tertiary) ponds

This type of oxidation pond is widely used throughout the world as a tertiary treatment process for improving the effluent quality from secondary biological processes, including facultative ponds. Effluent quality is improved by removing suspended solids, reducing ammonia, nitrate and phosphate concentration, and by reducing the number of pathogens (Maynard et al., 1999). They are the same depth as facultative ponds, 1.0–1.5 m, although shallower ponds are used where insolation is high to achieve maximum light penetration. The retention period is normally 10–15 days, although shorter periods can be used for either suspended solids (4 days) or phosphate removal (7–10 days) only (Gloyna, 1971). In Europe, maturation ponds are frequently used as a buffer between the secondary biological phase and the receiving water, balancing the effects of toxic loads and fluctuations in the performance of the secondary biological phase by dilution. They are also used to improve the quality of lowland river water prior to potable water treatment (Section 10.1).

Work on the maturation ponds at Rye Meads Sewage Treatment Works, which discharges to the River Lee in Britain, has shown that phosphate removal is directly correlated with the algal biomass reaching a maximum removal of 73% in May and a minimum of 2% in January (Toms et al., 1975). However, only a small proportion (<20%) of the phosphate removal can be attributed to nutritional uptake by phytoplankton, the majority being removed by precipitation at enhanced pH values (pH >8.2) due to photosynthetic activity (Section 6.4). For every unit increase in pH, the concentration of phosphate remaining decreased by a factor of 10. In the temperate climate of Britain it is only possible to maintain sufficient algal biomass to remove a reasonable proportion of the phosphate from the pond during the spring and autumn by having a long HRT. In winter this becomes unfeasible, although the higher dilution factor, reduced temperature and insolation all significantly reduce the risk of eutrophication in rivers. Removal of nitrate is not linked with the algal biomass and while small amounts are utilized by the algae, the majority is lost by denitrification at the sludge–water interface. Optimum nitrate removal is achieved by ensuring the development of a sludge layer and by using shallow ponds to give the maximum sludge–water surface contact. In this way up to $0.8\,\mathrm{g\,N\,m^{-2}day^{-1}}$ can be removed.

Maturation and facultative ponds have been used for fish production for many centuries. The nutrients released by heterotrophic bacteria are utilized by algae, which in turn are consumed by zooplankton and up the food chain to fish. They are widely used in China and Eastern Europe for coarse fish. In Western Europe, roach and carp are the prime fish species farmed, with chub and perch to a lesser extent. The main operational problem is disease due to the high density of fish stocked. Fishponds require careful management to ensure that the rate of organic loading does not disturb the ecological balance. Also, the inclusion of fish in a pond has significant effects on ecological balance as the fish are feeding on the zooplankton, which are feeding off the phytoplankton. If overstocked, the zooplankton will be overgrazed with a subsequent increase in phytoplankton resulting in algae in the final effluent (Fig. 18.4).

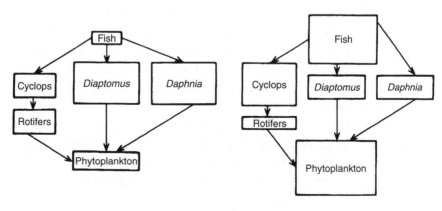

FIGURE 18.4 The effect of overstocking with fish on the food web in a fish pond system. (Reproduced from Hawkes (1983) with permission of Academic Press Ltd, London.)

River purification lakes

Diffuse pollution is extremely difficult to control and often results in serious degradation of water quality, especially in large rivers (Section 7.4). Polluted rivers can be treated by the development of large impounded lakes to remove and degrade this residual pollution. Such lakes are very similar to maturation ponds. Five such lakes have been constructed on the River Ruhr (Germany) with Lake Baldeney, south of Essen, the largest with a retention period of 60 h at low flows (Imhoff, 1984). Together they provide an area of 4488 km^2, with Lake Baldeney alone providing the organic treatment capacity equivalent to a population of 100 000 day^{-1}. Up to 170 tonnes of phosphorus per annum, together with heavy metals, is removed by sludge settlement in the lake. In Britain on the River Tame, a polluted tributary of the River Trent in the West Midlands, a purification lake has also been built. The Lea Marston Purification Lake was built at a disused gravel works after the last major conurbation and effluent input. It removes settleable solids from the river, which are especially a problem during storms. The solids are removed from the main body of water by settlement in the lake and which are subsequently removed by dredging. In contrast with the Rhur lakes, the Lea Marston Lake is continuously desludged although it takes about 12 months to cover the whole flow area of the lake. The effect on water quality is quite dramatic with the suspended solids reduced by 56% and the BOD by 34% overall (Woods et al., 1984).

High-rate aerobic stabilization ponds

High-rate aerobic stabilization ponds are not designed for optimum purification of wastewater but for algal production. The ponds are very shallow (20–50 cm deep), with an HRT of 1–3 days, and are mixed at night to prevent settlement and to maintain aerobic conditions. Algae are harvested for biomass or single-cell protein. Green algae, such as *Chlorella* and *Scenedesmus*, have a protein content of 50% (dry weight) compared to 60–70% for the blue-green alga *Spirulina*. It is possible to control the cellular constituents of algae by altering the growth conditions; for example, by limiting the nitrogen availability the protein content of the algae decreases while its lipid content increases. The cellular constituents of *Chlorella* can be controlled in this way over a range of 7–88% protein, 6–36% carbohydrate and 1–75% true fat. Production of algae can also be linked to biogas production with between 60% and 67% of the potential energy converted into methane. Light availability is the most critical factor controlling algal growth and so such ponds are restricted to those areas where there is plenty of sunshine. Recovery of the algae is extremely difficult because of their small size. The filamentous algae (blue-green) *Osciallatoria* and *Spirulina* are preferred as they are easier to harvest and slightly more digestible than single-celled algae, such as *Euglena* and *Scenedesmus*. Loading depends on insolation, and in California the average loading throughout the year is 134 kg BOD ha^{-1}day^{-1}, reaching an optimum summer loading of 366 kg BOD ha^{-1}day^{-1}.

18.1.4 AERATION LAGOONS

Aeration lagoons are used for the pre-treatment of industrial wastewaters prior to secondary treatment, such as a facultative pond or activated sludge. The lagoons are deep (3.0–4.0 m), with oxygen provided by aerators and not by the photosynthetic activity of algae. Oxygen is normally provided by mechanical pier mounted or floating surface aerators, which ensures that the microbial biomass is kept in suspension and that sufficient dissolved oxygen is provided for maximum aerobic activity. Bubble aeration, provided by compressed air pumped through plastic tubing laid across the bottom of the pond, is also widely used. A predominately bacterial microbial biomass develops and, as there is no provision for settlement or sludge return, the process relies on sufficient mixed liquor developing in the lagoon. For this reason they are ideal for high strength, highly degradable wastewaters such as sugar beet and food-processing effluents. Retention times vary between 3 and 8 days depending on the degree of treatment required, the temperature and strength of the influent wastewater. A retention time of 5 days at 20°C provides an 85% reduction in the BOD of domestic sewage, although a fall in temperature by 10°C results in a reduction in BOD removal to 65% (SCS, 1992).

18.2 PLANTS AND WETLANDS

18.2.1 PLANTS

The controlled culture of aquatic plants is becoming widely used in wastewater treatment. The plants employed can be classified into submerged algae and plants, floating macrophytes, and emergent vegetation (Dingles, 1982).

The role of algae has already been discussed in relation to stabilization ponds (Section 18.1). Floating macrophytes (duckweeds, water ferns and water hyacinth) use atmospheric oxygen and carbon dioxide, but obtain the remaining nutrients they require from the water (Table 18.3). It is the water hyacinth (*Eichhornia crassipes*), a rhizomatous plant with large glossy green leaves and a feathery unbranched root system that is most widely used. Hyacinths are grown in special lagoons and need to be harvested periodically. They are used mainly for tertiary treatment, although they are also employed for secondary treatment. The roots provide a substrate on which heterotrophic micro-organisms thrive. Most improvement in BOD and suspended solids is due to settlement rather than biological activity. The plants exhibit luxury uptake of nutrients, far in excess of normal requirement, and also accumulate other components in the wastewater, such as heavy metals and synthetic organic compounds. The uptake and accumulation of such compounds can be disadvantageous where the concentration in the plant tissue prohibits the subsequent use of the biomass for conversion into energy by digestion or its use as an animal feed supplement.

18.2.2 WETLANDS

The most widely used application of plants in wastewater treatment employs emergent vegetation in artificial wetlands or reed beds (Cooper and Findlater, 1990). Constructed wetlands are a low-cost alternative to conventional treatment processes in both capital and operational terms, often operating without any mechanical or electrical equipment. They can be used for secondary or tertiary treatment of wastewaters, the treatment of storm water and the stabilization of sludge. The process uses the root zone method of treatment. Emergent plants are grown in beds of soil or gravel retained by an impermeable sub-surface barrier (clay or synthetic liner). The base of the bed has a slope of 2–8% to encourage settled sewage to pass horizontally through the soil (Fig. 18.5). The outlet control pipework is designed to allow the

TABLE 18.3 The key functions of macrophytes in constructed wetland systems used for the treatment of wastewater

Macrophyte section	Function
Vegetation above water	Storage of nutrients
	Visual aesthetic quality
	Creation of microclimate
	Reduces effect of wind
	Suppresses algal growth by shading
	Reduces heat loss in winter
Submerged vegetation	Uptake of nutrients
	Provides surface area for biofilm development
	Photosynthesis leads to enhanced aerobic conditions and rapid oxidation
	Increases rate of sedimentation by reduction of current velocity
	Physical filtering of solids
Roots/rhizomes in media	Uptake of nutrients
	Release of oxygen leading to nitrification
	Increase of porosity due to roots
	Stabilization of deposited sediment preventing loss of solids

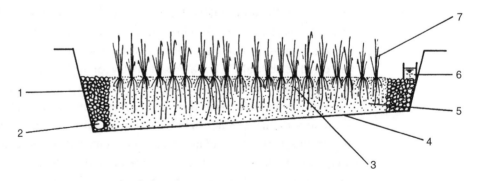

FIGURE 18.5 Basic features of a horizontal flow reed bed. (1) Drainage zone composed of large rocks, (2) water level control and drainage collection for treated effluent, (3) roots and rhizomes (root zone), (4) impervious liner, (5) gravel or soil, (6) influent sewage distribution system and (7) reeds.

water level in the bed to be adjusted to permit near saturation of the substrate during the establishment of the reeds, but lowering the level as the reeds mature. Flooding the bed encourages a horizontal flow of wastewater through the bed, the flooded conditions creating aerobic, anoxic and anaerobic treatment zones. The depth of the bed depends on the root penetration of the plants used, for example cattails 300 mm, reeds 600 mm, bulrushes 750 mm. The most widely employed emergent aquatic macrophytes are *Phragmites communis* (common reed), *P. australis* and *Typha latifola* (cattail). The substrate and the roots provide a large potential surface area for the growth of heterotrophic micro-organisms. The long roots of the vegetation penetrate deep into the substrate. The internal gas spaces of the plant, an adaptation to its partially submerged existence, transports oxygen to the saturated rhizomes and roots (i.e. aerenchyme tissue). Thus, an aerobic zone is created within an otherwise waterlogged sediment. Within the rhizosphere there are also anoxic and reduced areas (Fig. 18.6). This combination allows both aerobic and anaerobic bacteria to survive, allowing carbonaceous oxidation, nitrification, denitrification and anaerobic degradation to occur. The plants take up nutrients and store them in their tissues that are removed from the system when harvested. Finally, the substrate has a sorption potential to remove metals from solution. Removal potential exceeds 90% for most metals and these become immobilized in the anaerobic mud layer in the base of the filter as metal sulphides (Table 18.4). Excessive accumulation will eventually inhibit the process. During the winter the emergent part of the vegetation dies back and so

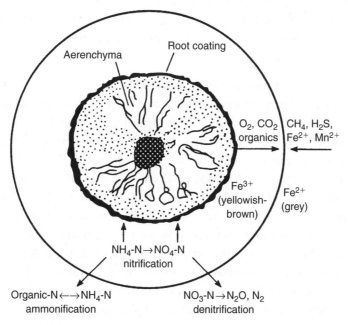

FIGURE 18.6 Cross section of an oxidizing root growing in a reduced sediment. The oxidized rhizosphere is depicted, along with some processes that occur as a result of this aerobic–anaerobic interface. (Reproduced from Mendelssohn and Postek (1982) with permission of the American Journal of Botany.)

TABLE 18.4   Predominant removal mechanisms in reed bed treatment systems

Wastewater constituent	Removal mechanism
Suspended solids	Sedimentation
	Filtration
Soluble organics	Aerobic microbial degradation
	Anaerobic microbial degradation
Nitrogen	Ammonification/nitrification
	Denitrification
	Plant uptake
	Matrix adsorption
	Ammonial volatilization
Phosphorus	Matrix sorption
	Plant uptake
Metals	Adsorption and cation exchange
	Complexation
	Precipitation
	Plant uptake
	Microbial oxidation/reduction
Pathogens	Sedimentation
	Filtration
	Natural die-off
	Predation
	UV irradiation
	Root excretion of antibiotics

treatment is restricted to the root zone only, but in practice performance is not severely affected. They do not require daily attention, although during the first year of operation beds do require weeding, either by hand or by flooding, before the reeds are fully established.

A minimum hydraulic conductance of 10^{-3}–$10^{-4}\,\mathrm{m\,s^{-1}}$ is required to prevent surface flow and channelling occurring, so the selection of the correct substrate or medium is vital. Once the reeds are established and the rhizomes have penetrated the entire root zone, the hydraulic conductance is secured by the network of micro-channels left after dead rhizomes have decayed (5–15 mm in diameter). Therefore, porosity increases with the age of the bed. The best results are achieved using gravel (5–32 mm) rather than soils, although this medium has a low potential for the removal of heavy metals and phosphorus. Different substrate compositions have different qualities. For example, substrates with a low nutrient content encourage direct uptake of nutrients from wastewater by the plants rather than from the substrate itself. A high Al or Fe content provides effective phosphorus adsorption, while a high organic or clay content ensures a good removal of heavy metals. Crushed limestone is used in the treatment of acidic wastewaters and also gives enhanced phosphorus removal.

The root zone method is becoming popular for the treatment of a range of non-domestic wastewaters including landfill leachate, acid mine drainage, pulp mill effluents, surface and urban run-off, and pharmaceutical wastewaters. Owing to their ability to withstand a wide range of operating conditions and their high potential for wildlife conservation, reed beds have become popular with the industrial sector as part of an overall environmental management strategy. However, their main application is in the treatment of domestic sewage from isolated developments, such as schools, housing estates, etc. Owing to the relatively low loading rates that can be applied, constructed wetlands require large areas of land compared to conventional processes, and so are normally restricted to treating domestic wastewater from small communities. Wastewaters require primary settlement before being applied to a reed bed. Average design loading for domestic wastewaters is $5\,\mathrm{m}^2$ per capita. The precise area of reed bed required is calculated using Equation (18.7):

$$A = \frac{Q(\ln C_0 - \ln C_1)}{K_1} \qquad (18.7)$$

where A is the area of the bed (m^2), Q the influent flow rate ($\mathrm{m}^3\,\mathrm{day}^{-1}$), C_0 the influent BOD concentration ($\mathrm{mg\,l}^{-1}$), C_1 the required effluent BOD concentration ($\mathrm{mg\,l}^{-1}$) and K_1 the BOD reaction rate constant (Equation (5.54)). Cooper (1990) has recommended that a K_1 value of 0.1 can be used in most European (temperate) situations. This equation can also be used for reed beds used for tertiary treatment.

Example 18.3 Calculate the area of a reed to treat a flow rate of $100\,\mathrm{m}^3\,\mathrm{day}^{-1}$ of influent wastewater with a BOD of $250\,\mathrm{mg\,l}^{-1}$ to a required BOD concentration of $20\,\mathrm{mg\,l}^{-1}$. Assume a K_1 value of 0.1.
Using Equation (18.7):

$$A = \frac{100(\ln 250 - \ln 20)}{0.1} = \frac{100(5.52 - 3.0)}{0.1} = \frac{252}{0.1} = 2520\,\mathrm{m}^2$$

During the growing season between 1.5 and $1.8\,\mathrm{m}^3$ of water is lost by evapotranspiration per square metre of fully developed *Phragmites*, significantly reducing the effluent volume. Reed beds can be designed to operate with the wastewater flowing in either horizontal or vertical direction, with the latter achieving better nitrogen removal overall.

Vertical flow reed beds, which consist of graded layers of gravel media topped with sharp sand, are also used. These beds are intermittently loaded with wastewater allowing complete discharge of treated effluent between each cycle. In this way oxygen is pulled into the interstices of the media as the effluent drains away ensuring aerobic conditions. Hybrid systems are generally constructed for treatment of settled sewage, which employ a combination of vertical and horizontal flow beds in series. Several vertical beds in series are used for high level BOD and COD removal followed by nitrification. Horizontal flow beds are used subsequently to polish the effluent by removing suspended solids and associated BOD, pathogens, carrying out denitrification and extra P removal. Design and operation overviews are given by Kadlec *et al.* (2000) and Gschlößl and Stuible (2000).

The reeds remove moisture from the sludge

FIGURE 18.7
Design of a sludge
mineralization reed
bed. (Reproduced
with permission of
TransForm ApS
Danish Rootzone,
Copenhagen,
Denmark.)

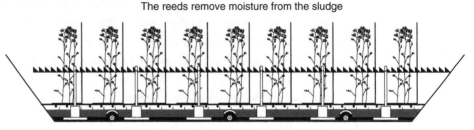

The reeds remove moisture from the sludge

The drains remove moisture from the sludge

An adaptation of the reed bed is used for the mineralization of sewage sludge. The reed bed is lined and a 500 mm layer of substrate, which is generally gravel, is laid on a drainage system to collect the excess moisture from the sludge which is then returned to the inlet of the wastewater treatment plant (Fig. 18.7). The reeds, usually *Phragmites australis*, are planted into the substrate and sludge is applied. The application rates and procedure for spreading the sludge depends on the type of sludge and its solids content. However, 60–70% of the organic matter is mineralized. Depending on the climate, the sludge is dewatered to 35–48% solids within 2 or 3 weeks, and in the long term the volume of sludge is reduced to 2–5% of the original. Loading rates vary from 10 to 20 $m^3 m^{-2} year^{-1}$ for sludges with a low dry matter content, to 3–4 $m^3 m^{-2} year^{-1}$ for sludges with a high dry matter content. Sludge mineralization beds should be rested every third season to allow for complete mineralization, so at least three beds are required at each plant. The residue from sludge mineralization has to be removed every 10–20 years and the reeds replanted. However, the beds must not be allowed to dry out and be kept partially flooded throughout the year.

18.3 NUTRIENT REMOVAL

18.3.1 DENITRIFICATION

There has been a serious increase in the concentration of nitrates in both surface and ground waters mainly from agricultural sources, although the contribution from treated sewage effluents is significant in many areas (Section 6.4). Nitrate can be converted via nitrite to gaseous nitrogen under low dissolved oxygen conditions by the process known as denitrification. The process occurs in any nitrified effluent when deprived of oxygen. So, unless a specific denitrifying reactor is constructed, denitrification is only likely to occur in the sludge layer of the sedimentation tank after an aerobic biological reactor where nitrification is occurring. The process can only proceed under anoxic conditions (i.e. when the dissolved oxygen conditions are very low but not necessarily zero) and when a suitable carbon source is available. Denitrification can occur at dissolved oxygen concentrations up to, but not exceeding, 2% saturation. The process is carried out by a wide range of facultative anaerobes, the most common being *Pseudomonas* spp. with *Achromobacterium* spp., *Denitrobacillus* spp., *Spirillum* spp.,

Micrococcus spp. and *Xanthomonas* spp. frequently present. Under anoxic conditions these bacteria use nitrate in place of oxygen as a terminal electron acceptor, although if oxygen is present then this will be used in preference. The overall stoichiometric equations for denitrification using methanol (CH_3OH) as the carbon source are:

(a) overall nitrate removal:

$$NO_3^- + 1.08CH_3OH + H^+ \rightarrow 0.065C_5H_7O_2N + 0.47N_2 + 0.76CO_2$$
$$+ 2.44H_2O \tag{18.8}$$

(b) overall nitrite removal:

$$NO_2^- + 0.67CH_3OH + H^+ \rightarrow 0.04C_5H_7O_2N + 0.48N_2 + 0.47CO_2$$
$$+ 1.7H_2O \tag{18.9}$$

(c) overall deoxygenation:

$$O_2 + 0.93CH_3OH + 0.056NO_3^- + 0.056H^+ \rightarrow 0.056C_5H_7O_2N$$
$$+ 0.65CO_2 + 1.69H_2O \tag{18.10}$$

From Equation (18.10), it can be seen that the denitrifying bacteria remove significant amounts of BOD during the reduction of nitrate. While some effluents will contain enough residual organic carbon to allow denitrification to proceed, if not untreated settled sewage can be used although the preferred source of organic carbon is methanol, which is also used for potable supplies. This is because the input of organic carbon into the reactor can be accurately controlled.

The amount of methanol, or organic carbon equivalent, can be estimated from:

$$C_m = 2.47N_0 + 1.52N_i + 0.87D_0 \tag{18.11}$$

where C_m is the concentration of methanol required ($mg\,l^{-1}$), N_0 the initial nitrate concentration ($mg\,N\,l^{-1}$) and N_i the initial nitrite concentration ($mg\,N\,l^{-1}$), while D_0 is the initial dissolved oxygen concentration ($mg\,l^{-1}$) as the reaction will take place under anoxic as well as anaerobic conditions.

The biomass produced can be estimated as:

$$C_b = 0.53N_0 + 0.32N_i + 0.19D_0 \tag{18.12}$$

where C_b is the biomass production ($mg\,l^{-1}$).

When N_i and D_0 are zero, the methanol:nitrate ratio is 2.47 and the biomass production is $0.53N_0$. In these calculations only the amount of nitrogen present is used, so for a sample containing $15\,mg\,l^{-1}$ of nitrate ions at $0\,mg\,l^{-1}$ nitrite ions and dissolved oxygen, then the amount of methanol required will be $2.47 \times 15 \times 14/62 = 8.4\,mg\,l^{-1}$ with $0.53 \times 15 \times 14/62 = 1.8\,mg\,l^{-1}$ of biomass produced. If, however, the sample contains $15\,mg\,l^{-1}$ of nitrate-nitrogen then $37\,mg\,l^{-1}$ of methanol would be required producing $8\,mg\,l^{-1}$ of biomass. It is important to provide the correct dose of methanol to the reactor as, if it is underestimated, denitrification will be incomplete and, if overestimated, methanol will remain in the final effluent and cause further pollution.

The overall effect of the denitrification reaction is to raise the pH by the formation of hydroxide ions. This replaces about 50% of the alkalinity consumed by the oxidation of ammonia during nitrification. For each mg of ammonia oxidized to nitrate, $7\,mg$ of alkalinity are utilized, and $3\,mg$ are produced during denitrification. The denitrification reaction is sensitive to pH with an optimum range between pH 6.5 and 7.5, but falling to 70% efficiency at pH 6 or 8. Temperature is also important with the reaction occurring between 0°C and 50°C but with optimum reaction rate between 35°C and 50°C. Denitrification and nitrification process kinetics are explained in detail in IAWQ (1995).

The denitrifying filter

The major denitrification process is the denitrifying filter which is a submerged filter containing an inert medium on which a film of denitrifying bacteria develops. This contact system provides the necessary MCRT for the development of these slow-growing bacteria and depending on the voidage of the medium the sludge residence time (SRT) can be very short, often <1 h. The most frequent operational problem is excessive film accumulation or entrapped gas bubbles which alter the internal flow pattern restricting the flow causing a significant head loss or reducing the HRT by channelling the effluent through the filter. Nitrate removal efficiency increases with the accumulation of biological solids; however, removal efficiency sharply falls off as the flow is restricted and the pressure builds up, until finally the liquid forces it way through the bed scouring much of the film away. Therefore, it is better to opt for a larger medium (15–25 mm diameter) with larger voids and a slightly reduced surface area so that film accumulation is less of a problem. In this way the filter will operate more consistently. Better film control has been achieved using fluidized bed systems, although this has been designed primarily for potable water treatment (Croll et al., 1985).

Activated sludge systems

In activated sludge systems denitrification occurs in anoxic zones or reactors with a simultaneous loss in BOD. For each 1 g of nitrate N utilized 2.9 g of BOD is also removed. Denitrification is carried out by facultative heterotrophic bacteria, such as *Pseudomonas*, *Achromobacter* and *Bacillus* spp. (Muyima et al., 1997). In plug-flow systems both BOD oxidation and nitrification occur in the same aerobic zone. A high degree of nitrification is achieved with only approximately 10% of the original influent BOD remaining. This will limit the degree of denitrification in the subsequent anoxic zone due to a lack of available organic carbon to just 20% of the available nitrate. In completely mixed systems this is overcome by constructing a separate anoxic zone in front of the aerobic zone (Figs 18.8 and 18.9) with the degree of denitrification dependent on the quantity of effluent and sludge returned to the anoxic zone. Denitrification efficiency (D_e) can be estimated by Equation (18.13):

$$D_e = \frac{Q_r + Q_{re}}{Q + Q_r + Q_{re}} \qquad (18.13)$$

where Q is the volume of influent wastewater ($m^3 h^{-1}$), Q_r the return sludge rate ($m^3 h^{-1}$) and Q_{re} the recirculated effluent rate ($m^3 h^{-1}$). Large quantities of treated effluent must be recirculated in order to achieve high denitrification efficiencies (Table 18.5). This is very expensive due to extra reactor volume and pumping costs.

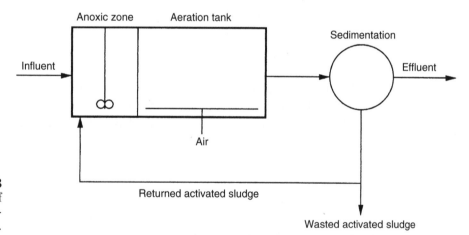

FIGURE 18.8
Schematic layout of a conventional pre-nitrification system.

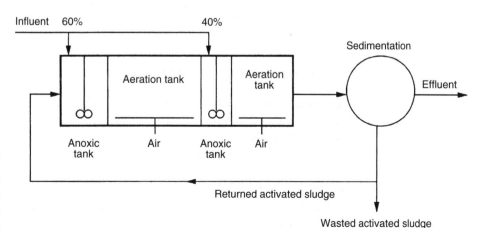

FIGURE 18.9
Schematic layout of a two-stage nitrification–denitrification system.

TABLE 18.5 Typical relationship between denitrification and the rate of recirculation

Recirculation factor	Denitrification (%)
1	50
2	67
3	75
5	83
10	91
20	95
30	97

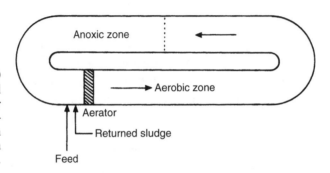

FIGURE 18.10 Aerobic and anoxic zone for dual nitrification–denitrification within single oxidation ditches.

However, in oxidation ditches simultaneous nitrification and denitrification occurs to a high degree achieved by natural internal recirculation (Fig. 18.10). The internal recirculation factor (R) can be calculated using Equation (18.14):

$$R = \frac{Q_r + (V/t_r)}{Q} \qquad (18.14)$$

where V is the volume of the oxidation ditch (m^3) and t_r the recirculation time (h). In most oxidation ditches R is >20 which according to Table 18.5 is equivalent to >95% denitrification.

18.3.2 PHOSPHORUS REMOVAL

Traditionally phosphorus is removed by the addition of coagulants to wastewater, although biological treatment is now widely used. Chemical removal of phosphorus is fully explored in Section 20.5, with the main design configurations for chemical removal shown in Fig. 20.10.

Conventional activated sludge mixed liquor contains 1.0–2.5% of phosphorus as dry weight. This can be increased to 5% or more in biological phosphorus removal processes due to enhanced biological (luxury) uptake by bacteria. Luxury uptake of

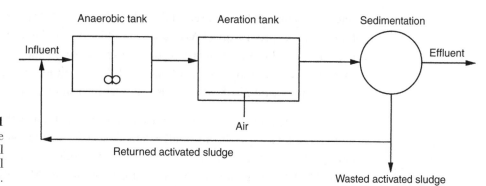

FIGURE 18.11
The two-stage
(A–O) biological
phosphorus removal
process.

phosphorus is achieved by subjecting the mixed liquor to a cycle of anaerobic and aerobic conditions. In an anaerobic environment the bacteria of the mixed liquor are conditioned or stressed so that once back in an aerobic environment with the stress removed the bacteria will rapidly take up phosphorus as long as an adequate supply of BOD is available. In the anaerobic environment volatile fatty acids (VFAs), which are products of fermentation, are rapidly absorbed by the mixed liquor, accumulating as storage products within bacterial cells while at the same time phosphorus is released from the cell into solution. Once in an aerobic environment, the bacteria oxidize the absorbed VFAs causing a simultaneous uptake of phosphorus from solution. The phosphorus is stored within the cell as polyphosphate granules. Phosphorus enters bacterial cells either by diffusion or biochemically via an active transport mechanism, such as adenosine triphosphate (ATP). The latter is more important in activated sludge as the concentration of phosphorus is normally higher in the cell than in the wastewater. Commonly anaerobic zones do not generate sufficient VFAs, so in-line fermentors may be employed. The genus *Acetinobacter* is especially successful at absorbing phosphorus, being able to accumulate polyphosphate by up to 25% of their cell mass (Muyima *et al.*, 1997). The accumulated phosphorus is removed from the system with the waste activated sludge stream. To be effective the final effluent suspended solids concentration must be kept as low as possible as suspended solids have a high phosphorus content. Also, if the excess sludge is not kept aerobic before wastage is complete then the phosphorus will be released back into the process water.

The main biological phosphorus removal system is the A–O process, which stands for anaerobic–oxic (Fig. 18.11). In the anaerobic tank the bacteria are conditioned and any phosphorus present becomes soluble. This is then absorbed by the bacterial biomass in the aeration tank. Biological phosphate removal is largely independent of temperature over the range of 5–20°C. The system depends on a BOD:P ratio >20 in order to achieve final effluent concentrations of $<1\,mg\,P\,l^{-1}$, and if the ratio falls below this critical limit then chemical coagulation will be required to remove the residual phosphorus from the final effluent. The anaerobic reactor should have an HRT of 1–2 h, although this can be reduced if the influent wastewater is either septic or has a high BOD:P ratio. If the BOD is not readily degradable then an extended retention time will be required in the anaerobic tank. Nitrate inhibits the fermentation processes producing VFAs in the anaerobic zone and must be either removed by denitrification before any recycling takes place or nitrification should not be allowed to occur. The aeration tank is of standard design, although if a long sludge age is required to permit nitrification a separate denitrification zone will be required.

The phostrip process combines biological and chemical removal. Phosphorus is accumulated biologically by luxury uptake then released back into solution in the anaerobic (stripper) tank (Fig. 18.12). The concentrated solution of phosphorus produced is treated by lime coagulation. The anaerobic tank with an HRT of 10 h also conditions the mixed liquor so that it is able to take up the phosphorus in the aeration tank as well as carrying out carbonaceous oxidation. The process is continuous with only 10–15% of the total wastewater flow passing through the anaerobic tank, so that only a small proportion of the flow is actually treated with the coagulant, resulting in a considerable saving. The higher phosphorus concentration of the liquid from the stripper makes chemical precipitation much more cost-effective. There is no nitrogen removal in this process. However, if nitrification does occur in the aeration tank due to temperature or sludge age then a separate anoxic zone is required for denitrification prior to the anaerobic tank.

18.3.3 BIOLOGICAL NUTRIENT REMOVAL SYSTEMS

BNR is widely used in Australia, South Africa and the USA, and in response to the stricter discharge limits for nutrients in the EU Urban Waste Water Treatment Directive such systems are becoming increasingly common in Europe. There are many different designs of BNR systems, and some of the commonest are described below.

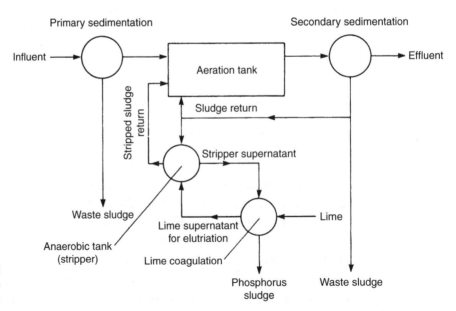

FIGURE 18.12
The phostrip
process.

The Bardenpho process

The Bardenpho process incorporates nitrification, denitrification, phosphorus removal and carbonaceous oxidation by providing a series of aerobic, anoxic and anaerobic zones (Fig. 18.13). Influent wastewater enters an anaerobic zone where the mixed liquor is conditioned for phosphorus uptake. It then flows into an anoxic zone where denitrification occurs and then enters the aeration tank where carbonaceous oxidation, nitrification and luxury uptake of phosphorus occurs simultaneously. Mixed liquor is constantly recycled back to the previous anoxic zone resulting in up to 80% denitrification before leaving the aeration tank to the final anoxic zone to complete denitrification. The final small aerobic zone is to prevent anaerobic conditions developing in the final sedimentation tank and to prevent loss of absorbed phosphorus from the mixed liquor. With a typical SRT of 18 days, an HRT of 22 h and an anaerobic zone HRT of 1.5 h, the phosphorus concentration in the effluent is $<1 \, \text{mg} \, \text{l}^{-1}$. The sludge can accumulate between 6% and 8% phosphorus dry weight, although the sludge must be rapidly dewatered before it becomes anaerobic and phosphorus is resolubilized. There are a large number of varieties of different BNR systems based on the Bardenpho concept (Figs 18.14 and 18.15) mostly of the AAO design, which is an acronym for anaerobic, anoxic and oxic (Fig. 18.16).

Sequencing batch reactors

Sequencing batch reactors have been described in Section 17.5.4. When used for nutrient removal all the processes (aerobic, anaerobic and anoxic) occur within a single

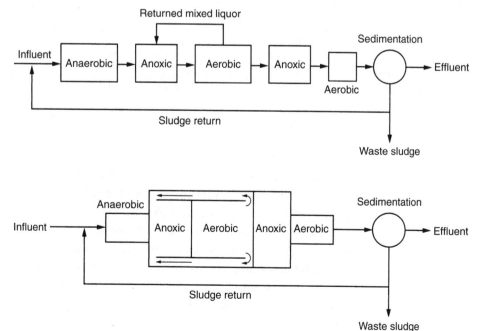

FIGURE 18.13
Typical schematic designs of the Bardenpho process.

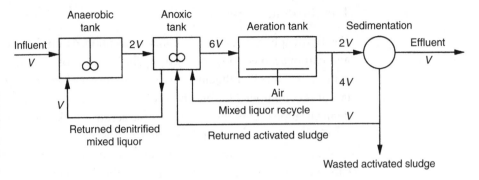

FIGURE 18.14
Schematic layout of
the University of
Cape Town–Virginia
Initiative
(UCT–VIP) process.

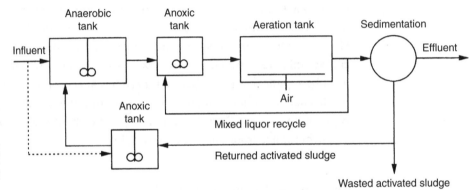

FIGURE 18.15
Schematic layout of
the Johannesburg
(JHB) process.

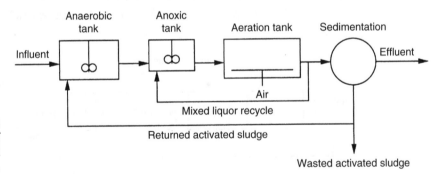

FIGURE 18.16
Schematic layout of
the AAO activated
sludge process.

reactor. The process cycle outlined in Fig. 17.16 is changed to: (I) wastewater addition, (II) anaerobic mixing, (III) aeration, (IV) anoxic mixing, (V) clarification, (VI) effluent removal and (VII) sludge wastage (Fig. 18.17). The batch nature of the process provides an almost ideal plug-flow configuration that prevents the development of filamentous bacteria, but such systems are not good at dealing with storm water flows. This design is particularly effective and is becoming increasingly popular. For example, the new plant serving North Brisbane in Queensland, Australia treats a population equivalent of 40 000 and is achieving a final effluent of $10\,\mathrm{mg\,l^{-1}}$ BOD, $10\,\mathrm{mg\,l^{-1}}$ suspended solids, $5\,\mathrm{mg\,l^{-1}}$ total nitrogen and $1\,\mathrm{mg\,l^{-1}}$ total phosphorus (Hayward, 1998).

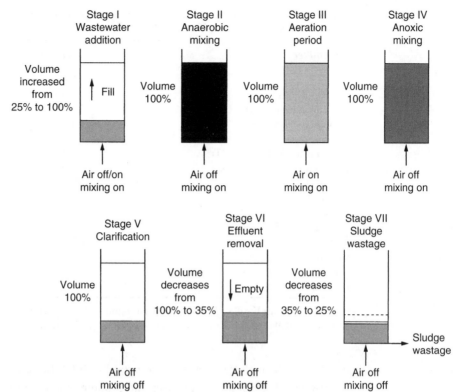

18.4 COMPOSTING

18.4.1 INTRODUCTION

Composting is a stabilization process used for sewage sludge and other organic wastes, including sewage sludge, animal and agricultural wastes, and household refuse (Gasser, 1985). Almost all the widely used composting systems are aerobic, with the main products being water, carbon dioxide and heat:

$$\text{Organic matter} + O_2 \xrightarrow{\text{aerobic bacteria}} \text{New cells} + CO_2 + H_2O + NH_3 + SO_4$$

$$(18.15)$$

Significantly less heat is produced during anaerobic composting with methane, carbon dioxide and low-molecular-weight organic acids being produced. Owing to the nature of the intermittent products of anaerobic breakdown there is also a high possibility that odours will be produced. Fermentation and anaerobiosis also lead to products which will undergo further decomposition when exposed to air (Table 18.6).

18.4.2 THE PROCESS

On average aerobic composting takes 30 days to produce a fully stabilized compost. The process is exothermic and proceeds from psychrophilic (ambient) temperatures

TABLE 18.6 Typical analysis of sewage-sludge-based compost produced at Ipswich Sewage Treatment Works, UK

Parameter	Concentration
pH	7.4
Dry solids (%)	75.9
Organic matter (% DS)	53.1
Ammonia–N (% DS)	0.1
Total nitrogen (% DS)	2.0
Phosphorus (as P_2O_5) (% DS)	1.2
Potassium (as K_2O) (% DS)	1.0
Mn ($mg\,kg^{-1}DS$)	273
Cu ($mg\,kg^{-1}DS$)	64
Zn ($mg\,kg^{-1}DS$)	222
Ni ($mg\,kg^{-1}DS$)	15
Pb ($mg\,kg^{-1}DS$)	137
Cr ($mg\,kg^{-1}DS$)	14
Cd ($mg\,kg^{-1}DS$)	0.5
Hg ($mg\,kg^{-1}DS$)	<0.1

Reproduced from Barnes (1998) with permission of The Chartered Institution of Water and Environmental Management, London.

of 10–20°C through the mesophilic range (20–40°C) into the thermophilic range up to 50–60°C. The compost can reach thermophilic temperatures within the first few days, and the longer these temperatures can be maintained the better will be the destruction of pathogens. For example, if compost temperature is maintained at >55°C for 10 days 99.9% pathogen destruction is achieved, which is equivalent to 60 min at >70°C. Composting also reduces the water content of the composted waste by evaporation thereby reducing the volume of material that has to be eventually disposed. Composting effectively recycles the organic matter and nutrients in wastewater into a useful material which can be used as a soil conditioner (i.e. to reduce the bulk density and to increase the water holding capacity of soil), and also a fertilizer containing nitrogen, phosphorus and trace elements. The nitrogen is organically bounded and so slowly released throughout the growing season, with minimum loses due to leaching in contrast to when soluble inorganic fertilizers are used.

18.4.3 OPERATIONAL FACTORS

To successfully compost dewatered sludge cake the bulk density needs to be reduced so that air can penetrate through the cake more efficiently. This is normally achieved either by the addition of an organic amendment that decomposes during composting or a bulking agent that is recovered after composting and reused. The most popular is the addition of an amendment to the sludge, as it allows greatest control over the quality of the final product as well as disposing of a secondary waste material. The amendment is an organic material which is added to the substrate to reduce its bulk density and to increase the voidage, so air can penetrate providing adequate aeration.

Sawdust, straw, peat, rice husks, manure, refuse and garbage, and lawn and tree trimmings have all been successfully used. The choice of amendment is generally limited to what is available locally, but ideally it should be dry, have a low bulk density and be degradable. Bulking agents can be either organic or inorganic particles of sufficient size and shape to provide structural support to the sludge cake as well as maintaining adequate aeration. The bulking agent provides a matrix of interstices between particles in which sludge is trapped and undergoes decomposition. Enough space is left between particles to ensure sufficient ventilation. After composting the bulking agent is recovered, normally by screening, and reused. Inert bulking agents have the longest life although degradable materials can be used to improve the organic quality of the compost and are less of a problem is not totally removed by the screens. The most widely used bulking agents are wood chips, although other suitable materials include pelleted refuse, shredded tyres, peanut shells, tree trimmings and graded mineral chips.

Since composting is an exclusively biological process, those factors which affect microbial metabolism either directly or indirectly are also the factors which affect the process as a whole. The most important operational factors are aeration, temperature, moisture, C:N ratio and pH (Finstein *et al.*, 1983).

Aeration

In the compost pile the critical oxygen concentration is about 15%, below which anaerobic micro-organisms begin to exceed aerobic ones. Oxygen is not only required for aerobic metabolism and respiration, but also for oxidizing the various organic molecules that may be present. During composting the oxygen consumption is directly proportional to the microbial activity, with maximum oxygen consumption rates occurring at temperatures between 28°C and 55°C.

Temperature

High temperatures in composting result from heat produced by microbial respiration. Composting material is generally a good insulator and as the heat is only slowly dispersed the temperature in the pile increases. Heat loss to the outside of the pile is a function of the temperature difference and the rate of microbial activity. As the rate of activity is limited by the rate that oxygen can enter the pile, heat production is affected by oxygen availability. In an aerated pile 60°C can be reached within 1–2 days while in unaerated windrow systems, where oxygen is more limited, similar temperatures are reached after 5 days. The compost will remain at this maximum operating temperature until all the available volatile solids have been consumed (Fig. 18.18).

Water content

Moisture content is linked to aeration of the pile. Too little moisture (<20%) will reduce or even stop biological activity, resulting in a physically stable but biologically

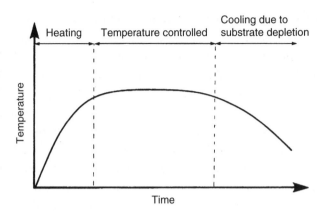

FIGURE 18.18
The three stages of
the composting
process in a static
pile system.
(Adapted from
Stentiford *et al.*
(1985) and repro-
duced with permis-
sion of Elsevier
Science Ltd,
London.)

unstable compost. Too much moisture (>60%) fills the interstices of the compost, excluding air, so that aerobic growth is replaced by anaerobic activity which will result in malodour production. The optimum moisture content for composting is between 40% and 50%.

C:N ratio

The optimum ratio of available carbon to nitrogen (C:N) should be about 12 for good microbial growth. However, where bulking agents, such as wood chips, are used all the carbon present is not readily available and will only be slowly degraded so only a small proportion of the carbon is available. In these circumstances an optimal C:N ratio for composting may be as high as 25–30 (De Bertoldi *et al.*, 1985). A higher C:N ratio will slow up decomposition until excess carbon is oxidized, while a low C:N ratio will also slow decomposition and increase the rate of nitrogen loss through ammonia volitization, especially at the higher pH and temperatures so typical of the composting process. The final compost product must have a C:N ratio <10–12 otherwise the soil microbial population will immediately tie up the available nitrogen, which will limit crop growth and defeat the purpose of using compost to improve soil quality. So the sewage sludge and bulking agent or amendment used must be blended to give an optimum C:N ratio as well as an adequate moisture content. For example, the approximate C:N ratio of sewage sludge (activated sludge) is 6:1, sewage sludge (digested) 16:1, straw 48:1, wood chips 500:1 and farm yard manure 14:1.

pH

The optimum pH for composting is between 5.5 and 8.0, although organic matter within the pH range of 3–11 can be successfully composted. The pH will begin to drop as soon as composting commences due to acid-forming bacteria breaking down complex carbohydrates to organic acids, which tends to favour fungal development. A high pH can result in nitrogen loss if the temperature is also high.

18.4.4 COMPOSTING SYSTEMS

There are two main methods of composting: open systems which do not require a reactor but employ piles and windrows; and closed systems which take place within a specially constructed reactor where the temperature and aeration are controlled.

Open systems

Open systems are of two types: turned piles or windrows (long rows of compost triangular or trapezoidal in cross section); and static piles. In the former oxygen is provided by natural diffusion and periodic turning, while in the latter oxygen is normally supplied by forced aeration.

Windrowing is more popular than piles as it is more efficient in terms of space utilization. They can be constructed either on a soil base or on concrete floors with underdrainage. Soil-based systems are generally used by farmers composting animal wastes using straw as a bulking agent. They are generally constructed at the periphery or headland of fields and are left to rot slowly without turning, which allows pathogens to survive in the outer part of the pile. The final quality of the compost produced in this way is much poorer compared to piles that have been turned periodically. At sewage treatment plants concrete bases are normally employed with long windrows 1–2 m high and 4–6 m wide constructed at regular intervals so that they can be regularly turned mechanically with ease. This ensures that all parts of the compost mixture, especially the sludge cake, spends an adequate period in the centre of the pile at 55°C, so that maximum pathogen destruction occurs. Composting is complete within 12–14 weeks although the finished material improves if left to cure for a further 6–9 months. Oxygen is mainly supplied by gas exchange during turning, although oxygen also enters the pile by diffusion and natural ventilation caused by the warm gases escaping which induces convection currents through the pile. The pile is only fully oxygenated periodically when turned which results in a cyclic variation in the oxygen concentration within the pile. In practice, this means that biological oxidation can never be maintained at maximum efficiency as the oxygen concentration is normally limiting (Fig. 18.19). Also, pile turning requires more space than static pile systems especially if they are moved laterally. There are a variety of turned systems in use, employing a variety of amendments and bulking agents. The Bangalore process was developed in 1925 in India and is used extensively there today. A 0.5–1.0 m deep trench is dug in the ground and filled using alternative layers of refuse, night soil, earth and straw. The material is then turned by hand as frequently as possible with composting completed after 120–180 days.

During the final stages of composting when the material is nearly dry, turning releases quantities of dust containing spores of *Aspergillus fumigatus* which is hazardous to the operator. The spores or conidia are present in soil at very low densities and rarely constitute a hazard. Within the compost vast densities of spores accumulate, which are thermo-tolerant, and are released when the pile is turned. There does not

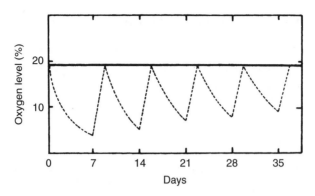

FIGURE 18.19 Cyclic variation of oxygen during composting within a pile that is periodically turned. The straight line represents the oxygen demand for a composting mass, and the dotted line the oxygen availability within the pile during each period between being turned. (Reproduced from Stentiford *et al.* (1985) with permission of Elsevier Science Ltd, London.)

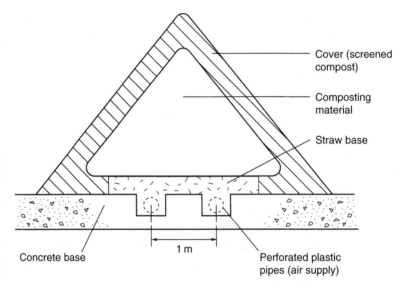

FIGURE 18.20
The most effective pile structure. (Reproduced from Stentiford *et al.* (1985) with permission of Elsevier Science Ltd, London.)

appear to be a problem with aerated static piles or closed systems of composting, with the release of spores being worst when the compost is dry. Health-associated problems are restricted to plant operators who, if they are hypersensitive or suffer from pulmonary disease, may display a severe allergic response. It is possible that the spores will germinate after inhalation and that the fungus may invade lung-parenchyma to produce typical aspergillosis (Vincken and Roels, 1984).

Static piles are not turned for aeration, instead oxygen is supplied by natural ventilation or more usually by forced aeration (Fig. 18.20). By using forced aeration the amount of oxygen entering the pile can be controlled, as can the moisture content and temperature of the compost. Static piles take up less area than turned piles and savings are made in manpower as the compost is not touched until the process is completed. Air can be sucked or blown through the pile or used, alternatively, to give

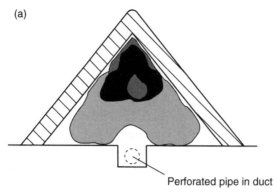

(a)

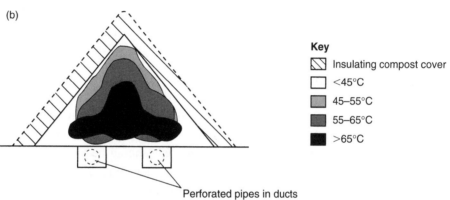

Perforated pipe in duct

(b)

Key

◩ Insulating compost cover

☐ <45°C

▨ 45–55°C

▨ 55–65°C

■ >65°C

Perforated pipes in ducts

FIGURE 18.21
Temperature profile of composting sewage sludge where (a) is positive pressure (blowing) aeration and (b) negative pressure (sucking) aeration. (Reproduced from Stentiford *et al.* (1985) with permission of Elsevier Science Ltd, London.)

overall oxygen and temperature control (Haug, 1980; De Bertoldi *et al.*, 1985). The system can be controlled using thermocouples placed within the pile which can activate the blower automatically if the temperature reaches critical levels. Air continues to be blown through the pile until the temperature returns to the required optimum. Unlike suction methods, blowing air through the compost enhances evaporation so producing a highly stabilized end product within 4–6 weeks with a low moisture content (Fig. 18.21). Most micro-organisms do not survive at temperatures >60°C, especially fungi which degrade cellulose and lignin; by controlling the temperature in this way composting can continue at maximum microbial efficiency. Once the temperature rises above 60°C microbial degradation begins to fall off rapidly and eventually stops, so composting will not continue until the temperature falls and the micro-organisms can reinvade the affected area. The most widely used system of this type is the Beltsville process, which is popular in the USA.

Closed systems

Closed-system composting takes place in specially constructed reactors or vessels. In these systems the mixture is fed into a reactor where air is forced through the compost under controlled conditions. Often compost is pre-treated by windrow composting and finished using an incubator box system. Small plants use a batch system of insulated vessels in series, while larger plants utilize continuous flow-through systems, usually of the rotating drum type. Additives, such as caustic soda, are often used in such systems to control the pH. In addition, some compost may be recycled in order to seed the incoming mixture with the correct micro-organisms.

18.4.5 STRAW COMPOSTING

In Europe, research has centred on open systems of composting. For example, in the UK much research has been conducted into composting sewage sludge with straw (Audsley and Knowles, 1984). The straw can be used whole or chopped and the sludge can be either raw or digested with the dry solids (DS) content between 4% and 20%, so dewatering is not essential although preferable. The process is cheap both in capital and operational terms and, because the straw decomposes during composting, there is no need for screening the finished compost to remove a bulking agent.

Field-based composting systems compost sewage sludge where it is to be used the following year. Straw is taken to the headland of each field while the sludge is stored in a lagoon until harvest. An open windrow is constructed on the headland, which is not turned. As the pile is not turned pathogens will not be killed in the outer surface of the pile, so digested sludge only should be used. If dewatered sludge is used the volatile solid content may be too low to maintain the critical temperature. On-farm systems use a central base for composting. The straw is baled and stored centrally, with windrows constructed on a concrete base as the sludge is delivered. The hard base allows the farmer to use his tractor to turn the windrows regularly, so maximum pathogen removal is achieved allowing raw sludge to be used. Central-site composting is carried out at the sewage treatment plant where the straw is bailed and stored at the farm and brought to the treatment plant as required. As with the previous system, windrows are constructed on a concrete base and turned regularly, with the compost returned to the farm for spreading. This system should be more carefully managed and is the most expensive due to the transportation costs.

Maintaining an adequate C:N ratio is difficult as straw has a typical C:N ratio of 48 and sewage sludge 5–9. To obtain the optimum C:N ratio of 12 for good microbial growth, it is necessary to mix sludge dry matter to straw dry matter in a ratio of 1.3:1, 0.7:1, 4.7:1 and 4.0:1 for raw (4% DS), digested (4% DS), dewatered raw (20% DS) and dewatered digested sludges (20% DS), respectively. More dewatered sludge is required as it contains significantly less nitrogen than non-dewatered sludges. To ensure that a high temperature is maintained for a sufficient time large windrows with a minimum cross-sectional area of $5\,m^2$ are required to ensure that the heat generated exceeds the heat lost. As heat is produced from microbial degradation of volatile solids, a high volatile solids content in the compost mix is desirable.

Outdoor composting requires a high degree of operational care and is liable to fail if the winter is very cold. Forced aeration improves the chance of successful composting, while enclosed (covered) forced aeration generally always succeeds.

18.4.6 PRODUCT QUALITY

Fully composted sludge closely resembles peat, being dark brown in colour with a crumbly texture and with a pleasant earthy smell. The fertilizer value of such compost is generally low compared with inorganic fertilizers, although it does reflect the original fertilizer value of the substrate and amendments used. It is possible to fortify the compost to increase its nutrient value by adding nitrogen, phosphorus and potassium, although the cost of this probably exceeds its value. Heavy metals may be a problem with composted sludge as with sewage sludge when disposed to land. Therefore, compost must be treated as sewage sludge under the existing legislation and guidelines (Section 21.3). Composts are generally pathogen free. The major cause of inactivation of pathogens in composting is exposure to high temperatures (>55°C) for a number of days. In the USA, the Environmental Protection Agency has produced standards to ensure pathogen destruction during composting. In aerated piles the coolest part of the pile must be at least 55°C for 3 days. While in the windrow system the centre must be at least 55°C for 15 days out of the total of a 21–30-day composting period. During this time the pile must be turned five times (Haug, 1980). In forced aeration systems the volatile solids can be utilized very quickly, so although a high temperature is achieved it may not last as long as the slower and slightly cooler turned system, and so may give poor pathogen inactivation (Audsley and Knowles, 1984). Pathogen inactivation is greater if the pile is turned as cool spots develop within the pile, especially in the corners where pathogens can survive and even increase in numbers. Regular turning ensures all parts of the pile are subjected to the high temperatures. It is now common practice in forced aeration piles to build the new compost mixture on top of a layer of mature compost and for the pile to be sealed by a thick layer of mature compost that acts as insulation and reduces odour. Under these conditions high pathogen kills are possible without turning and there is no outer layer in which the pathogens can survive.

REFERENCES

Audsley, E. and Knowles, D., 1984 *Feasibility Study – Straw/Sewage Sludge Compost*, NIAE Report R45, National Institute of Agricultural Engineering, Silsoe, UK.

Barnes, L. M., 1998 Tunnel composting at Ipswich. *Journal of the Chartered Institution of Water and Environmental Management*, **12**, 117–23.

Cooper, P. F., 1990 *European Design and Operation Guidelines for Reed Bed Treatment Systems*, Report No. U117, Water Research Centre, Stevenage.

Cooper, P. F. and Findlater, B. C. (eds), 1990 *Constructed Wetlands in Water Pollution Control*, Pergamon Press, Oxford.

Croll, B. T., Greene, L. A., Hall, T., Whitford, C. J. and Zabel, T. F., 1985 Biological fluidized bed denitrification for potable water. In: Tebbutt, T. H. Y., ed., *Advances in Water Engineering*, Elsevier, London, pp 180–7.

De Bertoldi, M., Vallini, G., Pera, A. and Zucconi, F., 1985 Technological aspects of composting including modelling and microbiology. In: Gasser, J. K. R., ed., *Composting of Agricultural and Other Wastes*, Elsevier, London, pp 27–41.

Dingles, R., 1982 *Natural Systems for Water Pollution Control*, Van Nostrand Reinhold, New York.

Finstein, M. S., Miller, F. C., Strom, P. F., MacGregor, S. T. and Psarianos, K. M., 1983 Composting ecosystem management for waste treatment, *Biotechnology*, **1**, 347–53.

Gasser, J. K. R. (ed.), 1985 *Composting of Agricultural and Other Wastes*, Elsevier, London.

Glancer, M. and Soljan, V., 1998 Beyond biological treatment, *Water Quality International*, November/December, 19–20.

Gloyna, E. F., 1971 *Waste Stabilization Ponds*, WHO Monograph Series No. 60, World Health Organization, Geneva.

Gray, N. F., 2004 *Biology of Waste Water Treatment*, 2nd edn, Imperial College Press, London.

Gschlößl, T. and Stuible, H., 2000 Reed bed systems: design, performance and maintainability, *Water Science and Technology*, **41** (1), 73–6.

Haug, R. T., 1980 *Compost Engineering: Principles and Practice*, Ann Arbor Science, Michigan.

Hawkes, H. A., 1983 Stabilization ponds. In: Curds, C. R. and Hawkes, H. A., eds, *Ecological Aspects of Used Water Treatment: 2. Biological Activities and Treatment Processes*, Academic Press, London, pp 163–217.

Hayward, K., 1998 BNRs control key, *Water Quality International*, September/October, 14–15.

Huban, C. M. and Plowman, R. D., 1997 Bioaugmentation: put microbes to work, *Chemical Engineering*, March, 74–84.

IAWQ, 1995 *Activated Sludge Model No. 2*, Scientific and Technical Report No. 3, International Association on Water Quality, London.

Imhoff, K. R., 1984 The design and operation of the purification lakes in the Ruhr Valley, *Water Pollution Control*, **83**, 243–53.

Kadlec, R. H., Knight, R. L., Vymazal, J., Brix, H., Cooper, P. and Haberl, R., 2000 *Constructed Wetlands for Pollution Control: Processes, Performance, Design and Operations*, Scientific and Technical Report No. 8, International Water Association, IWA Publishing, London.

Mara, D. D., 1997 *Sewage Treatment in Hot Climates*, 2nd edn, John Wiley and Sons, Chichester.

Maynard, H. E., Ouki, S. K. and Williams, S. C., 1999 Tertiary (maturation) lagoons: a review of removal mechanisms and performance, *Water Research*, **33**, 1–13.

Mendelssohn, I. A. and Postek, M. T., 1982 Elemental analysis of deposits on the roots of *Spartina alterniflora* Loisel, *American Journal of Botany*, **69**, 904–12.

Metcalf and Eddy, 1991 *Wastewater Engineering: Treatment, Disposal and Reuse*, McGraw-Hill, New York.

Muyima, N. Y. O., Momba, M. N. B. and Cloete, T. E., 1997 Biological methods for the treatment of wastewaters. In: Cloete, T. E. and Muyima, N. Y. O., eds, *Microbial Community Analysis: The Key to the Design of Biological Wastewater Treatment Systems*, Scientific and Technical Report No. 5, International Association on Water Quality, London, pp 1–24.

Powledge, T. M., 1983 Prospects for pollution control with microbes, *Biotechnology*, November, 743–755.

Minto, T., Van Loosdrecht, M. C. M. and Heijnen, J. J., 1998 Microbiology and biochemistry of the enhanced biological phosphate removal process, *Water Research*, **32**, 3193–207.

SCS, 1992 *Agricultural Waste Management Field Handbook*, US Department of Agriculture, SCS, Washington, DC.

Stentiford, E. I., Mara, D. D. and Taylor, P. L., 1985 Forced aeration co-composting of domestic refuse and sewage sludge in static piles. In: Gasser, J. K. R., ed., *Composting of Agricultural and Other Wastes*, Elsevier, London, pp 42–55.

Toms, I. P., Owens, M., Hall, J. A. and Mindenhall, M. J., 1975 Observations on the performance of polishing lagoons at a large regional works, *Water Pollution Control*, **74**, 383–401.

Vincken, W. and Roels, P., 1984 Hypersensitivity pneumonitis due to *Aspergillus fumigatus* in compost, *Thorax*, **39**, 74–75.

Woods, D. R., Green, M. B. and Parish, R. C., 1984 Lea Marston Purification Lake: operational and river quality aspects, *Water Pollution Control*, **83**, 226–42.

FURTHER READING

Barnard, J. L., 1998 The development of nutrient-removal processes, *Journal of the Chartered Institution of Water and Environmental Management*, **12**, 330–7.

Cooper, P. and Green, B., 1995 Reed bed treatment systems for sewage treatment in the UK: the first ten years experience, *Water Science and Technology*, **32** (3), 317–27.

De Bertoldi, M., Ferranti, M., Hermite, P. L. and Zucconi, F., 1988 *Compost Production, Quality and Use*, Elsevier, London.

Harberl, R., Perfler, R., Laber, J. and Cooper, P., 1997 Wetland systems for water pollution control 1996, *Water Science and Technology*, **35** (5), 1–347.

Mara, D. D., 1996 *Low Cost Sanitation*, John Wiley and Sons, Chichester.

Moshire, G. A. (ed.), 1993 *Constructed Wetlands For Water Quality Improvements*, Lewis Publishers, Boca Raton, FL.

Nuttall, P. M., 1999 *Constructed Wetland Technology: A Global Perspective*, John Wiley and Sons, Chichester.

Reddy, K. R. and DeBuske, T. A., 1978 State of the art utilization of aquatic plants in water pollution control, *Water Science and Technology*, **19** (10), 61–79.

Sidwick, J. M. and Holdom, R. S. (eds), 1987 *Biotechnology of Waste Treatment and Exploitation*, Ellis Horwood, Chichester.

Introduction

Anaerobic processes are used to treat strong organic wastewaters (biological oxygen demand (BOD) $>500\,mg\,l^{-1}$), and for further treatment of primary and secondary sludges from conventional wastewater treatment. Liquid wastewaters rich in biodegradable organic matter are generated primarily by the agricultural and food-processing industries. Such wastewaters are difficult to treat aerobically mainly due to the problems and costs of satisfying the high oxygen demand and maintaining aerobic conditions. The majority of the organic carbon removed from solution during conventional wastewater treatment is converted into waste sludge (Section 14.2). This can be further stabilized either aerobically or, more conventionally, using anaerobic digestion. Anaerobic treatment, although slow, offers a number of attractive advantages in the treatment of strong organic wastes. These include a high degree of purification, ability to treat high organic loads, production of a small quantity of excess sludge that is normally very stable and the production of an inert combustible gas (methane) as an end product (Sterritt and Lester, 1988). Unlike aerobic systems, complete stabilization of organic matter is not possible anaerobically and so subsequent aerobic treatment of anaerobic effluents is normally necessary. The final effluent produced by anaerobic treatment contains solubilized organic matter, which is amenable to aerobic treatment indicating the potential of using combined anaerobic and aerobic units in series. The advantages and disadvantages of anaerobic treatment are outlined in Table 19.1.

Anaerobic degradation occurs in the absence of oxygen. The basic difference between aerobic and anaerobic oxidation is that in the aerobic system oxygen is the ultimate hydrogen acceptor with a large release of energy, but in anaerobic systems the ultimate hydrogen acceptor may be nitrate, sulphate or an organic compound with a much lower release of energy (See 'Introduction' in Chapter 5). The process of anaerobic decomposition involves four discrete stages (Fig. 19.1). First is the hydrolysis of high-molecular-weight carbohydrates, fats and proteins, which are often insoluble, by enzymatic action into soluble polymers. The second stage involves the acid-forming bacteria that convert the soluble polymers into a range of organic acids (acetic,

TABLE 19.1 The advantages and disadvantages of anaerobic treatment compared to aerobic treatment

Advantages	Disadvantages
Low operational costs	High capital costs
	Generally require heating
Low sludge production	Low retention times required (>24 h)
Reactors sealed giving no odour or aerosols	Corrosive and malodorous compounds produced during anaerobiosis
Sludge is highly stabilized	Not as effective as aerobic stabilization for pathogen destruction
Methane gas produced as end product	Hydrogen sulphide also produced
Low nutrient requirement due to lower growth rate of anaerobes	Reactor may require additional alkalinity
Can be operated seasonally	Slow growth rate of anaerobes can result in long initial start-up of reactors and recovery periods
Rapid start-up possible after acclimation	Only used as pre-treatment for liquid wastes

Reproduced from Wheatley et al. (1997) with permission of the Chartered Institution of Water and Environmental Management, London.

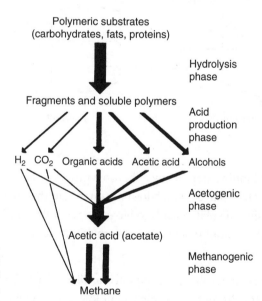

FIGURE 19.1 Major steps in anaerobic decomposition.

butyric and propionic acids), alcohols, hydrogen and carbon dioxide. Acetic acid, hydrogen and carbon dioxide are the only end products of the acid production, which can be converted directly into methane by methanogenic bacteria. So a third stage is present where the organic acids and alcohols are converted to acetic acid by aceto-genic bacteria. In the final phase, which is the most sensitive to inhibition, methanogenic bacteria convert the acetic acid to methane. Although methane is also produced from hydrogen and carbon dioxide, in practice about 70% of the methane produced is from acetic acid. Obviously the methanogenic stage is totally dependent on the production of acetic acid and so it is the third stage, the acetogenic phase,

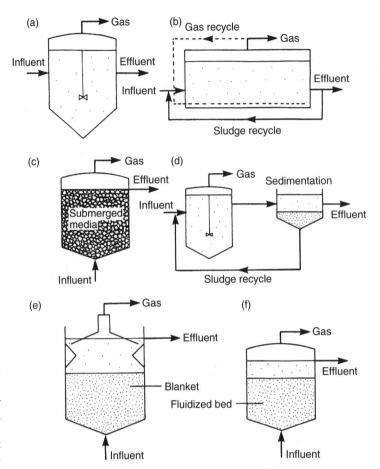

FIGURE 19.2
Schematic representation of digester types. Flow-through digesters (a and b) and contact systems (c–f).

which is the rate-limiting step in any anaerobic process. Although in practice, as the methanogenic bacteria have a much lower growth rate than the acid-producing bacteria, the conversion of volatile acids to biogas is generally considered to be the rate-limiting step of the overall reaction with the methanogenic phase normally used for modelling purposes.

A large number of anaerobic processes are available including anaerobic lagoons, digesters and filters. Present anaerobic technology can be divided into two broad categories. *Flow-through systems* for the digestion of concentrated wastes, such as animal manures or sewage sludges that have a solids concentration in the range of 2–10%. These include completely mixed reactors (Fig. 19.2a), which are used primarily for sewage sludges, and plug-flow reactors used to a limited extent for the digestion of animal manures (Fig. 19.2b) (Section 19.1). Wastewaters with a lower solids concentration are treated anaerobically by *contact systems* in which the wastewater is brought into contact with an active microbial biomass that is retained within the system (Section 19.3). In flow-through systems the residence time of the waste (i.e. the hydraulic retention time (HRT)) and of the microbial biomass (i.e. the mean cell retention time (MCRT)) in the reactor are the same, while in a contact system the MCRT is far greater than the HRT of the wastewater. This is achieved by:

(a) recycling of effluent biomass,
(b) biofilm development on internal support materials, or
(c) by the formation of well-settling flocs or granules. In this way the HRT can be reduced to several days or even hours.

The basic kinetics of anaerobic treatment are different to those of aerobic treatment (Section 15.2). Relating substrate concentration to specific growth rate (i.e. the Monod function) is invalid for anaerobic reactors, since the volatile acids not only act as substrate for methanogenic bacteria but are also inhibitory at higher concentrations. The Monod function is therefore replaced by an inhibition function so that:

$$u = \frac{u_m}{(1 + K_s/s + s/K_i)} \tag{19.1}$$

where K_i is the inhibition function (Section 15.2). The dynamic models available all use Monod kinetics and normally consider the interactions between several substrates and bacterial populations.

19.1 FLOW-THROUGH SYSTEMS (DIGESTION)

Digestion is carried out either in large open tanks or lagoons (Section 18.1) at ambient temperatures of 5–25°C (psychrophilic digestion) or more rapidly in covered tanks heated to between 25°C and 38°C (mesophilic digestion) or at 50–70°C (thermophilic digestion) (Koster, 1988). Psychrophilic digesters are usually unmixed and as they are uncovered any gas produced is dispersed to the atmosphere. In the winter the rate of digestion will be extremely low or even zero while in the summer digestion will be rapid. Therefore, long retention times are required of between 6 and 12 months to stabilize sludges under these conditions and to balance between sludge accumulation and sludge degradation. Such digesters are restricted to smaller works where the output of sludge is low and land is readily available. Both septic tanks and Imhoff tanks fall within the psychrophilic range (Section 22.2).

Heated digesters are much more cost effective at larger treatment plants where there is sufficient sludge available to ensure a continuous operation, with gas collected and used to either directly or indirectly heat the digesters. Conventional digestion is usually a two-stage process with the primary digester heated to the desired temperature to allow optimum anaerobic activity with acid formation and gas production occurring simultaneously (Fig. 19.3). The primary reactor is continuously stirred, unlike the earlier stratified digesters, which reduces the retention time from 30–60 days to <15 days. The secondary digester is unheated and can be used for two functions.

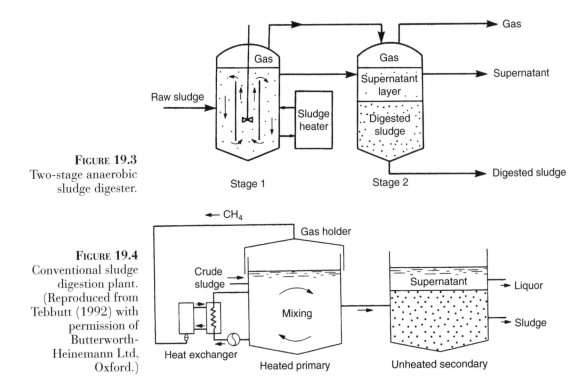

FIGURE 19.3
Two-stage anaerobic
sludge digester.

FIGURE 19.4
Conventional sludge
digestion plant.
(Reproduced from
Tebbutt (1992) with
permission of
Butterworth-
Heinemann Ltd,
Oxford.)

Either to continue digestion under psychrophilic conditions in which case it will be stirred to ensure complete mixing as in the primary digester, or it can be used for sludge separation. However, as the evolution of gas interferes with settlement, the second reactor cannot be used for both functions. When used for sludge separation the secondary digester is not stirred and is generally uncovered (Fig. 19.4). Under quiescent conditions the solids separate leaving a strong supernatant liquor on top of the stabilized sludge. The liquor has a high soluble organic content with a high BOD, which needs to be recirculated back to the inlet of the treatment plant. The settled solids are removed, dewatered and disposed of (Chapter 21).

Digestion is used to stabilize both primary and secondary sludges with a solids content of between 20 000 and 60 000 mg l^{-1} (2–6%). About 70% of the sludge is degradable and up to 80% of this will be digested reducing the solids content by about 50%. The mixture of primary and secondary sludges fed into a digester is rich in carbohydrates, lipids and proteins which are ideal substrates for microbial degradation, and although the sludge is already rich in a variety of anaerobic bacteria, in order to obtain the correct balance of hydrolytic acid producers and methanogenic bacteria, the raw sludge needs to be seeded. This is particularly important, as the main bacterial groups responsible for digestion are dependent on the end product of the other. The particulate and high-molecular-weight organic matter is broken down to lower fatty acids, hydrogen and carbon dioxide by facultative anaerobic bacteria (first stage). The end products are converted to acetic acid by acetogenic bacteria (second stage), which is subsequently converted in the final stage by methanogenic bacteria to methane and carbon dioxide. Obviously anaerobic digestion will proceed most efficiently when the rates of reaction at each stage are equal. However, if the first stage is limited then the nutrient supply to the other stages will be reduced with the overall effect of suppressing the rate of digestion and biogas formation, but not inhibiting it. If either of the subsequent stages are restricted then the first stage products will accumulate causing a gradual rise in the carbon dioxide fraction in the biogas (>30%) and a gradual fall in the pH as volatile acids accumulate until the pH falls below 7.0 and the whole process becomes stressed. All three stages proceed simultaneously within the same reactor with the bacterial groups in close proximity to each other. However, as the anaerobic chain of reaction proceeds the bacterial groups involved become progressively more sensitive to their environmental conditions. Hydrolysis and acid formation are carried out by a diverse and large group of facultative bacteria, which can tolerate a wide variation in temperature, pH, and a range of inhibitory substances. In contrast the acetogenic and methanogenic bacteria are far more sensitive micro-organisms which are highly specialized and severely inhibited by even minor changes in operating conditions. So operational problems generally involve the third and fourth stages of digestion.

19.1.1 DESIGN

Primary digesters are normally covered with a fixed or floating top for gas collection (Fig. 19.5). While floating covers rise and fall according to the volume of gas and sludge, being of the same design as conventional gas holders with the weight of the floating cover providing the gas pressure, fixed covers require a separate gas holder which allows the gas to move freely in both directions. This is particularly important if a vacuum is not to be produced when sludge is removed from the digester, which would not only affect operation but could affect the structural stability of a reactor with a fixed cover. Methane is biochemically inert and so it can be stored above the sludge within the reactor without affecting or inhibiting the process. Digesters are mostly circular in plan with maximum diameters of approximately 25 m, and to ensure the minimum depth the ratios of reactor height to diameter is normally between 1:3 and 1:2, although smaller digesters tend towards a ratio of 1:1. The entire contents of the tank needs to be turned over once every 3–4 h, with the sludge warmed by heat exchangers using hot water from gas boilers or the cooling systems of gas engines. Heating and mixing of digester contents can be achieved by a number of techniques, most favoured being a central screw mixing pump with an external heat exchanger, a circulating pump and exchanger housed in a projecting chamber or internal gas-lift pumps (IWPC, 1979).

The amount of heat required to raise the temperature of the incoming sludge to the optimum temperature and maintain it is calculated as:

$$H = WC \, \Delta T + UA \, \Delta T \tag{19.2}$$

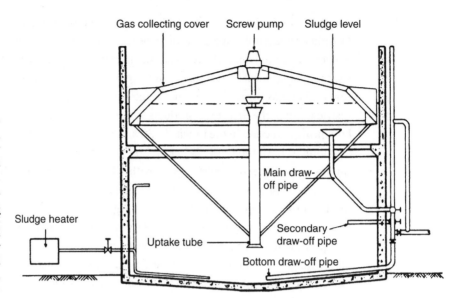

FIGURE 19.5
Primary digestion
tank with screw
mixing pump and
external heater.
(Reproduced from
IWPC (1979) with
permission of
the Chartered
Institution of Water
and Environmental
Management,
London.)

where H is the heat required to heat the incoming sludge and compensate for heat losses ($kg\,cal\,h^{-1}$), W the mass rate of influent sludge ($kg\,h^{-1}$), ΔT the difference between the digester temperature and influent sludge temperature, U the heat transfer through the digester walls ($kg\,cal\,m^{-2}h^{-1}°C$), A the surface area of digester losing heat and C the mean specific heat of feed sludge ($1\,kg\,cal\,kg^{-1}°C$).

The basic design criterion is the provision of sufficient capacity to ensure an adequate sludge retention time. It is common practice to use 25–30 days as the design retention period to allow for variation in daily sludge loading. However, as the theoretical retention time can be as low as 7–10 days, with the improvements in mixing and heating digesters now available, 15 days possibly allows an adequate safety margin to cover most operational difficulties. Digester capacity can also be calculated on population equivalents (pe). Assuming a per capita sludge production of 1.8 l of raw sludge containing 4.5% dry solids (DS), the capacity of a digester with a 25 days retention time is:

$$\text{Digester capacity} = \frac{1.8 \times 25 \times \text{pe}}{1000} \; m^3 \tag{19.3}$$

The volume of a batch digester can be more accurately calculated using Equation (19.4):

$$V_s = [V_1 - 0.66(V_1 - V_2)]t \tag{19.4}$$

where V_s is the digester volume (m^3), V_1 the initial sludge volume (m^3), V_2 the final sludge volume (m^3) and t the retention time (days).

EXAMPLE 19.1 Calculate the digestion tank volume required to treat waste activated sludge from a treatment plant treating a population equivalent of 35 000. The sludge production rate is 0.1 kg (DS) ca^{-1}day^{-1}, the sludge has a solids content of 3.5% and a volatile content 78% DS, a specific gravity of wet sludge of 1.017. After digestion the sludge is 6% DS with a specific gravity of 1.025 and with 65% of the volatile solids destroyed.

$$\text{Influent sludge production} = 35\,000 \times 0.1 = 3500 \text{ kg day}^{-1}$$
$$\text{Volatile suspended solids (VSS)} = 3500 \times 0.78 = 2730 \text{ kg day}^{-1}$$
$$\text{Fixed suspended solids (FSS)} = 3500 \times 0.22 = 770 \text{ kg day}^{-1}$$
$$\text{VSS destroyed} = 2730 \times 0.65 = 1774.5 \text{ kg day}^{-1}$$

$$\text{Remaining VSS in sludge after digestion} = 2730 - 1774.5$$
$$= 955.5 \text{ kg day}^{-1}$$
$$\text{Remaining FSS in sludge after digestion} = 700 \text{ kg day}^{-1}$$
$$\text{Total solids in sludge after digestion} = 955.5 + 770$$
$$= 1725.5 \text{ kg day}^{-1}$$

$$\text{Influent sludge volume} = 3500 \times \frac{100}{3.5} \times \frac{1}{1.017} = 98.3 \text{ m}^3 \text{ day}^{-1}$$
$$\text{Digested sludge volume} = 1725.5 \times \frac{100}{6} \times \frac{1}{1.025} = 28.1 \text{ m}^3 \text{ day}^{-1}$$

Volume of digested sludge (V_s) using Equation (19.4) is using a sludge residence time of 25 days:

$$V_s = [98.3 - 0.66(98.3 - 28.1)]25 = 1299.2 \text{ m}^3$$

As the sludge in the digester only occupies two-thirds of the tank, the total tank volume required is:

$$1299.2 \times 1.5 = 1948.8 \text{ m}^3$$

Although often quoted, organic loadings expressed as kg organic or volatile matter m^{-3} are not appropriate design parameters for digester design. With the standard 25 days retention time, the typical sludge feed of 4.5% total solids with an 85% organic (volatile) matter content results in an organic loading equivalent to 1.5 kg m^{-3}day^{-1}. However, in the UK, organic loading ranges from 0.3 to 2.8 kg m^{-3}day^{-1} indicating the varying water contents of the sludges being digested. So in order to achieve the optimum organic loading of 1.5 kg m^{-3}day^{-1} with a thin sludge, a much shorter retention time will result with incomplete digestion occurring. For this reason sludges should be characterized and loadings specified as a concentration of total (dry) solids (%).

Secondary digestion tanks are generally used for storage and separation, although they can be used for the further digestion and gas collection. Sludge is passed as frequently as possible from the primary digester into the uncovered tank where it cools and allows the liquor to separate from the solids so that each can be withdrawn separately. Temperature differences between the cool and warm sludge can cause convection currents within the digester which will hinder settlement, as will gas production, so that it becomes difficult to obtain a solids-free liquor. To overcome this, secondary digesters are relatively shallow with a maximum depth of 3.5 m. The capacity of older tanks were between 50% and 70% of the primary digester providing retention times of 15–20 days. Newer tanks are approximately the same size as the primary tanks providing similar retention times for the sludge. Tank volume can be estimated on a pe basis using Equation (19.5):

$$\text{Secondary digester volume} = 0.035 \times \text{pe m}^3 \qquad (19.5)$$

In order to ensure adequate operating conditions for digestion the pH value, the concentration of carbon dioxide in the biogas and the volatile acid concentrations should be continuously monitored. The normal operating conditions should be a pH of 7.0–7.2, alkalinity (as $CaCO_3$) of 4000–5000 mg l^{-1} and a concentration of volatile acids (as acetic acid) of <1800 mg l^{-1}. The carbon dioxide content of the biogas should not exceed 30%, and once any of these values are exceeded then immediate remedial action is required.

19.1.2 OPERATION

Like aerobic micro-organisms, those anaerobes responsible for digestion require certain substrates, growth factors, trace elements and nutrients for successful development. Nitrogen and phosphorus are both vital for bacterial growth and are required at minimum concentrations of 2.5% and 0.5% of the dry organic matter content of the sludge. Anaerobic processes are far less demanding in terms of nitrogen and phosphorus than aerobic systems. The optimum BOD:N:P (mass) ratio is 100:0.5:0.1 while the COD:N:P ratio ranges from 42:0.7:0.1 to 150:0.7:0.1; COD: chemical oxygen demand.

Non-ionic and cationic detergents have little effect on anaerobic digestion, even at high concentrations while anionic detergents inhibit the process. Household washing powders, which contain alkyl benzene sulphonates (ABS), both hard and soft, are non-degradable anaerobically. Being strongly adsorbed onto organic solids they are invariably present in sewage sludge. Anionic detergents at concentrations (expressed as Manoxol OT) in excess of 1.5% of the dry raw sludge solids inhibit anaerobic digestion even in heated digesters with relatively long retention times of up to 40 days, with gas production particularly affected.

The most quoted cause of inhibition of sewage sludge digesters are heavy metals and in particular chromium, copper, nickel, cadmium and zinc, although particularly high concentrations of metals are required to have a significant effect. Synergistic effects have been noted with heavy metals and a number of other inhibitory substances, so under certain conditions even low concentrations of heavy metals may cause problems. Apart from the concentration of heavy metals in the raw sludge other factors, such as solubility, pH and the concentration of sulphide present, will all affect their concentration in the digester. There is always more than one heavy metal present in sewage sludge and the inhibitory effects are generally additive on an equivalent weight basis. Where the milligram equivalent weight (meq) per kg dry sludge solids (K) exceeds $400\,\mathrm{meq\,kg^{-1}}$ there is a 50% chance of digester failure which rises to 90% when K exceeds $800\,\mathrm{meq\,kg^{-1}}$. To ensure a 90% probability that digestion will not be affected, the value of K should be $<170\,\mathrm{meq\,kg^{-1}}$ (Mosey, 1976).

K is measured using the concentrations of the most abundant heavy metals ($\mathrm{mg\,l^{-1}}$) in sewage sludge, excluding chromium, using the equation:

$$K = \frac{(\mathrm{Zn})/32.7 + (\mathrm{Ni})/29.4 + (\mathrm{Pb})/103.6 + (\mathrm{Cd})/56.2 + (\mathrm{Cu})/47.4}{\text{Sludge solids concentrations (kg l}^{-1})}\ \mathrm{meq\,kg^{-1}} \quad (19.6)$$

Sulphate can be a particular problem in the digestion process if present in sufficient quantities. The sulphate is reduced to sulphide by bacterial action with hydrogen sulphide eventually being formed. Sulphate concentrations $>500\,\mathrm{mg\,l^{-1}}$ can reduce methane production and generate up to 4% hydrogen sulphide in the biogas. Nitrates can also prove problematic because if denitrification occurs within the digester there will be a shift in the redox potential, which will suppress methane production. Methanogens are strict anaerobes and can be completely inhibited by a dissolved oxygen concentrations as low as $0.01\,\mathrm{mg\,l^{-1}}$. They require a reduced environment with a redox potential within the range -200 to $-420\,\mathrm{mV}$. This has been a serious problem at a number of sewage treatment plants in the UK and has been remedied by providing an anoxic step to permit denitrification to occur before the sludge enters the digester with the nitrate converted to nitrogen gas. Pilot trials to test the biodegradability of wastewaters and assess the possible inhibitory effects of wastewaters and chemicals should always be carried out prior to anaerobic digestion (HMSO, 1987).

Anaerobic digestion can occur over a wide temperature range. The rate of anaerobic digestion and gas production is temperature dependent with optimum gas production at the higher temperature ranges, so the warmer the reactor the shorter the MCRT needs to be for complete digestion (Fig. 19.6).

Digestion is very sensitive to low pH, with a declining pH indicating the process is unbalanced. Most anaerobic treatment systems have problems with pH control that arise from differences in the growth rate of the synergistic bacterial populations. The activity of the acid-producing bacteria tends to reduce the pH of digesting sludge from

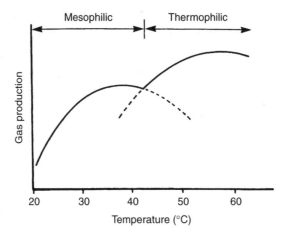

FIGURE 19.6
Effect of temperature
on gas production.

the optimal 7.0–7.5 range required by methanogenic bacteria. Under normal operating conditions, once a stable population of each of the groups has been established, an equilibrium is maintained by the buffering action of ammonium bicarbonate (the bicarbonate alkalinity) and so no external pH control is required. The bicarbonate ions are derived from carbon dioxide in the digester gas while the ammonium ions derive from the degradation of proteins in the raw sludge. However, the digesting sludge does have a tendency to become acidic especially if the methanogenic bacteria are inhibited, or the digester is overloaded, resulting in an excessive accumulation of volatile acids. Under these conditions the buffering capacity may be exceeded with the pH rapidly decreasing to 6.0 causing the process to fail. Methanogenic bacteria exhibit a negative response when the pH shifts towards the acid region as they do when the temperature falls. Growth of methanogens is inhibited below pH 6.6, while the fermentative bacteria will continue to function until the pH has dropped to 4.5–5.0. The measurement of pH must be done quickly as samples of digesting sludge once exposed to the atmosphere will rapidly lose carbon dioxide causing erroneously high pH values. So long as the sludge has a fairly high alkalinity an increase in acid production will initially produce little effect on pH, so in practice the measurement of volatile acids is a better control factor of the buffering capacity within a digester. Any change in the loading of the digester must be gradual to ensure that the concentration of volatile acids does not exceed the normal buffering capacity of system. Normal volatile acids concentrations in sewage sludge digesters are between 250 and 1000 $mg\,l^{-1}$ but values in excess of 1500 ml^{-1} result in a decline in pH, while values in excess of 1800–2000 $mg\,l^{-1}$ indicate serious problems. Lime (calcium hydroxide) is normally used to control the pH value within the reactor and preserve the buffering capacity of the sludge. If the sludge contains heavy metals in solution at concentrations likely to cause inhibition, the addition of an alkali to raise the pH to 7.5–8.0 will precipitate most of the metal ions out of solution usually as carbonates, reducing the inhibitory effect. Sulphide can also be used to precipitate the heavy metal ions without causing a significant pH change (Mosey, 1983).

On start-up there is a rapid decline in pH due to faster-growing acid-forming bacteria. The methane bacteria build up slowly gradually reducing the need for acid control and increasing gas production. To avoid excess pH on start-up reactors should be seeded with a good anaerobic sludge, or operated initially at only 10% of the normal loading, gradually increasing the loading to maximum, to prevent excessive acid production inhibiting the growth of methanogenic bacteria.

19.2 BIOGAS AND OTHER PRODUCTS OF DIGESTION

There are three products of anaerobic digestion: the digested sludge, a waste liquor and the sludge gas. Digested sludge is different to either primary or secondary sludges in a number of ways. It is pathogen free, stabilized and far less offensive with a characteristic tarry smell and drying to an inert friable condition making it ideal for disposal to agricultural land (Section 21.3). The BOD of the liquor can be very high reaching up to $10\,000\,mg\,l^{-1}$, although up to 60% of the BOD is due to the suspended solids fraction ($500–1000\,mg\,l^{-1}$). Owing to the degradation of organic nitrogen the liquor may have high concentrations of soluble nitrogen present. The characteristics and strength of the liquor make it difficult to dispose of or treat separately, and so it is returned to the works inlet where it is diluted by the incoming sewage and treated in admixture.

Sludge or digester gas is widely known as biogas. Gas production is the most direct and sensitive measure of the rate of anaerobic digestion, a decrease in production being the first indication that the process is unstable. Apart from trace amounts of water vapour, hydrogen sulphide, hydrogen, nitrogen, unsaturated hydrocarbons and other gases, biogas is essentially a mixture of carbon dioxide and methane, the exact proportions of which determine its calorific value. A typical biogas contains between 65% and 70% methane by volume with the remaining 30–35% carbon dioxide. When burnt, biogas produces water which complicates the determination of its calorific value (Equation (19.7)). The net calorific value for combustion is where the water formed remains in the vapour phase, while the gross calorific value is where the water formed is condensed. The calorific values are calculated using the percentage of methane present (M) as the net calorific value = $334 \times M\,kJ\,m^{-3}$ (saturated with water) or the gross calorific value = $370 \times M\,kJ\,m^{-3}$ (saturated with water), both at 15.5 °C and at 1 atmosphere. So for normal biogas the gross calorific value is between $24\,000$ and $26\,000\,kJ\,m^{-3}$ with the net value some 10% less

$$\text{Organic matter} \xrightarrow{\text{anaerobic digestion}} CH_4 + CO_2 + H_2 + NH_3 + H_2S \quad (19.7)$$

The exact gas yield per kg of volatile solids removed depends on the composition of the waste but is approximately $1.0\,m^3$. Some food-processing industries produce a pure substrate waste with 80–95% of the organic (volatile) matter removed, whereas only 40–50% of the organic matter in sewage sludge and even slightly less in animal slurries will be utilized with a proportionately lower gas yield in terms of m^3 of gas produced per kg of substrate supplied. So for a mixed primary and secondary sludge from a purely domestic sewage treatment works the gas production will be of the order of $0.5\,m^3\,kg^{-1}$ organic (volatile) matter, or about $0.375\,m^3\,kg^{-1}$ of total (dry) solids, added to the digester (Table 19.2). The composition and quantity of gas produced by complete digestion can be theoretically determined using the equation:

$$C_cH_hO_oN_nS_s + 1\frac{1}{4}(4c - h - 2o + 3n + 2s)H_2O = \frac{1}{8}(4c - h + 2o + 3n + 2s)CO_2$$
$$+ \frac{1}{8}(4c + h - 2o - 3n - 2s)CH_4 + nNH_3 + sH_2S \qquad (19.8)$$

where c, h, o, n and s are the number of atoms of carbon, hydrogen, oxygen, nitrogen and sulphur, respectively (Table 19.3).

A more accurate estimation of methane production per unit time can be obtained by using:

$$G = 0.35(L - 1.42\,S_t) \qquad (19.9)$$

where G is the volume of methane produced ($m^3\,day^{-1}$), L the mass of ultimate BOD removed ($kg\,day^{-1}$) and S_t the mass of volatile solids accumulated ($kg\,day^{-1}$).

TABLE 19.2 Typical expected gas yields and methane content from the anaerobic digestion of various wastes

Material	Gas yield ($m^3\,kg^{-1}DS$)	Methane (% volume)
Sewage sludge (municipal)	0.43	78
Dairy waste	0.98	75
Abattoir: paunch manure	0.47	74
blood	0.16	51
Brewery waste sludge	0.43	76
Potato tops	0.53	75
Beet leaves	0.46	85
Cattle manure	0.24	80
Pig manure	0.26	81

TABLE 19.3 Gas yield and composition of biogas produced by the digestion of carbohydrates, proteins and lipids

Substrate	Gas yield ($m^3\,kg^{-1}$)	Composition	
		% CH_4	% CO_2
Carbohydrates	0.8	50	50
Proteins	0.7	70	30
Lipids	1.2	67	33

S_t can be estimated as:

$$S_t = \frac{aL}{1 + bt_s} \tag{19.10}$$

where a is the mass of volatile solids synthesized per kg of ultimate BOD removed, b is the endogenous respiration constant and t_s is the solids retention time. Methane is the most valuable by-product of anaerobic digesters and so it is useful to express the yield as $m^3 CH_4 kg^{-1}$ organic matter removed or $m^3 CH_4 kg^{-1}$ COD removed. Under continuous reactor operation a mesophilic digester will be producing between 12 and 16 l of methane per capita per day for primary sludge rising to about $20 l day^{-1}$ for combined primary and secondary sludges (Speece and McCarty, 1964).

There are three main options for the use of biogas. Burn it to produce heat or to generate electricity, or to fuel vehicles. The energy can be supplied to the National Grid, either in the form of gas or electricity, when sufficient biogas is produced. Each of these options do require further intermediate processing after production, ranging from simple storage until required to cleaning and compression in the production of fuel gas (liquefied petroleum gas (LPG)). Methane has a high calorific value of $35.8 MJ m^{-3}$ at standard temperature and pressure compared with $37.3 MJ m^{-3}$ for natural gas, which is a mixture of methane, propane and butane. However, digester gas can be only 65% methane, so without cleaning to remove the carbon dioxide the biogas has a typical calorific value of $23 MJ m^{-3}$. At the sewage treatment plant the gas is generally used in gas-heated boilers to heat sludge digesters with the excess used to generate electricity for use on the plant (Section 14.4.2). The electricity production at some intensive agricultural plants may be so high that they are able to sell the surplus energy via the National Grid. Smaller digestion units are used widely in the Third World for cooking and lighting.

19.3 CONTACT ANAEROBIC SYSTEMS

Contact anaerobic systems are specifically designed to treat weaker sludges with low solids concentrations and strong effluents. Unlike the conventional continuously stirred digester, the HRT and MCRT are independent of each other with the biomass either retained within the reactor or recycled after separation. Four major contact systems have been developed: anaerobic activated sludge, sludge blanket, static media filters and fluidized media.

19.3.1 ANAEROBIC ACTIVATED SLUDGE

The principle of operation is similar to that of the activated sludge process. The active anaerobic bacteria are separated from the final effluent by sludge settlement or other separation process (e.g. filtration or centrifugation). A portion of the sludge is then returned to the reactor to maintain a high biomass content (Fig. 19.2d). These systems are widely known as anaerobic contact digesters and operate at relatively short HRTs of between 2.0 and 5.0 days. This system is particularly suited to warm, dilute effluents especially food-processing wastewaters that have a high suspended solids content (e.g. abattoir wastes, distillery slops, starch waste, sugar beet processing wastes, yeast fermentations, and so on). The major operational problem appears to be poor settlement in the secondary reactor due to gas bubbles becoming attached to the solids. This is generally overcome by vacuum degassing followed by settlement using inclined plates within the tank. Settlement tanks often need to be covered to prevent odour problems.

19.3.2 UP-FLOW ANAEROBIC SLUDGE BLANKET (UASB) PROCESS

Up-flow anaerobic sludge blanket (UASB) process relies on the propensity of anaerobic biomass to aggregate into dense flocs or granules over a long period. This heavily flocculated sludge develops within the special reactor (Fig. 19.2e) able to withstand high mixing forces. Mixing is achieved by pumping the feed wastewater through the base of the reactor up through the sludge blanket. Above the blanket finer particles flocculate and in the upper settlement zones they settle as sludge back to the blanket, thus preventing washout of biomass. The settlement zones occur between the gas collection bowls which slope at an angle of 50° which promotes the return of the sludge from the settlement areas. The process is characterized by very high MCRT, which is even higher than for anaerobic filters, and a low HRT (usually <1 day). Compared to the other contact processes, it is very difficult to operate and to maintain the structure of the sludge blanket. It will take up to 6 months for a suitable granular sludge to develop from a soluble wastewater, so a suitable inoculum is required for rapid start-up. The process is dependent entirely on the ability of the biomass to form granules using particular wastewaters and is widely used for strong industrial wastewaters with a low suspended solids content (e.g. creamery wastes and the waste from the manufacture of soft drinks).

19.3.3 STATIC MEDIA FILTER

Anaerobic reactors using fixed or static media to retain the active biomass are generally known as anaerobic biofilters (Fig. 19.2c). In design they are essentially percolating filters that are completely flooded so that no oxygen can enter the reactor. They are operated utilizing a wide variety of filter media from mineral to random and ordered packed plastics. The biomass comprises a thin bacterial film that is firmly attached to the support media. They can be operated in either an up-flow or down-flow mode, although there are distinct advantages and disadvantages to each (Hobson and Wheatley, 1993). Different media result in very different performance characteristics. In up-flow filters a significant portion of the active biomass will be present not as attached film, but as unattached dispersed growths in the interstices (voids) of the medium. The suspended solids flocculate as they travel up through the filter forming larger particles that finally settle back down the filter column. The movement of solids and gas bubbles being released introduces an element of mixing into this essentially plug-flow system. When operated in the down-flow mode all the dispersed solids are washed out of the reactor leaving only the attached biomass. In these reactors the media must ensure stable film development as well as preventing excessive solids accumulation. So channelled or ordered media as opposed to random media is best. As all the biomass is attached in down-flow anaerobic biofilters, the performance is directly related to the specific surface area of the medium, which is not the case in up-flow filters.

Anaerobic biofilters are ideal for relatively cold and dilute wastes as they have an extremely high MCRT:HRT ratio. This gives the process a high degree of stability, an excellent resistance to inhibitory compounds and a satisfactory performance even at low temperatures. At present the major use of anaerobic biofilters has been for denitrification, to remove nitrates from sewage effluents (Section 18.4.1). However, the process is becoming increasingly popular due to its simplicity and robustness of operation. It is used mainly for industrial wastewaters with a low suspended solids content (e.g. food and distillery wastes and some pharmaceutical wastewaters).

19.3.4 FLUIDIZED AND EXPANDED MEDIA

The biomass in an anaerobic fluidized bed reactor (Fig. 19.2f) develops on tiny-grained inert (sand) or reactive (activated carbon) random medium. The medium, which is very light, becomes coated with bacteria and is mixed within the reactor by applying high rates of gas and effluent recycle. The particles can be fully mixed (fluidized) which increases the volume of the reactor occupied by the medium by 20–25%, or less vigorously mixed (expanded) by <10% of its volume. Particle size ranges from 0.3 to 1.0 mm depending on the type of medium used. The decrease in media particle size has resulted in a significant increase in the specific surface area available for biological growth making such systems highly efficient. Smaller sizes are readily lost from the reactor while larger sizes are difficult to fluidize. It is also difficult to maintain the optimum mixing velocity to ensure the medium remains in suspension without causing shear forces strong enough to strip off the accumulated biomass, so there is increasing interest in expanded beds. Fluidized beds are normally tall, with a height to diameter ratio of 5–6:1. Although difficult to start up, fluidized beds are highly resistant to temperature, toxic compounds and high organic loadings, tolerant to wide fluctuations in influent quality and also tolerant to a high solids content in the influent. It has proved very difficult to scale up this system from pilot- to full-scale operation, resulting in very few full-scale plants worldwide.

REFERENCES

Hobson, P. N. and Wheatley, A. D., 1993 *Anaerobic Digestion: A Modern Theory and Practice*, Elsevier Applied Science, London.

HMSO, 1987 *Determination of the Inhibitory Effects of Chemicals and Waste Waters on the Anaerobic Digestion of Sewage Sludge*, Methods for the Examination of Waters and Associated Materials, Department of the Environment, HMSO, London.

IWPC, 1979 *Sewage Sludge. I. Production, Preliminary Treatment and Digestion*, Institute of Water Pollution Control, Maidstone.

Koster, I. W., 1988 Microbial, chemical and technological aspects of the anaerobic degradation of organic pollutants. In: Wise, D. L., ed., *Biotreatment Systems: 1*, CRC Press, Boca Raton, FL.

Mosey, F. E., 1976 Assessment of the maximum concentration of heavy metals in crude sewage which will not inhibit the anaerobic digestion of sludge, *Water Pollution Control*, **75**, 10–20.

Mosey, F. E., 1983 Anaerobic processes. In: Curds, C. R. and Hawkes, H. A., eds, *Ecological Aspects of Used-Water Treatment: 2. Biological Activities and Treatment Processes*, Academic Press, London, pp 219–60.

Speece, R. E. and McCarty, P. L., 1964 Nutrient requirements and biological solids accumulation in anaerobic digestion. In: Ekenfelder, W. W., ed., *Advances in Water Pollution Research. Proceedings of the First International Conference on Water Pollution Research, 2*, Pergamon Press, Oxford, pp 305–33.

Sterritt, R. M. and Lester, J. N., 1988 *Microbiology for Environmental and Public Health Engineers*, E. & F. N. Spon, London.

Tebbutt, T. H. Y., 1992 *Principles of Water Quality Control*, 4th edn, Pergamon Press, London.

Wheatley, A. D., Fisher, M. B. and Grobicki, A. M. W., 1997 Applications of anaerobic digestion for the treatment of industrial waste waters in Europe, *Journal of the Chartered Institution of Water and Environmental Management*, **11**, 39–46.

Further Reading

HMSO, 1989 *The Assessment of Biodegradability in Anaerobic Digesting Sludge*, Methods for the Examination of Water and Associated Materials, Department of the Environment, HMSO, London.

La Farge, B., 1998 *Biogas: Methanic Fermentation Procedures*, John Wiley and Sons, Chichester.

Malina, J. F. and Pohland, F. G., 1992 *Design of Anaerobic Processes for the Treatment of Industrial and Municipal Wastes*, Technomic Publishing, London.

van Haanel, A. C. and Lettinga, G., 1994 *Anaerobic Sewage Treatment: A Practical Guide for Regions with a Hot Climate*, John Wiley and Sons, Chichester.

Zehnder, A. J. B. (ed.), 1988 *Biology of Anaerobic Micro-organisms*, John Wiley and Sons, New York.

INTRODUCTION

There are a wide range of non-biological unit processes, some of which are applicable to both water and wastewater treatment. Advanced processes are those not normally employed in conventional treatment but are required to conform to ever stricter water quality and discharge standards (Sections 8.1 and 8.3). Advanced water treatment processes include iron and manganese removal, softening, ion exchange, adsorption and chemical oxidation. Advanced (tertiary) wastewater treatment is more varied and includes biological processes for nitrification, denitrification and phosphorus removal (Section 18.4), as well as a range of physico-chemical processes (Table 20.1).

20.1 EQUALIZATION

While sewage displays a diurnal variation in flow (Fig. 13.3), wastewater treatment plants are designed to handle normal variations of up to 3 times the dry weather flow (3DWF). The composition of the sewage entering large treatment plants also varies very little during this period. The exception is storm water which can result in an initial toxic shock loading as the stored contents of road-side gully pots are displaced followed

TABLE 20.1   Principal advanced physico-chemical wastewater treatment processes

Process	Removal function
Filtration	Suspended solids
Air stripping	Ammonia
Breakpoint chlorination	Ammonia
Ion exchange	Nitrate, dissolved inorganic solids
Chemical precipitation	Phosphorus, dissolved inorganic solids
Carbon adsorption	Toxic compounds, refractory organics
Chemical oxidation	Toxic compounds, refractory organics
Ultrafiltration	Dissolved inorganic solids
Reverse osmosis	Dissolved inorganic solids
Electrodialysis	Dissolved inorganic solids
Volatilization and gas stripping	Volatile organic compounds

by increasing dilution of the wastewater as the hydraulic flow increases (Section 13.3). In domestic and municipal plants it is unusual for equalization to be used if the peaking factor (i.e. peak flow/average flow) is <2. Flows in excess of 3DWF are dealt with by the storm water tanks, which offer a degree of equalization (Section 14.2). Industrial wastewaters are subject to significant variation in both flow and composition due to the nature of the processes used and the work patterns operated. For this reason equalization or balancing may be required before entry to a sewer in order to avoid problems at the treatment plant. Equalization ensures a constant and continuous flow of wastewater and ensures that changes in wastewater composition occur slowly causing minimum disruption to treatment processes, especially the biological systems.

Four main designs are used for equalization (Fig. 20.1). In-line equalization is most widely used with all the flow passing through the balancing tank resulting in significant flow and load (organic, toxicity, etc.) dampening (Fig. 20.1a). Alternating equalization uses two separate tanks so that while one is filling the other is being discharged after the maximum storage period for equalization (Fig. 20.1b). Where there are a number of

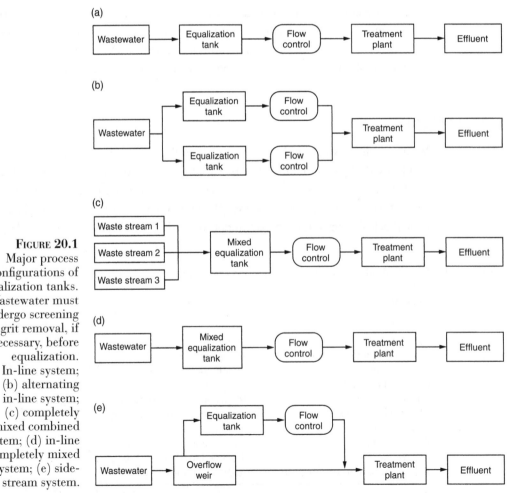

FIGURE 20.1
Major process configurations of equalization tanks. Wastewater must undergo screening and grit removal, if necessary, before equalization.
(a) In-line system;
(b) alternating in-line system;
(c) completely mixed combined system; (d) in-line completely mixed system; (e) side-stream system.

compatible process streams these can be fed simultaneously into a single balancing tank which must be mixed to ensure full equalization (Fig. 20.1c). Where settleable solids are present or where the wastewater is readily degradable then mixing is required to prevent settlement, septicity or both (Fig. 20.1d). Where sufficient mixing is available the tank also acts as a pre-aeration unit resulting in improved settling characteristics due to pre-flocculation of particles as well as some biochemical oxygen demand (BOD) removal. Side-line equalization is used only periodically when the flow exceeds the normal design flow rate. Excess flow is diverted into a balancing tank and then allowed to discharge slowly back into the main process stream at a controlled rate (Fig. 20.1e). Pumping costs are significantly cheaper with side-line systems as only a small portion of the total flow is diverted. Equalization is carried out after preliminary treatment but prior to primary sedimentation. Equalization tanks can be a major source of odour generation, especially if volatile compounds are being stored and mixed. In such cases

TABLE 20.2 Calculation of equalization tank volume for an industrial wastewater with a large diurnal variation

Time	(i) Volume $(m^3 h^{-1})$ (h)	(ii) Cumulative actual flow (m^3)	(iv) Cumulative equalized flow (m^3)	(v) Difference in flow (m^3)
0–1	372	372	574	202
1–2	292	664	1148	484
2–3	384	1048	1722	674
3–4	472	1520	2296	776
4–5	482	2002	2870	868
5–6	535	2537	3444	907
6–7	558	3095	4018	923
7–8	562	3657	4592	935
8–9	593	4250	5166	916
9–10	624	4874	5740	866
10–11	641	5515	6314	799
11–12	658	6173	6888	715
12–13	749	6922	7462	540
13–14	765	7687	8036	349
14–15	808	8495	8610	115
15–16	810	9305	9184	−121
16–17	831	10 136	9758	−378
17–18	677	10 813	10 332	−481
18–19	681	11 494	10 906	−588
19–20	624	12 118	11 480	−638
20–21	559	12 677	12 054	−623
21–22	516	13 193	12 628	−565
22–23	296	13 489	13 202	−287
23–24	287	13 776	13 776	0

(iii) Average (equalized) flow = $13\,776/24 = 574\,m^3 h^{-1}$.
Cumulative equalized flow exceeds cumulative actual flow by a maximum of $935\,m^3 h^{-1}$. Allowing a 20% safety margin the equalization tank volume at the equalized flow rate of $574\,m^3 h^{-1}$ is $935 \times 1.20 = 1123\,m^3$.

tanks should be covered and the air space continuously vented through a scrubber or biofilter. Enclosed tanks can be extremely hazardous to workers and may even result in the build up of explosive or flammable gases, so care must be taken (Section 13.6).

The size of an equalization tank can be calculated by the following method:

(i) Calculate the actual flow rate ($m^3 h^{-1}$) each hour over at least a 24-h period.
(ii) Calculate the cumulative (actual) flow in m^3 over 24 h using hourly increments.
(iii) The average flow ($m^3 h^{-1}$) is calculated by dividing the cumulative (actual) flow by 24.
(iv) Using the average flow per hour the cumulative (equalized) flow in m^3 over 24 h using hourly increments is calculated.
(v) The maximum hourly difference between the cumulative (equalized) flow and the cumulative (actual) flow is the size of the equalization tank required. As a safety margin, the design volume should be 20–25% larger than the actual volume calculated.

A worked example is given in Table 20.2.

20.2 COAGULATION

Colloidal particles cannot be removed by normal filtration or sedimentation processes due to their small size and negligible settling velocities. They have a very high specific surface area and as a consequence their behaviour is governed largely by surface properties. Hydrophobic particles (e.g. metal oxides, clays, etc.) derive stability due to having charges that repel each other. These charges are derived from either their chemical structure or by the adsorption of a single ion type. All particles similarly charged result in a stable suspension preventing particles colliding under the influence of Brownian motion or van der Waal forces. Particles have a solid boundary layer covered by a rigid solution boundary layer. Each has a charge resulting in a potential gradient between the layers. Only the potential at the outer rigid solution boundary can be measured and is known as the zeta potential (ζ) and is related to the particle charge and the thickness between the two boundary layers (Sayer et al., 1994):

$$\zeta = \frac{4\pi\delta q}{D} \qquad (20.1)$$

where π is 3.1416, δ is the thickness between the two boundary layers, q is the particle charge and D is the dielectric constant for the liquid. The zeta potential is estimated experimentally by measuring the movement of colloidal particles in an electric field. The addition of coagulants results in a decrease of ζ leading to a decrease in the repulsion between particles. Normal values are between 10 and 200 mV, while optimum coagulation occurs at 0.0–0.5 mV. However, measurements of ζ vary with the nature of the solution components making measurements non-repeatable and so in practice it is of limited value.

Hydrophilic particles (organic residues, macromolecules such as proteins, starch and detergents) have a high affinity for water and their stability is due to bound water layers that prevent particles from coming into close contact. They also have weak charges from the ionization of attached functional groups (e.g. carboxyl, hydroxyl groups, etc.).

The application of coagulation is discussed in Section 10.1. Metal salt coagulants react with the alkalinity in the water to produce insoluble metal hydroxide precipitates that enmesh the colloidal particle in water and adsorbs other material including dissolved organic matter present. The hydroxide flocs carry a small positive charge that attracts the negative particles in the water. Polyelectrolytes used as coagulant aids are available with either a positive or negative charge and are used to optimize flocculation. The most widely used coagulant is aluminium sulphate (alum) $(Al_2(SO_4)_3 \cdot 14H_2O)$ but it is highly corrosive and requires careful handling and storage. Alum reacts with the alkalinity in the water to produce insoluble aluminium hydroxide floc $(Al(OH)_3)$ according to Equation (20.2):

$$Al_2(SO_4)_3 \cdot 14H_2O + 3Ca(HCO_3)_2 \rightarrow 2Al(OH)_3 + 3CaSO_4 + 14H_2O + 6CO_2$$

$$(20.2)$$

where alkalinity is insufficient in the water lime in the form of calcium hydroxide is added (Equation (20.3)):

$$Al_2(SO_4)_3 \cdot 14H_2O + 3Ca(OH)_2 \rightarrow 2Al(OH)_3 + 3CaSO_4 + 14H_2O \quad (20.3)$$

In the production of the finished drinking water minimum residuals of coagulant metals are required in order to conform to drinking water quality standards (Section 8.1). The solubility of iron and aluminium are related to pH with a residual of $<200\,\mu g\,l^{-1}$ for iron occurring between pH 3.7 and 13.5 and for aluminium between pH 5.2 and 7.6 at 25°C (Fig. 20.2).

The rate of addition of coagulant is governed by many factors that can alter very quickly. So in practice very careful control is required over the addition of the coagulant. The optimum conditions for coagulation are determined as often as possible using a simple procedure known as the jar test (Solt and Shirley, 1991). This measures the effect of different combinations of the coagulant dose and pH, which are the two most important factors in the process, as well as the dosage of coagulant aid if required. The jar test allows a comparison of these different combinations under standardized conditions after which the colour, turbidity and pH of the supernatant (clarified water) are measured. The jar test is a simple piece of equipment that comprises of six standard 600 ml beakers each containing an identical flat stainless steel stirring paddle. Once the coagulant doses and pH to be tested have been selected each beaker is then subject

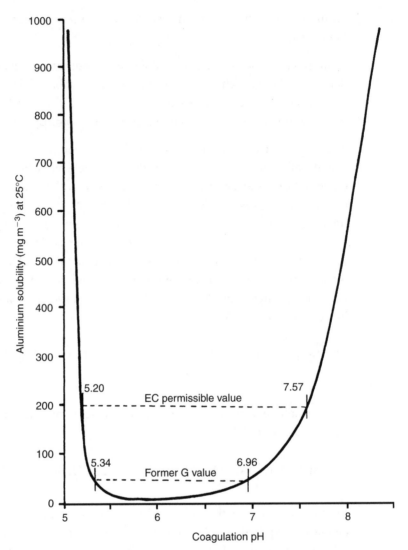

to identical mixing. First a fast stir (e.g. 60–$80 \, \text{rev min}^{-1}$ for 1–$4 \, \text{min}$) followed by a slow stir (e.g. 5–$20 \, \text{rev min}^{-1}$ for 15–$20 \, \text{min}$). The stirrers are switched off and the samples allowed to settle and the water tested for best result. This involves three separate sets of jar tests:

1. The test is first carried out on the raw water without altering the pH. The coagulant dose is raised over a suitable range, so if five jars are used then five different doses can be tested. From this a simple curve can be drawn which shows the turbidity against the coagulant dose (Fig. 20.3). From this the best coagulant dose can be calculated.

2. The next stage is to alter the pH of the raw water by adding either an alkali or acid, and then to repeat the test using the best coagulant dose determined in the first set of tests. The range selected is normally between 5.5 and 8.5, and if possible 0.5 pH

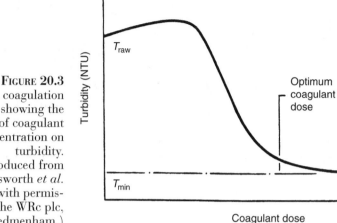

FIGURE 20.3
Typical coagulation curve showing the effect of coagulant concentration on turbidity. (Reproduced from Ainsworth *et al.* (1981) with permission of the WRc plc, Medmenham.)

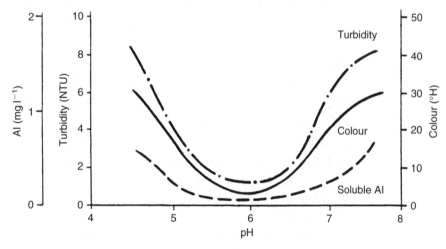

FIGURE 20.4
Example of a pH optimization curve for a hard lowland water showing the effect on turbidity, colour and the soluble aluminium concentration. (Reproduced from Ainsworth *et al.* (1981) with permission of the WRc plc, Medmenham.)

increments are used. From this a plot of final colour and turbidity is plotted against pH, allowing the optimum pH for coagulants to be selected (Fig. 20.4).

3. Finally, the test is repeated again using fresh raw water but this time corrected to the optimum pH and tested at various coagulant doses again. This determines the exact coagulant dosage at the optimum pH. These values are then used to operate the coagulant process.

The curve produced from a lowland river in Fig. 20.5 is typical for hard waters of its type, and from this it is clear that few problems should occur in operational management. However, soft upland waters, which are usually highly coloured due to humic material, are far more difficult to treat. Figure 20.5 shows that there is only a very narrow band of pH at which to achieve optimum removal of colour and turbidity. So residual colour and turbidity will always be much higher when compared to lowland water or groundwater supplies. The hydroxide flocs formed are small and weaker causing problems at

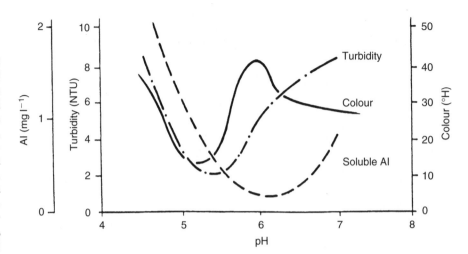

FIGURE 20.5
Example of a pH optimization curve for a soft, coloured water showing the effect on turbidity, colour and the soluble aluminium concentration. (Reproduced from Ainsworth *et al.* (1981) with permission of the WRc plc, Medmenham.)

later stages in treatment. Another problem is that as conditions change quite rapidly, these tests need to be repeated, certainly daily and sometimes more frequently. At small plants where such waters are frequently treated, this is just not possible so quality in terms of turbidity and colour is more likely to fluctuate.

Changes in quality can occur rapidly, so if the operator sees a change in the colour or turbidity he or she may increase the dosage rate of coagulant, but optimum conditions may not be maintained so that excess coagulant will enter the distribution system. One of the major causes of seasonal and often daily changes is the presence of algae in reservoirs, which can significantly alter the pH of the water. The jar test does not indicate how much insoluble coagulant will pass through the clarification and filtration stages. It does, however, give an idea of how much soluble coagulant will be in the final water as this depends solely on pH and the dosage rate. If the curves obtained from the jar test are overlaid by the amount of aluminium that has been taken into solution, a new and more serious problem arises (Figs 20.4 and 20.5).

The curve of soluble aluminium in water that is abstracted from the hard lowland river follows the same curve as both turbidity and colour, so that the amount of soluble aluminium coagulant in the treated water will be acceptably low at the optimum conditions for removal of turbidity and colour (Fig. 20.4). With soft coloured waters a dilemma occurs, for the best conditions for the removal of turbidity and colour do not coincide with those for minimum coagulant solubility. So in the example shown the optimum removal of turbidity and colour occurred at a coagulant dose of 3 mg Al l^{-1} at pH 5.3, which results in 0.7 mg l^{-1} of soluble aluminium in the water compared to the EC limit of 0.2 mg l^{-1}. So the water is clear and colourless but contains 3.5 times more than the legal limit of aluminium. Where this occurs much more careful analysis of the dosage rates must be carried out to find the optimum conditions to achieve minimum aluminium in the water. In this case it turned out to be a coagulant dose of 5 mg l^{-1} of Al, nearly twice as much as before at pH 6.6. This resulted in satisfactory aluminium (0.06 mg l^{-1}), colour (5.2 mg l^{-1}) and turbidity (1.4 FTU), although far more expensive, and resulted in larger quantities of aluminium sludge being produced. Because the operational range is so narrow, any minor change in water quality will result in rapid deterioration in colour, turbidity and, of course, soluble aluminium. With higher residual levels of aluminium in soft waters anyway, any random adjustment by the operator, no matter how minor, could cause a significant increase in the amount of aluminium in the water (Ainsworth *et al.*, 1981).

Synthetic organic polymers, in particular polyacrylamides, are used extensively as coagulant aids to produce stronger flocs and more stable floc blankets during sedimentation (Section 10.1). In the UK, the use of polyelectrolytes is very widespread with up to 70% of all plants using such chemicals. Polyacrylamide is primarily used in the treatment of sludge, but as the polymer-rich supernatant is normally returned to the inlet of the plant it is inevitable that polyelectrolytes will find their way in to drinking water (Sections 8.1 and 10.1.5).

20.3 SEDIMENTATION

Sedimentation is the most widely used unit operation in water and wastewater treatment. The terms sedimentation and settling are the same and can be used interchangeably. Sedimentation is the process by which the suspended solids, which are heavier than water and commonly referred to as the settleable fraction, are removed from suspension by allowing the particles to gravitate to the floor of a tank to form a sludge under near quiescent conditions. The process is used in primary treatment to remove settleable organic and inorganic material to reduce the organic load to the secondary treatment processes. It is also used for secondary sedimentation, to remove material converted to settleable solids during the biological phase of treatment, such as the removal of humus from filter effluents and the recovery of activated sludge (Section 14.2).

The settleable solids fraction in surface waters used as sources of drinking water are normally extremely low so that plain sedimentation is normally unnecessary. When large rivers are used for supply purposes the solids loading can become high during flooding so that intermittent storage in reservoirs or primary sedimentation is required.

20.3.1 SEDIMENTATION THEORY

The way in which particles settle out of suspension is a vital consideration in the design and operation of sedimentation tanks as well as other unit processes. Four types of settling can be identified. These are type I (discrete), type II (flocculant), type III (hindered or zonal settling) and, finally, type IV (compression) settlement. Where particles are dispersed or suspended at a low solids concentration then type I or II settlement occurs. Type III or IV settlement only occurs when the solids concentration has increased to such an extent that particle forces or particle contact affects normal settlement processes. During settlement it is common to have more than one type of settling occurring at the same time and it is possible for all four to occur within the settlement tank simultaneously.

Type I or discrete particle settling

Discrete particles settle out of a dilute suspension as individual entities. Each particle retains its individual characteristics and there is little tendency for such particles to flocculate, so settlement remains solely a function of fluid properties and particle characteristics. Settling of heavy inert particles, such as grit and sand particles, is an example of type I settlement.

Type I settlement is analysed by means of the classic laws of sedimentation formed by Newton and Stokes. Newton's law yields the terminal particle velocity by equating the gravitational force of the particle to the frictional resistance or drag. The rate of settlement of discrete particles varies with the diameter of the particle, the difference in density between the particles and fluid in which it is suspended and the viscosity, as shown by Stokes' law (Equation (20.4)):

$$V_c = \frac{g(p_s - p)d^2}{18\eta} \qquad (20.4)$$

where V_c is the terminal (Stokes) settling velocity ($\mathrm{m\,s^{-1}}$), g is the gravity constant ($9.80\,\mathrm{m\,s^{-1}}$), p_s is the density of particle ($\mathrm{kg\,m^{-3}}$), p is the density of the fluid ($\mathrm{kg\,m^{-3}}$), d is the particle diameter (m) and η is the viscosity of the suspending fluid ($\mathrm{kg\,m^{-1}s^{-1}}$) (Metcalf and Eddy, 1991).

Under quiescent conditions discrete particles settle out at their terminal velocity, which will remain, constant provided the fluid temperature does not alter. Knowledge of this velocity is fundamental in the design of sedimentation tanks and in the evaluation of their performance. A sedimentation tank is designed to remove all particles that have a terminal velocity equal to or greater than V_c. The selection of V_c will depend on the specific function of the tank and on the physical characteristics of the particles to be removed. The rate at which clarified water is produced (Q') in a sedimentation tank is given by:

$$Q' = V_c A \ (\mathrm{m^3\ day^{-1}}) \qquad (20.5)$$

where V_c is the terminal velocity of the particle and A is the surface area of the settling chamber. Therefore in theory, the rate of clarification by type I settlement is independent of depth. The surface overflow rate or surface loading, which is the average daily flow in cubic metres per day divided by the total surface area of the sedimentation tank in square metres and so equivalent to the terminal settling velocity, is the basis of sedimentation tank design (Equation (20.6)):

$$V_c = \frac{Q'}{A} = \text{Surface overflow rate (m}^3 \text{ m}^{-2} \text{ day}^{-1}) \qquad (20.6)$$

Effluent weir loading is also a widely used design criterion and is equal to the average daily overflow divided by the total weir length, being expressed as cubic metres of effluent flowing over per metre of weir per day $(m^3 m^{-1} day^{-1})$.

Modern sedimentation tanks are operated on a continuous-flow basis. Therefore, the hydraulic retention time (HRT) of the tank should be long enough to ensure that all particles with the desired velocity V_c will settle to the bottom of the tank. HRT is calculated by dividing the tank volume by the influent flow:

$$\text{HRT} = 24\frac{V}{Q} \text{ (h)} \qquad (20.7)$$

where V is the tank volume (m^3) and Q is the mean daily flow $(m^3 day^{-1})$. The depth of a sedimentation tank is taken as the water depth at the sidewall, measuring from the tank bottom to the top of the overflow weir. This excludes the additional depth resulting from the slightly sloping bottom in both rectangular and circular sedimentation tanks. The terminal velocity (V_c), tank depth and HRT are related as:

$$V_c = \frac{D}{\text{HRT}} \text{ (m s}^{-1}) \qquad (20.8)$$

where D is the depth (m).

Type II flocculant settling

Type II particles in relatively dilute solutions do not act as discrete particles but coalesce or flocculate with other particles during gravitational settlement. Subsiding particles coalesce with smaller particles falling at lower velocities to form larger particles which then settle faster than the parent particle. Settlement of solids in the upper layer of both primary and secondary sedimentation tanks are typical of type II settlement, and flocculation is particularly important in the separation of the sludge in the activated sludge process where rapid separation is required, so that the sludge can be returned to the aeration basin. The degree of flocculation is dependent on the opportunity for particle contact, which increases as the depth of the settling tank increases. So the removal of suspended material depends not only on the clarification rate (Q) but also on depth, which is a major difference between type I and type II settlement. Surface overflow rate, the velocity gradients in the system, the concentration of particles and the range of particle sizes are also important considerations in flocculant settling. As there is no adequate mathematical relationship to determine the effect of flocculation on sedimentation, the effects of all these variables can only be determined by settling column analysis.

Settling column analysis is carried out in a cylinder of the same depth as the proposed tank and with a minimum width of 150 mm to minimize wall effects. Sample ports spaced at 0.5 m intervals allow samples to be taken for analysis from all depths. The wastewater is poured rapidly into the column and mixed to ensure that there is a uniform distribution of particles throughout the column at the beginning of the analysis. Care must also be taken to ensure that uniform temperature is maintained throughout the column to avoid convection currents affecting the settlement rate. Water samples are taken from the ports at regular intervals and are analysed for suspended solids. The percentage removal from the original suspended solids concentration of the wastewater is plotted against time and depth, which allows curves of equal percentage removal, iso-concentration curves, to be plotted (Fig. 20.6). These curves allow depth and retention time to be compared so providing the optimum design and operating conditions for particular wastes. The degree of flocculation is indicated by the slope of the iso-concentration curve, increasing as the slope increases. Overall removal of solids can be estimated by reading values directly from the plotted curves. For example, in Fig. 20.6 the percentage removal after 25 min retention is 40% at 1 m and 36% at 2 m. The curves also represent the settling velocities of the particles for specific concentrations in a flocculant suspension (Equation 20.8). So after 25 min 40% of the particles in suspension at 1 m will have average velocities of at least $6.7 \times 10^{-4} \mathrm{m\,s^{-1}}$ and 36% at least $1.3 \times 10^{-3} \mathrm{m\,s^{-1}}$ at 2 m.

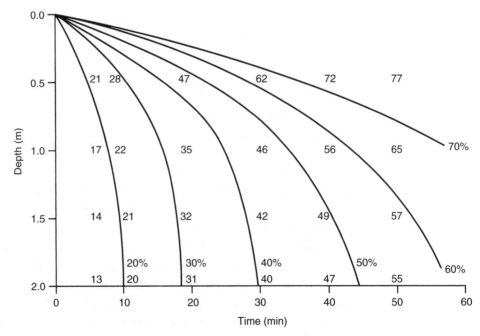

FIGURE 20.6
Iso-concentration curves for solids removal from a flocculent suspension. The numbers plotted represent the percentage removal of solids based on settling column analysis.

Type III or hindered (zonal) settling

In dilute suspensions particles settle freely at their terminal velocity until they approach the sludge zone when the particles decelerate and, finally, become part of the sludge. In concentrated suspensions ($>2000\,\mathrm{mg\,l^{-1}}$ of suspended solids) hindered settlement occurs. Owing to the high concentration of particles there is a significant displacement of liquid as settlement occurs which moves up through the interstices between particles reducing their settling velocity and forming an even more concentrated suspension (Fig. 20.7). The particles are close enough for interparticulate forces to hold them in fixed positions relative to each other so that all the particles settle as a unit or blanket. A distinct solids–liquid interface develops at the top of the blanket with the upper liquid zone relatively clear and free from solids. This type of settling is normally associated with secondary settlement after a biological unit, in particular the activated sludge process. As settlement continues a compressed layer of particles begin to form at the base of the tank, where the particles are in physical contact with one another. The solids concentration increases with depth and this gradation of solids is even apparent throughout the hindered settling zone from the solids–liquid interface to the settling–compression zone.

Hindered settling and compression can be clearly demonstrated using a graduated cylinder, by measuring the height of the interface between the settling particles and the clarified liquid at regular intervals (Fig. 20.8i). This is known as a batch-settling test and can be used to plot a settlement curve (Fig. 20.8ii). On the curve, A–B shows hindered settling of the solids–liquid interface, B–C deceleration as there is a transition between hindered and compressive settling and, finally, C–D which represents compression. Further settlement in the compression zone is due to physical compression of the particles. The surface area required in a continuous-flow system designed to handle concentrated suspensions depends upon the clarification and thickening capabilities of the system. Batch settling tests are used to estimate these factors using the method developed by Talmadge and Finch (1955).

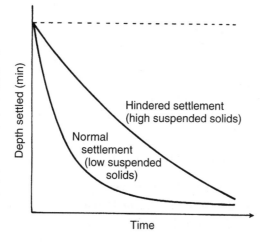

FIGURE 20.7 Comparison of the rate of settlement of solids for a wastewater with a high suspended solids concentration showing hindered settlement compared with a normal wastewater containing a low suspended solids concentration.

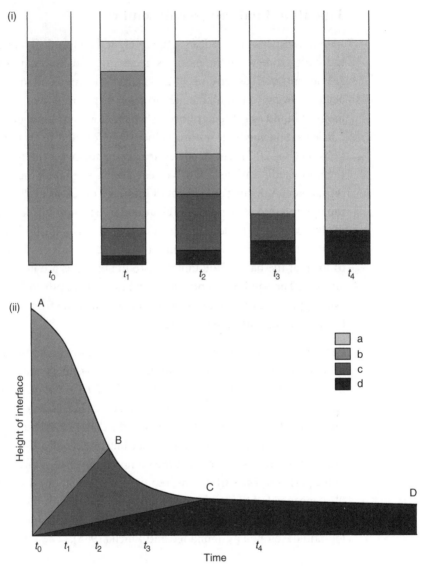

FIGURE 20.8 (i) The development of discernible settlement zones at various times during the settling test on a flocculant suspension, where (a) is the clarified effluent; (b) the wastewater with solids at its initial concentration and distributed evenly throughout its volume; (c) the zone where particle concentration is increasing and compaction is prevented by constant water movement up through the sludge blanket due to water being displaced in the lower zone; (d) where the particles have compacted. (ii) The data derived from the settling test is used to plot a settlement curve, where A–B shows hindered settling of the solids–liquid interface, B–C deceleration as there is a transition between hindered and compressive settling, and C–D, which is compression.

Type IV or compressive settling

Consolidation of particles in the compressive zone is very slow as can be seen in Fig. 20.8ii. With time, the rate of settlement decreases as the interstices between the particles become progressively smaller until eventually the release of liquid from the zone is inhibited. The particles are so concentrated that they are in physical contact, forming a thick sludge. Further settling is only possible by the weight of the new particles physically compressing the sludge layer. Gentle agitation enhances further compaction of the sludge by breaking up the flocs and releasing more liquid. Similarly, stirring has also been shown to increase the rate of settlement in the hindered zone. Compression normally occurs in sludge thickening tanks where specifically designed stirrers (e.g. picket-fence thickeners) are used to encourage consolidation (Fig. 21.4), or in the bottom of deep sedimentation tanks.

20.3.2 DESIGN OF SEDIMENTATION TANKS

Sedimentation tanks have two functions, the removal of settleable solids to produce a clarified effluent and the concentration of solids to produce a handleable sludge. The design of a sedimentation tank takes both of these functions into consideration and the eventual size will depend on whichever function is limiting. For example, in the design of tanks for activated sludge settling, the limiting factor is usually sludge thickening. The general criteria for sizing sedimentation tanks are the overflow rate, tank depth and HRT. If wastewater was only composed of discrete particles then primary sedimentation tanks would be constructed as shallow as possible to optimize removal efficiency, as the rate of type I settlement is independent of depth. However, wastewater also contains some flocculant particles and so the depth of primary sedimentation tanks is roughly the same depth as secondary tanks, which normally have a minimum depth of 2.1 m. Although increasing the depth does not increase the surface-loading rate, it does however increase the HRT, which allows a greater degree of flocculation and so enhances settlement. In recent years HRT has been used as the primary design criterion instead of surface-loading or overflow rate. However, whether HRT or surface-loading rate is used, the other is normally kept at an accepted value, and since the depth varies only between narrow limits the choice of criterion has had little influence in practice on the dimensions of the tanks constructed.

20.4 FLOTATION

Where suspended particles have a specific gravity less than water they can be removed by allowing them to float to the surface and then removed mechanically by skimming. This can be done passively using gravity flotation or assisted using air and dissolved air flotation (DAF).

Gravity flotation units or grease traps, as they are widely known, are employed to remove fats, oils and grease (FOG) from individual premises such as restaurants and garages. The FOG present rise slowly to the surface of the water where they can be collected manually. To be effective a minimum HRT of 30 min is required with a maximum flow-through velocity of $<5\,m\,h^{-1}$ (Section 22.2.5) (Fig. 22.5). The amount of FOG in domestic and municipal wastewaters is generally low enough to be removed by the skimming device on primary sedimentation tanks. However, in some industrial processes large amounts of FOG are produced (e.g. dairy industry) and if not removed would block pipework and cause problems in other unit processes. So they must be removed by a special separation process. Air bubbles are introduced at the base of a flotation tank and become attached to the suspended particles buoying them up to the surface of the water where they are removed by skimming. DAF is the most widely used process for removing large quantities of FOG. Effluent is recycled to a pressure vessel where it is saturated with air. It is then added to the incoming waste-water where under atmospheric pressure the dissolved air comes out of solution forming minute bubbles that rise to the surface bringing all the FOG with them (Fig. 20.9) (Haarhoff and van Vuuren, 1993). Alternatively, air can be introduced directly into the flotation tank using an impeller, a process known as air flotation.

Performance of DAF systems depends on the ratio of the volume of air to mass of solids (a/s) to give a specific clarification rate. This ratio is calculated experimentally for specific wastewaters and sludges using a DAF test apparatus. From this value the design parameters for full-scale systems can be calculated (Metcalf and Eddy, 1991). For a particle to be removed by flotation its density (p_s) must be reduced by the attachment to a bubble (p_c) which is less than that of the liquid in which it is suspended (p):

$$p_c = \frac{V_a p_a + V_s p_s}{V_a + V_s} \qquad (20.9)$$

where V_a is the volume of air attached to a particle of volume V_s, while p_a is the density of air. The ratio a/s is:

$$\frac{a}{s} = \frac{V_a p_a}{V_s p_s} \qquad (20.10)$$

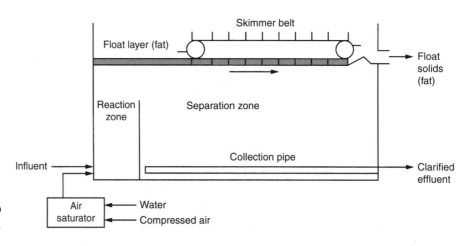

Skimmer belt

Float layer (fat)

Float solids (fat)

Reaction zone

Separation zone

Collection pipe

Influent

Clarified effluent

Air saturator ← Water
← Compressed air

FIGURE 20.9
Typical DAF unit.

By combining Equations (20.9) and (20.10) then:

$$p_c = \frac{(1 + a/s)}{(1/p_s + 1/p_a \cdot a/s)} \qquad (20.11)$$

in order for flotation to occur $p_c < p$. The minimum a/s ratio to achieve flotation is when $p_c = p$. So by modifying Equation (20.11):

$$\text{Minimum } a/s = \frac{(1 - p/p_s)}{(p/p_a - 1)} \text{ ml air mg}^{-1} \text{ solids} \qquad (20.12)$$

Apart from removing FOG from wastewater, flotation is also used for a wide variety of other applications including the thickening of sewage sludges, the recovery of paper fibres from pulp and paper mill wastewaters, removing coagulant hydroxide floc from potable water, and concentrating metallic ores in the mining industry.

20.5 CHEMICAL PRECIPITATION

20.5.1 REMOVAL OF CALCIUM AND MAGNESIUM

Chemical precipitation is more widely known as precipitation softening. It is used primarily to remove or reduce the hardness in potable waters caused by excessive salts of calcium and magnesium. Precipitation softening converts the soluble salts into insoluble ones, so that they can be removed by subsequent flocculation and sedimentation. Lime or soda ash are normally used to remove the hardness, although the exact method of addition depends on the type of hardness present (Table 20.3). Lime is most widely used, but like soda ash, produces a large volume of sludge, which has to be disposed (American Water Works Association, 1990). Softening using ion exchange is becoming increasingly common.

Single-stage lime softening uses hydrated lime ($Ca(OH)_2$) to remove calcium bicarbonate from solution by forming insoluble calcium carbonate ($CaCO_3$) with the amount of lime required equivalent to the amount of bicarbonate present:

$$Ca(HCO_3)_2 + Ca(OH)_2 \rightarrow 2CaCO_3 \downarrow + 2H_2O \qquad (20.13)$$

TABLE 20.3   Selection of chemical precipitation process for the removal of water hardness

Process	Selection criteria by type of hardness		
	Ca^{2+}	Mg^{2+}	Non-carbonate
Single-stage lime	High	Low	None
Excess lime	High	High	None
Single-stage lime–soda ash	High	Low	Some
Excess lime–soda ash	High	High	Some

While the saturation concentration of $CaCO_3$ is very low ($8.4\,mg\,l^{-1}$ at 20°C) the saturation concentration of $MgCO_3$ is relatively high ($110\,mg\,l^{-1}$ at 20°C) resulting in little precipitation and removal. This is overcome by ensuring excess lime to raise the pH to about 11 when the $MgCO_3$ is converted to $Mg(OH)_2$ which has a similar saturation concentration to $CaCO_3$. This is known as excess lime softening:

$$Mg(HCO_3)_2 + Ca(OH)_2 \rightarrow MgCO_3 + CaCO_3 \downarrow + 2H_2O \qquad (20.14)$$

$$MgCO_3 + Ca(OH)_2 \rightarrow + Mg(OH)_2 \downarrow + CaCO_3 \downarrow \qquad (20.15)$$

Lime soda softening is used for both carbonate and non-carbonate hardness with Ca^{2+} and Mg^{2+} exchanged for Na^+:

$$Na_2CO_3 + CaSO_4 \rightarrow Na_2SO_4 + CaCO_3 \downarrow \qquad (20.16)$$

$$Na_2CO_3 + MgSO_4 + Ca(OH)_2 \rightarrow Na_2SO_4 + Mg(OH)_2 \downarrow + CaCO_3 \downarrow \qquad (20.17)$$

Owing to the short contact time available during treatment and the variability in the solubility of calcium carbonate with temperature there is always a residual amount of calcium carbonate in treated water (up to $70\,mg\,l^{-1}$). This can subsequently precipitate out in the distribution system causing scaling. Carbonation is used to convert the remaining calcium carbonate present back to the soluble bicarbonate form:

$$CaCO_3 + CO_2 + H_2O \rightarrow Ca(HCO_3)_2 \qquad (20.18)$$

Precipitation results in large volumes of sludge. Calcium carbonate sludge can be calcined and then slaked with water to produce lime solving the disposal of the sludge, as well as producing excess lime for sale:

$$CaCO_3 \xrightarrow{\text{heat}} CaO + CO_2 \qquad (20.19)$$

$$CaO + H_2O \rightarrow Ca(OH)_2 \qquad (20.20)$$

Chemical precipitation can also be used to recover metals from industrial effluents. For example, hexavalent chromium (Cr^{6+}), which is found in wastewaters from metal plating and anodizing processes, is highly toxic. In a two-stage process Cr^{6+} is first reduced to Cr^{3+} and then precipitated out of solution as a hydroxide (Sayer et al., 1994). The major use of chemical precipitation in wastewater treatment is the removal of phosphate as a precipitate of calcium, magnesium or iron.

20.5.2 PHOSPHATE REMOVAL

Phosphorus is present in wastewater as orthophosphate (PO_4^{3-}, HPO_4^{2-}, $H_2PO_4^-$ and H_3PO_4), polyphosphates and organic phosphate. The average phosphorus concentration in sewage is between 5 and $20\,mg\,Pl^{-1}$ as total phosphorus, of which 1–$5\,mg\,Pl^{-1}$ is organic the rest being inorganic. Normal secondary treatment can only remove 1–$2\,mg\,Pl^{-1}$ and so there is a large excess of phosphorus that is discharged in the final effluent that gives rise to eutrophication in surface waters. New legislation requires effluents discharged into sensitive waters to incorporate phosphate removal to bring final effluent phosphorus concentrations to below $2\,mg\,Pl^{-1}$ (Section 8.3). Chemical precipitation is used to remove the inorganic forms of phosphate by the addition of a coagulant (lime, aluminium salts or iron salts).

The selection of a coagulant depends on a number of factors:

(a) influent P concentration;
(b) influent suspended solids concentration and alkalinity;
(c) cost of the chemical used at treatment plant;
(d) availability and reliability of chemical supply;
(e) sludge handling facilities at plant or cost of new system;
(f) disposal method of sludge and cost. For example, if sludge is disposed to agricultural land then iron or aluminium sludges are less favoured than those treated with lime;
(g) compatibility with other unit processes used at plant;
(h) the potential environmental impact of use coagulants.

Lime addition removes calcium ions as well as phosphorus from wastewater, along with any suspended solids. The lime ($Ca(OH)_2$) reacts first with the natural alkalinity in the wastewater to produce calcium carbonate which is primarily responsible for enhancing suspended solids removal:

$$Ca(HCO_3)_2 + Ca(OH)_2 \rightarrow 2CaCO_3 \downarrow + 2H_2O \tag{20.21}$$

After the alkalinity is removed calcium ions combine with the orthophosphate present under alkaline conditions (pH 10.5) to form insoluble and gelatinous calcium hydroxyapatite ($Ca_5(OH)(PO_4)_3$):

$$5Ca^{2+} + 4OH^- + 3HPO_4^{2-} \rightarrow Ca_5(OH)(PO_4)_3 \downarrow + 3H_2O \tag{20.22}$$

The lime dose required can be approximated at 1.5 times the alkalinity as $CaCO_3$. Neutralization may be required to reduce the pH before the aeration tank to protect the micro-organisms. This is done by injecting carbon dioxide to produce calcium carbonate although the carbon dioxide produced by the micro-organisms themselves is often sufficient to maintain a neutral pH in the aeration tank. A disadvantage is that large quantities of lime sludge are produced possibly doubling the normal volume of sludge requiring disposal.

Alum or hydrated aluminium sulphate ($Al_2(SO_4)_3 \cdot 14.3H_2O$) is widely used precipitating phosphates as aluminium phosphate ($AlPO_4$):

$$Al_2(SO_4)_3 \cdot 14.3H_2O + 2PO_4^{3-} \rightarrow 2AlPO_4 \downarrow + 3SO_4^{2-}\ 14.3H_2O \quad (20.23)$$

This reaction causes a reduction in the pH and the release of sulphate ions into the wastewater. Although the natural alkalinity accounts for some of the coagulant, the dosage rate of alum is a function of the degree of phosphorus removal required. The efficiency of coagulation falls as the concentration of phosphorus decreases. So for phosphorus reductions of 75%, 85% and 95% the aluminium to phosphorus ratio will have to be increased to approximately 13:1, 16:1 and 22:1, respectively. In practice, an 80–90% removal rate is achieved at coagulant dosage rates of between 50 and 200 mg l^{-1}. Sodium aluminate ($NaAlO_2$) can be used instead of alum but causes an increase in the pH and the release of sodium ions. Aluminium coagulants can adversely affect the microbial population in activated sludge, especially protozoa and rotifers, at dosage rates greater than 150 mg l^{-1}. However, in practice such high dosage rates have little effect on either BOD or suspended solids removal as the clarification function of the protozoa and rotifers is largely compensated by the enhanced removal of suspended solids by chemical precipitation.

Ferric chloride ($FeCl_3$) or sulphate ($Fe_2(SO_4)_3$) and ferrous sulphate ($FeSO_4 \cdot 7H_2O$), also known as copperas, are all widely used for phosphorus removal, although the actual reactions are not fully understood. For example, ferric ions combine to form ferric phosphate. Ferric chloride reacts slowly with the natural alkalinity and so a coagulant aid, such as lime, is normally added to raise the pH in order to increase the hydroxyl ion concentration as well as enhance coagulation overall. The ferric ion reacts with both the natural alkalinity and the lime to precipitate out as ferric hydroxide. The overall reaction is:

$$FeCl_3 + PO_4^{3-} \rightarrow FePO_4 \downarrow + 3Cl^- \quad (20.24)$$

The main design modifications to the activated sludge process that permit the addition of coagulants for the removal of phosphorus are shown in Fig. 20.10. In terms of efficiency, post-coagulation is by far the most effective phosphorus removal system. The coagulant is added after biological treatment, so that nearly all the phosphorus present has been hydrolysed to orthophosphate and so is potentially removable, but before final sedimentation which removes the chemical precipitate. A portion of the chemical floc is returned to the aeration tank along with the returned sludge. This results in the mixed liquor having a greater inorganic content (i.e. lower mixed liquor volatile

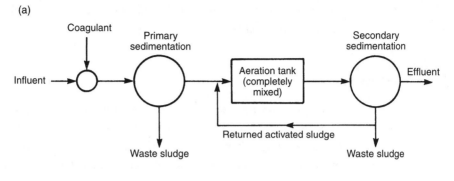

(a)

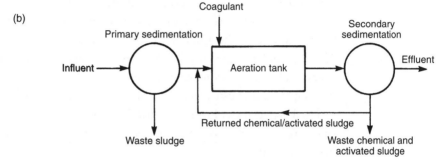

(b)

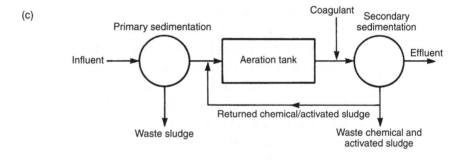

(c)

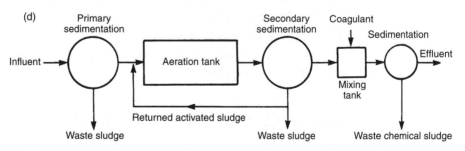

(d)

FIGURE 20.10 Phosphorus removal using coagulants. (a) Primary coagulation; (b) simultaneous coagulation; (c) post-coagulation and (d) post-precipitation.

suspended solids (MLVSS)), so the operational mixed liquor suspended solids (MLSS) must be increased to maintain adequate BOD removal. The problem of returning chemical sludge back to the aeration tank can be overcome by post-precipitation. The coagulant is added after final sedimentation but requires an extra mixing tank and sedimentation tank, making it essentially an advanced treatment process. Post-precipitation will also remove fine suspended solids, thereby increasing clarity of the effluent and reducing the final BOD.

20.6 ADSORPTION

Adsorption is a physical process where soluble molecules (adsorbate) are removed by attachment to the surface of a solid substrate (absorbent) primarily by van der Waals forces, although chemical or electrical attraction may also be important. Adsorbents must have a very high specific surface area and include activated alumina, clay colloids, hydroxides and adsorbent resins, with the most widely used being activated carbon. To be effective the surface of the adsorbent must be largely free of adsorbed material, which may require the adsorbent to be activated before use (McKay, 1996). Used both for water and wastewater treatment a wide range of organic materials are amenable to removal by adsorption, including detergents. In water treatment it is used to remove taste and odour-causing trace organic compounds as well as colour and other organic residual especially chlorination disinfection by-products, such as THMs. In waste-water treatment it is used to improve settleability of activated sludge and to remove toxic compounds.

The most widely used adsorbent is activated carbon, which can be produced by pyrolytic carbonization using a number of different raw materials including bituminous coal, lignite, peat and wood. Activated carbon has a highly porous structure resulting in specific surface areas of between 600 and 1500 $m^2 g^{-1}$. Particles are irregular in shape with a highly porous internal structure providing the large available surface area for adsorption. The rough external surface is ideal for the attachment of micro-organisms which can enhance adsorption by biological removal mechanisms. The overall rate of adsorption is dependent on particle size, which varies reciprocally with the square of the particle diameter. The adsorption rate also increases with increasing molecular weight of solute and decreases with increasing pH with adsorption very poor at pH >9.0. The rate of adsorption is also proportional to the square root of the time of contact (Casey, 1997). Activated carbon is used either as a powder being added to water as a slurry, or as granules which are housed in a simple filter or column through which contaminated water is passed.

20.6.1 POWERED ACTIVATED CARBON (PAC)

Powered activated carbon (PAC) offers unique design flexibility. It can be added to process water at various stages within the water and wastewater treatment processes. It can also be used on an intermittent basis that is particularly useful in controlling sea-sonal taste and odour problems in drinking water caused by algae, actinomycetes or fungi (Table 11.5). In water treatment PAC is fed directly into the water stream as a slurry either at the rapid mix stage of chemical coagulation or immediately prior to sand filtration. As the PAC must be removed from the treated water its addition at the coagulation stage ensures maximum contact time and mixing to occur with removal of the PAC by sedimentation and sand filtration. Not only does PAC result in additional sludge production but filter efficiency is also reduced due to increased head loss and reduced filter run times between backwashing. Unlike granular activated carbon (GAC) it is not possible to regenerate PAC due to contamination with other sludge materials. Typical dosage rates are between 5 and $10\,\mathrm{g\,m^{-3}}$, although at dosages $>20\,\mathrm{g\,m^{-3}}$ on continuous basis GAC columns become increasingly more economical. A key factor in the selection of PAC is bulk density as the higher the bulk density the greater the absorption capacity. Applications in wastewater treatment are mainly as an additive to activated sludge to improve floc structure to increase settleability. The influent wastewater is dosed at a rate of $50–300\,\mathrm{g\,m^{-3}}$, although the exact dosage has to be determined experimentally using laboratory or pilot-scale simulation. It is also used to enhance the natural removal of toxic compounds by the activated sludge microbial biomass. Micro-organisms will become acclimated to certain toxic compounds removing them primarily by adsorption. Trace residuals of these compounds can be removed from the final effluent by activated carbon treatment.

20.6.2 GRANULAR ACTIVATED CARBON (GAC)

GAC is easier to handle than PAC and is used for continuous applications in either a packed or expanded bed. Packed beds are only suitable for low turbidity waters due to clogging and can be designed either as a downward or upward flow system. Downward flow systems are operated as batch processes with all the carbon replaced once breakthrough occurs. Saturation of the carbon starts at the top of the bed and gradually moves downwards. Where they are also used for filtration as well as adsorption then the bed must be backwashed in the same way as a rapid sand filter. When this is done then the stratification of the activated carbon seen in other downward flow systems is lost.

The mass transfer zone (MTZ) is the area of the filter where adsorption occurs. As the adsorbent becomes saturated the MTZ is displaced downwards (Fig. 20.11). After contaminated water has passed the zone the concentration of solute remaining will have reached its minimum concentration with no further adsorption occurring below the MTZ within the bed. The MTZ continues to move downwards until breakthrough of the solute occurs. If the hydraulic loading is too high (i.e. the empty bed contact time (EBCT) is too short) then the depth of the MTZ will exceed the depth of the filter and the solute will escape.

Adsorption can be modelled using isotherms to predict the mass of solute removed per mass of adsorbent used versus concentration (e.g. the Freundlich and Langmuir isotherms). The Freundlich equation is most widely used:

$$\frac{X}{M} = k_{\mathrm{f}} C_{\mathrm{e}}^{1/n} \tag{20.25}$$

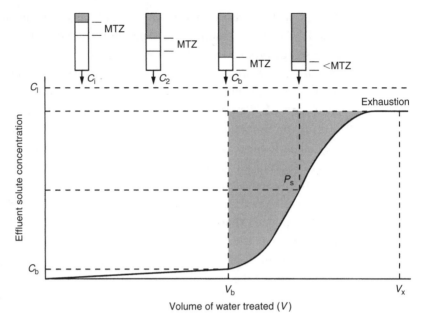

This can be linearized as:

$$\log \frac{X}{M} = \log k_f + \frac{1}{n} \log C_e \tag{20.26}$$

where X is the mass of adsorbate, M is the mass of adsorbent, C_e is the concentration of solute remaining at equilibrium, while k_f and n are constants derived from the adsorption isotherm by plotting X/M against C_e on log–log paper which produces a straight line with a slope $1/n$ while the y intercept is k_f. Remember the intercept of a log–log plot occurs at $x = 1$, not zero. This is normally done experimentally in the laboratory using PAC and for a given temperature relates the mass of solute adsorbed per unit mass of adsorbent to give the concentration of solute remaining in the final effluent. Typical values for key organic compounds at pH 7 are given in Table 20.4. The Freundlich isotherm can also be used to describe the adsorption of pathogens, such as viruses, onto soil or onto the aquifer matrix (Powelson and Gerba, 1995).

The Langmuir equilibrium adsorption isotherm is based on the concept of equilibrium in a monomolecular surface layer. It is expressed as:

$$\frac{X}{M} = \frac{abC_e}{1 + aC_e} \tag{20.27}$$

where a and b are constants. Equation (20.27) is more commonly expressed as:

$$\frac{1}{X/M} = \frac{1}{b} + \frac{1}{ab} \cdot \frac{1}{C_e} \tag{20.28}$$

where b is the amount of adsorbate needed to form a complete monolayer on the adsorbent surface and so increases with molecular size. The Langmuir equilibrium

TABLE 20.4   Values of K_f and $1/n$ for some common organic pollutants at neutral pH used in the calculation of the Freundlich isotherm

Organic pollutant	K_f (mg/g)	$1/n$
Chlorobenzene	93	0.98
Dibromochloromethane	63	0.93
Hexachlorobutadiene	360	0.63
Hydroquinone	90	0.25
α-naphtol	180	0.31
Nitrobenzene	68	0.43
Pentachlorophenol	150	0.142
p-xylene	85	0.16

Reproduced from Bitton (1998) with permission of John Wiley and Sons Inc., New York.

adsorption isotherm is produced by plotting $(C_e/(X/M))$ against C_e. As with the Freundlich isotherm, the Langmuir isotherm is only valid if the plotted line is straight. In reality the equation is of limited value due to adsorption of many different organics at the same time. These isotherms provide the basic data from which the volume of carbon required and the breakthrough times for GAC beds can be calculated (Bryant *et al.*, 1992). Organic compounds in drinking water rarely exist on their own, resulting in competition between adsorbable compounds for adsorption sites. Competition depends not only upon the strength of adsorption of the competing molecules but also their relative concentrations and type of activated carbon used. So in practice the volume of activated carbon required to remove a set amount of a particular compound is much greater when it is present in a mixture of adsorbable compounds (American Water Works Association, 1990).

Upflow systems are widely used and offer continuous operation. Saturated GAC is removed periodically from the base of the bed and fresh carbon added to the surface. This ensures continuous operation and maximum utilization of the adsorbent. Saturated carbon also tends to migrate to the base of expanded beds as the particles become denser allowing selective replacement of carbon. Expanded beds are self-cleansing and so can be used to treat turbid waters. With an upflow velocity >15 m h^{-1} and a grain size of 0.8–1.0 mm then a 10% expansion is obtained. Regardless of configuration, minimum contact times for GAC beds should be between 5 and 20 min when the media is clean (i.e. the EBCT). The design of GAC beds depends on having as low a turbidity in the water as possible. The volume required is calculated by the rate of carbon saturation which can only be carried out by pilot plant trials. In practice, a downflow batch GAC bed should have a minimum operational life of between 3 and 12 months before the carbon is removed for regeneration. Thermal regeneration is most widely employed, although other methods include solvent washing, acid or caustic washing, or steam treatment. The process is generally used after sand filtration to produce a very high quality water.

Microbial development on GAC is common and generally beneficial by controlling many taste and odour compounds by biodegradation. An advantage is that this biological action continues after the physical adsorption capacity of the carbon is exhausted. Where microbial activity is important then GAC is referred to as biological activated carbon (BAC). Pre-treatment of drinking water by ozonization reduces the size of large-molecular-weight organics making them generally more readily adsorbable and biodegradable by the attached microbial flora. However, in practice biological growth in GAC filters result in an increased rate of head loss build-up requiring an increased backwashing frequency. Where GAC is used to partially replace conventional media in rapid sand filters the GAC media can be rapidly worn down by the sharp sand during backwashing, especially if this includes air scouring. Selection of the correct size and type of sand is essential to achieve adequate media separation during backwash, which will also reduce erosion of the GAC particles.

20.7 ION EXCHANGE

Ion exchange is an adsorption process that employs the reversible interchange of ions of the same charge between a solid ion-exchange medium and a solution. Used primarily for water softening Ca^{2+} and Mg^{2+} cations are exchanged by Na^+ where the Na_2R is the ion-exchange medium with R the negatively charged polymer:

$$Ca^{2+} + 2NaR \leftrightarrow CaR_2 + 2Na^+ \tag{20.29}$$

$$Mg^{2+} + 2NaR \leftrightarrow MgR_2 + 2Na^+ \tag{20.30}$$

When all the exchange sites have been used the medium is regenerated by flushing with a 5–10% NaCl solution:

$$CaR_2 + 2Na^+ \leftrightarrow Ca^{2+} + 2NaR \tag{20.31}$$

$$MgR_2 + 2Na^+ \leftrightarrow Mg^{2+} + 2NaR \tag{20.32}$$

Ion-exchange water softening is carried out in a downflow fixed-bed reactor. The granular resin is housed in an enclosed metal tank rather similar to a pressurized rapid sand filter in design. The depth of the bed is between 0.8 and 2.0 m to avoid short-circuiting of the water, with flow rates $<1\,m^3min^{-1}m^{-3}$ of medium. Once performance falls off (breakthrough) the bed is backwashed with clean water to remove solids and then regenerated. Water used for softening must have a low turbidity and be free from organics which become adsorbed onto the medium inactivating sites.

Most exchange media are synthetic polymer resins, although naturally occurring zeolites which are sodium-alumino-silicates (e.g. analcite, clioptilolite and montmorillonite) are also used (Hill and Lorch, 1987). Exchange resins are available for the removal of a wide range of cations and anions. Cations are normally exchanged for Na^+ or H^+ ions while anions are exchanged for OH^- ions.

Hydrogen cation exchange media

$$2HR + Ca^{2+} \leftrightarrow CaR_2 + 2H^+ \tag{20.33}$$

Regeneration is by 2–10% sulphuric acid:

$$CaR_2 + 2H^+ \leftrightarrow 2HR + Ca^{2+} \tag{20.34}$$

Hydroxide anion exchange media

$$2ROH + SO_4^{2-} \leftrightarrow R_2SO_4 + 2OH^- \tag{20.35}$$

Regeneration is by 5–10% sodium hydroxide:

$$R_2SO_4 + 2OH^- \leftrightarrow 2ROH + SO_4^{2-} \tag{20.36}$$

While water softening remains the largest application of ion exchange, the process is also used to remove cations such as chromium, barium, strontium and radium, and anions such as nitrate, fluoride cyanide and humates. Different resins have a different order of affinity based on the relative concentrations of the ions. For example, quaternary ammonium resin ($NO_3^- > CrO_4^{2-} > Br^- > Cl^-$) and sulphonic acid resin ($Fe^{2+} > Al^{3+} > Ca^{2+}$) (Casey, 1997; Solt and Shirley, 1991).

Ion-exchange capacity of media is measured by the number of charges it can replace per unit volume expressed as equivalents ($eq\,m^{-3}$). This is done experimentally using a simple column containing a known volume of medium. The total quantity of ions exchanged is measured until the medium is exhausted. Normal capacities for water softening exchange resins are between 100 and $1500\,eq\,m^{-3}$. Design details and calculation of operational periods between regeneration are given in American Water Works Association (1990).

20.8 MEMBRANE FILTRATION

Membrane filtration is a highly sophisticated process that employs primarily synthetic polymeric membranes to physically filter out of solution under pressure minute particles including viruses and some ions. Conventional filtration can only deal effectively with particles larger than $10^{-2}\,mm$ so a range of synthetic membranes with very small pores is used to remove from water particles of any size down to $10^{-7}\,mm$ (Fig. 20.12). As with

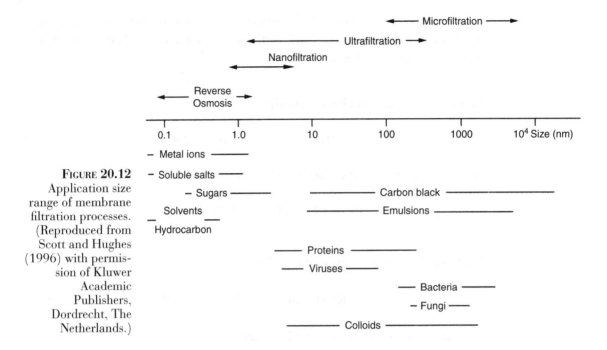

FIGURE 20.12 Application size range of membrane filtration processes. (Reproduced from Scott and Hughes (1996) with permission of Kluwer Academic Publishers, Dordrecht, The Netherlands.)

all filtration processes the size of the particles retained are approximately an order of magnitude smaller than the pore size of the filter. Membrane filtration is widely used for the advanced and tertiary treatment of potable and wastewaters (Fig. 20.13), and as the technology develops it will gradually replace many existing conventional treatment processes. Currently membrane filtration is only widely used as a replacement for the secondary settlement step after the activated sludge aeration tank (Gander *et al.*, 2000), and for separating the solid and liquid phases in the anaerobic digestion process (Wen *et al.*, 1999) (Section 20.8.5).

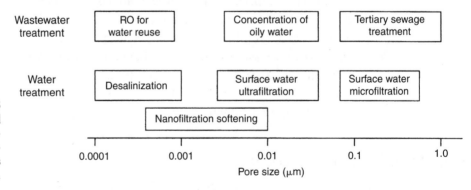

FIGURE 20.13 Principal water and wastewater treatment membrane filtration applications by pore size.

20.8.1 MICROFILTRATION

Microfiltration (MF) is a physical process marketed in a ready to use cartridge or modular housing that can remove particles between 0.05 and 5 μm in size. The membranes

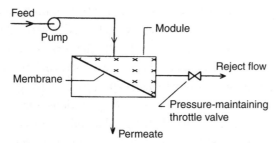

FIGURE 20.14 Schematic layout of a membrane filtration module. (Reproduced from Casey (1997) with permission of John Wiley and Sons Ltd, Chichester.)

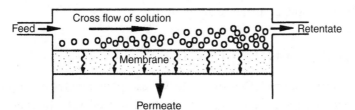

FIGURE 20.15 Cross-flow membrane separations. (Reproduced from Scott and Hughes (1996) with permission of Kluwer Academic Publishers, Dordrecht, The Netherlands.)

come in a variety of forms, such as tubular, capillary, hollow fibre and spirally wound sheets. The untreated water is pressurized to 100–400 kPa forcing the permeate (clarified liquid) through the membrane with the particulate matter retained (Fig. 20.14). Membranes used in MF are generally made from a thin polymer film with a uniform pore size and a high pore density (75–80%). The high density of pores results in a low hydrodynamic resistance allowing high flow rates. Most MF systems are not continuous but batch processes with the permeate produced continuously during the period of operation but the concentration of the retained solids or solutes increasing over time. Membranes require periodic backwashing with either water or gas under pressure to remove trapped solids from the micropores. There are many different designs of MF systems, but these can be classified according to the way the untreated water is introduced to the membrane. In dead-end systems the flow is perpendicular to the membrane with the retained particles accumulating on the surface of the membrane forming a filter cake. The thickness of the cake increases over time with the rate of permeation decreasing accordingly. When the filtration rate becomes too low the membrane module or cartridge is replaced or backwashed. Cross-flow is where the flow is introduced tangentially to the membrane surface that promotes self-cleansing ensuring longer operation periods between replacement or backwashing (Fig. 20.15).

MF can remove microbial cells, large colloids and small particles (<0.5 μm). It is widely used to remove chlorine-resistant pathogens, in particular *Cryptosporidium* oocysts and *Giardia* cysts. Bacteria are also removed depending on the particle size cut-off limit, while viruses are not removed unless they are associated with larger particulate matter. There is much interest in the use of MF as a replacement for conventional potable water treatment using processes, such as coagulation, flocculation, sedimentation and filtration processes (Yuasa, 1998). The process is already widely used as a point of use water treatment system in private homes. An important industrial application is the treatment of low-volume high-value process streams. For example, the clarification and sterile filtration of heat-sensitive solutions, such as removing yeast from cider and wine. Other uses include the production of pure water for the electronics, chemical, pharmaceutical and food industries; process solvent recovery in the chemical industry; removing heavy metal precipitates, small inert particles from process waters (e.g. glass) and algae. When used to reduce the suspended solids concentration in treated wastewaters then the BOD, chemical oxygen demand (COD) and turbidity of the effluent are all reduced.

20.8.2 ULTRAFILTRATION

Ultrafiltration (UF) is similar to MF except membrane micropores are much smaller and the pressure required to force the untreated water through is much greater (up to 3000 kPa). Widely used polymers for UF membranes include polysulphone, polyacrylonitrite, polyimide and cellulose acetate. Inorganic ceramic membranes are also used for certain industrial processes. The membranes are very thin (0.1–1.0 μm thick) and so must be supported on a stronger but highly porous layer (50–250 μm thick) that takes the form of either tubular, capillary or spirally wound. Each module is composed of a series of hollow membranes with the untreated water fed tangentially at high velocity to minimize blinding or fouling of the membrane surface. The permeate passes through the membrane and the concentrated particles are pumped to a waste tank. Owing to the smaller particles removed (0.001–0.02 μm) UF has different applications to MF with solvents and salts of low molecular weight able to pass through the membrane while larger molecules are retained. For that reason its major application is in the separation of macromolecules with molecules with a molecular mass of <1000 passing the membrane. In industry they are used to concentrate proteins, enzymes and hormones, while UF is used with food-processing waters to remove bacteria, viruses and cellular fragments (pyrogens). Colour and humic substances are removed from drinking water by UF, while macromolecules, such as dyes, can be recovered as can oil which is effectively separated from oil emulsions.

20.8.3 REVERSE OSMOSIS

Reverse osmosis (RO) uses a semi-permeable membrane that, unlike MF and UF membranes, does not have pores. Removal of salts and metal ions occur by diffusion-controlled transport, a mechanism where one molecule at a time diffuses through vacancies in the molecular structure of the membrane material. Osmosis is the movement of water (or any solvent) from a weak solution to a strong solution through a semi-permeable membrane. So if the membrane was placed between freshwater and saltwater, the solvent (i.e. pure water) will move through the membrane until the salt concentration on either side is equal. Only water can pass the membrane so the salts are retained. The movement of water across the membrane is caused by a difference in pressure and continues until the pressure in both solutions is equal limiting further passage. The pressure difference, which causes osmosis to occur, is known as the osmotic pressure. RO uses this principle to make the solvent move from the concentrated solution to the weak by exerting a pressure higher than the osmotic pressure on the concentrated solution, thus reversing the direction of flow across the membrane (Fig. 20.16). The osmotic pressure ψ of a solution is calculated as:

$$\psi = 1.12 \, (T + 273) \sum mi \quad \text{psi} \tag{20.37}$$

where ψ is the osmotic pressure (psi), T is the temperature (°C) and Σmi is the summation of the molarities of all the ionic and non-ionic constituents in solution. So by subjecting saline water to pressures greater than the osmotic pressure, pure water passes through the semi-permeable membrane and can be collected for use as drinking water. Commercial units use membranes made from cellulose acetate, triacetate and polyamide polymers. Multi-stage units are generally used with the best results from the use of brackish water, rather than seawater. There is a production of wastewater in which all the various contaminants are concentrated which has to be discharged and may be very polluting. For example, the disposal of concentrated brine solution is a major problem for desalinization systems.

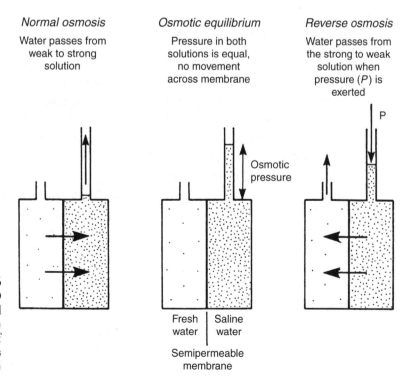

Normal osmosis
Water passes from weak to strong solution

Osmotic equilibrium
Pressure in both solutions is equal, no movement across membrane

Reverse osmosis
Water passes from the strong to weak solution when pressure (P) is exerted

P

Osmotic pressure

Fresh water | Saline water

Semipermeable membrane

FIGURE 20.16 Principles of the RO process. (Reproduced from Gray (1994) with permission of John Wiley and Sons Ltd, Chichester.)

The performance of the membrane depends very much on the quality of the raw water. For example, its life is prolonged if chlorine is removed by activated carbon pre-treatment, also if iron is removed from raw water using a catalysed manganous dioxide membrane filter, which oxidizes the iron to an insoluble form that can then be removed by a simple fibre filter. Performance of RO is defined in terms of solute retention R:

$$R = 100(1 - C_p/C_r)\% \tag{20.38}$$

where C_p is the solute concentration of the permeate and C_r is the solute concentration in the retentate (Scott and Hughes, 1996). RO is ideal for removing most inorganic and organic compounds with particles 0.0001–0.003 μm (1–30 Å) being removed. Metals, in particular aluminium, copper, nickel, zinc and lead, are removed with on average between 94% and 98% of the total dissolved solids also removed. It can also remove 85–90% of all organics including trihalomethanes (THMs), polychlorinated byphenyls (PCBs), pesticides, benzene, etc. So when used downstream of an activated carbon system then organics are essentially eliminated. Nitrate removal is generally poor (60–75%) so in order to remove all the nitrate present (99%) ion exchange is used upstream of the RO unit. The installation of ion-exchange systems will increase the capacity of the RO 10-fold, also making it much more efficient. Some problems have been reported due to excessive water pressure, or microbial degradation of the membrane material, causing membrane fracture and so treatment failure.

20.8.4 NANOFILTRATION

Nanofiltration (NF) is similar to RO and is a pressure-driven process that employs membranes that are capable of both physical sieving and diffusion-controlled transport. This is due to the recent development of thin film non-cellulose membranes. Compared to RO, NF systems operate at much lower pressures but yield higher flow rates of permeate. So while NF is not able to produce permeate of the same quality as RO, it is rapidly taking over many areas where UF was not sufficient. The molecular weight cut-off for NF is about 200 when a high sodium rejection (typical of RO) is not required but where other salts, such as Ca^{2+} and Mg^{2+}, and other divalent ions need to be removed. The particle size removal range for NF is 0.0005–0.005 μm. Other specific uses include the removal of colour, total organic carbon (TOC), humic acids and organic molecules.

20.8.5 PROCESS DESIGN

There are many different configurations of membrane filters used in the design of water and wastewater treatment systems. Most of the plant configurations described can be used with any of the membrane types. *Tapered plants* are widely used for industrial

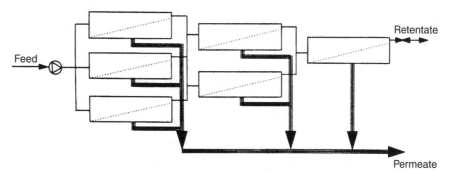

FIGURE 20.17 Schematic diagram of the arrangement of modules in a taper plant. (Reproduced from Scott and Hughes (1996) with permission of Kluwer Academic Publishers, Dordrecht, The Netherlands.)

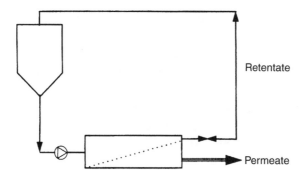

FIGURE 20.18 Schematic diagram of a simple batch plant arrangement. (Reproduced from Scott and Hughes (1996) with permission of Kluwer Academic Publishers, Dordrecht, The Netherlands.)

water treatment and comprise of a decreasing number of parallel modules known as an array (Fig. 20.17). The untreated liquid is fed into the first array and the permeate drained off while the retentate is passed onto a second, third and even fourth array. The retentate is finally discharged from the system via a pressure control value. The simplest type of configuration is the batch plant (Fig. 20.18). Here the permeate is continuously removed while the retentate is returned to the batch tank. The system continues to operate until the retentate concentration increases to such a concentration that filtration falls below the minimum acceptable rate. While this design is mainly used to clarify fruit juices by UF, it is also used to concentrate weak effluents with the retentate taken away for treatment. *Feed and bleed* systems allow a portion of the retentate to pass (bleed) into the next tank and so on until it leaves the plant via a control value. So this configuration allows the constant removal of both retentate and permeate (Fig. 20.19). It is widely used to process wastewater from paper pulp manufacture.

Two reactor designs are used in the activated sludge process (Section 17.1) to replace the post-settlement stage, which separates the final effluent from the microbial biomass, with membranes. External membranes are known as sidestream systems and behave exactly as the clarification step being an external unit (Fig. 20.20); while internal or

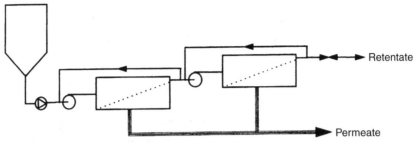

FIGURE 20.19 Schematic diagram of a food and bleed plant. (Reproduced from Scott and Hughes (1996) with permission of Kluwer Academic Publishers, Dordrecht, The Netherlands.)

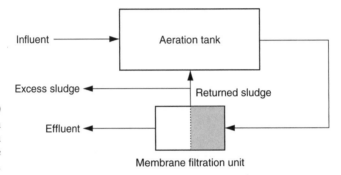

FIGURE 20.20 MBR system employing an external membrane filtration unit.

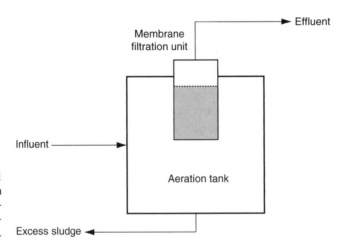

FIGURE 20.21 MBR system employing an internal membrane filtration unit.

submerged membranes are submerged within the aeration tank (Fig. 20.21 and Table 20.5). Membranes made from polymeric materials, such as acrylonitrite, polyethylene or polysulphone, and with a pore size of 0.1–0.4 μm are generally used. The membranes are configured as hollow fibre or flate sheet ensuring high flow rates per unit area of membrane (i.e. flux) of 0.5–1.0 m^{3}m^{2}day^{-1} at pressures as low as 0.1 bar (10^4 Pa). Known as membrane bioreactors (MBRs) these systems are very compact as they are operated at high MLSS concentrations (16 000–20 000 mg l^{-1}), with f/m ratios typically 0.02–0.4 and with long sludge ages. The long sludge age ensures

TABLE 20.5 Comparison of the relative advantages and disadvantages of submerged (internal) and sidestream (external) MBR systems

	Submerged MBR	*Sidestream MBR*
Capital costs	Higher	Lower
Aeration costs	Higher	Lower
Operating costs	Lower	Higher
Cleaning frequency	Higher	Lower
Footprint	Higher	Lower
Flux	Lower	Higher
Liquid pumping cost	Lower	Higher

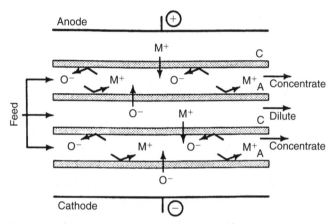

FIGURE 20.22 Separation by an electrodialysis cell. *C*: cation transfer membrane, *A*: anion transfer membrane. (Reproduced from Scott and Hughes (1996) with permission of Kluwer Academic Publishers, Dordrecht, The Netherlands.)

maximum endogenous respiration resulting in relatively low sludge production rates of $<0.3\,kg\,kg^{-1}$ BOD. MBR systems produce very high quality effluents with BOD $<5\,mg\,l^{-1}$, suspended solids $<2\,mg\,l^{-1}$ and a high removal efficiency of pathogens including bacteria and viruses. From an operational perspective the replacement of the secondary settlement step means that traditional separation problems associated with floc morphology and denitrification (Table 17.3) are no longer relevant.

20.8.6 ELECTRODIALYSIS

Electrodialysis is a membrane process that uses an electric field to separate ions of one charge from ions of an opposite charge. Ion selective membranes (ion-exchange membranes) have fixed charge groups bound into the polymer matrix to which mobile ions with the opposite charge are attracted and attached. The unit consists of numerous flat membrane sheets (up to 400) alternatively arranged as cation- and anion-exchange membranes forming cells 0.5–2.0 mm wide and sandwiched between an anode and cathode (Fig. 20.22). Anion membranes are permeable to anions and impermeable to cations, while cation membranes are permeable to cations and

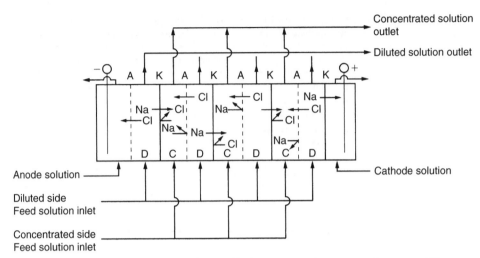

FIGURE 20.23 A typical electrodialysis cell. A: anion-exchange membrane and K: cation-exchange membrane. (Reproduced from Scott and Hughes (1996) with permission of Kluwer Academic Publishers, Dordrecht, The Netherlands.)

impermeable to anions. When an electrical current is applied the cations attempt to move towards the cathode and the anions towards the anode. The ion selective membranes restrict the movement of the charged ions with both cations and anions being captured resulting in alternating cells of ion-enriched solution (concentrate) and ion-depleted solution (diluate). This process has been widely used for the treatment of brackish groundwater (Fig. 20.23), although it is not suitable for strong saline solutions, such as seawater. It is also used in metal plating and pharmaceutical industries for the pre-treatment of wastewaters.

REFERENCES

Ainsworth, R. G., Calcutt, T., Elvidge, A. F., Evins, C., Johnon, D., Lack, T. J., Parkinson, R. W. and Ridgway, J. W., 1981 *A Guide to Solving Water Quality Problems in Distribution Systems*, Technical Report 167, Water Research Centre, Medmenham.

American Water Works Association, 1990 *Water Quality and Treatment: A Handbook For Community Water Supplies*, McGraw-Hill, New York.

Bitton, G., 1998 *Formula Handbook for Environmental Engineers and Scientists*, Wiley-Interscience, New York.

Bryant, E. A., Fulton, G. P. and Budd, G. C., 1992 *Disinfection Alternatives for Safe Drinking Water*, Van Nostrand Reinhold, New York.

Casey, T. J., 1997 *Unit Treatment Processes in Water and Waste Water Engineering*, John Wiley and Sons, Chichester.

Gander, M., Jefferson, B. and Judd, S., 2000 Aerobic MBRs for domestic wastewater treatment: a review with cost considerations. *Separation and Purification Technology*, **18**, 119–30.

Gray, N. F., 1994 *Drinking Water Quality: Problems and Solutions*, John Wiley and Sons, Chichester.

Haarhoff, J. and van Vuuren, L., 1993 *Dissolved Air Flotation: A South African Design Guide*, Water Research Commission, Pretoria.

Hill, R. and Lorch, W., 1987 Water purification. In: Lorch, W., ed., *Handbook of Water Purification*, Ellis Horwood, Chichester.

McKay, G. (ed.), 1996 *Use of Adsorbents for the Removal of Pollutants from Wastewaters*, CRC Press, Boca Raton, FL.

Metcalf and Eddy, 1991 *Wastewater Engineering*, McGraw-Hill, New York.

Powelson, D. K. and Gerba, C. P., 1995 Fate and transport of micro-organisms in the vadose zone. In: Wilson, L. G. *et al.*, eds, *Handbook of Vadose Zone Characterisation and Monitoring*, Lewis Publishers, Boca Raton, FL.

Sayer, C. N., McCarty, P. L. and Parkin, G. F., 1994 *Chemistry for Environmental Engineers*, McGraw-Hill, New York.

Scott, K. and Hughes, R. (eds), 1996 *Industrial Membrane Separation Technology*, Blackie Academic and Professional, London.

Solt, G. S. and Shirley, C. B., 1991 *An Engineer's Guide to Water Treatment*, Avebury Technical, Aldershot.

Talmadge, W. P. and Finch, E. B., 1955 Determining thickener unit areas, *Industrial Engineering Chemistry*, **47**, 38–41.

Wen, C., Huang, X. and Qian, Y., 1999 Domestic wastewater treatment using an anaerobic bioreactor coupled with membrane filtration. *Process Biochemistry*, **35**, 335–40.

Yuasa, A., 1998 Drinking water production by coagulation-microfiltration and adsorption-ultrafiltration, *Water Science Technology*, **37** (10), 135–46.

FURTHER READING

Adham, S. S., Jacangelo, J. G. and Laine, J. M., 1996 Characteristics and costs of MF and UF plants, *Journal of the American Water Works Association*, **88**, 22–31.

Hyde, R. A., 1989 Application of granular activated carbon in the water industry, *Journal of the Institution of Water and Environmental Management*, **3**, 174.

Kunikane, S., Magara, Y., Itoh, M. and Tanaka, O., 1995 A comprehensive study on the application of membrane technology to the public water supply, *Journal of Membrane Science*, **102**, 149–54.

Madaeni, S. S., 1999 The application of membrane technology for water disinfection, *Water Research*, **33**, 301–8.

Najim, I. M., Soenyink, V. L., Lykins, B. W. and Adams, J. Q., 1991 Using powered activated carbon: A critical review, *Journal of the American Water Works Association*, **83**, 65.

Newton, D. and Solt, G., 1994 *Water Use and Reuse*, Institution of Chemical Engineers, Rugby.

Chapter 21 Sludge Treatment and Disposal

21.1 Sludge Characteristics

The removal of the settleable fraction of raw sewage at the primary settlement stage, and of the settleable solids produced by biological conversion of dissolved nutrients into bacterial cells at the secondary stage, continuously produces a large quantity of concentrated sludge. While the liquid fraction of the wastewater can be fully treated and disposed of safely to surface waters, the accumulated sludge has to be further treated and finally transported from the site for disposal. Sludge separation, treatment and disposal represents a major capital and operational cost in sewage treatment, with dewatering and disposal costs for a medium-sized activated sludge plant representing as much as 50% of the initial capital and 65% of the operating costs.

At moisture contents >90% sludges behave as liquids, while below 90% they behave as non-Newtonian fluids exhibiting plastic rather than viscous flow (Fig. 21.1). In

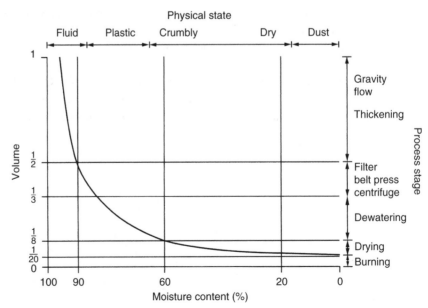

FIGURE 21.1 Relationship between moisture content and the volume of sewage sludge as produced by various stages of sludge processing. (Reproduced from Best (1980) with permission of the Chartered Institution of Water and Environmental Management, London.)

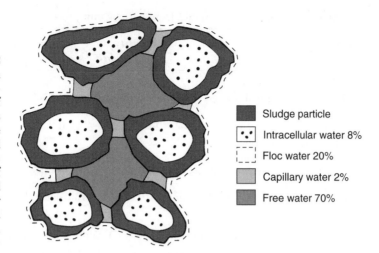

FIGURE 21.2
Schematic diagram of a sludge floc showing the association of the sludge particles with the available water. (Reproduced from Best (1980) with permission of the Chartered Institution of Water and Environmental Management, London.)

■ Sludge particle

Intracellular water 8%

Floc water 20%

Capillary water 2%

Free water 70%

sludges with a moisture content of >95% most of the water is in a free or readily drained form (70%) while the remainder is bound to the sludge and more difficult to remove (Fig. 21.2). While most of the free water can be removed by gravity settlement, some of the bound water must also be removed if the sludge is to be mechanically handled rather than pumped. This is normally done by coagulation using chemicals and dewatering equipment which removes moisture by altering the particle formation of flocs and the cohesive forces that bind the particles together, thus releasing floc and capillary water.

The type of sludge produced depends on a number of factors, such as the type of sludge separation and the treatment processes employed, which are really a function of the size of the treatment plant and wastewater characteristics. Industrial wastewaters produce very different types of sludge which can be broadly categorized as either organic or inorganic but both of which may contain toxic materials, especially heavy metals. The presence of gross organic solids in sludge (e.g. offal from meat-processing wastes), pathogenic organisms or toxic compounds (e.g. pharmaceutical and metal industries) all contaminate the sludge and so restrict the disposal options. Primary or raw sludge is the settleable fraction of raw or crude sewage and is a rather unpleasant smelling thick liquid, which is highly putrescible with a moisture content of between 94% and 98%. Secondary sludges contain the solids washed out of the biological treatment units, so either comprise of wasted activated sludge or sloughed microbial film. The secondary sludges are more stabilized than primary sludge with percolating filter sludge (humus) far more stabilized than wasted activated sludge having a dry solids (DS) content of 6–8% compared with 1.5–2.5% for activated sludge (Table 21.1). It is normal practice to mix primary and secondary sludges together for treatment and disposal. The mean flow of wet sludge produced at a conventional treatment plant and requiring treatment will be between 1% and 2% of the flow of raw sewage. The daily production of sludge can be estimated as:

$$Q(K_1 S + (1 + K_2) Y b K_3) \text{ kg solids day}^{-1} \qquad (21.1)$$

TABLE 21.1 Sludge production characteristics from various unit processes

Process stage	Quantity (g cap^{-1} day^{-1})	Solids content (% by weight)	Volume (l cap^{-1} day^{-1})
Primary sedimentation	44–55	5–8	0.6–1.1
Biofiltration of settled sewage	13–20	5–7	0.2–0.4
Standard-rate activated sludge pre-settled sewage	20–35	0.75–1.5	1.3–4.7
Extended aeration raw sewage	22–50	0.75–1.5	1.7–6.7
Tertiary sand filtration	3–5	0.01–0.02	15.0–50.0
Phosphorus precipitation (Al or Fe)	8–12	1–2	0.4–1.2

Reproduced from Casey and O'Conner (1980) with permission of Enterprise Ireland, Dublin.

where Q is the mean flow of sewage (m^{3}d^{-1}), S and b the mean concentration of suspended solids and biological oxygen demand (BOD), respectively (mg l^{-1}), K_1 and K_2 the fraction of suspended solids (e.g. 0.6) and the fraction of BOD (e.g. 0.3) removed at the primary sedimentation stage, respectively, and K_3 the fraction of BOD removed at the biological oxidation stage (e.g. 0.90–0.95), while Y is the sludge yield coefficient for the conversion of BOD to secondary sludge in the biological oxidation unit (i.e. Y is 0.5–1.0 for activated sludge and 0.3–0.5 for single-pass percolating filter systems).

In Germany alone some 100 million tonnes of liquid sewage sludge has to be processed annually producing 2.75 million tonnes of dry solids. Although raw liquid sludge can be disposed directly either to land or sea, it is normal practice for the sludge to receive further treatment in the form of thickening and dewatering, which may include chemical conditioning, to reduce the volume of the sludge. The putrescible nature of sludge is also a problem especially if it has to be stored before final disposal. This is overcome by digestion or lime stabilization to make the sludge more stable (i.e. less likely to undergo further decomposition on storage) before final disposal (Fig. 21.3). Such treatment will alter the physical and chemical nature of the sludge which is characterized by its water content and stability.

21.2 SLUDGE TREATMENT

The treatment of sewage sludge follows a general sequence. The water content is reduced by some form of thickening which is either followed by a stabilization process and secondary thickening, and/or chemical conditioning and a dewatering process to reduce the water content even further (Fig. 21.3). The sludge is then ready for disposal (Metcalf and Eddy, 1991).

21.2.1 THICKENING

Untreated sludges from the primary and secondary sedimentation tanks have high water contents (Table 21.1) and in order to reduce the volume of sludge handled in the stabilization or dewatering processes the sludge needs to be concentrated or thickened.

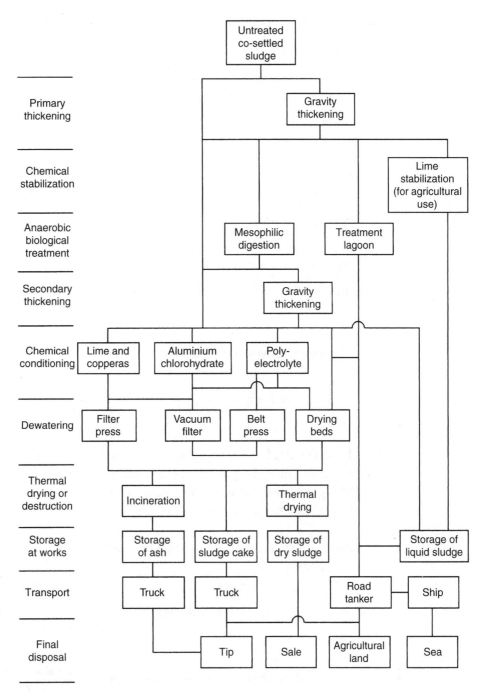

FIGURE 21.3
Process selection
for the disposal of
sewage sludge.
(Reproduced from
Hall and Davis
(1983) with permis-
sion of the Water
Research Centre plc,
Medmenham.)

Thickening is achieved by physical means, such as flotation, centrifugation, lagooning but most commonly by gravity settlement. Gravity thickeners can increase the sludge concentration in raw primary sludge from 2.5% to 8.0% resulting in a three-fold decrease in sludge volume, while a five-fold decrease in the volume of wasted activated sludge is not uncommon being thickened from 0.8% to 4.0% solids (Table 21.2).

TABLE 21.2 Typical concentrations of unthickened sludges and solids loadings for gravity thickeners

Type of sludge	Sludge concentration (%)		Solids loading for gravity thickeners (kg m^{-2} day^{-1})
	Unthickened	Thickened	
Separate			
Primary sludge	2.5–5.5	8–10	100–150
Percolating filter sludge	4–7	7–9	40–50
Activated sludge	0.5–1.2	2.5–3.3	20–40
Pure oxygen sludge	0.8–3.0	2.5–9.0	25–50
Combined			
Primary and percolating filter sludge	3–6	7–9	60–100
Primary and modified aeration sludge	3–4	8.3–11.6	60–100
Primary and air-activated sludge	2.6–4.8	4.6–9.0	40–80

Reproduced from Metcalf and Eddy (1991) with permission of McGraw-Hill Inc., New York.

The specific gravity (sg) of sludge is estimated using Equation (21.2):

$$\text{Sludge sg} = \frac{100}{(\% \text{ sludge solids/solids sg}) + (\% \text{ water/water sg})} \tag{21.2}$$

EXAMPLE 21.1 A sewage sludge has a solids content of 2%, with the solids having a specific gravity (sg) of 1.35. So 1 tonne of sludge will contain 20 kg of solids and 980 kg of water. Calculate the volume occupied by 1 tonne of sludge with 2% DS before and after dewatering to 35% DS.

(a) Calculate the volume occupied by 1 tonne of sludge at 2% DS using Equation (21.2):

$$2\% \text{ DS sludge (sg)} = \frac{100}{(2/1.35) + (98/1)} = 1.005$$

$$\text{Volume } (V) \text{ occupied by sludge (m}^3) = \frac{\text{Total weight of sludge (kg)}}{(\text{sg of sludge} \times 1000)} \tag{21.3}$$

So the volume occupied by 1 tonne of sludge at 2% DS

$$V = \frac{1000}{1.005 \times 1000} = 0.995 \, \text{m}^3$$

(b) Once dewatered to 35% DS, the 20 kg of solids would now be associated with 57.14 kg of water making a total sludge weight of 77.14 kg.

$$35\% \text{ DS sludge sg} = \frac{100}{(35/1.35) + (65/1)} = 1.100$$

So the volume occupied by 77.24 kg of sludge at 35% DS

$$V = \frac{77.14}{1.100 \times 1000} = 0.07\,\text{m}^3$$

The volume now occupied by the dewatered sludge (35% DS) is only 7% of the volume occupied by the sludge at 2% DS.

Gravity thickening takes place in a circular tank that is vaguely similar in design to a primary sedimentation tank (Fig. 21.4). The dilute sludge is fed into the tank where the particles are allowed to settle over a long period, which may be several days. The concentrated sludge is withdrawn from the base of the tank and pumped to the digesters or dewatering equipment. Sewage sludge is very difficult to dewater due to the small size of the particles resulting in small interstices that retain the water. Some of the water in the sludge exists as a gel and combined with strong capillary and electrostatic forces, water separation is difficult. Settlement is enhanced by slowly stirring the sludge with a rotating set of vertical blades or rods, which are spaced approximately 100 mm apart. The rate of rotation is fairly critical with optimum thickening occurring between peripheral blade velocities of 0.5 and 3.0 m min^{-1}. The stirrer, which looks like a garden fence, gives its name to the unit that is universally known as a 'picket-fence thickener'. The sludge is thickened by the weight of the particles compressing the sludge at the base of the tank forcing the water out, while the picket-fence stirrer enhances particle settlement by encouraging particle size to increase by enhanced flocculation. The stirrer enhances sludge consolidation by aiding water release by forming channels, preventing solids forming bridges within the sludge and by releasing gas bubbles formed by anaerobic microbial activity. Such units can either be operated as a batch or continuous process.

The liquid released from the sludge will contain more organic material than the incoming raw sewage and so it must be returned to the inlet of the treatment plant and be recycled. Because of the strength and the volume of this liquid, it must be taken into account in the organic loading to the plant.

21.2.2 STABILIZATION

Once the sludge has been thickened two options are available. It can be further dewatered to a solids content of between 20% and 35% or it can undergo stabilization

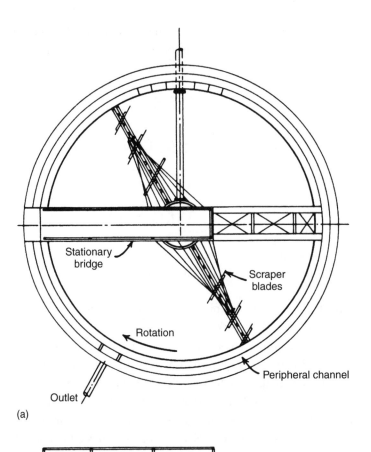

(a)

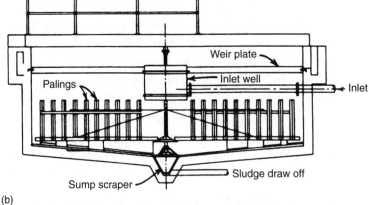

FIGURE 21.4
Picket-fence thickener in (a) plan and (b) cross section. (Reproduced from Hammer (1977) with permission of John Wiley and Sons Inc., New York.)

(b)

before the dewatering stage. Sludges are stabilized to prevent anaerobic breakdown of the sludge on storage (i.e. putrefaction) so producing offensive odours. There are other advantages depending on the stabilization process selected, including destruction of pathogens, partial destruction of sludge solids, increase in the concentration of soluble nitrogen and improved flow characteristics. There are three categories of stabilization process: biological, chemical and thermal. Each process prevents the utilization of the volatile and organic fraction of the sludge during storage by different effects. There is

no standardized test for sludge stability and the European Union (EU) defines a stabilized sludge as simply as one that has undergone biological or chemical treatment,
or long-term storage. However, sewage sludge can only be considered fully stabilized
when it has been humified (i.e. fully decomposed to humic substances which are non-
putrescible, odourless and which only degrade further very slowly) (Bruce, 1984).

Biological stabilization processes are the most widely practiced, resulting in the utilization of the volatile and organic fraction of the sludge. The resultant sludge has a
reduced volume, is odourless and has a higher solids content. The most common stabilization method for medium- to large-sized treatment plants is heated anaerobic
digesters, while at small works cold anaerobic digestion in tanks or lagoons is most
common (Section 19.1). Anaerobic digestion utilizes up to 40% of the organic matter present in raw sludge resulting in an increased nitrogen concentration of up to 5%
of the dry solids, of which 70% is in the form of ammonical nitrogen. The sludge
entering digesters does not have to be pre-thickened, but if it is then the resultant
digested sludge will have a higher dry solids and nitrogen content and there will be a
saving in digester capacity. Thickening is required after digestion and can be
achieved by gravity settlement. Lagoons can also be used with settlement taking
place over long periods with further digestion taking place resulting in a highly humified sludge with 8–10% DS. The manurial value of lagooned sludge is reduced in terms
of nitrogen and phosphorus content, which is leached into the supernatant, with the
proportion of total nitrogen present as ammonical nitrogen being reduced at <25%.

Sludge can also be stabilized by continuing oxidation of waste activated sludge over a
prolonged period. Similar to the activated sludge process, the sludge is mixed and aerated until fully stabilized. The basic objective of aerobic digestion is to produce a biologically stable sludge, and at the same time to reduce its mass and volume. The final
product should be a mineralized sludge with good settleability characteristics that
can be easily thickened and dewatered. The process is most widely used at smaller
treatment plants, as unlike anaerobic digestion there is no recovery of energy making
aerobic digestion comparatively expensive to operate due to the high energy costs
associated with aeration and mixing, although in capital terms it is relatively cheap
and easy to operate. Aerobic digestion is most cost-effective at plants with a design
treatment capacity of $<20\,000\,\text{m}^3\,\text{day}^{-1}$ (Water Environment Federation, 1992). The
process is inserted between the activated sludge tanks and dewatering processes, so
that it also provides a degree of sludge equalization giving a more uniform product
and allowing better control over sludge production. It is a low odour process while
aerobic conditions are maintained producing a supernatant that is largely oxidized,
although often with a high suspended solids concentration, that is returned to the
inlet of the plant. Stabilization in aerobic digestion is defined as a specific oxygen
uptake rate (SOUR) $\leqslant 1\,\text{mg}\,O_2\,h^{-1}\,g^{-1}$ (EPA, 1989). There is a decrease in pH during
aerobic digestion due to nitrification that can inhibit digestion so that pH control to
6.5 may be necessary. Performance is directly related to microbial activity which in
turn is largely dependent on temperature. So the oxidation rate falls dramatically

during the winter unless the digester is heated. The reduction in the biodegradable solids concentration follows a first-order reaction rate according to Equation (21.4):

$$\frac{dM}{dt} = -k_d M \tag{21.4}$$

where M is the concentration of biodegradable volatile solids (BVS) at time t, dM/dt the rate of change of the BVS and k_d the reaction rate constant (day^{-1}), which approximates to $0.1 \, \text{day}^{-1}$ at 20°C and $0.15 \, \text{day}^{-1}$ at 30°C. The size of an aerobic digestion tank can be approximated from Equation (21.5):

$$V = \frac{Q_0 X_i}{X(k_d + 1/\phi_c)} \tag{21.5}$$

where V is the aerobic digestion tank volume (l), Q_0 the influent sludge flow (l day^{-1}), X_i the influent BVS (mg l^{-1}), X the average BVS in the digester (mg l^{-1}) and ϕ_c the sludge residence time (SRT) (day^{-1}).

Another biological stabilization process is composting, with or without a bulking agent or recycled material, which can convert sewage sludge into a useful soil conditioner and fertilizer (Section 18.5).

Lime stabilization is the commonest form of chemical stabilization. Lime is added to untreated sludge until the pH is raised to above 11 creating an environment unsuitable for the survival of micro-organisms. Pathogens are readily killed by the process with lime stabilization at pH 12 for 3 h giving a higher reduction in pathogens than anaerobic digestion. Lime stabilization does not reduce the organic matter or provide permanent stabilization, but prevents putrefaction so long as the high pH is maintained. However, while the high pH suppresses the emission of volatile sulphides and fatty acids, the emission of amines and ammonia is increased making lime-treated sludge less offensive than raw sludge but certainly not odourless. The elevated pH is normally maintained for several days or until incorporated into the soil. Obviously lime addition has advantages to the farmer if sludge is disposed to agricultural land, while the addition of lime can help in the dewatering process. Two approaches to lime addition are used, the addition of hydrated lime to non-dewatered sludge and the addition of quicklime to dewatered sludge. The dosage is expressed as grams of lime per kg dry sludge solids (g kg DS^{-1}). The average dosage for a primary sludge is between 100 and 200 g Ca(OH)$_2$ kg DS^{-1} so the mass of solids will increase significantly after treatment. The pH of lime-treated sludges falls with time if the initial dosage is not sufficient, so it is important that sufficient lime is added to maintain the pH for the required period.

Thermal or heat treatment is a continuous process that both stabilizes and conditions sludges by heating them for short periods (30 min) under pressure. This releases

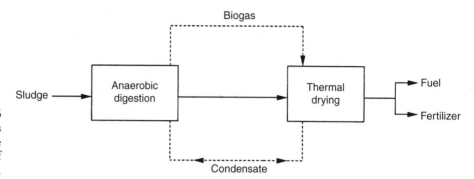

FIGURE 21.5
Schematic process
flow diagram for the
thermal drying of
digested sludge.

bound water allowing the solids to coagulate, while proteinaceous material is hydro-lysed resulting in cell destruction and the release of soluble organic compounds and ammonical nitrogen. The Zimpro process heats the sludge to 260°C in a reactor ves-sel at pressures up to 2.75 MN m^{-2}. The process is exothermic which results in the operating temperature rising. The solids and liquid separate rapidly on cooling with up to 65% of the organic matter being oxidized. The process sterilizes the sludge, practically deodorizes it and allows dewatering to occur mechanically without the use of chemicals. Owing to the disadvantages of capital and operating costs, operational difficulties and the large volume of very strong waste liquor produced, there are very few such plants operating in Europe. There are many improved designs becoming available. Thermal drying is normally now combined with anaerobic digestion. The digestion process produces the biomass and energy to run the drier, and the energy recovered from the drying process is used to heat the digester (Fig. 21.5). Modern thermal drying units produce small quantities of final product in the form of a stable granular material similar in size to artificial fertilizer allowing it to be easily handled by standard agricultural spreading equipment. With a dry solids content of between 90% and 95% it can be bagged and stored for long periods without problems. The product is odourless and can be used for a wide range of disposal options primarily as a fertilizer but also as a low sulphur (0.5%) and greenhouse gas neutral (0.5% CO_2) fuel. The main disadvantages are the high capital cost of thermal drying and the risk of combustion if operating conditions are not carefully controlled.

21.2.3 DEWATERING

Dewatering is a mechanical unit operation used to reduce the water content of sludge to obtain a solids concentration of at least 15%, usually much more. It is nor-mally preceded by thickening and conditioning, which is the addition of chemicals to aid flocculation and water separation, and may be followed by further treatment (Section 20.2). This reduces the total volume of sludge even further so reducing the ultimate transportation cost of disposal. The resultant sludge is a solid, not a liquid, and so can be easily handled by conveyers or JCB tractors although experience has shown that the dried sludge, known as cake, is more easily handled at solids concentrations of >20%. Its solid nature makes it suitable for many more disposal options than liquid sludge. The ratio of primary to secondary sludges is a major factor controlling dewatering, with primary sludge on its own capable of being dewatered to 35–45% DS but when mixed with secondary sludge this falls to a maximum of 17–20% DS.

The process is a physical one involving filtration, squeezing, capillary action, vacuum withdraw, centrifugal settling and compaction. The most widely used European methods include drying beds, lagoons, filter and belt presses, vacuum filtration, centrifugation, and reed beds.

Sludge drying beds

Drying beds are the cheapest and simplest form of dewatering. They are mainly restricted to smaller treatment plants due to the area of land required and the problems with odours and flies. Sludge drying beds are shallow tanks with a system of underdrains covered with layers of graded filter medium, normally a 100 mm of pea gravel covered with a 25 mm layer of sand (Fig. 21.6). The floor of the bed slopes towards the outlet at a minimum gradient of 1 in 60, and the sludge is pumped onto the bed to a depth of between 150 and 300 mm. Dewatering occurs mainly by drainage which is rapid for the first day or two and then progressively decreases until the solids have become so compacted on to the surface of the medium that drainage ceases. The surface liquid is decanted off and returned to the inlet of the treatment plant, so that evaporation can take over as the major dewatering process. Evaporation is affected by the weather in particular the wind, humidity and solar radiation, and at this stage rainfall will dilute the solids concentration. As the sludge dries it cracks which increases the surface area of the sludge thus increasing the evaporation. When the cracks reach the support medium, rainfall is no longer a problem as it can drain through the sludge via the cracks to the medium below. At this stage the sludge can be lifted either manually or by machine. The performance of drying beds depends on the type of sludge, the initial solids content, the period of drying, the porosity of the medium, loading rate and weather conditions. Solids contents of up to 40% are possible, although 35% is a more commonly attained value. They can only be used when

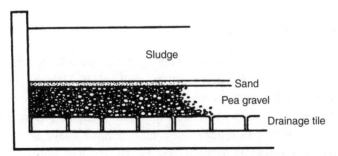

FIGURE 21.6 Cross section through a typical sludge drying bed. (Reproduced from Institution of Water Pollution Control (1981) with permission of the Chartered Institution of Water and Environmental Management, London.)

drying conditions are good, which in the UK limits their use to 6–7 months of the year. In hot weather 40% solids can be obtained after storage for 10–15 days but periods of 30–40 days are more usual.

Sludge is not usually conditioned by the addition of coagulants prior to application to drying beds, although aluminium chlorohydrate and polyelectrolytes do improve drainage characteristics, so that more water is able to escape via the underdrains during the first few days after application. The total surface area of bed required is related to the population although local factors, such as average rainfall, the dry solids concentration of the sludge and its filterability are also important. As a rough guide 0.12–0.37 m^2 of bed are required per head of population for primary sludge with 50–60% less area required for digested sludge. Secondary sludges (humus and activated sludge) are more difficult to dewater, and conditioning is required before application to drying beds.

Filter presses

Filter presses are comprised of a series of metal plates, between 50 and 100 in number, suspended either from side bars (side-bar press) or an overhead beam (overhead beam press). Each plate is recessed on both sides and supported face-to-face so that a series of chambers are formed when the plates are closed. Filter cloths are fitted or hung over each plate which are then closed with sufficient force, using either hydraulic rams or powered screws, to form sealed chambers. The sludge is conditioned before being pumped into the chamber under pressure ($\sim$700 kN m^{-2}) via feed holes in each plate. The pressure builds up and filtration occurs with the filtrate passing through the filter cloths and leaving the chamber via ports in the plates (Fig. 21.7). Pumping of sludge into the chambers continues until the chamber is filled with cake and the flow of filtrate has virtually stopped. Filter presses are normally situated on the first

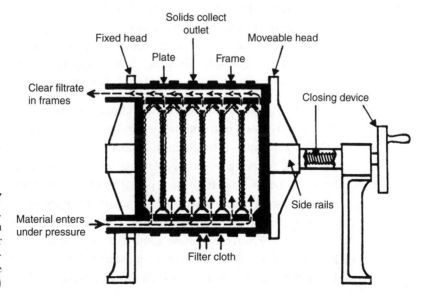

FIGURE 21.7
Sludge filter press. (Reproduced from Casey and O'Conner (1980) with permission of Enterprise Ireland, Dublin.)

floor of a specially constructed building that allows access for trailers below. The pressure is released to the press and the plates separated allowing the cake to drop out into the trailer parked directly below the press. In this batch process the operational cycle can vary from 3 to 14 h with filling and pressure build-up taking between 0.2 and 1.5 h, filtration under maximum pressure from 2 to 12 h and cake discharge 0.1–0.5 h, although cycles normally take between 3 and 5 h to complete. Much work has been done to try and reduce the time taken for the second phase of the cycle, which is filtration under pressure. Important factors affecting the rate of filtration include the choice of filter cloth, pump pressure, filterability of dry solids content of sludge and condition of filter cloth.

The process is primarily used for raw sludge, being equally successful with digested, mixed and thermally conditioned sludges. The final cake is quite thick, 25 or 38 mm, with a solids concentration of 30–45%. The advantage of the process is that it produces a drier cake than other dewatering devices, which results in a lower volume of sludge cake and reduced handling costs. The filtrate contains a relatively small concentration of suspended solids, although this is dependent on the type of sludge and filter cloth used. Filter cloths need to be washed occasionally between cycles using a high-pressure water jet. The frequency depends on the suitability of the cloth used and the chemical conditioner selected. For example, cloths used to filter sludge conditioned with lime and copperas should give up to 20 cycles between washing, while organic polyelectrolytes permit 25–30 cycles.

Conditioners

Conditioners aid the dewatering process by improving the filtration characteristics of sludges by increasing the degree of flocculation of the sludge particles so that the absorbed water can be more easily removed. It also prevents small particles clogging the filter cloths in filter presses. Although expensive, the use of chemical conditioners are cost-effective due to the increased solids content of the sludges produced which reduces the sludge volume that has to be disposed. There are numerous chemical conditioners available including lime, copperas, aluminium chlorohydrate, organic polymers or polyelectrolytes, ferric chloride and many more (Section 20.2).

Lime ($Ca(OH)_2$) is nearly always used in conjunction with copperas (ferrous sulphate $FeSO_4 \cdot 7H_2O$) unless iron salts are present in the sludge when lime can be used on its own. The combination is used almost exclusively in association with filter presses, not being suitable for belt presses or centrifuges which required a more rapid flocculation that can only be obtained from using polyelectrolytes. Copperas is kept in a crystalline form and is dissolved as required. It is an acidic and highly corrosive liquid and special precautions have to be taken with its use. The lime is delivered by bulk tanker and is normally kept in a silo ready for use. Lime is usually added after the copperas with doses of 10% copperas and 20% lime (dry solids) used for raw sludges, and 40% copperas and 30% lime for digested sludges. The large weights of chemicals used results in a significant increase in the weight of solid cake that has to be disposed, in the case of lime and copperas this can be as much as 50% of the solids in the final cake. Aluminium chlorohydrate ($Al_2(OH)_4Cl_2$) is just one of the aluminium salts used for conditioning sludge prior to dewatering, and is delivered to the site as a concentrated solution which needs to be diluted before use. It is only suitable for filter presses and drying beds, and has the advantage over lime and copperas of not significantly adding to the mass of sludge cake produced. Lime addition (5–10%) is sometimes used to control the emission of odours, especially if the sludge has been stored for longer than 4–5 days.

The mode of action of these conditioners, that is the charges on the metallic ions neutralizing the surface charges on the small sludge particles, is more controllable when using polyelectrolytes. These macro-molecular organic polymers are water soluble and like the other conditioners are able to flocculate dispersed particles. They are available in a very large range of molecular weights and charge densities, which allows the conditioner to be changed to cope with even subtle changes in sludge character. Polyelectrolytes are available either as a liquid, granules or powder, with the dry forms requiring to be dispersed in water before use so that they contain between 0.10% and 0.25% of active ingredient. Dispersion can take up to 2 h and as their ability to flocculate particles deteriorates with storage beyond this period, they are usually made up as required. Dosage is expressed as either volumetric ($mg\,m^{-3}$ of sludge) or more commonly as the weight of active ingredient added to a unit weight of dry sludge solids ($kg\,tonne^{-1}\,DS$). Normal doses for raw sludge are between 1 and 4 $kg\,tonne^{-1}$ for belt presses, so there is no significant addition to the mass of sludge. Although they can be used in conjunction with filter presses, polyelectrolytes are most efficient when used with belt presses and centrifuges. They are particularly useful for use with dewatering systems that utilize shear force to aid water release. As a general rule, the higher the shear force required, low in drying beds and filter presses but high in belt presses and centrifuges, then the greater the molecular weight of the polymer used. Lime can be used with polyelectrolytes to suppress odours, but hydrogen peroxide is more efficient for this purpose and easier to inject into the sludge. The optimum dosage of conditioner for a specific sludge is calculated by measuring its filterability after conditioner is added. Numerous methods of assessing the dosage are used including visual observation the specific resistance to filtration or the capillary suction time (CST) (Department of the Environment, 1985). The CST is measured using a small portable battery-operated unit. A small sample of sludge is placed in a central well (either 10 or 18 mm in diameter) that stands on a sheet of absorbent chromatography paper (Whatman No. 17) (Fig. 21.8). The capillary suction exerted by the paper pulls the water from the sludge and the rate at which it wets a standard area of the paper is dependent on the filterability of the sludge. The CST is the time to wet a standard area of paper which is measured automatically by electrodes that are turned on and off by the water as it passes. Other factors, in particular floc strength, are also important in the selection of dewatering equipment and conditioners. The effect of the shear exerted by

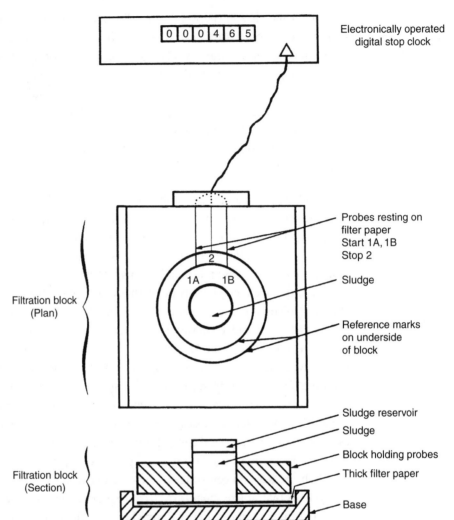

Electronically operated
digital stop clock

Probes resting on
filter paper
Start 1A, 1B
Stop 2

Sludge

Reference marks
on underside
of block

Filtration block
(Plan)

Sludge reservoir
Sludge
Block holding probes
Thick filter paper
Base

Filtration block
(Section)

FIGURE 21.8
The capillary
suction test (CST)
apparatus.
(Reproduced from
the Department of
the Environment
(1985) with permis-
sion of Her Majesty's
Stationery Office,
London.)

pumps and presses can be simulated by using a standard laboratory stirring apparatus prior to CST measurement, and can have a significant affect on filterability (Department of the Environment, 1985). Typical CST values for conditioned sludges with a solids concentration of 5% using an 18 mm well are 10 s for belt presses, 15 s for filter presses, 40 s for vacuum filters and centrifuges, and 300 s for drying beds.

Belt presses

Unlike filter presses, belt presses are a continuous process. Instead of porous sheets of filter mesh being squashed between plates, the belt press comprises of two continuous belts (one porous filter belt and an impervious press belt), between which sludge is added, which are driven by a series of rollers which also compress them squeezing excess water from the sludge.

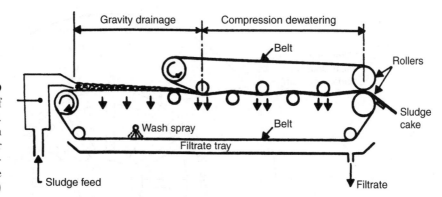

FIGURE 21.9
Schematic outline of
a sludge belt press.
(Reproduced from
Casey and O'Conner
(1980) with permis-
sion of Enterprise
Ireland, Dublin.)

Conditioned sludge is fed evenly onto a continuously moving porous filter belt that acts as a drainage medium. Dewatering then occurs in three distinct phases. First, before any pressure is added, water drains rapidly from the sludge due to the action of the poly-electrolyte. Sometimes a vacuum is used to increase the water removal at this stage. The second phase is the initial compression of the sludge as a low pressure is applied by a series of rollers as the belt press converges with the lower filter belt. Water is squeezed out of the sludge as the pressure increases. In the final phase the sludge is subjected to increased pressure and a shearing effect, which rips open the sludge between the two belts producing new channels for more water to escape, while the increased pressure forces out the remaining free water. As the belts part, the sludge cake falls from the machine onto a conveyer belt or into a trailer ready for transportation to the disposal site. The filtrate from the process falls from the filter belt to drains under the machine and is pumped back to the inlet of the treatment plant (Fig. 21.9).

Filter presses are all the same in principle, although the configuration of the rollers and the complexity of the machines can differ dramatically. The efficiency of these presses depends on the pressure applied to the sludge (i.e. clearance between rollers) and the retention time (i.e. length of compressed filter belt and belt speed). Other factors are also important, for example with primary sludges fresh material dewaters more successfully than stored material or if insufficient polyelectrolyte is used then large conglomerate flocs are not formed so that complete free drainage does not occur. Organic polyelectrolytes are normally used with doses ranging between 1.3 and 7.7 kg tonne^{-1} (DS). Final solids concentrations of between 20% and 25% are normal, although surplus activated sludge and aerobically digested sludge are far less amenable to dewatering. New machine designs with vertical belts and spring-loaded rollers (e.g. multi-stage and incremental pressure type filter-belt presses) can pro-duce cake with up to 35% solids.

Other systems

Vacuum filtration and centrifugation are also used for dewatering sludges, although their use is largely restricted to industrial wastewaters. In vacuum filtration water is

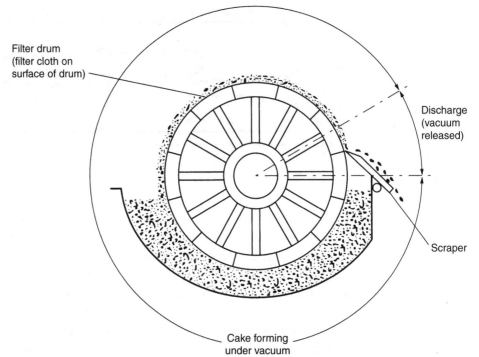

Filter drum
(filter cloth on
surface of drum)

Discharge
(vacuum
released)

Scraper

Cake forming
under vacuum

FIGURE 21.10
Principle of
vacuum filtration for
dewatering sludge.
(Reproduced from
Casey and O'Conner
(1980) with permis-
sion of Enterprise
Ireland, Dublin.)

sucked from the sludge under vacuum through a filter cloth that is carried on a slowly revolving drum partly immersed in the sludge (Fig. 21.10). Centrifuges consist of a rotating bowl into which sludge and polyelectrolytes are added. Centrifugal forces enhance the settling rates of the particles causing solids to separate out at its periphery from where it is removed (Fig. 21.11). Often called decanters, in order to operate continuously they need a mechanism to scrape the centrifugally deposited solids from the cylinder walls and convey them to a discharge point. This is done using a scroll running coaxially with the main bowl that acts as a screw conveyor.

21.3 SEWAGE SLUDGE DISPOSAL

With the introduction of the EU Urban Waste Water Treatment Directive (91/271/EEC) the amount of sewage sludge produced in Europe will almost double due to the increase in treatment provision (Section 8.3). In Ireland, where there has not been the same tradition of constructing sewage treatment plants as in other parts of Europe, sludge production will treble over the same period. This will mean new challenges in disposing of sewage sludge. In order to enhance the perception of sludge as a valuable resource, the EU is promoting a new name for the material, biosolids, however in this text the term sewage sludge will be used to avoid confusion.

The disposal option dictates the degree of treatment sludges receive (e.g. dewatering, stabilization). Other factors are also involved in selection including the presence of toxic compounds (e.g. heavy metals, organics), pathogens, volume for disposal,

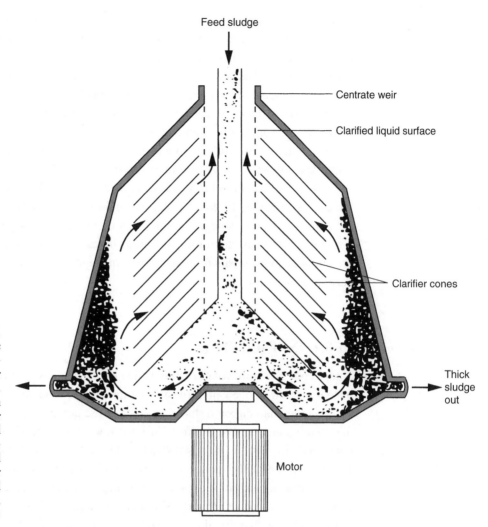

Feed sludge

Centrate weir

Clarified liquid surface

Clarifier cones

Thick sludge out

Motor

FIGURE 21.11 Schematic diagram of a nozzle-bowl centrifuge for dewatering sludge. (Reproduced from Institution of Water Pollution Control (1981) with permission of the Chartered Institution of Water and Environmental Management, London.)

TABLE 21.3 Expected trend in sewage sludge disposal in the European Union

Year	1995	2005
Volume (tDS year^{-1})	6.5×10^6	10.1×10^6
Land disposal (%)	37	45
Sea (%)	6	0
Fuel/incineration (%)	11	38
Landfill (%)	40	17
Other (%)	6	Unknown

location of treatment plant and cost. In practice, it is the *cost* and *need for environmental protection* that are the key factors in selection, and normally in that order.

Disposal options, and these do not include transformation processes such as anaerobic digestion and composting, fall into five main categories: landfill, disposal to agricultural land, disposal to sea, incineration, and other processes (Table 21.3). Other processes include sacrificial land use, forestry, land reclamation, use as building materials and as feedstuff. These other processes currently utilize a very small percentage of the total volume of sludge produced. The EU Directive on hazardous wastes (91/689/EEC) and its amendment (94/31/EEC) categories wastes as either hazardous or non-hazardous, which then affects how it may be disposed of. It categories wastewater sludges in the European Waste Catalogue (EWC) under the general code 19-08. Sludge from industrial wastewater (EWC code No. 19-08-04) and domestic wastewater treatment (19-08-05) are not included in the list of hazardous wastes. However, some types of water and wastewater treatment sludges are classified as hazardous (e.g. 19-08-03: Fat and oil mixtures from oil separators; 19-08-06: Saturated or used ion-exchange resins; 19-08-07: Solutions and sludges from the regeneration of ion-exchangers). Where sludges contain metals or metal compounds, making them unsuitable for use in agriculture, or contain listed toxic substances, then these sludges are automatically classified as hazardous and cannot then be mixed with other non-hazardous wastes.

21.3.1 INCINERATION

Sludge contains more volatile combustible matter and less fixed carbon than coal, so once it is dried sludge will burn to generate considerable heat. Raw sludge contains more energy than digested sludge as much of the potential energy is released during digestion as methane. Calorific values for raw sludge are $16\,000$–$24\,000\,kJ\,kg^{-1}\,DS$ compared to just $11\,000$–$14\,000\,kJ\,kg^{-1}\,DS$ for digested (Water Environment Federation, 1991). Incineration destroys the organic and volatile components of sludge including any toxic organic compounds, leaving a sterile ash in which all the toxic metals are concentrated. Although the weight of sludge cake is very much reduced, the ash that remains is a hazardous waste that must be disposed of at a regulated site. Alternatively it can be used as an amendment in cement or aggregate manufacture. If the solids concentration of sludge is <30% then additional fuel is required to burn the sludge because the amount of energy released during combustion is less than that required to evaporate the water present. At solids concentrations >30% the reaction is autothermic. This problem is generally overcome by mixing sludge with refuse to increase combustibility of sludge.

Incineration requires a high capital cost, so is not widely adopted except for large cities where other disposal routes are not available or metal contamination is high. With the introduction of the European Union EU Urban Waste Water Treatment Directive and the ban on sewage sludge disposal to sea, incineration has been selected by most large cities. For example, Thames Water have installed two fluidized bed incinerators to dispose of the sludge produced at their Crossness and Beckton Treatment Plants treating the wastewater from London. Operating at 860°C the dewatered sludge is instantly vaporized.

21.3.2 SEA DISPOSAL

Prior to 1998, over 25% of the total sludge produced in the UK was discharged from boats with a further 2% discharged via pipelines. Apart from Ireland, which disposed some 45% of its sludge to sea by both methods, no other EU country was so highly dependent on this route for disposal. With the implementation of the EU Urban Waste Water Directive (Section 8.3) sewage sludge disposal to sea ended on 31 December 1998. This affected 9 out of the 10 main water companies in England and Wales, with only Severn and Trent unaffected. Some water companies were very severely affected, for example Strathclyde Regional Council in Scotland (80% of sludge to sea), North-West Water (59%) and Thames Water (30%). In the UK alone, there will be an extra 10.5×10^6 tonnes of wet sludge per annum from over 30 locations due to the cessation of sludge dumping. There are numerous small towns throughout Europe with no sewage treatment, or just basic screening to remove floatable sanitary material, discharging to coastal waters via short or long sea outfalls. These will be required to have full primary and secondary treatment over the next decade, which will result in even more sludge being produced. All this sludge will now require a land-based disposal site.

While dumping of sludge at sea has now been made illegal within Europe there is very little evidence to show that it is in fact detrimental to marine organisms, including fish embryos and larvae if properly regulated. Mesocosm and field trials have shown that acute toxicity due to sewage sludge dumping is unlikely to occur at ecologically significant or detectable levels at dump sites. The main impact is organic enrichment leading to changes in species richness, relative abundance and the biomass of macro-invertebrate species (Costello and Read, 1994).

21.3.3 LANDFILL

The disposal of sewage sludge to landfill sites is undesirable due to:

(a) contamination of leachate with metals and organics,
(b) pathogen transfer risks,
(c) enhanced methane production on decomposition.

The addition of sludge to landfill is widely practiced and is considered a major route in all European countries (e.g. Luxembourg 10%, UK 26%, The Netherlands 32%, Denmark 45%, West Germany 49%, France 50%, Ireland 51%, Belgium 53%, Italy 55% and Greece 100%). The draft EU Directive on landfill requires sludge to have a minimum solids content of 35% while most technologies are achieving only 12–25%. So for many treatment plants the required solids level can only be achieved by sludge drying which is extremely expensive. In the long term there may be a prohibition on organic matter disposal to landfill altogether in order to reduce carbon emissions. There is a low level of acceptance of sludge disposal to landfill sites by communities with problems relating to odour, flies, public health and increased transport. So in most instances there are only two major disposal routes left, incineration and disposal to agricultural land. All the UK water companies have indicated that incineration will be the preferred option and with growing concerns about trace organic contamination in sludges, several European countries have indicated a wish to significantly reduce the application of sludge to farmland.

21.3.4 DISPOSAL TO AGRICULTURAL LAND

Sewage sludge is rich in organic matter and nutrients, especially N and P. It is also rich in trace elements, so it is useful both as a soil conditioner and as a fertilizer (Younas, 1987). There is considerable variation in the agricultural value of sludge, which depends largely on the treatment it receives (Table 21.4). The main problems are that a significant portion of the N is lost in the final effluent, and that sludge has a low K concentration as the K present is mainly soluble and so is also lost in the final effluent. Therefore, when sludge is used as a fertilizer a supplementary source of K (e.g. potash) may be required. For example, arable crops require a K supplement, while a low K is desirable for dairy farmers as a high concentration of K in the herbage can cause hypomagnesaemia in cows. On average a single application of $100 \, m^3$ of sludge (at 5% DS) provides $10–20 \, kg \, K \, ha^{-1}$.

The content of phosphorus is unaffected by treatment as it is generally present in insoluble forms, and so retained in the sludge. It is constant at between 1.0% and 1.8% of DS, giving a low N:P ratio compared to artificial fertilizers. So sewage sludges are rather like a superphosphate fertilizer with 50–60% of the P readily available. Phosphorus as a plant nutrient is less important than N, as most soils have adequate reserves, also the extent to which P is used is controlled by N availability. Phosphorus is a major eutrophication nutrient so the disposal of sewage sludge to agricultural land must be carefully managed to prevent surface-water pollution.

Organic matter increases the quality of soil and its productivity by binding the soil particles forming aggregates, which creates more air spaces so that air and water can penetrate, by reducing soil bulk density and increasing its water-holding capacity. Trace elements are plentiful in sludge. For example, 95% of grassland soils in the UK are deficient in zinc and 81% deficient in Cu, and sludge is an excellent source of most vital plant elements. In practical terms disposal to grassland used for silage

TABLE 21.4   Concentration of nutrients vary with sludge treatment

%	Liquid undigested	Liquid digested	Liquid cake
Dry solids	2–7	2–8	20–50
Organic matter (DS)	50–70	50–60	50–70
Total N (DS)	2.1–7.6	0.9–6.8	1.5–2.5
Total P (DS)	0.6–3.0	0.5–3.0	0.5–1.8
Total K (DS)	0.1–0.7	0.1–0.5	0.1–0.3
Total Ca (DS)	1.4–2.1	1.5–7.6	1.6–2.5
Total Mg (DS)	0.6–0.8	0.3–1.6	0.1–0.5

appears the best option. Disposal is possible from March to September and there are long no-grazing periods. Also the nutrient balance of sewage sludge is generally best suited to grass production. Throughout Europe farmers' attitudes to using sewage sludge appears to be hardening due to increasing concern over food quality, and increasing regulation and control that are making the disposal of their own animal wastes more difficult. For example, in 1993 some 191 million tonnes of animal manure required disposal in the UK compared with 1.5 million tonnes of sewage sludge.

21.3.5 HEAVY METAL CONTAMINATION

All sludges contain metals to a greater or lesser degree. Sludges can retain up to 96% of all the metals entering the treatment plant in sewage, therefore, where sludge is disposed of to land, heavy metals will accumulate in the soil. Effects of heavy metal accumulation are long-lasting and even permanent. Phytotoxicity is the main problem, although metals can be transferred directly to man via vegetables and other crops or indirectly via animals, primarily cattle, eating herbage (zootoxic). The commonest metals in sludge are Zn, Cu, Ni, Pb and Cr, with Cd generally the most toxic metal found in high concentration (Table 21.5). For example, levels $>10\,mg\,Cd\,kg^{-1}\,DS$ are common in municipal sludges.

The accumulation of metals in the sludge can be estimated from the normal concentration of metals in the wastewater. For example, if Cd is present in crude sewage at $0.008–0.01\,mg\,l^{-1}$ and we assume 85% removal efficiency of metal during treatment, and 350 mg of primary and secondary sludge solids generated per litre of sewage treated. Then the sludge will contain $20\,mg\,Cd\,kg^{-1}\,DS$. The normal range of metals found in soil and sludge are given in Table 21.6. What are the safe level of metals in soil (i.e. the acceptable level for plant growth) and the levels accumulated in plants that present a danger to livestock or humans? These questions are difficult to answer as the availability of metals to plants will vary according to soil and plant type, the

TABLE 21.5   Concentration of metals in sewage sludges disposed to agricultural land in the UK in 1981

Metal	Concentration of metals (mg kg⁻¹ DS)				No. of works	No. of samples
	Minimum	Median	Mean	Maximum		
Zinc	199	1270	1820	19 000	193	2386
Copper	36	546	613	2889	193	2379
Nickel	5	94	188	3036	192	2343
Zinc eq.	507	3440	4550	40 502	192	2343
Chromium	7	335	744	10 356	188	2310
Cadmium	0.4	17	29	183	180	2319
Lead	19	324	550	3538	164	2189

Note: Medians and means are weighted according to population served by works.
Reproduced from National Water Council (1981) with permission of Her Majesty's Stationery Office, London.

metal considered (e.g. some elements, such as Cd and Zn, are rapidly absorbed while Pb is largely unavailable), and finally the environmental factors that affect plant growth. For example, if the organic content and pH of a soil is high then availability of most metals is reduced. Most metals are absorbed to soil or organic matter, or form relatively insoluble precipitates at pH values normally associated with agricultural soils. Therefore, heavy metals are largely immobile and are not readily leached from the soil into ground water. So lime-treated sludges help reduce the availability of metals. The exception to this is molybdenum, which becomes available to plants at high pH and causes copper deficiency symptoms in livestock. Threshold toxicity concentrations in plants and animals are given in Table 21.7.

Metals come from other sources, and in the UK only 2% of agricultural land receives sewage sludge. All land receives atmospheric deposition and most agricultural land receives inputs from phosphate fertilizers with the Cd levels in phosphate fertilizers ranging from 1 to $160\,\text{mg}\,\text{Cd}\,\text{kg}^{-1}$, depending on the origin of the rock phosphate. The estimated average input of Cd to UK soils is $8\,\text{g}\,\text{Cd}\,\text{ha}^{-1}$ per annum (5 g from fertilizer

TABLE 21.6 The range (R) and common values (CV) of the major elemental contaminants in sewage sludge and soil in $\text{mg}\,\text{kg}^{-1}\text{DW}$

Contaminant	Sludge		Soil	
	R	CV	R	CV
Ag	5–200	25	1–3	1
AS	3–30	20	1–50	6
B	15–1000	30	2–100	10
Be	1–30	–	0.1–40[a]	3[a]
Cd	2–1500	20	0.01–2.4	1
Co	2–260	15	1–40	10
Cr	40–14 000	400	5–1000	100
Cu	200–8000	650	2–100	20
F	60–40 000	250	30–300	150
Hg	0.2–18	5	0.01–0.3	0.03
Mo	1–40	6	0.2–5.0	2
Ni	20–5300	100	10–1000	50
Pb	50–3600	400	2–200	20
Sb	3–50	12	2–10[a]	–
Se	1–10	3	0.01–2	0.2
Sn	40–700	100	2–200	10
Tl	–	1[a]	–	0.1[a]
V	–	15[a]	–	100
W	1–100	20	–	1[a]
Zn	600–20 000	1500	10–300	50
Dieldrin	<0.03–300[a]	0.4[a]	–	–
PCB (Aroclor 1254)[b]	<0.01–20[a]	3[a]	–	–

[a]Tentative data.
[b]A Water Research Centre survey of 11 UK sludges found PCB (Aroclor 1260) concentrations of $006–2.4\,\text{mg}\,\text{kg}^{-1}\text{DS}$.
Reproduced from Davis (1980) with permission of the Water Research Centre plc, Medmenham.

TABLE 21.7 Background and upper critical concentrations of elemental contaminants in plant tissue in $mg\,kg^{-1}\,DW$

Contaminant	Background concentrations	Upper critical concentrations	
		Phytotoxic thereshold	*Zootoxic threshold*[a]
Ag	0.06	4	–
As	<1	20	–
B	30	80	–
Be	<0.1	0.6	–
Cd	<0.5	10	–
Co	0.5	6	50
Cr	<1	10	50
Cu	8	20	30
F	8	>2000[b]	30
Hg	0.05	3	1
Mo	1	135	10
Ni	2	11	50
Pb	3	35	15
Sb	<0.1	–	–
Se	0.2	30	5
Sn	<0.3	60	–
Tl	<1[a]	20	–
V	1	2	–
W	0.07	–	–
Zn	40	200	500

[a]Tentative data; it is particularly difficult to assign zootoxic thresholds to Cd, Pb and Hg.
[b]Applies only to F taken up from soil.
Reproduced from Davis (1980) with permission of the Water Research Centre plc, Medmenham.

and 3 g from atmospheric deposition). So the contribution from sludge disposal overall is low. As a general guide $1\,kg\,ha^{-1}$ of metal added in the sludge to the soil will raise the soil metal concentration by $0.5\,mg\,l^{-1}$ if the plough depth is 20 cm or 0.25–$0.30\,mg\,l^{-1}$ if this is increased to 30 cm.

21.3.6 GENERAL SLUDGE APPLICATION GUIDELINES

A general guide to the toxicity of sludge can be calculated using the zinc equivalent from which application rates of sludge to land can be calculated. The system is based on the assumption that Cu is twice as toxic as zinc and Ni is eight times more phytotoxic. The effects of these metals are considered additive. Therefore, the toxicity of the sludge is calculated as a single value, the zinc equivalent (Zn eq.), using the concentration of the three metals in the sludge in $mg\,kg^{-1}$ in the equation below:

$$Zn\ eq. = Zn + (2 \times Cu) + (8 \times Ni)\ mg\ kg^{-1} \qquad (21.6)$$

Where no previous contamination in the soil exists, then 250 mg of Zn eq. kg^{-1} DS per annum over a 30-year period is considered safe, provided the pH is close to neutral.

TABLE 21.8 General maximum permissible addition of metals by sewage sludge and maximum soil concentrations

Element	Maximum permissible addition (kg ha^{-1})	Provisional maximum soil concentration (mg l^{-1})
Cd	5	3.5
Cu[a]	280	140[b]
Ni[a]	70	35[b]
Zn[a]	560	280[b]
Pb	1000	550
Hg	2	1
Cr	1000	600
Mo	4	4
As	10	10
Se	5	3
B	3.5	3.25[a]
F	600	500

[a]These elements are assumed to have additive toxic effects in the ratio Zn + 2(Cu) + 8(Ni) = zinc eq. The limits for the zinc eq. are as for zinc.
[b]Extractable.
Reproduced from National Water Council (1991) with permission of Her Majesty's Stationery Office, London.

However, on a routine basis the application rate is calculated for all major metals. The lowest application rate calculated indicates the metal that is limiting and gives the recommended annual loading.

Application rate (R) is calculated as:

$$R = \frac{A - B}{C} \times \frac{1000}{D} \quad \text{tonnes DS ha}^{-1} \text{ year}^{-1} \tag{21.7}$$

where A is the recommended limit for addition of specific metal (kg ha^{-1}) (Table 21.8), B the soil metal level (kg ha^{-1} or 2.2 × mg kg^{-1} DS), C the sludge metal level (mg kg^{-1} DS) and D the application period (e.g. 30 years). An example is given in Table 21.9 where the limiting metal is cadmium so the maximum permissible application rate of that particular sludge to that specific soil is 2.7 tonnes DS ha^{-1} year^{-1}.

21.3.7 EU SLUDGE APPLICATION GUIDELINES

Controls have been based on the concentration of metals in the soil or sludge. So soil quality controls are in effect environmental quality standards, while sludge quality controls are in effect heavy metal emission standards. Because of the variable nature of sludge, guidelines based on soil are thought to give a more reliable estimate of heavy metal loading. The EU Directive on the protection of the environment, and in particular the soil, when sewage sludge is used in agriculture (86/278/EEC) incorporates both

TABLE 21.9 Example of the calculation of the elemental application rates for a specific soil and sludge using the data in Table 21.8. Where national standards have been set (e.g. EC, USEPA, etc.) then these should be used instead

Parameter	Concentration in sludge $(mg\,kg^{-1})$	Concentration in soil $(mg\,kg^{-1})$	Elemental application rate $(tDS\,ha^{-1}year^{-1})$
Cd	35	1	$\dfrac{5-(2.2\times1)}{35}\times\dfrac{1000}{30}=2.7$
Cu	380	22	$\dfrac{280-(2.2\times22)}{380}\times\dfrac{1000}{30}=20.3$
Pb	200	40	$\dfrac{1000-(2.2\times40)}{200}\times\dfrac{1000}{30}=152.0$
Ni	27	8	$\dfrac{70-(2.2\times8)}{27}\times\dfrac{1000}{30}=64.7$
Zn	1200	62	$\dfrac{560-(2.2\times62)}{1200}\times\dfrac{1000}{30}=11.8$
Zn eq.	2176	170	$\dfrac{560-(2.2\times170)}{2176}\times\dfrac{1000}{30}=2.8$

TABLE 21.10 EU maximum limit values of heavy metals in sludge and soil, and maximum permitted application rates as a 10-year average, based on the EU Sewage Sludge Directive (86/278/EEC)

Parameter	Sludge $(mg\,kg^{-1})$	Soil $(mg\,kg^{-1})$	Maximum application $(kg\,ha^{-1}year^{-1})$
Cd	40	3	0.15
Cu	1750	140	12
Ni	400	75	3
Pb	1200	300	15
Zn	4000	300	30
Hg	25	1.5	0.1

Reproduced with permission of the European Commission, Luxembourg.

standards. It defines: (a) limit values (G and I) for metals in soil; (b) limit values (G and I) for metals in sludge; and (c) limit values on application rates (Table 21.10).

There are other requirements:

(a) sludge must be treated (biological, chemical, heat treatment), otherwise no grazing limits must be set to protect animal and public health, also untreated sludge must be injected or worked into soil;
(b) sludge must be analysed every 6–12 months depending on variability;
(c) soil must be analysed before and during application;
(d) records must be kept.

The suitability of sludges for agricultural use depends on factors other than metal concentrations such as the presence of animal-processing waste (e.g. danger of beef tapeworm transmission), contamination by tannery wastes (e.g. if foreign hides are processed then there is a danger of anthrax transmission) and contamination from vegetable processing (e.g. cyst eelworm transfer to agricultural land). There has been a significant reduction in the use of sewage sludge in agriculture in many European countries since the late 1990s due to consumer worries about direct and indirect pathogen contamination of food. In the UK, this has led to the introduction of a new code of practice, the Safe Sludge Matrix, which sets new levels of treatment based on pathogen removal rates for sludges used on agricultural land (ADAS, 2002). Significant changes have been proposed for the Sewage Sludge Directive with stricter limits for Cd and Hg, adoption of the current G values as future limit values for the other metals listed, short- and long-term reduction targets for all listed metals, including Cr for which a limit is set, and the inclusion of limit values for a wide range of organic compounds (Spinosa, 2001).

Rates of addition can be calculated using simple formulae, as can the calculation of the amount of sludge required to satisfy the N and P requirement of plants or crops.

(a) The capacity C ($mg\,l^{-1}$) of a given soil to receive a potentially toxic compound is calculated as:

$$C = \text{EU guide limit (mg } l^{-1}) - \text{initial soil concentration (mg } l^{-1}) \qquad (21.8)$$

(b) The permissible load L ($kg\,ha^{-1}$) of a potentially toxic compound is calculated using Equation (21.9), where 1 ha at a plough depth of 25 cm has a volume of $2500\,m^3$.

$$L = \frac{C \times 2500}{1000} \qquad (21.9)$$

(c) The permissible rate of addition, R, in tonnes of dry solids (tDS) $ha^{-1}\,year^{-1}$ is calculated using Equation (21.10) using a 30-year application period:

$$R\,(\text{tDS ha}^{-1}\,\text{year}^{-1}) = \frac{L \times 1000}{\text{Sludge quality } Q\,(\text{mg kg}^{-1}) \times 30} \qquad (21.10)$$

(d) In order to calculate the weight of sludge dry solids that must be applied to satisfy crop requirements for N, Equations (21.11) and (21.12) are used: a crop requires a certain number of units of nitrogen per hectare per annum (U) and this can be converted to crop requirement P ($kg\,ha^{-1}year^{-1}$) by

$$P = 0.51 \times U \qquad (21.11)$$

The application rate T (tDS ha^{-1}year^{-1}) is calculated as:

$$T = \frac{P \times 1000}{\text{Available N in sludge (N mg kg}^{-1})} \quad (21.12)$$

(e) If the concentration of a potentially toxic compound is high, then T will be greater than R and, to avoid contamination of the soil, the application rate determined by R will have to be used and the crop requirement for N will not be satisfied. To meet the crop requirement the concentration of the toxic compound must be reduced in the sludge so that sludge quality Q (mg kg^{-1}) is now:

$$Q = \frac{L \times 1000}{T \times 30} \quad (21.13)$$

21.3.8 INTERNATIONAL SLUDGE APPLICATION GUIDELINES

International regulations of sludge disposal all aim to control the same factors (i.e. pathogen transfer, soil pollution, surface- and ground-water pollution), however, due to the different approach to risk assessment very different standards are applied. Most regulations are based on keeping below known critical harmful limits based on current knowledge (risk assessment approach) while others believe on achieving the highest possible reduction of contaminants in sludges with maximum soil metal standards as close to natural levels as possible (precautionary approach). In Germany and The Netherlands, the maximum permissible soil concentrations are so close to natural background soil levels that land application of sludges is extremely difficult (Table 21.11).

In the USA, sludges are classified using pathogen reduction criteria. Rule 503 of the US Environmental Protection Agency (USEPA) standards for the use and disposal of sewage sludge specifies pathogen limits and acceptable treatment procedures. Class A sludges have a highly reduced pathogen transfer risk due to thermal drying or lime stabilization largely eliminating all pathogens. In contrast Class B sludges are untreated and so have a high pathogen transfer risk. Class A sludges have unrestricted use and have led to the widespread adoption of Class A process technologies, while Class B sludges are subjected to permitting and extensive regulation (Water Environment Federation, 1993).

TABLE 21.11   Comparison of maximum soil concentrations (mg kg^{-1}DS) for metals for sewage sludge disposal to agricultural land

Parameter	USA	UK	Ireland	Germany	Australia (NSW)	The Netherlands
Cd	39	3	1	1.5	3	0.8
Cu	1500	135	50	60	100	36
Pb	300	300	50	100	150	85
Zn	2800	200	150	200	200	140

There are many different methods of applying sludge (Table 21.12). The most effect-ive is the direct injection system, which employs a chisel plough and injects the sludge 25–50 mm below the surface of the soil. This reduces odours, reduces the risk of disease transmission and allows animals back on the land more quickly. High application rates can be achieved at a single pass at up to 180 tonnes ha^{-1}. Groundwater contam-ination is unlikely due to attenuation. However, groundwater contamination can occur if sludge is applied to dry clay soils which have cracked, allowing the sludge to drain rapidly deep into the soil. Surface contamination of watercourses is most likely due

TABLE 21.12   Comparison of the various methods employed to spread liquid sewage sludge onto an agricultural soil in terms of operational cost and performance

Application method	Prevention of odours and disease transmission	Effective on wet soil and for many land uses	Evenness of spread[a]	Economy	General comments
Tanker direct	+	+	+	+++	Widely used. Economic. Allows operational flexibility
Field tanker	+	++	++	+	High capital cost. Should be kept continu-ously supplied with sludge
Tractor-drawn tanker	+	++	++	+	Widely used
Movable rain gun	+	+++	+	++	Uneven application, problems of moving dirty pipes
Travelling irrigator	+	+++	+++	++	No risk of damage to soil
Subsoil injection using field tanker or tractor-drawn	+++	++	+++	+	Avoids odour and disease transmission problems

+: fair; ++: good; +++: excellent.
[a]Evenness of spread does depend considerable on the type of ancillary equipment used and driver skill.
Reproduced from Hall and Davis (1983) with permission of the Water Research Centre plc, Medmenham.

to run-off when the land is wet. Soil damage is a major problem and extra-wide tyres (flotation type) are required if soil is damp.

21.3.9 OTHER USES

Sludge is also becoming more widely used for land reclamation, forestry and a wide variety of other uses. Among the more unusual are brick manufacture, cement production (ash is used as a filler and the dried sludge as a fuel supplement), formation of aggregates, fibre production, wall board, top soil, in vermiculture (worm biomass and compost), fish and animal food, and for protein and vitamin (B_{12} in particular) recovery (Hall, 1991).

Europe is also currently requiring governments to develop local, regional and national sludge disposal strategies to ensure maximum recycling of sludge and minimum environmental damage. This is in conjunction with a requirement to reduce metal concentrations in sludge, and to licence sludges and record its movement and disposal, rather like hazardous/toxic waste.

REFERENCES

ADAS, 2002 *Guidelines for the Application of Sewage Sludge to Agricultural Land*, 2nd edn, Agricultural Development and Advisory Service, London.

Best, R., 1980 Want not, waste not! Sensible sludge recycling, *Water Pollution Control*, **79**, 307–21.

Bruce, A., 1984 *Sewage Sludge Stabilization and Disinfection*, Ellis Horwood, Chichester.

Casey, T. J. and O'Connor, P. E., 1980 Sludge processing and disposal. In: J. Ryan, ed., *Today's and Tomorrow's Wastes*, National Board for Science and Technology, Dublin, pp 67–80.

Costello, M. J. and Read, P., 1994 Toxicity of sewage sludge to marine organisms: a review, *Marine Environmental Review*, **37**, 23–46.

Davis, R. D., 1980 *Control of Contamination Problems in the Treatment and Disposal of Sewage Sludge*, Technical Report 156, Water Research Centre, Stevenage.

Department of the Environment, 1985 *The Conditionability, Filterability, Settleability and Solids Content of Sludges 1984. A Compendium of Methods and Tests*, Methods for the Examination of Waters and Associated Materials, Department of the Environment, Her Majesty's Stationery Office, London.

EPA, 1989 Standards for the disposal of sewage sludge: Proposed rule. 40 CFR Parts 257 and 503, *Federal Register*, **54**, 23–48.

Hall, J. E. (ed.), 1991 *Alternative Uses for Sewage Sludge*, Pergamon Press, London.

Hall, E. R. and Davis, R. D., 1983 *Sludge Utilization to Farmland*. WRc Regional Seminar, February, 1983, Dublin, Water Research Centre, Stevenage.

Hammer, M. J., 1977 *Water and Wastewater Technology*, John Wiley and Sons, New York.

Institution of Water Pollution Control, 1981 *Sewage Sludge. II. Conditioning, Dewatering and Thermal Drying*, Institution of Water Pollution Control, Maidstone.

Metcalf and Eddy, 1991 *Wastewater Engineering*, 3rd edn, McGraw-Hill Inc., New York.

National Water Council, 1981 *Report of the Sub-committee on the Disposal of Sewage Sludge to Land*, Standing Technical Committee Report 20, Department of the Environment, London.

Spinosa, L., 2001 Evolution of sewage sludge regulations in Europe, *Water Science and Technology*, **44** (10), 1–8.

Water Environment Federation, 1991 *Sludge Incineration: Thermal Destruction of Residues*, Manual of Practice FD-19, Water Environment Federation, Alexandria, VA.

Water Environment Federation, 1992 *Design of Municipal Waste Water Treatment Plants: II*, Water Environment Federation, Alexandria, VA.

Water Environment Federation, 1993 Biosolids and the 503 standards, *Water Environment and Technology*, May, 1993, Water Environment Federation, Alexandria, VA.

Younas, T. M. (ed.), 1987 *Land Application of Wastewater Sludge*, American Society of Civil Engineers, New York.

FURTHER READING

Cheremisinoff, P. N., 1994 *Sludge Management and Disposal*, Prentice-Hall, Englewood Cliffs, New Jersey.

Dirkzwager, A. H. and L'Hermite, P. (eds), 1989 *Sewage Sludge Treatment and Use: New Developments, Technological Aspects and Environmental Effects*, Elsevier, Amsterdam.

Gray, N. F., 2004 *Biology of wastewater treatment*, 2nd edn, Imperial College Press, London.

Lescher, R., 2002 International report: Sludge management and related legislation, *Water Science and Technology*, **46** (4–5), 367–71.

L'Hermite, P. (ed.), 1991 Treatment and Use of Sewage Sludge and Liquid Agricultural Wastes, Elsevier, Amsterdam.

Snaith, S. R., 1998 *Agricultural Recycling of Sewage Sludge*, CAB International, London.

Chapter 22 Household and Small-scale Treatment Systems

Introduction

Wastewater treatment systems for individual houses are required to be cheap, robust, compact, hygienic at site, odourless, require little maintenance and be installed, and operated, by unskilled labour. Their designs are based on full-scale units but are highly modified. Selection depends on: (a) the loading to the unit (Table 22.1), (b) the treatment objectives, (c) site conditions and location within the site, (d) effluent discharge requirements, (e) unit requirements (e.g. power, maintenance, etc.) and finally (f) cost. The most commonly employed systems are cesspools, septic tanks, rotating biological contactors (RBCs), constructed wetlands and small complete treatment systems (National Water Council, 1980). Some basic guidelines to the process selection of treatment systems for small communities up to a population equivalent of 50 are given in Fig. 22.1.

TABLE 22.1 Guide to per capita volumes of wastewater from different types of accommodation

Type of accommodation	Volume of sewage $(l\,ca^{-1}\,day^{-1})$
Domestic house (basic)	120
Domestic house (luxury)	200
Hotel (shared bathroom)	200
Hotel (private bathroom)	350
Restaurant	35[a]
Camping site (toilets only)	70
Camping site (toilets and showers)	110
Caravan parks	140
School (day with meals)	60
School (boarding)	200

[a] Grease problem probable.

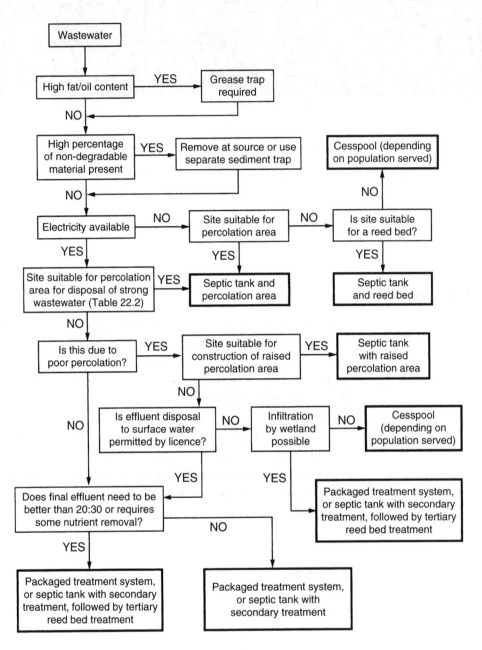

FIGURE 22.1 Process selection algorithm for small sewage treatment plants.

22.1 CESSPOOLS

These are underground chambers constructed solely for the reception and storage of wastewater with no treatment taking place. Cesspools are required to be water tight and constructed so as not to overflow, have a minimum capacity of $18\,m^3$ or 45 days retention for two people assuming a per capita water usage of $180\,l\,day^{-1}$, to be constructed so that it can be completely emptied, and be adequately vented and covered. They are not septic tanks (Section 22.2) and so no treatment occurs. Normally only used for single dwellings or small groups of houses (three to four maximum), they can be constructed from a variety of different materials including concrete, plastic and fibreglass, although prefabricated units must be set in concrete. The maximum prefabricated unit available is the Klargester-sealed cesspool with a capacity of 54 550 l. It is cylindrical in design (12 000 × 2740 mm) and must be embedded in 150 mm of concrete, which is equivalent to 25 m³, and then backfilled. However, where necessary units can be used in series to increase overall capacity. Construction is covered by BS 6297:1983. Clearly cesspools are only used where no other form of treatment is possible, and the need to have them regularly emptied means that they are amongst the most expensive form of treatment for domestic dwellings in terms of both capital and operational costs. The only improvement in such a system is achieved by adopting water-saving improvements that reduce the volume of wastewater discharged. There are advantages, however, such as no power requirement, no quality control required, no mechanism that can go wrong, the process is not injured by intermittent use, and as there is no effluent discharge there is no immediate environmental impact. The major limitation on the use of cesspools is the cost of emptying, although the construction of large underground storage tanks can be both difficult and expensive.

EXAMPLE 22.1   The land in a recent housing development was found after the construction of the houses to be unsuitable for septic tank usage, so cesspools were considered. The development had a persons equivalent of 58, so assuming a daily per capita water usage of 180 l the daily flow was:

$$\frac{180 \times 58}{1000} = 10.44 \, \text{m}^3\text{day}^{-1}$$

In this case the council was faced with three tank sizes, 31 m³ to give a minimum retention time of 3 days to cover a weekend, 73 m³ for 7 days retention to cover the Christmas period or 475 m³ for the recommended minimum 45 days retention period. While the land proved ultimately unsuitable for the construction of such a tank, the removal of 3810 m³ of sewage each year from the site using a 9 m³ tanker would have required over 420 trips per annum and at an average cost of £50 per load, this would have amounted to £21 000 each year. The comparable cost of collecting and disposing of 9 m³ of effluent for a private individual would cost in excess of £200 per load. In this case the problem was resolved by the construction of a small package sewage treatment plant comprising two septic tanks, a RBC and a tertiary reed bed (Upton and Green, 1995).

One area where storage tanks are still widely used is on farms. Storage of farm wastewaters is expensive and often dangerous, so ideally raw wastewaters should be spread immediately onto the land. However, this is impossible due to seasonal crop cycles, and at certain times of the year due to the risk of surface run-off causing pollution. The use of slurry tanks, both of underground and above-ground construction, is widespread and the design of such systems has been reviewed by Weller and Willetts (1977).

22.2 SEPTIC TANKS

Septic tanks are often confused with cesspools and it is important to differentiate between the two as in many areas these two terms are incorrectly used interchangeably. A cesspool is solely a storage tank, which requires regular emptying and is not intended as a septic tank in which the settleable solids are separated from the liquid fraction that is continuously discharged to a percolation area. The accumulated solids retained within the tank undergo anaerobic decomposition (Section 19.1).

22.2.1 THE PROCESS

Septic tanks can provide only partial treatment to wastewaters (Table 22.2), so further treatment is required either by a secondary treatment system (Section 14.2) or by the provision of a percolation area in which the sewage percolates into the soil via a system of underground distribution pipes (Fig. 22.2). If the effluent from a septic

TABLE 22.2 Variation in strength of the effluent from 28 Irish septic tanks ($n = 84$)

	Mean	Standard deviation	Minimum	Maximum
pH	7.0	0.24	6.3	7.5
BOD $(mg\,l^{-1})$	264	94.1	110	510
COD $(mg\,l^{-1})$	500	175.7	225	900
Suspended solids $(mg\,l^{-1})$	124	53.4	40	260
NH_4-N $(mg\,l^{-1})$	55	21.4	16	106
Total phosphorus $(mg\,l^{-1})$	16.2	6.4	2.0	29.0
COD:BOD	1.9	0.21	1.3	2.4

COD: chemical oxygen demand.

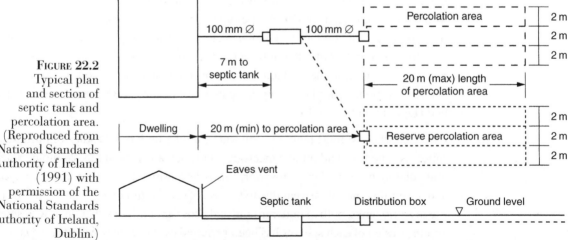

FIGURE 22.2
Typical plan and section of septic tank and percolation area. (Reproduced from National Standards Authority of Ireland (1991) with permission of the National Standards Authority of Ireland, Dublin.)

tank is discharged directly to surface waters then a minimum dilution factor of 300–400 is required subject to licensing. Although most commonly used for individual houses, septic tanks can be used to serve small communities of up to 500 and are commonly used in combination with a percolating filter or reed bed for small rural villages. Septic tanks require no power and where sub-surface drainage is used quality control can be confined to suspended solids removal only. They are not adversely affected by intermittent use, have a very small head loss and can achieve a 40–50% biochemical oxygen demand (BOD) and 80% suspended solids removal. In its simplest form the septic tank consists of a single chamber (Fig. 22.3a) with a single input and output. They can be of any shape although they are normally rectangular in plan, with two chambers and made out of concrete (Fig. 22.3b).

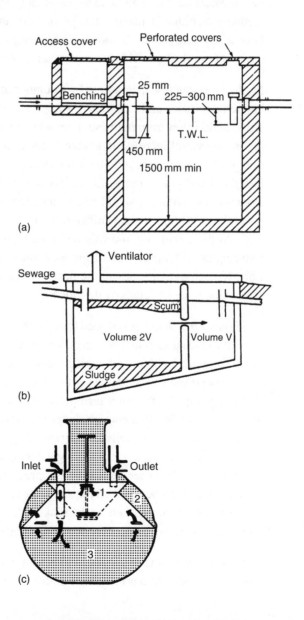

FIGURE 22.3
Examples of manufactured septic tanks. (a) Original single-chamber system; (b) more common two-chamber system; (c) GRP septic tank. (Reproduced from Mann (1979) with permission of the WRc plc, Medmenham.)

The tank consists of three separate zones, a scum layer on top of the clarified liquid with a sludge layer in the base. Wastewater enters and leaves the tank via T-shaped pipes, which prevents disturbing the scum layer or allowing solids to be carried out of the tank. As the wastewater moves through the chamber settleable solids gradually form a sludge, while any fats or buoyant material floats to the surface where it is retained by the baffle edge of the T pipe and forms the scum layer. The scum layer, although not vital to the successful operation of the tank, helps its operation in three ways. It prevents oxygen transfer through the air–water interface, insulates the anaerobic chamber by preventing heat loss, and finally attracts and retains fats, floating material and any solids risen from the sludge layer. It is in the sludge layer where anaerobic decomposition occurs slowly degrading the organic fraction (Section 19.1). Decomposition is often incomplete and the settled material is only hydrolysed and not broken down to methane, resulting in intermediate products of anaerobic digestion such as short-chain fatty acids being produced and slowly diffusing back into the clarified liquid to be discharged from the tank. This incomplete anaerobic activity results in unpleasant odours.

Methane production is inhibited by a number of factors in septic tanks, most notably low temperatures and insufficient sludge storage capacity. Methane production in septic tanks is low even in the summer while in winter it may cease completely resulting in an overall increase in sludge accumulation. However, gas formation occurs even in the absence of methane formation, and the rising gases can carry solids back into the liquid phase and cause an increase in the rate of scum accumulation. Most wastewaters contain a significant proportion of non-degradable solids, so even if anaerobic digestion is highly efficient there will be a gradual increase in solids in the tank. As these build up, the volume of the liquid zone in the chamber is reduced, so reducing the retention time of the wastewater and the degree of settlement is possible. If discharge of solids to the next treatment phase is to be minimized, then the septic tank must be regularly desludged.

The floor of the chamber is sloped towards the inlet end above which a manhole is situated into which a suction pipe can be lowered. Ideally septic tanks should be desludged every 12 months, with only the sludge removed. About 5–10% of the sludge should be left to reseed the new sludge layer to ensure rapid anaerobic activity. In practice the sludge is pumped out until it turns into water and then 5–10 l are returned. However, many operators feel that enough sludge will be left behind in the corners of the chamber and adhering to the walls to ensure seeding. The scum should be disturbed as little as possible and the tank, which is designed to operate full, should be refilled with water as soon as possible after desludging. In practice, the scum collapses as the sludge is pumped out. If this occurs then the scum must also be pumped out to prevent the overflow pipe becoming blocked as the tank refills. Also, an excessively thick scum layer should also be removed to prevent the inlet and outlet pipe becoming blocked.

When commissioning septic tanks they should be filled with water before use and if possible seeded with a few litres of sludge. Commercial seeds are now available that are easier to handle and more convenient. There is no shortcut to effective tank care and the addition of yeast additives, enzymes or bacteria are not necessary for digestion within the tank. It is a common fallacy that odours produced by septic tanks can be reduced by cleaning the chamber using a proprietary cleaner. Under no circumstances should a tank be disinfected, as this will only result in even worse odours being produced on recovery and inhibiting the degradation processes already taking place in the chamber. If odours are a problem then the tank should be desludged.

22.2.2 DESIGN

The design and constructional details of septic tanks are given by the British Standards Institution BS 6297 (British Standards Institution, 1983) and National Standards Authority of Ireland SR6 (National Standards Authority of Ireland, 1991). The two most important design criteria for septic tanks being the suitability of land for percolation and the capacity of the chamber. Many by-laws insist on minimum distances for the siting of tanks and percolation areas from houses, wells, streams and boundaries. These vary, but in Ireland no part of the percolation area should be closer than 20 m from the nearest habitable building, or 10 m from the nearest road boundary, stream or ditch, or 3 m from the boundary of the adjoining site. Distances from wells vary from 30 to 60 m depending on the percolation characteristics, the minimum depth of subsoil above rock and the minimum depth to the water table. The capacity and hence the retention time for settlement and sludge digestion is generally calculated according to the number of people discharging. The minimum capacity of a septic tank must not be <2720 l with the actual capacity calculated as:

$$C = (180\,P + 2000)\,\text{l} \qquad (22.1)$$

where C is the capacity in litres and P is the number of people discharging into the system. A common problem is when an older property in which an elderly couple have lived for many years is purchased and the septic tank system is found not to be able to cope once it has been modernized with mains water, a new bathroom and a kitchen full of labour-saving devices. For example, a disposal unit for household kitchen waste (garbage grinder) results in a 30% increase in the BOD and a 60% increase in suspended solids. Where these are installed the capacity of the septic tank must be calculated as:

$$C = (250\,P + 2000)\,\text{l} \qquad (22.2)$$

So in the design of a septic tank the future size of the family, number of bathrooms, visitors and other potential developments must be included in the design. It is unacceptable to use half values for children in such calculations as their water demands are just as great as adults, and they do not stay as children for long. In Ireland, the capacity of tanks are based on the number of bedrooms in the house served, where the

TABLE 22.3 Calculation of septic tank capacity and size of two-chambered tank in Ireland

Number of bedrooms in house	Minimum capacity (l)	Internal tank dimensions (m)			
		Chamber length		Width	Depth
		First	Second		
3	3000	2.0	1.0	1.0	1.0
4	3500	2.2	1.0	1.0	1.1
5	4000	2.4	1.0	1.0	1.2

Reproduced from National Standards Authority of Ireland (1991) with permission of the National Standards Authority of Ireland, Dublin.

minimum allowable capacity is 3000 l (Table 22.3). Some wastewaters are not suitable for discharge to septic tanks. For example, one lady who started a pottery at her country retreat discharged the wash water from her wheel directly to the septic tank. Within a year the excess clay had compacted so hard in the base of the chamber that the tank had to be drained and manually dug out, so beware. Septic tanks are generally unsuitable for industrial wastewaters or agricultural uses.

While the majority of septic tanks are made from precast concrete or are cast *in situ*, new prefabricated units in plastic or fibreglass are now widely available and come in a variety of shapes, designs and even colours. The Klargester septic tank has a characteristic onion shape (Fig. 22.3c). It is a three-chamber design which must positioned vertically in the ground. To prevent unalignment it must be set in concrete and the immediate site protected from traffic. The wastewater enters at the base (zone 3) that also stores $12 \, m^3$ of sludge. Settled wastewater is displaced upwards into the second chamber (zone 2) where the inclined base encourages further settlement with solids falling into the lower chamber through the peripheral slots. The wastewater passes into the final chamber (zone 1) from where it is discharged via the outlet to the percolation area.

Septic tanks are easy to up-rate by adding extra chambers in series. This reduces the effects of surge flows and excessive sludge accumulation in the first chamber. For populations in excess of 60 duplicate tanks and percolation areas, each of half the total required capacity should be provided and operated in parallel. This allows tanks to be desludged or percolation areas to be serviced without shutting down the entire system. The commonest faults associated with septic tanks include the following.

1. *Leaking joints* when tanks have been constructed from concrete panels or concrete rings. It is important that septic tanks should be water tight to prevent contamination of the groundwater and so they should be constructed, if possible, without joints.
2. *Non-desludging* is the commonest fault resulting in a reduced retention time and a stronger effluent due to less settlement. The loss of sludge under these circumstances can block pipes, percolation areas or filters.

3. *Blocked outlet pipes* are frequently a problem due to the scum layer becoming too thick or sludge physically blocking the outlet pipe which causes the tank to overflow. Many systems have inadequate percolation areas and access to the tanks for the sludge tanker is often inadequate, with the desludging pipe occasionally having to pass through the owners' house to reach the tank.

22.2.3 OPERATION

Little information is available on the operation of septic tank systems and on the whole they are very robust. However, care should be taken when using the following:

1. *Disinfectants* should be used moderately as their bactericidal properties kill off the anaerobic bacteria which can result in awful odours being produced during recovery. It is best to use disinfectants using free chlorine as it reacts with the organic matter in the sewage rendering it harmless by the time it reaches the tank.
2. *Caustic soda* which is so often used to remove grease from drains can cause the sludge to flocculate and rise. It can result in sludge passing out of the tank and blocking the percolation area. Small amounts of acidic or alkaline cleaners do no lasting harm and should be used in preference. Some strong cleaners can upset the pH of the tank which should be as near to neutral as possible. The pH of the tank can be controlled by the addition of hydrated lime, but this should only be used as a last resort. Under good operational practice the tank will buffer itself.
3. *High sodium* concentrations in the water do not affect the septic tank system directly but can impair the drainage properties of the soil. Where water softeners are used advice will be required from a hydrogeologist to ensure that the soil percolation area will not be damaged.
4. *Detergents*, especially alkyl benzene sulphonate, are known to inhibit the digestion process, although providing the tank capacity is sufficient, and normal concentrations of detergents used, the performance of the system will not be impaired. Enzyme-based washing powders have no effect on septic tank systems. Excessive use of shampoos, washing liquids and powders can prevent scum formation and cause fats and oils to be carried over to the percolation area with subsequent blockages occurring.
5. *Large flushes of water* to the tank should be avoided, if possible, to prevent scouring and resuspension of the sludge. Unless the tank is large enough to withstand such flushes a dual- or triple-chamber system must be installed. Where permissible bathwater should be diverted to a soakaway. Surface run-off including rainwater and melted snow from roofs and paved areas should not be disposed to a septic tank system.
6. *Solid material* such as disposable nappies, tampons, sanitary towels, bones, cigarette butts and cat litter will not degrade in the tank and should not be discharged to the septic tank as they will reduce the volume of chamber very quickly and can

be difficult to remove. Tampons remain partially buoyant and expand to a large size occasionally passing out of the tank and blocking percolation pipes. Fats and cooking oils should also be disposed of separately whenever possible, and it may be necessary to incorporate a grease trap in non-domestic situations. Mineral and fuel oils must not be discharged to a septic tank under any circumstances.

22.2.4 MAINTENANCE

Maintenance of the septic tank, like all sewage treatment units, is important in maximizing treatment efficiency and preventing pollution. The scum and depth of sludge accumulated in the tank should be inspected twice a year, depending on its volume. The depth of the sludge and the thickness of the scum should be measured in the vicinity of the outlet pipe. Records should be kept so that desludging frequency can be accurately predicted. The tank should be cleaned whenever the bottom of the scum layer is within 7.5 cm of the bottom of the outlet pipe or the sludge level is within 25–30 cm of the bottom of the outlet device. Scum thickness is measured using a hinged flap device which is a weighted flap attached to long rod (Fig. 22.4). Although, any device can be used which allows the bottom of the scum mat to be felt. The measuring device is pushed through the scum layer until the hinged flap falls into the horizontal position. It is then gently pulled upwards until the flap engages against the bottom of the scum layer. The handle is marked to correspond to a reference point on top of the tank. The same procedure is used to locate the lower end of the outlet pipe. The difference in height on the handle corresponds to the distance the scum is from the outlet. The depth of sludge is measured using a portable sludge depth indicator. An alternative method is to wrap a long stick in rough white towelling, which is tied securely, and then is slowly lowered into the tank through the vertical piece of the outlet pipe to the bottom of the tank to avoid the scum. Wait a few minutes and then slowly remove the stick. The depth of the sludge can be distinguished on the towelling by black particles clinging to it. If the depth of the sludge is more than one-third of the total liquid and sludge depth, desludging should be arranged.

22.2.5 SPECIAL WASTEWATERS

If laundry wastewater exceeds 10% of the total flow to the septic tank then this should be treated separately using a dedicated septic tank with a separate percolation area. Alternatively, it can be stored and sprayed onto lawns during dry periods. Catering facilities such as restaurants produce a large quantity of grease that cannot be dealt with by the system. So fats must be kept to a minimum, especially cooking oils. A grease trap should be installed on the kitchen waste line as close to the source as possible, and before any foul water enters. Any settleable solids will also be retained in the trap, which reduces its efficiency and rapidly blocks it. So where necessary a dedicated disposal line should be installed for grease and oils, or solids should

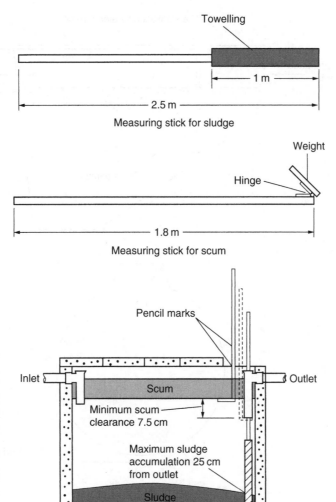

not be allowed to enter the drain. It is important that oils must be cooled before entering the trap. Grease traps are available in a wide range of materials (e.g. plastic, fibreglass, etc.) and must have a high surface area (Fig. 22.5). Collected grease must be regularly skimmed off the surface manually using a scoop.

Petrol and oil from car parking areas, garages and petrol station forecourts must be removed prior to septic tank or secondary treatment. This is best done by the use of oil interceptors (Fig. 22.6), which are installed before the foul water from paved areas enters the system. They are designed to ensure maximum separation with retention times for adequate separation ranging from 6 to 20 min depending on the density of the oil or petrol. Units are normally prefabricated in plastic or glass fibre and vary in size from 2000 to 54 000 l. The design of the unit depends on the area served and the maximum expected rainfall per hour. They are also useful for containing spillages of light-density liquids.

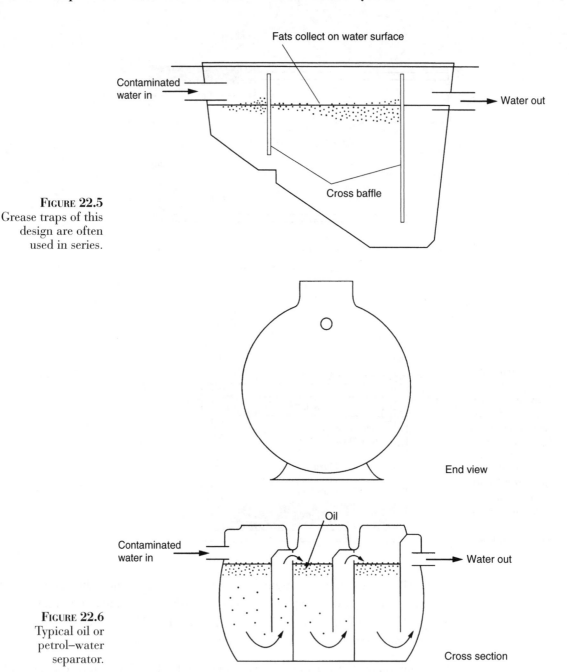

Fats collect on water surface

Contaminated water in

Water out

Cross baffle

FIGURE 22.5
Grease traps of this design are often used in series.

End view

Oil

Contaminated water in

Water out

FIGURE 22.6
Typical oil or petrol–water separator.

Cross section

22.3 PERCOLATION AREAS

Where treated wastewater cannot be discharged to surface water a percolation area is constructed to allow the effluent to seep through the soil and into the water table. This is achieved by a system of perforated or open jointed distribution pipes which are laid on 20–30-mm diameter gravel or clean crushed stone to allow the water to freely drain away (Fig. 22.7a). In Ireland the area required for percolation is determined by

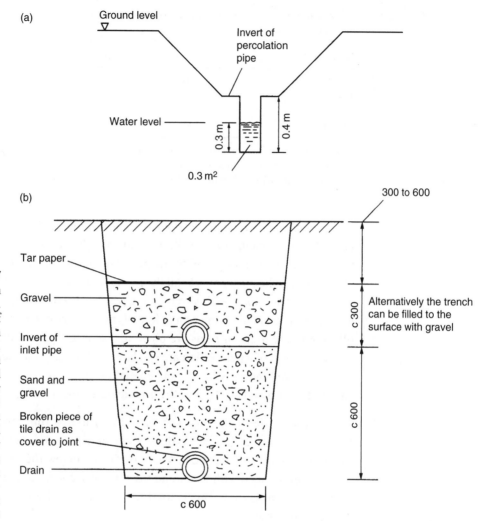

FIGURE 22.7
(a) Irish percolation test hole dimensions. (b) Cross section of typical underdrain used in the UK. (Reproduced from (a) National Standards Authority of Ireland (1991) with permission of the National Standards Authority of Ireland, Dublin; and (b) British Standards Institution (1983) with permission of the British Standards Institution, London.)

carrying out soil percolation tests on the site designated for the percolation area. Test holes 0.3 × 0.3 m and at least 0.4 m deep are dug below the proposed depth of the percolation pipe. The sides and bottom of the hole is scratched with a knife to ensure free drainage and clean water is then poured into the hole to a minimum depth of 300 mm. It is necessary to ensure the surrounding soil is saturated to simulate conditions below the percolation area. For that reason it is best to leave the hole until the following day. The test is carried out once the ground is saturated by adding water to a depth of 300 mm and measuring the time in minutes for the water level in the hole to drop by 100 mm. The result is divided by four to give the time in minutes for the water level to drop 25 mm, which is the test value t. An average of all the t values is taken (T) which is used to calculate the total length of percolation pipe required (Table 22.4).

TABLE 22.4 Length of piping per capita for construction of percolation area

Value of T	Length of pipe (m)
5–10	30
11–15	36
16–30	48
31–60	96

Reproduced from National Standards Authority of Ireland (1991) with permission of the National Standards Authority of Ireland, Dublin.

Values of $T < 5$ may indicate a percolation rate that is too fast which may lead to groundwater contamination, while a T value of >30 indicates a percolation rate that is too slow and will possibly lead to the percolation area becoming flooded. Percolation areas with T values >60 or <5 fail the test and are not suitable for use. In most cases a reserve percolation area is also required which is used in rotation with the main area and can be used in the event of a failure of the primary percolation area. Surface run-off from paved areas or roofs are piped to a separate percolation area or soakaway. A typical layout of a percolation area is shown in Fig. 22.2. Full details of site suitability assessment and the construction of percolation areas are given in National Standards Authority of Ireland (1991) and British Standards Institution (1983).

The British Code of Practice (BS 6297:1983) uses a different percolation test to determine the value V_p, which is the average time required for the water level to drop 1 mm in seconds. The test hole is dug to 250 mm below the proposed invert level of the land drain, filled with water and the time in seconds for the water to completely seep away is measured. This is used to calculate the floor area of the sub-surface drainage trench A_t in m^2

$$A_t = P \times V_p \times 0.25 \tag{22.3}$$

where P is the population equivalent. For effluents that have received secondary treatment followed by settlement this area is reduced by 20%:

$$A_t = P \times V_p \times 0.20 \tag{22.4}$$

In the UK shallow soakaways are permitted where the geology and hydrogeology is suitable. Drainage ditches should have a uniform gradient no $>1:200$, and should be from 300 to 900 mm wide with a minimum of 2 m between parallel trenches. Perforated or porous pipes (75–110 mm diameter) are laid on a 150–250 mm layer of clean gravel or crushed stone (20–50 mm grade) and then covered by 50–150 mm of the same material. The trench is sealed by a strip of air-permeable sheeting to prevent the entry of silt and filled with soil. Pipes should be a minimum of 500 mm below the surface. Where high percolation values ($V_p > 140$ s) are obtained the soil is not suitable for the construction of a percolation area. In this case a complete treatment system must be employed or a percolation area constructed above the ground level. Where V_p is between 100 and 140 s then underdrains are required to prevent flooding. In this system a second system of drainage pipes are laid on the bottom of the trench to convey the surplus water to an outfall to a ditch or watercourse (Fig. 22.7b).

22.4 Small Complete Treatment Systems

Secondary treatment systems are often used with septic tanks to treat the final effluent before disposal, and include percolating filters (Section 16.2), RBCs (Section 16.4) and increasingly reed beds (Section 18.3) (Upton and Green, 1995). However, there are a wide range of small packaged systems available for the treatment of domestic effluent from single or several houses. These are largely based on fixed-film or activated sludge systems. Three popular systems are *Biocycle®*, *Bioclere®* and *Puroflow®*, although there are many different designs.

Biocycle® is a complete treatment system that incorporates a septic tank and a submerged aerated filter for secondary treatment within a single tank (Fig. 22.8). The final effluent passes through a simple secondary settlement tank with the settled sludge returned by a simple airlift system to the primary chamber. The final effluent passes through a chlorinator that uses sodium hypochlorite tablets and is then pumped out from the holding tank section when full by a submersible pump. The unit requires electricity to run a small air blower and the submersible pump, but is capable of producing a very clean ($10 \, \mathrm{mg \, l^{-1}}$ BOD, $10 \, \mathrm{mg \, l^{-1}}$ suspended solids), disinfected and nitrified effluent. The company provides a variety of options for final effluent disposal via surface or sub-surface percolation. The standard unit serves a population equivalent (pe) of eight, while the system is available in larger sizes to treat up to 150 pe. Unlike many other systems, *Biocycle®* provide a comprehensive maintenance service that checks the system twice a year and ensures optimum operation. The system also has a simple diagnostic alarm system that can alert the householder, or the company directly, that the system is malfunctioning or needs attention. A similar version is the *HiPAK®* system, although this is simpler in design and the final effluent flows passively to a percolation area with no disinfection option. The standard system treats a population equivalent of five, while the largest version can treat a population equivalent of 55.

Bioclere® also incorporates a filter system, but this is not a submerged system but a mini-percolating filter using a small random plastic filter medium. It requires a separate septic tank for primary settlement and like the other systems it requires electricity to power a small irrigation pump to draw the wastewater from the sump and to spray it continuously over the filter medium (Fig. 22.9). A 20:30 effluent is possible but nitrification is poor. Secondary sludge is removed from the base of the tank. Small RBC systems are also available but also require a septic tank for primary settlement.

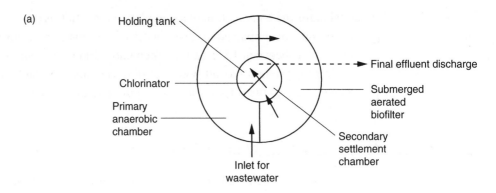

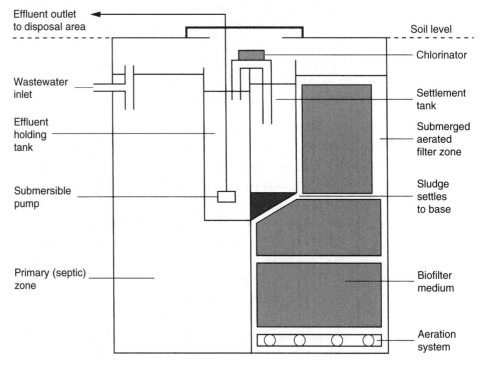

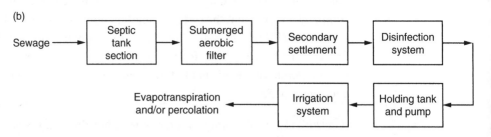

FIGURE 22.8
(a) Schematic plan and cross section of the *Biocycle®* complete wastewater treatment system for a single dwelling. (b) Flow diagram of process.

Puroflow® is a packaged peat bed system. It is a shallow filter filled with a coarse peat mixture. Wastewater is spread evenly over the surface by means of a static distribution system. It requires a separate septic tank and, where there is not sufficient hydraulic head, a sump from where the tank effluent is pumped to the distribution system above the peat filter (Fig. 22.10). The filters are efficient at BOD and suspended solids

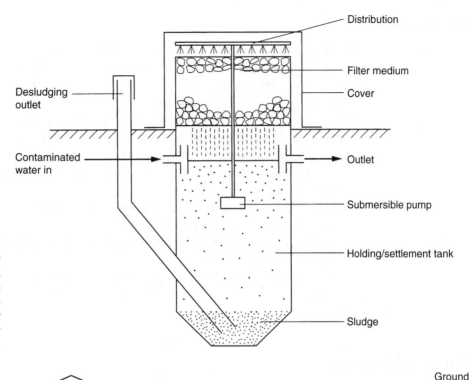

FIGURE 22.9
Schematic cross section of the *Bioclere*® complete wastewater treatment system for a single dwelling.

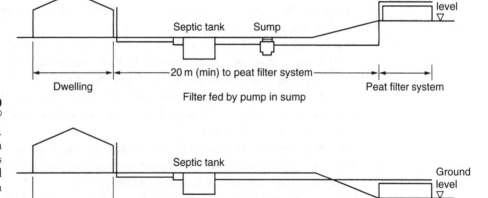

FIGURE 22.10
Layout of *Puroflow*® peat filter system. (Reproduced from National Standards Authority of Ireland (1991) with permission of the National Standards Authority of Ireland, Dublin.)

removal, and as peat has a high adsorptive capacity cations and anions are also effectively retained within the system.

Disposal of the final effluent from all of these systems is normally to a percolation area. However, as the final effluent has been treated then discharge to a watercourse is possible where sufficient dilution is available. This will, however, require a discharge licence from the relevant regulatory authority. The design and specification of small package systems are constantly changing so reference should always be made to the manufacturer before system selection.

REFERENCES

British Standards Institution, 1983 *Design and Installation of Small Sewage Treatment Works and Cesspools*, Code of Practice, CP 6297, British Standards Institution, London.

Mann, H. T., 1979 *Septic Tanks and Small Sewage Treatment Plants*, Technical Report 107, Water Research Centre, Medmenham.

National Water Council, 1980 *The Operation and Maintenance of Small Sewage Works*, Occasional Technical Paper No. 4, National Water Council, London.

National Standards Authority of Ireland, 1991 *Septic Tank Systems Recommendations for Domestic Effluent Treatment and Disposal from a Single Dwelling House*, SR6. National Standards Authority of Ireland, Dublin.

OSU, 1999 *Septic Tank Maintenance*, Oklahoma Co-operative Extension Service Fact Sheet F-1657, Oklahoma State University, Stillwater, OK.

Upton, J. and Green, B., 1995 A successful strategy for small treatment plants, *Water Quality International*, **4**, 12–4.

Weller, J. B. and Willetts, S. L., 1977 *Farm Wastes Management*, Crosby Lockwood Staples, London.

FURTHER READING

Boller, M., 1997 Small treatment plants – a challenge to wastewater engineers, *Water Science and Technology*, **35** (6), 1–12.

Crabtree, H. E. and Roswell, M. R., 1993 Standardization of small wastewater treatment plants for rapid design and implementation, *Water Science and Technology*, **28** (10), 17–24.

EPA, 2000 *Treatment Systems for Small Communities, Business, Leisure Centres, and Hotels*, Wastewater Treatment Manual, Environmental Protection Agency, Wexford.

EPA, 2000 *Treatment Systems for Single Houses*, Wastewater Treatment Manual, Environmental Protection Agency, Wexford.

Gray, N. F., 2004 *Biology of Waste Water Treatment*, 2nd edn, Imperial College Press, London.

Green, B. and Upton, J., 1995 Constructed reed beds: appropriate technology for small communities. *Water Science and Technology*, **32** (3), 339–48.

Nicoll, E. H., 1992 *Small Water Pollution Control Works: Design and Practice*, Ellis Horwood, Chichester.

Part V The Future

Chapter 23 Sustainability Principles in Water Management

23.1 THE GLOBAL PERSPECTIVE

Everyone of the current world population of 6 billion requires clean water and a safe disposal route for their wastewater. In developing countries waste disposal was traditionally achieved locally while in developed countries waste disposal has become increasingly centralized resulting in large and damaging point sources of pollution. The trend of centralization is no longer a developed world phenomenon, with increasingly fewer urbanized people with satisfactory drinking water or adequate sanitation. The provision of water services simply cannot cope with the rate of population growth, especially where such services are prohibitively expensive. Currently in excess of 1 billion people lack access to clean water and 1.7 billion have no sanitation. Of the 1.75 million tonnes of faeces and 7 million tonnes of urine produced each day, only 15% is currently treated. This, coupled with unregulated industrial development in many parts of the world, has left global water quality in the worse state ever recorded. Thus with the world population predicted to double by the end of the century the water crisis is set to worsen with potentially catastrophic ramifications.

This rapidly increasing demand for water and sanitation, especially in developing countries, is an obvious obstacle to sustainability. Conversely the urgent necessity for its provision is similarly an obstacle with short-term solutions often leading to serious long-term problems. The water industry is struggling to cope with current water demands with little or no opportunity to look to the future. Yet, if we are to preserve what we have and find ways of providing for future generations then radical action is required. A sustainability approach to these problems seems an obvious solution yet Mitcham (1995) suggests that many developing countries see sustainable development as a Western idea meant to maintain the Western way of life. In developed countries population growth is often seen as the major threat to sustainability while in developing countries over-consumption is more frequently cited as the major problem. However, to be successful sustainability will eventually require global compliance; so sustainability in developing countries cannot be ignored with action required at both a local and national level in order to achieve global stability. At the heart of the sustainability concept is self-regeneration and self-reliance. Communities in which people are denied opportunity, face poverty and exclusion, and those whose economies are in long-term recession are not sustainable (Petts, 2001). This is the major challenge for the future.

23.2 CLIMATE CHANGE

The critical impacts induced by climate change over the next 20–50 years are changes in temperature, precipitation and sea levels, all of which will have varying consequences not only for ecosystems but on the availability of water supplies around the world. The timescale for climate change is hotly debated but only minor impacts are expected up to 2025 becoming increasing serious over the subsequent decades. Climate change is predicted using general circulation models (GCMs) that currently are unable to predict regional or local responses of aquatic ecosystems. While some countries will experience positive impact effects, for most countries the impacts will be negative although the effects will vary. What is certain, however, is that there will be greater warming in higher latitudes resulting in colder regions warming faster. This will alter run-off patterns by decreasing the amount of winter precipitation that falls as snow resulting in earlier run-off into river basins. For example, Canada is a large land mass in the northern mid- to high latitudes so larger than average increases in temperature are predicted in its interior with the north having a greater increase in winter temperature and more precipitation than at present. Warming in northern areas is also expected to melt permafrost areas, allowing shallow summer groundwater tables to drop. This is causing concern that the subsequent drying of wetlands will result in the enormous quantities of organic carbon stored as peat being released into the atmosphere as CO_2 or methane. The most significant problems will occur in those areas of the world where water supplies are already vulnerable due to shortages, a problem that will be exacerbated by climate change. Average temperatures in the UK have risen by approximately 1°C over the past 100 years with further temperature increases of between 1.2°C and 3.4°C predicted by 2080. This will result in drier summers, especially in the southeast and wetter winters, especially in the northwest, with severe storms becoming more frequent (Petts, 2001).

The average global surface temperature is expected to increase by between 1.5°C and 5.8°C by 2100, although it has been projected that it may be greater in the US (Wigley, 1999; Houghton *et al.*, 2001). Temperature is a critical factor in determining the species diversity of surface waters (Chapter 3). Even small increases or decreases in temperature can cause significant changes in the metabolic rates of organisms, overall productivity of ecosystems, survival of macro-invertebrates and fish, and the vulnerability of waters in terms of oxygen solubility and maintaining sufficient dissolved oxygen to maintain ecological status. So temperature increase will alter the geographical distribution of aquatic species as well as fundamental ecological processes. While species will attempt to adapt by migrating to a more suitable habitat, the ability to do this is limited in freshwaters where distribution is physically restricted to within individual catchments or water bodies. In larger catchments, where there is a possibility that migration northwards is feasible, this may be restricted or prevented by human alteration of potential migratory corridors or by pollution. The overall effect of an increase in temperature will be a change of species diversity and community structure at the local level, with an increasing likelihood of species extinction and loss of biodiversity.

Climate change will alter the hydrological characteristics of surface water due to changes in seasonal rainfall patterns and surface run-off. This will affect river and reservoir yields and also the recharge of groundwater aquifers, making water resources more difficult to manage in order to maintain abstraction rates for water supplies. The loss of winter snowpack will also greatly reduce a major source of groundwater recharge and spring run-off, resulting in a lowering of water levels in streams, rivers, lakes and wetlands adversely affecting species during the growing season. Alteration in discharge regimes in rivers will affect both species composition and productivity, with species varying in their ability to cope with the frequency, duration, timing and magnitude of extreme rainfall events that includes both floods and droughts. The life cycles of many aquatic species are closely associated with river discharge regimes that are controlled by the availability and seasonal timing of water from precipitation and run-off. Temperature effects will act synergistically with changes in the seasonal timing of run-off to freshwater and coastal systems resulting in falling water quality. Freshwater wetlands are particularly vulnerable as they are already under stress being transitional between aquatic and terrestrial, often upland, ecosystems. This makes them particularly sensitive to changes in temperature and precipitation. Wetlands fed primarily by precipitation are already vulnerable to dry periods, so any changes in precipitation patterns could lead to their contraction in size or elimination. Groundwater-fed wetlands have the greatest ability to withstand climate change as they have large reserves of water in the underlying aquifer. If the climate becomes wetter water tables will rise expanding such wetlands, but if water tables fall then they will contract in area (Winter, 2000). Other protected water bodies, such as turloughs, which are temporary lakes characteristic of some limestone areas and protected under the European Union (EU) Habitats Directive (92/43/EEC), are also vulnerable to water table variability brought about by climate change.

Rising sea levels will have significant effects on aquatic systems by extending estuaries inland, increasing the possibility of flooding, and by increased saline intrusion of coastal aquifers. Coastal wetlands provide unique but sensitive habitats that are especially important to fisheries. As they are already under increasing pressure from a wide range of human activities they are particularly vulnerable to climate change. Sea level rise will change both water quality and flow regimes in such systems and as associated species are poorly adapted for survival under other environmental conditions, then impacts will be severe. Coastal wetlands play important roles in nutrient cycling and in capturing sediment transported downstream from catchments, functions that if lost will significantly alter the nature of estuaries.

In the US and Canada climate change is predicted to seriously disrupt the current distribution of plants and animals in fresh and coastal waters. Particularly vulnerable are the cold-water fish species, such as salmon and trout, which are expected to disappear from large areas of the continent when water temperatures exceed their thermal tolerance limits. In contrast fish, such as largemouth bass and carp, that prefer warmer water, will expand their ranges as temperatures increase. It is probable that the ranges of cold-water species will expand northward or to higher elevations, where the very cold temperatures currently limit growth (Clark *et al.*, 2001). Without intervention by man, this natural colonization of isolated northern catchments with very limited productivity and diversity may take a considerable time. Surface water productivity increases with water temperature allowing undesirable species to dominate, as in the case of algal blooms that are associated with algal toxins (Section 6.4.1). Lakes are especially vulnerable with species eliminated from smaller lakes as the surface water temperature rises and algal development increases.

Apart from concerns as to how the biota will respond to climate change, of equal concern is how will humans react to a rapidly changing and largely unpredictable climate, which will introduce a high degree of uncertainty about future water availability and quality. Inevitably there is going to be increased pressure to control water resources by modifying existing water bodies by river transfer, reservoir construction and compensation, increased groundwater abstraction and better surface water run-off systems. So how humans respond by altering important water resources is a critical factor in predicting how aquatic ecosystems will respond to climate change. In practice freshwater ecosystems have a limited ability to adapt to climate change so alteration in species diversity and community structure is inevitable. However, it is possible to slow the effects or even mitigate the effects to a certain degree, allowing populations longer periods to possibly acclimatize. As climate change will make individual species more vulnerable to pollutants and other sources of environmental stress, reducing or eliminating these contaminants and stresses will minimize the effects induced by climate change. Restoration of damaged ecosystems, the improvement and proper maintenance of riparian and bank side vegetation, better land use through catchment (river basin) management, habitat restoration, controlling the spread of exotic species, the strategic use of compensation reservoirs, controlling surface water abstraction and minimizing groundwater withdrawal can all minimize the effects of climate change on both the ecosystem and our utilization of resources for other needs.

While the predicted changes in temperature will occur too rapidly for many species to adapt, some species may benefit. However, it is clear that such changes must be considered in terms of overall planning and management of the water cycle in the future. Increased incidence of flooding, higher water demands in dry areas with fewer resources to meet such demands, reduced flows in some rivers, changes in the impacts on lakes and rivers, and increased demand for water for irrigation, are major issues that will need to be addressed if a sustainable water cycle is to be achieved.

23.3 SUSTAINABILITY

Although there are many different definitions of sustainable development the broad definition proposed by the World Commission on Environment and Development (1997) remains the most appropriate:

> *Development that meets the needs of the present generation without compromising the ability of future generations to meet their own needs.*

This definition is also about equity between all inhabitants of the planet and equity among generations to come. So the term *needs* in the definition refers in particular to the essential needs of the world's poor. So the concept outlined in the above definition is about realising that we cannot isolate ourselves on the planet and that the actions of every individual impinges of the health and well-being of everyone else. So the concept of sustainability affects the very basis of how we live. Therefore the remediation of current social and economic problems is crucial for global sustainability to be achieved.

The fundamental problem is that while most people agree in principle to the idea of sustainable development, few are able to agree on how it can be achieved. Some see it as a slow evolution of current technologies leading to greater efficiency in the use of energy and better controls on the emission of wastes, while others consider that sustainability requires fundamentally different solutions and technologies. What is apparent is that water cycle sustainability is going to require significant changes for both industry and consumers alike resulting in a new partnership incorporating novel and radical changes in our use and management of water.

A major step towards sustainability in Europe is that water and wastewater treatment are no longer seen in isolation but as an integral part of the urban water cycle (Fig. 1.2), which itself forms part of the natural hydrological cycle (Fig. 1.1). In a sense water is just intercepted from the hydrological cycle, treated, used and finally returned after wastewater treatment back into the cycle, often at almost the same point it was originally removed. The water invariably becomes contaminated during use and so will change catchment water quality once it has been returned. Van der Graaf *et al.* (1997) examined the urban water cycle in detail and found that at every stage within the cycle significant improvements could be made to reduce overall contamination. So in terms of sustainability the best scenario is to ensure that the urban water cycle becomes a closed system isolated from the natural water cycle, to protect resources and their ecology, with only treated effluents of the highest quality returned to the catchment (Van der Graaf, 2001). This concept is encapsulated in the Water Framework Directive (WFD), which has introduced integrated and holistic management of water resources on a catchment basis.

Earlier in the text when the WFD (Section 7.3) was introduced the bold statement that introduces the legislation was quoted. It is quoted again here to give a starting point for the development of the concept of a sustainable water cycle. It says, 'Water is not a commercial product like any other but rather a heritage which must be protected, defended and treated as such'. This declaration is a powerful synthesis as to what the water industry, and all who use water, should be striving to achieve, and offers us a valuable insight into what water cycle sustainability is all about. The EU introduced the WFD (2000/60/EC) in response to a number of specific concerns:

1. the need for action to avoid long-term deterioration of surface freshwater and coastal quality;
2. the dwindling supplies of water suitable for supply proposes and the preservation of drinking water quality;
3. the growing demand on water resources for a wide variety of uses;
4. groundwater quality and increasing abstraction;
5. the discharge of dangerous substances to surface waters and their subsequent accumulation in the marine environment;
6. the preservation of wetlands that play a vital role in protection of water resources;
7. resolving transboundary water quality and quantity problems;
8. the lack of an integrated approach to water management.

The objective of the WFD is to create a suitable mechanism that can establish the basic principles of sustainability in water policy and subsequently water management. To do this the WFD will create a management system for the control of the water cycle based on river basin areas (catchments), including ground and coastal waters, with legislation that controls both the sources of contaminants and their impacts on receiving waters. The WFD incorporates an integrated and holistic approach to water cycle management using the precautionary principle for wastes, encouraging the principle that preventive action should always be taken, that environmental damage should be rectified at source and that the polluter should pay.

By setting a minimum quality status for all water bodies with the long-term objective of improving the ecological quality of all water bodies, the WFD will gradually create the closed water system described by Van der Graaf (2001). For example, the stated ultimate aim of the Directive is to eliminate priority (hazardous) substances so that the concentrations of these substances in the marine environment will ultimately fall to zero or to near background concentrations for naturally occurring substances. To do this River Basin Management Plans will need to control all discharges at source by setting emission limit values and environmental quality standards. Thus the WFD is about the sustainable and equitable exploitation of water resources by providing a policy framework to achieve a sustainable water cycle supported by integrated, but separate, legislation dealing with drinking water, treatment of wastewaters, eutrophication from agriculture, groundwater protection and bathing water quality (Fig. 7.4). Water is a sustainable resource that is recycled naturally. So it is not a question of exhausting a limited resource, but rather the optimization of available supplies, preservation and enhancement of quality, and the protection of water environments.

Traditional wastewater treatment has several disadvantages:

1. it requires too much water on a *per capita* basis;
2. it results in dilute wastewater;
3. it raises levels of nutrients, metals, and organic contaminants via effluent and sludge disposal;
4. it facilitates the spread of disease organisms and encourages the development of antibiotic resistance during treatment;
5. is often prohibitively expensive both in energy, capital, and operational costs.

Current trends in design and practice are to look for marginal reductions or improvements in these areas. This fine-tuning of existing technology, while improving the current situation, cannot lead to a sustainable water cycle. The fact that global water quality has never been poorer, implies the need for more radical action to preserve what we currently have and what we will require in the future. So the challenge is to translate sustainability into real environmental objectives from which operational systems can be developed. Before objectives can be formalized then it must be accepted that the water issue is not only a problem of water supply and wastewater treatment, it engulfs the whole socio-economic basis of all societies. Therefore mechanisms for the achievement of sustainability will vary between rural, suburban and urban environments, as well as between climatic zones, and economic zones.

Current wastewater treatment plant design has evolved very little since its inception with emphasis still on end of pipe solutions. Development is primarily led by the need to conform to rapidly changing legislation rather than the need to achieve sustainability. Wastewater treatment is an integral part of the urban water cycle where new criteria are needed to visualize and create new sustainable systems. In order to achieve water cycle sustainability future treatment objectives must be:

1. The elimination of all material in used water that was not present before its use, regardless of concentration (i.e. using the precautionary principle there can be no acceptable level for an introduced substance).
2. The protection and if necessary the reclamation of water resources to a condition that can be deemed to be wholly natural. To achieve this will require a unique partnership between users and the industry as well as fundamental changes in wastewater generation and treatment.

23.4 CONCLUSION

The old ways of waste generation and disposal are now just too expensive both economically and environmentally. The risks to our health are also rapidly increasing which will threaten our current way of life, a problem exacerbated by the effects of climate change. New attitudes amongst governments and the public, new technological developments and new management frameworks are all enabling the concepts of sustainability to become gradually accepted. However, a sustainable water cycle cannot be achieved without active support of consumers. This is particularly true of waste minimization and the exclusion of certain wastes from the water cycle. Therefore it would appear that a sustainable water cycle is unlikely to succeed in isolation and must be part of an overall community acceptance of the sustainability principles into every aspect of our lives.

In 1992 a blueprint for future survival on plant earth was established at the United Nations Conference on Environment and Development held in Rio de Janeiro. This blueprint has become known as Agenda 21, and in this a number of key principles for integrated water management are listed. These are:

1. That water is a scarce resource.
2. That all those interested in water allocation and use should be involved in decision making.
3. Water should be managed within a comprehensive framework including water supply and waste management.

Integrated catchment management leading to sustainable agricultural is now widely accepted as the key to environmental protection of terrestrial and aquatic resources and has been widely adopted worldwide (Shields and Good, 2002). In Europe this has been achieved through the introduction of the WFD. The term stakeholder is widely used in catchment management and implies an investment either in terms of financial commitment or effort (Nimmo, 2001). However, for sustainability to be fully achieved then everyone living within a catchment must be seen as a stakeholder with a responsibility to preserve the water quality within it through whatever means are necessary (Carroll *et al*., 2002). So everyone has a part to play. Interdisciplinarity in the water industry is also vital for the future success in sustaining and expanding water supplies while at the same time maintaining ecological quality of our water bodies. Scientists and engineers must work closely together to make the concept of a sustainable water cycle a reality.

REFERENCES

Carroll, C., Rohde, K., Millar, G., Dougall, C., Stevens, S., Richie, R. and Lewis, S., 2002 Neighbourhood catchments: a new approach for achieving ownership and change in catchment and stream management. *Water Science and Technology*, **45** (11), 185–91.

Clark, M. E., Rose, K. A., Levine, D. A. and Hargrove, W. W., 2001 Predicting climate change effects on Appalachian trout: combining GIS and individual-based modeling. *Ecological Applications*, **11**, 161–78.

Houghton, J. T., Ding, Y., Griggs, D. J., Noguer, M., van der Linden, P. J. and Xiaosu, V. (eds), 2001 *Climate Change 2001: The Scientific Basis*, Intergovernmental Panel on Climate Change: Working Group I, Cambridge University Press, Cambridge, UK.

Mitcham, C., 1995 The concept of sustainable development: its origins and ambivalence. *Technology in Society*, **17** (3), 311–26.

Nimmo, J., 2001 From river to ridge: a blueprint for sustainable management in the Georges River Catchment. *Water Science and Technology*, **43** (9), 251–62.

Petts, G., 2001 Sustaining our rivers in crisis: setting the international agenda for action. *Water Science and Technology*, **43** (9), 3–16.

Shields, J. and Good, R., 2002 Environmental water in a regulated river system: the Murrumbidgee River planning approach to the determination of environmental needs. *Water Science and Technology*, **45** (11), 241–9.

Van der Graaf, J. H. J. M., 2001 What to do after nutrient removal? *Water Science and Technology*, **44** (1), 129–35.

Van der Graaf, J. H. J. M., Meester-Broertjes, H. A., Bruggeman, W. A. and Vles, E. J., 1997 Sustainable technological development for urban water cycles. *Water Science and Technology*, **35** (10), 213–20.

Wigley, T. M. L., 1999. *The Science of Climate Change: Global and U.S. Perspectives.* Pew Center on Global Climate Change, Arlington, VA.

Winter, T. C., 2000 The vulnerability of wetlands to climate change: a hydrologic landscape perspective. *Journal of the American Water Resources Association*, **36**, 305–11.

World Commission on Environment and Development, 1997 *Our Common Future*, Oxford University Press, Oxford.

FURTHER READING

Balkema, A. J., Preisig, H. A., Otterpohl, R. and Lambert, F. J. D., 2002 Indicators for the sustainability assessment of wastewater treatment systems. *Urban Water*, **4**, 153–61.

Burton, J. M., 1998 Sustainable waste management a case study from the UK water industry. *Water Science and Technology*, **38** (11), 51–8.

Greenfield, P. and Oliver, G., 2001 River symposium 2000. *Water Science and Technology*, **43** (9), 1–284.

Serageldin, I., 1994 *Water Supply, Sanitation and Environmental Sustainability,* The International Bank for Reconstruction and Development, The World Bank, Washington DC.

Index

Page numbers for figures have suffix **f**, those for tables have suffix **t**